# Environmental and Economic Considerations In Energy Utilization

## Proceedings of the Seventh National Conference on Energy and the Environment

# Environmental and Economic Considerations In Energy Utilization

## Proceedings of the Seventh National Conference on Energy and the Environment

November 30–December 3, 1980, Phoenix, Arizona

Edited by
**Joseph P. Reynolds**
**William N. McCarthy, Jr.**
**Louis Theodore**

**Conducted by** Manhattan College Chemical Engineering Department and the Energy-Environmental Interactions Committee of the Air Pollution Control Association

**Sponsored by** the U.S. Department of Energy and the U.S. Environmental Protection Agency

**Hosted by** American Institute of Chemical Engineers (Arizona Chapter), Arizona Department of Health Services, Arizona Public Service, Arizona State University Department of Chemical Engineering, Salt River Project

ANN ARBOR SCIENCE
PUBLISHERS INC / THE BUTTERWORTH GROUP

# PREFACE

The National Conference on Energy and the Environment was established in 1973 in response to a need for a national meeting to explore and call attention to the rather delicate balance between energy and environmental issues. It was in 1973 that the energy-environment balance exploded into the national consciousness as a result of the Arab oil embargo, which, by chance, coincided with the scheduling of the first of the conferences. Since that time, public interest and concern have mounted steadily; in 1980, the Seventh National Conference was one of over 75 to examine this important topic. In this ever-increasing field, however, this conference continues to maintain its uniqueness. While most of the other conferences focus mainly on one of the two areas (either on energy or on the environment), this conference continues to emphasize the critical balance between the two.

The program for the Seventh National Conference (and the contents of these proceedings) should be of interest not only to the technologist, the engineer and the scientist, but also to a host of others involved or interested in the interactions among the areas of energy, environment and economics. For example, several of the topics are directed toward the public official, the administrator, the manager and the concerned citizen.

The financial support of both the U.S. (DOE) and the U.S. (EPA), without which this undertaking would have been impossible, is gratefully acknowledged. One of the unique features of the conference is that it provides an opportunity for individuals and organizations not under contract to DOE and EPA (and who, therefore, are not participants in the DOE Environmental Conference or the EPA Interagency Conference on Energy and the Environment) to present and publish their findings. Many persons and organizations, also deserving of recognition and gratitude for help and support in the organization and running of the conference, are listed in the Acknowledgments section.

Joseph P. Reynolds
William N. McCarthy, Jr.
Louis Theodore

v

**Joseph Reynolds**       **William N. McCarthy, Jr.**       **Louis Theodore**

**Joseph Reynolds** is a Professor of chemical engineering and Chairman of the Chemical Engineering Department at Manhattan College. He received a BA in chemistry from Catholic University and his PhD in chemical engineering from Rensselaer Polytechnic Institute.

Dr. Reynolds' current research interests are in the air pollution control area. He is responsible in part for the development of a new model for predicting the collection efficiencies of electrostatic precipitators by employing a computer-based Monte-Carlo technique for parameter assignment. He has also worked on a model to describe the pressure drop and compartment flow rates in a pulse-jet filler baghouse.

Dr. Reynolds has presented many papers and chaired sessions at technical meetings, has coordinated continuing education programs in connection with these meetings, has published numerous articles and has been chairman of several functions of the American Institute of Chemical Engineers (New York Section). He is presently the General Chairman of the Seventh National Conference on Energy and the Environment.

Dr. Reynolds is a member of the American Institute of Chemical Engineers, Air Pollution Control Association, and the American Society for Engineering Education, as well as Sigma Xi, Phi Beta Kappa, Phi Lambda Upsilon and Tau Beta Pi.

**William N. McCarthy, Jr.** is Manager of the Flue-Gas Desulfurization Program in the Office for Environmental Engineering and Technology of the U.S. Environmental Protection Agency. He received his MS and BChE in chemical engineering from the University of Maryland and The Catholic University of America, respectively.

Mr. McCarthy is currently responsible for energy-related environmental research and development in a program which is a coordinated interagency effort. Some of his programs have included fluidized-bed combustion coal processing, oil shale and flue-gas desulfurization, management information and quality assurance of management information systems.

Previous experience has been with the U.S. Food and Drug Administration, the National Security Agency and the U.S. Bureau of Mines.

**Louis Theodore** is Professor of chemical engineering and Director of the Graduate Program for the Chemical Engineering Department at Manhattan College. He received a BChE from The Cooper Union and his MChE and EngScD from New York University.

Dr. Theodore has taught courses in Transport Phenomena, Kinetics, Stattistics, Mathematics for Chemical Engineers, and Air Pollution and Its Control. He has lectured internationally in Transport Phenomena for industry and education, and has regularly presented invited lectures and seminars in the air pollution control area. He has consulted for industry in the field of computer applications and pollution control, served as research consultant, and supervised National Science Foundation environmental projects.

Dr. Theodore is a member of numerous professional and scholastic organizations including the American Chemical Society, Air Pollution Control Association and the New York Academy of Sciences.

# ACKNOWLEDGMENTS

Support for this conference was provided by the U.S. Department of Energy, Office of Energy Research, through Grant No. DE-FG02-80ER10205 and the U.S. Environmental Protection Agency, Office of Research and Development, through Grant No. R-807528-01. Thanks are due to the Project Officers for each grant: Richard E. Stephens (DOE) and William N. McCarthy, Jr. (EPA). Any opinions, findings, conclusions or recommendations are those of the authors and do not necessarily reflect the views of DOE or EPA.

Without the tireless efforts of the following persons, the Seventh Conference would never have become a reality:

### General Chairman

Joseph Reynolds
Chairman and Professor
Chemical Engineering Department
Manhattan College

### Conference Committee

Anthony Buonicore
Michael Deutch
Duane Earley
Ann Kaptanis,
    Finance Chairperson
Edward Lynch

Roger Raufer
Edmund Rolinski
Richard Stephens
Louis Theodore,
    Conference Preplanning
Craig Toussaint

### Technical Chairman

William N. McCarthy, Jr.
Senior Chemical Engineer
U.S. Environmental Protection Agency

### Program Chairman

R. Bruce Scott
Assistant Director
Arizona Department of Health Services

**Program Committee**

Anthony Buonicore
Michael Deutch
Paul Farber
Gary Foley
Ray Kary,
    Local Arrangements

Nils Larsen,
    Treasurer
Edmund Rolinski
Edwin Roberts,
    Registration

**Workshop Coordinator**

Louis Theodore
Graduate Program Director and Professor
Chemical Engineering Department
Manhattan College

Thanks also to the chairpersons of the following organizations that cooperated with Manhattan College in presenting the Conference:

Air Pollution Control Association
Energy-Environmental Interactions Committee
Paul Farber, Chairman

and

American Institute of Chemical Engineers
Arizona Chapter
Imre Zweibel, Chairman

Acknowledgment is also due to the following experts who conducted the workshops on November 30:

- *Rulemaking and Its Effect on Energy Development*, Ray Kary, Arizona Public Service

- *Resource Conservation and Recovery—An Industrial Perspective*, Lori Spencer, Western Electric

- *Fine Particulate Control for Industrial and Utility Boilers*, Louis Theodore, Manhattan College; Anthony Buonicore, York Research

- *Environmental Control Implications of Coal Use*, Paul Farber and David Livengood, Argonne National Laboratory

A final and special recognition must be given to William N. McCarthy, Jr. of EPA who, as technical chairman, was responsible for putting together an informative and relevant technical program.

# TABLE OF CONTENTS

## ENVIRONMENTAL ISSUES REGARDING
## THE BACA GEOTHERMAL PROJECT

**Harold Sando**
All Indian Pueblo Council, Inc.

The Department of Energy along with the Public Service Company and Union Geothermal Company of California is jointly funding Unit 1 of the Baca Project, a 50 megawatt geothermal power plant in the Valles Caldera of North Central New Mexico. The Caldera, 12 miles in diameter, is a collapsed volcanic crater, at an elevation of 9000 to 11000 feet. Eight Indian Pueblos are situated around the Southern and Eastern flanks of the mountain mass where the Caldera is located and religious leaders of those tribes have expressed concern that the project will impair important religious areas and adversely impact the water rights of several Pueblos. (Figure 1)

Although the project is being constructed on private land adjacent to Forest Service land, the area is highly referred as culturally significant by the Rio Grande Pueblos and other Pueblos. It is the ancestral and contemporary homelands of many Pueblos. These lands have been occupied since the 1200's. The Pueblos feel that the area impacted by the development is highly incompatible with the many current land uses.

Our cultural resources are being threatened and exploited in many ways. The religious and cultural elements are many. In a background study done by the Mitre Corporation under contract by the Department of Energy, it is stated that the Geothermal Demonstration Program would alter the environment in many ways that may pose a threat to the integrity of the system of beliefs and or set of practices which compromise the sacred aspects of Pueblo Culture. This study further states that the sources of infringement include the reduction or relocation of population of plants and animals gathered by the Pueblos for use in religious rituals and ceremonies.

The plants and medicines gathered there in the Redondo Peak area are of the highest order and importance and no other place in the Jemez Region is valued as such. They cannot be collected elsewhere because they would

1

not be valued as such.  It is pointed out by Dr. Florence Hawley Ellis and Mr. Richard W.Hughes in a report done for the All Indian Pueblo Council that the Redondo Peak is the most sacred place and the spot from which strength, psychological aid and cures come directly.  Like the shrines at Lourdes, Mecca and Jerusalem, it cannot be moved and a substitute be devised.

From time immemorial, the Pueblo People have practiced and maintained their native religion.  Pueblo religion has bonded the people together as a way of life.  The Mitre study states that in spite of contact with the other cultures in modern times, the Pueblo Culture has remained fairly intact.  The Pueblos have managed to maintain many of their traditional ways.

The All Indian Pueblo Council, in a position paper regarding the American Indian Religious Freedom Act, 1980, essentially states that because the Pueblos have never been relocated by the Government nor established on artificial "reservations", the Pueblo People, more than any other American Indian group have formed a special attachment for their aboriginal lands.

In the Final Environmental Impact Statement it states that a great secrecy is associated with the Pueblo religion.  This secrecy evolved partly in response to religious harrassment and precaution since the time of the Spanish Conquest.  As a rule, non-Indians are forbidden from viewing most Pueblo religious ceremonies, especially those taking place in the Jemez Mountains.  In the same report done by Dr. Ellis and Mr. Hughes it is further stated that the cornerstone on which the perpetuation of Pueblo religion has rested its security since Spanish times has been the secrecy from the non-Indians.  This concept is not easily understood  by the Whites and misconcepts have arisen because of it.

This classic example is further demonstrated by the Department of Energy in their Record of Decision from the Baca Geothermal Project, D.O.E. determined  that; 1) The Pueblos do not possess property rights in the Baca sufficient to support a valid claim of infringement on any specific religious activities that occur on the Baca, 2) There has been no showing by the Pueblos that the Project will infringe their religious freedom.

As far as the Pueblos are concerned, the Baca Project is another example of their rights of sovereign government being ignored.  Like so many actions by the Federal Government asserting dominion over lands, the Baca location grant was made in disregard of, and probably ignorance of Indian Rights.

The Pueblo of Jemez alone utilizes these sacred springs which are geographically located on all four sides of the Redondo Peak which are used throughout the ceremonial cycle.  The Mitre study further states that these waters are used for medicinal and ritual purposes and changes in them will degrade their meaning and utility.

The All Indian Pueblo Council also said in a Statement on Water Policy Implementation, The Pueblo Position, April 1980, that water has a deep spiritual meaning.  Flowing water has a spiritual meaning and cleanses not only the physical being, but the spiritual as well.  Where to the dominant society, it is only a physical commodity to exploit for commercial gains without regard to its spiritual value.

Because the Indian People do not use water as avariciously as non-Indians nor do they have the same dedication to the so-called progress as do the non-Indian water users, the Indian might suffer.  We Pueblos settled centuries ago where we are located, and we have lived there ever since.  Water Rights along the Jemez River have never been adjudicated.  The State Engineer considers the surface waters to be fully appropriated; that is no new Water Rights are available.

As reported by Mr. Elmer G. Hall, Special Assistant Attorney General, Water Resources Division, a hearing was held before the State Engineer to transfer surface water rights from the San Antonio Creek Tributary to the Jemez River.  A decision was rendered by the State Engineer contrary to the position of the Pueblos, although the ruling by the State Engineer also points out the geothermal development will exceed the permitted limit within 5 years of operation.  He further concludes that the State Engineer should, at his discretion, either deny this application or condition its approval in the entirement by applicant of sufficient rights, if they exist, to offset the most conservative, reasonable estimate of the effect of this application on the Jemez Stream system.

At any rate, the Pueblos distinctly feel that the development will impair the Water Rights that are rightfully theirs.

We are led to believe that the development will terminate at 50 megawatts.  Let me assure you this is not so.  The companies involved in the project have made clear their intentions to develop up to 400 megawatts of geothermal power at the Baca site, if Unit 1 is successful.  Recently, the Public Service Company in witness before the Public Service Commission Petition  admits the project to have a "cascading effect" and they have discovered reserves of Union Geothermal, to date could produce 400,000 KW of electricity for 30 years.

For example, among the critical provisions contained in the Union Baca Land and Cattle Company Lease (Union/PNM, 1978) are the following: Union may continue to hold the lease each year by pursuing a development program.  Failure to comply with the development obligation will result in a cancellation of the lease...the lease will be considered fully developed when Union has developed 500,000 kilowatts of geothermal energy and generating facilities capable of producing the amount of electricity has been installed.

Each 50 MWE portion of the entire program is expected to require about 740 acres of land for the plant and well sites.  This means we will have over a dozen power plants in the Jemez Mountains.  We asked ourselves who will benefit from all of this.  We have sufficient electricity at the nearby Pueblos.  The rate of development the energy companies are planning they will virtually destroy everything that we have, our culture, our water resources, springs, the ecology, and importantly, our way of life.  In addition, the geothermal leases around the Baca Land Grant held by Amax, Chevron and Sunoco surround the area.

The attitude of the Federal Government is a feeling of no guilt.  The Indian People are aware that they represent progress,  but for whom?  As we know and have heard, progress is also defined as the enemy of traditional culture and vice versa.  Out of the many thousands of acres that we used to own, and we stand to lose more because it directly affects our water shed, which is our livelihood.  The lack of knowledge, unawareness, insensitivity and neglect are the keynotes to the Federal Government's interaction with traditional Indian Religion and Culture.

Then we look at the transportation needs and the impacts it will present.

State Road 44 and State Road 4 are considered the best routes to the development site which traverses right through several Pueblos. At one Pueblo, the road dissects the residential area and houses are within 20 to 30 feet from the State Road. The road has been given unfavorable rating from the State Highway Department. No adequate consultation to the Pueblos has been provided either as to the future plans about construction or improvements.

Air quality changes of great significance are those resulting from operation of the plants. The Mitre study reveals that the cooling towers will emit moisture and gases having the following characteristics: During normal operations, the cooling towers at the 50 megawatt will result in a vapor plume of 800 feet in height. This plume will contain concentrations of hydrogen sulfide gas. These vapor plumes consist of droplets of water containing impurities. Many of these droplets can affect the plants and herbs highly valued in the area which are used for our ceremonies. The degregation of the air quality will certainly endanger the wildlife in the area.

In the Final Environmental Impact Statement a mitigation plan was formulated to provide an early warning system if there occurred a blowout which could release chemicals on the Jemez River. However, the Department of Energy failed to recognize the Pueblos downstream. The dangers of water contamination are real and we are certainly very concerned.

The overall impacts are unknown. It is our expressed feeling that the project be relocated elsewhere in the country where there will be less impacts on the environment and importantly, Indian Religious elements.

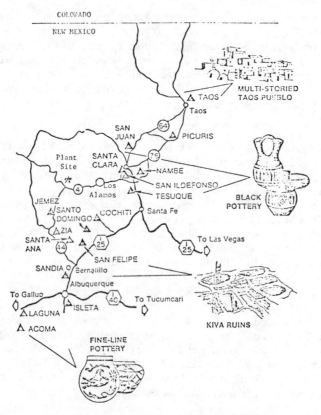

Figure 1. PRESENT DAY RIO GRANDE PUEBLOS

## SOCIOECONOMIC IMPACT ASSESSMENT
## OF ENERGY DEVELOPMENT ON INDIAN RESERVATIONS

**A. T. Anderson**
American Indian Science and Engineering Society

The mission was to evaluate the cost of reconstruction of the Tribal society and the elements which would guarantee self-sufficiency. The reconstruction became necessary for at least three unfortunate developments: (1) the involuntary relocation of all of the Tribal families from their traditional homes to a new settlement in a relatively small area, thereby disturbing the traditional clan structure and placing them within a new and unfamiliar kind of community; (2) the gradual loss in recent time of their freedom to hunt, fish and trap in areas to which they had enjoyed unrestricted access for many generations; and (3) the pollution of their waters by mercury and other effluents, which interrupted their own food supply, limited their commercial fishing and in many different ways had a highly demoralizing effect on individuals and on their society.

The problems of underdevelopment which prevent these Indian people from achieving the level of protection and good management required for survival in the modern world are so severe and wide ranging that massive, long-term restoration efforts are needed. It has been apparent for some time that a comprehensive redevelopment plan will necessitate major changes affecting the protection and development of the available natural resources.

It is believed that changes in structure and budget processes can be brought about by governmental action. We concentrated on the potential ability of the Tribe to govern itself and concluded that they can proceed with an orderly development of their resources.

It has become clear from experience of recent years that piecemeal attempts to resolve Indian problems have consistently failed, testifying to the failure of piecemeal vision. This case is no exception.

It is also clear, however, that the immediate needs of these Indian people must be satisfied by way of assistance from government and other sources, which should coincide with the progressive development of the appropriate natural resources. The issue is survival.

Looking to the future, an increasing population must be assured of food, shelter and energy, which can result only from innovative management together with a higher level of resource development.

Apart from the technological complexity or sophistication of possible solutions, development issues are human issues and involve human dilemmas. Thus, the best solutions are to be found in deliberation, debate and effort among fellow human beings who care and feel strongly about the need to succeed.

Our recommendations respond to unprecedented modern demands for a wide variety of skills, disciplined ethics, unrelenting integrity, constant diligence and total dedication.

The protection of both the Indian cultural structure and the integrity of Indian resources has been paramount in our deliberation. In the past, protection has been inadequate, resulting in the erosion of the original Indian homeland with the resources of forest, water and wild life which at one time enabled the inhabitants to support themselves. If the pittance that remains is to be preserved, there must be major changes in existing structures and mechanisms.

DEVELOPMENT

The strategy for development of resources must be continual in scope and generational in time. Only with vision and vigor, and the commitment of the several governments, can these native people hope to close the gap between a developed affluent society and an underdeveloped, poverty-ridden one.

Our analysis indicates that endorsement and authorization of governments for its creation, scope, budget and function is needed. A fifty-year program is the minimum required to achieve a sense of self-sufficiency and security.

We continue to believe that the past and present piecemeal approach has not succeeded and cannot succeed. These Indian reservation people are not greatly different from under-developed nations. No underdeveloped nation can achieve economic and social self-sufficiency and security with the building of a lodge here, an agricultural project there, and a gas station/store further down the road, with no eye for the comprehensive development of a plan--a strategy. The Indian reservation must be treated as part of a long-term regional program to create an appropriate response.

In large measure, the constraints and restrictions on Indian development have permitted poor access to financial resources for appropriate development. We believe that this limiting approach ought to be discontinued and that governments approve a truly comprehensive plan. Together with a

mediated settlement with the private sector, an acceptable
long range strategy is indicated.

LONG-TERM GOALS

The long-term goal is to overcome the dependence of the
Tribe on the federal government for survival and welfare.

In just a few years, this plan would begin to reverse a
tragic situation which took four centuries to create.  The
initial budgeting to accomplish the task will require substantial funding over a short period of time.  From this
positive, comprehensive and visionary development program
will emerge an Indian community, self-sufficient by 2025.

A REGIONAL MASTER PLAN

A master plan for resource development of the region is
essential for progress of this Indian reservation.  These
people cannot reach an acceptable level of social and
economic self-sufficiency in isolation.  Their condition is
deeply entwined with the land and water of the region and has
a historical background through past centuries.  The
unnaturally high concentration of methyl mercury and other
soluble organic and inorganic compounds in the surface
waters is a catastrophic regional problem.

The water level control is primarily directed towards generation of electrical power for use hundreds of miles distant.
The natural vegitation, animal, fish and fowl which form the
basis of self-sufficiency is in the hands of people who have
other objectives.  Even the need to move a whole people from
their motherland and sacred areas was determined by needs
other than those of the Indian people themselves.

A master plan ought to consider the needs of all of the
people and all of the institutions in the area.  The administrative approach of the several governments must turn to
a development management approach.  These resources of land,
water and people have an exciting potential for the twenty-
first century if they are managed with regional self interests
as the driving force for development.

An outline for the process of creating a regional master plan
is not presented here but the importance of a plan cannot be
over-emphasized.  Out of this planning process will emerge a
development plan by each Indian reservation in the region.
These plans for development will complement and supplement
the future development efforts of the nation and all of its
people.

The case described here is a tragic human drama which could
have been prevented if a harmonious social impact assessment
had been completed before the power dam was built.

A social impact assessment in the terms used by Anderson
Research Consultants, Inc. is described thusly:

SOCIAL IMPACT ASSESSMENT

OBJECTIVE

A Social Impact Assessment attempts to predict and evaluate
the social effects of a policy, program or project while in
the planning stage, and should be carried out concurrently
with the economic and technical investigations.  An assessment should: (1) describe the people likely to be affected
by some new development; (2) outline what positive and negative impacts are likely to affect people; (3) summarize the
acceptability of each alternative proposal; and (4) specify
the steps to be taken to minimize the negative effects and
maximize the positive effects.

Planners and decision makers have always given some consideration to the social effects of new developments, but recent
legislation is forcing a more systematic approach.

An Assessment requires a positive partnership between technical people and those who will be affected by the project.


METHODOLOGY

The steps in an Assessment are:  SUMMARIZE - Describe the
present situation, including the economy, government, education, physical and social services, community organization,
and personal and community goals; SURVEY - Estimate the
changes which may occur in the future, with and without the
proposed development.  (Community people have an important
contribution to make in defining what things are likely to
happen, and what they want to occur in their community.);
APPRAISE - Assess the alternatives with special focus on
the side effects of each choice.  Clearly, local people
have a major part to play in assessing the future;
PRIORITIZE - Evaluate each change and judge the importance
of each group of people affected by the change; and
DETERMINE - Reconcile each alternative and harmonize the
negative and positive impacts.

APPRAISAL

The study area must be large enough to include all of those
affected, including those who stand to benefit as well as
those who may suffer from the project.

Cost of an assessment is proportional to the size of the
population affected, the complexity, size and newness of
the project.

RESULT

A good social impact assessment can result in a positive
effect to the community without economic and legal entangle-
ment.

The following charts graphically describe the problem and
suggest solutions.  They are not annotated and are used as
a basis for the verbal presentation.

Table 1.  An evaluation profile of a case study based on a
scale of 1 (least important) to 10 (most important).

## EVALUATION PROFILE

| | Need | Ease of Doing | Probable Success | Jobs | Band Approval |
|---|---|---|---|---|---|
| Alcohol Treatment | 10 | 5 | 4 | 4 | 9 |
| Arts & Crafts | 7 | 7 | 9 | 8 | 8 |
| Blueberry Farming | 6 | 7 | 7 | 5 | 6 |
| Care for Elders | 9 | 2 | 7 | 9 | 10 |
| Chicken Farming | 7 | 2 | 5 | 4 | 7 |
| Child Care - Home | 8 | 2 | 4 | 8 | 10 |
| Commercial Fishing | 4 | 3 | 5 | 6 | 7 |
| Community Garden | 9 | 4 | 8 | 2 | 9 |
| Compensation | 1 | 9 | 1 | 1 | 9 |
| Credit Union | 1 | 1 | 1 | 5 | 2 |
| Day Care | 10 | 8 | 8 | 9 | 5 |
| Fish to Fertilizer | 3 | 2 | 3 | 4 | 8 |
| Government Jobs | 3 | 8 | 6 | 5 | 3 |
| Guiding | 9 | 9 | 5 | 6 | 8 |
| Health Center | 9 | 3 | 6 | 9 | 8 |
| Housing Construction | 10 | 4 | 7 | 7 | 9 |
| Hunting and Trapping | 9 | 6 | 7 | 5 | 6 |
| Life Scientist | 10 | 8 | 5 | 1 | 6 |
| Logging | 10 | 4 | 5 | 7 | 8 |
| Mechanical Repair Shop | 9 | 5 | 7 | 6 | 7 |
| Mercury Monitoring System | 10 | 5 | 5 | 1 | 6 |
| Paved Road | 4 | 3 | 6 | 4 | 7 |
| Police | 4 | 4 | 5 | 4 | 3 |
| Private Sector Jobs | 5 | 7 | 3 | 6 | 5 |
| Residential School | 2 | 3 | 3 | 6 | 7 |
| Rice Processing | 8 | 2 | 5 | 5 | 9 |
| Ricing | 10 | 7 | 7 | 10 | 10 |
| Saw Milling | 10 | 3 | 7 | 8 | 9 |
| Service to Old Reserve | 2 | 1 | 5 | 4 | 7 |
| Socio-Economic Fund | 10 | 2 | 4 | 3 | 6 |
| Store | 10 | 8 | 7 | 8 | 9 |
| Tourism | 4 | 3 | 2 | 8 | 7 |
| Water (for consumption) | 10 | 4 | 9 | 2 | 10 |

Table 2.  Number of part-time and full-time jobs available by occupation and the level of education
necessary to perform the task.

## EDUCATION - NEEDS ASSESSMENT

| GRADES | 0 - 8 | 10 | 12 | 14 | 16 |
|---|---|---|---|---|---|
| Alcohol Treatment | 1 | | 1* | 1* | |
| Arts & Crafts | (20) | | | | |
| Blueberry Farming | (20) | | | | |
| Care for Elders | 1 | 1 | 1* | 1* | |
| Chicken Farming | 2 | 1* | 1* | | |
| Child Care - Home | 1 | 1* | 1* | | |
| Commercial Fishing | (10) | 1 | 1* | | |
| Community Garden | (40) | 1* | | | |
| Compensation | | | 1* | 1* | |
| Credit Union | | | 1* | 1* | |
| Day Care | 1 | 1 | 1* | | |
| Fish to Fertilizer | (5) | 1 | 1* | | |
| Government Jobs | | 1* | 1* | | 1 |
| Guiding | (20) | | 1* | | |
| Health Center | 1 | 1 | 1* | 1* | 1* |
| Housing Construction | 10 | 1* | 1* | | |
| Hunting and Trapping | (20) | 2* | 1* | | |
| Laundromat | 1 | 1* | | | |
| Life Scientist | | | | | (1) |
| Logging | (20) | 1 | 1* | | |
| Mechanical Repair Shop | | 1* | 1* | | |
| Mercury Monitoring System | | | | 1* | (1) |
| Paved Road | (10) | 1 | | | |
| Police | | | 2* | | |
| Private Sector Jobs | 1 | 1 | 2* | | |
| Residential School | 1 | 2 | 2* | 1* | |
| Rice Processing | 1 | 1 | 1* | | |
| Ricing | (200) | | 1* | | (1) |
| Saw Milling | 20 | 1* | 1* | | |
| Service to Old Reserve | (10) | 1 | | | |
| Socio-Economic Fund | 1 | 2 | 1 | | |
| Store | (2) | 1 | 1* | 1* | |
| Tourism | (2) | | 1* | 1* | |
| Water (for consumption) | (1) | | | | |

( ) = Part-time    * = Additional Training

9

## INDIAN'S LANDS

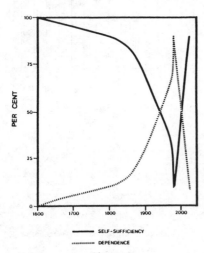

**1492**

**1820**  **1840**

**1860**  **1977**

Figure 1.
The disappearance of Indian occupied lands from the arrival of the European societies in 1492 through 1977.

## POPULATION DENSITIES

1979    2000    2027

FEDERAL

STATE

RESERVATION

*EXCLUDING CITIES WITH A POPULATION OF 100,000 OR MORE.

ONE PERSON PER SQUARE MILE.

Figure 2.
The effect of restricted land use on Indian population densities compared with non-Indian land population densities projected to the year 2027.

## SELF-SUFFICIENCY VS DEPENDENCE

PER CENT

——— SELF-SUFFICIENCY
·········· DEPENDENCE

Figure 3.
The result of a dominant society (non-Indian-European) on a previously self-sufficient society (Indian) and the dependence that results projected to show the improbable task of reversal during the next 50 years.

## SOCIO-ECONOMICS · INDIAN

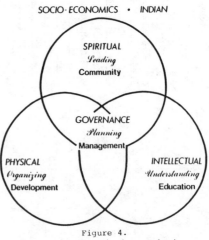

SPIRITUAL
*Leading*
Community

GOVERNANCE
*Planning*
Management

PHYSICAL
*Organizing*
Development

INTELLECTUAL
*Understanding*
Education

Figure 4.
Indian socio-economic interrelationships within self-sufficient communities.

## SOURCES OF HOUSEHOLD INCOME – 1977

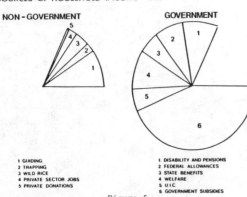

NON-GOVERNMENT

GOVERNMENT

1 GIRDING
2 TRAPPING
3 WILD RICE
4 PRIVATE SECTOR JOBS
5 PRIVATE DONATIONS

1 DISABILITY AND PENSIONS
2 FEDERAL ALLOWANCES
3 STATE BENEFITS
4 WELFARE
5 U I C
6 GOVERNMENT SUBSIDIES

Figure 5.
The overwhelming financial burden of government in a dominant/dependent relationship.

## EFFECT OF SOCIAL TRAUMA

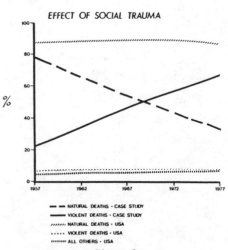

%

—— — NATURAL DEATHS · CASE STUDY
———— VIOLENT DEATHS · CASE STUDY
//////// NATURAL DEATHS · USA
····· VIOLENT DEATHS · USA
········· ALL OTHERS · USA

Figure 6.
Case study of the dramatic rise in the percentage of violent deaths when a previously isolated self-sufficient society is forced to move and assimilate into the rules and social structures of a dominant society.

10

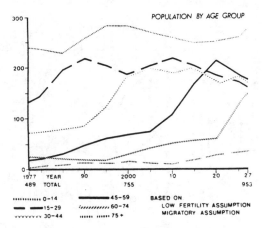

POPULATION BY AGE GROUP

| | | | |
|---|---|---|---|
| 1977 | YEAR | 90 | 2000 | 10 | 20 | 27 |
| 489 | TOTAL | | 755 | | | 953 |

............ 0-14          ――――― 45-59          BASED ON
――  ― 15-29          /////////// 60-74          LOW FERTILITY ASSUMPTION
vvvvvvvv 30-44          ⅢⅢⅢⅢ 75 +          MIGRATORY ASSUMPTION

Figure 7.
A case study of projected population
changes by age groups of an isolated/
dependent society from 1977 to 2027.

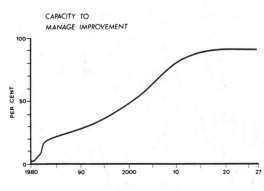

CAPACITY TO
MANAGE IMPROVEMENT

Figure 8.
Assuming that enormous assistance
would be forthcoming, the chart demonstrates the increase in capacity
to manage improvement necessary on the
part of the community in order to successfully reach the goals.

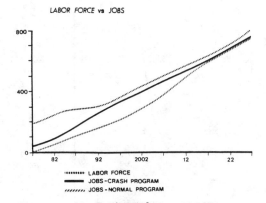

LABOR FORCE vs JOBS

............ LABOR FORCE
――――― JOBS-CRASH PROGRAM
/////////// JOBS-NORMAL PROGRAM

Figure 9.
A 50 year projection of the labor
force in a case study and two possible job programs to lower the high
unemployment to 10%.

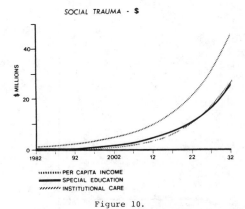

SOCIAL TRAUMA - $

............ PER CAPITA INCOME
――――― SPECIAL EDUCATION
/////////// INSTITUTIONAL CARE

Figure 10.
A case study of the amount of money
necessary to overcome existing social
trauma through special education and
institutional care and the projected
increase in per capita income.

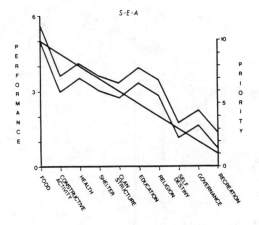

S·E·A

Figure 11.
A socio-economic assessment of the
minimum and maximum performance
levels of a prioritized list of essential activities necessary for
survival.

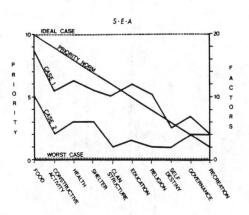

S·E·A

Figure 12.
A socio-economic assessment of two
case study communities and their
level of success in relation to a
prioritized list of activities necessary for survival.

11

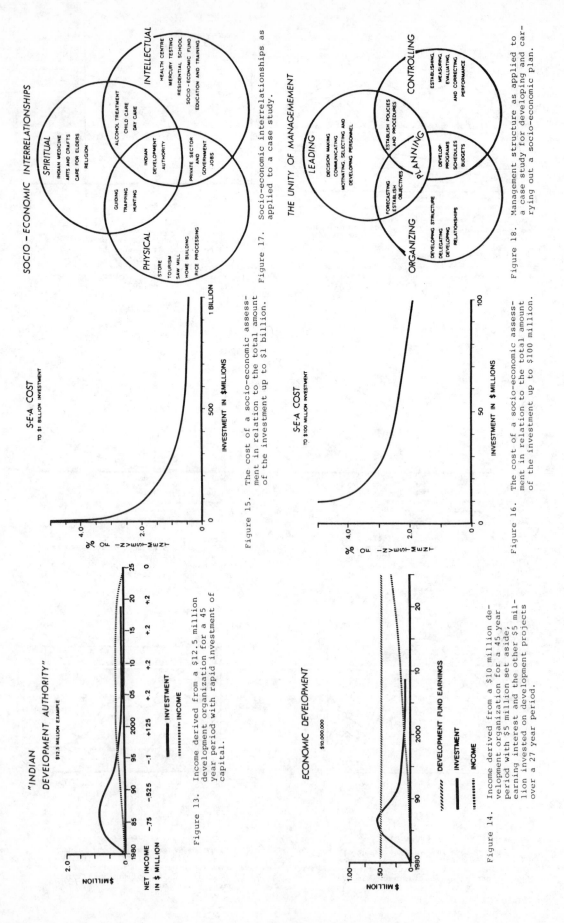

SOCIO – ECONOMIC INTERRELATIONSHIPS

SPIRITUAL
INDIAN MEDICINE
ARTS AND CRAFTS
CARE FOR ELDERS
RELIGION

INTELLECTUAL
HEALTH CENTRE
MERCURY TESTING
RESIDENTIAL SCHOOL
SOCIO-ECONOMIC FUND
EDUCATION AND TRAINING

ALCOHOL TREATMENT
CHILD CARE
DAY CARE

INDIAN
DEVELOPMENT
AUTHORITY

PRIVATE SECTOR
AND
GOVERNMENT
JOBS

GUIDING
TRAPPING
HUNTING

PHYSICAL
STORE
TOURISM
SAW MILL
HOME BUILDING
RICE PROCESSING

Figure 17. Socio-economic interrelationships as
applied to a case study.

THE UNITY OF MANAGEMEMENT

CONTROLLING
ESTABLISHING
MEASURING,
EVALUATING,
AND CORRECTING
PERFORMANCE

LEADING
DECISION MAKING
COMMUNICATING
MOTIVATING, SELECTING AND
DEVELOPING PERSONNEL

ESTABLISH POLICIES
AND PROCEDURES

PLANNING
DEVELOP
PROGRAMS
SCHEDULES
BUDGETS

FORECASTING
ESTABLISH
OBJECTIVES

ORGANIZING
DEVELOPING STRUCTURE
DELEGATING
DEVELOPING
RELATIONSHIPS

Figure 18. Management structure as applied to
a case study for developing and carrying out a socio-economic plan.

S-E-A COST
TO $1 BILLION INVESTMENT

% OF INVESTMENT

INVESTMENT IN $MILLIONS

Figure 15. The cost of a socio-economic assessment in relation to the total amount
of the investment up to $1 billion.

S-E-A COST
TO $100 MILLION INVESTMENT

% OF INVESTMENT

INVESTMENT IN $ MILLIONS

Figure 16. The cost of a socio-economic assessment in relation to the total amount
of the investment up to $100 million.

"INDIAN
DEVELOPMENT AUTHORITY"
$12.5 MILLION EXAMPLE

$ MILLION

INVESTMENT
INCOME

NET INCOME
IN $ MILLION

Figure 13. Income derived from a $12.5 million
development organization for a 45
year period with rapid investment of
capital.

ECONOMIC DEVELOPMENT
$10,000,000

$ MILLION

DEVELOPMENT FUND EARNINGS
INVESTMENT
INCOME

Figure 14. Income derived from a $10 million de-
velopment organization for a 45 year
period with $5 million set aside, $5 million
earning interest and the other $5 mil-
lion invested on development projects
over a 27 year period.

12

## HOW VISIBILITY REGULATIONS
## WILL AFFECT ENERGY DEVELOPMENT IN THE WEST

**Terry L. Thoem** and **David B. Joseph**
U.S. Environmental Protection Agency

Introduction

The Rocky Mountain West has been blessed with at least two forms of
natural resources - - scenic beauty and energy resources.  This dichotomy
provides a potential conflict between visibility protection and energy
resource development.  The magnitude and extent of this conflict will
only be defined through implementation of the EPA visibility and PSD
regulations.  Because of the form of the regulations prescribed by the
Clean Air Act, it is difficult, if not impossible, to forecast the degree
to which energy resource development will be constrained and/or the
amount of visibility degradation that will occur in Federal Class I areas.

EPA Region VIII States (Colorado, Montana, North Dakota, South Dakota,
Utah, and Wyoming) will be particularly stressed by this dichotomy.  The
six states contain half of the Nation's coal reserves (200 billion tons),
forty percent of the Nation's uranium reserves (240,000 tons of $30 per
pound $U_3O_8$), and essentially all of the Nation's economically attractive
oil shale resource (731 billion barrels).  In addition, two of the
Nation's hottest oil and gas prospects lie within the Region's boundaries-
the Overthrust Belt in Utah, Wyoming and Montana, and the Williston
Basin in North Dakota.  Coal fired generating capacity is expected to
reach 32,000 MW by 1985 ( a doubling from 1979 and a quadrupling from
1975.)  Shale oil production could be about 500,000 barrels per day in
the early 1990's.  Uranium yellow cake production is expected to double
between now and 1985.  (24,000 tons per year by 1985)

On the other side of the dichotomy is the existence of 40 mandatory
Federal Class I areas in the Region VIII States totalling over 10 million
acres.  This acreage represents nearly one-third of the Class I area
acreage of the U.S.  This substantial acreage emphasizes the importance
of good visibility to the Region vis-a-vis the tourism and recreational
industry.

13

Background of Regulatory Requirements

Section 169A of the Clean Air Act Amendments of 1977 established a number of visibility requirements. Sections 164(e) and 165(d) and (e) contain additional visibility requirements. In Section 169A Congress established as a national goal "the prevention of any future and the remedying of any existing impairment of visibility in mandatory Class I Federal areas which impairment results from man-made air pollution". To work toward achievement of this goal Section 169A requires...

- DOI to determine which mandatory Federal Class I areas have visibility as an important value.

- EPA to promulgate a list of these areas.

- EPA to provide a Report to Congress on methods for achieving progress toward the National goal. This report addresses methods for monitoring, characterizing, modelling, and controlling visibility impairment.

- EPA to promulgate regulations which will provide guidance to the States for including visibility protection requirements in their State Implementation Plans (SIP).

In addition, Congress provided visibility requirements in the Prevention of Significant Deterioration (PSD) section of the Act. The Federal Land Manager was mandated to have "an affirmative responsibility" to protect the visibility values of his Class I area.

EPA published its list of 156 mandatory Class I areas with visibility as an important value (40 CFR Part 81) and its Report to Congress (EPA-450/5-79-008) in late 1979. EPA is under court order to propose regulations by May 18, 1980 and to promulgate by November 15, 1970. Proposed regulations were signed by Administrator Costle on May 15, 1980 and published in the Federal Register on May 22, 1980 ( 45 FR 34762). The promulgated regulations were signed on November 21, 1980 following a court extension of the final deadline from the 15th to 26th of November. States have nine months following promulgation to submit a revised SIP.

Three major issues associated with the proposed regulations were 1) adequacy of the technical and scientific understanding of the cause and effect relationships of anthropogenic visibility impairment, 2) role of the Federal Land Manager, and 3) protection of integral vistas. Probable resolutions are discussed below. The visibility protection requirements have been designed to be implemented as a phased program. Existing impairment which can be reasonably attributed to a single source or small group of sources will be addressed first. Regional haze conditions will be addressed in a later phase as better understanding of cause/effect and improved monitoring and modelling techniques become routinely available. Thus, the first phase will probably be limited to control of particulate and $NO_2$ plumes. A clearly identifiable $SO_2$ to $SO_4$ condition resulting in adverse impairment may dictate control of an isolated source.

The role of the Federal Land Manager is one of consultation and recommendation. The State has the primary responsibility for all aspects of the program. It has been suggested that the "affirmative responsibility" language in the PSD section requires the FLM to pursue all possible avenues to protect visibility. These could include non-issuance of

14

right-of-way approvals if a State issued a permit over the objection of the FLM.

Integral vistas are defined as vistas outside the Class I area which are important to the visitor's visual experience of the area. The FLM recommends integral vistas to the State; the State identifies those integral vistas to be protected in the SIP.

Regulatory Impact on Energy Facilities

Contrary to a mistake-filled and misinterpreted contractor report prepared for EPA, BART requirements will impact only a very few(if any) existing energy facilities. It is probable that less than five power plants in the U.S. will be affected (none of which are in Region 8). At this time no other energy facilities have been identified as possible BART candidates. The process for imposing BART is as follows:

1.  The FLM identifies mandatory Federal Class I areas where visibility is being adversely impaired. This includes impairment of any integral vistas. Factors including geographic extent, intensity, duration, frequency, and time of impairment must be considered by the FLM in making the determination of adverse impairment. The FLM may provide advice to the State on the suspected cause of the impairment.

2.  The State identifies the cause of the impairment. If the cause may be reasonably attributed to a single source or small group of sources, the State identifies the source(s) as a possible BART candidate. The impairment would most likely be an identifiable brownish $NO_2$ plume or a grayish-white particulate plume.

3.  The State evaluates control options available to the source for the pollutant causing the impairment. If a reasonable control option will result in a perceived improvement in visibility, the State will establish an emission limitation and a compliance schedule for that source. The emission limit is established taking into consideration the technology available, cost of compliance, existing control equipment in use at the source, remaining useful life of the source, and the degree of visibility improvement which may result.

4.  The above process applies to impairment of vistas in a Class I area. If there is visibility impairment of State identified integral vistas, i.e., vistas outside the Class I area, additional factors including impacts on energy development and land use considerations may be balanced against the need for control.

5.  The BART limitation is adopted by the State as a SIP revision. The source must be in compliance with the BART limit in at least 5 years after SIP plan approval. Continuous source emisssion monitoring may be required by the State.

6.  Any existing major source required to install, operate and maintain BART may apply to the Administrator of EPA for an exemption from that requirement provided the source can demonstrate that it does not or will not by itself or in combination with other sources emit any air pollutant which

causes or contributes to significant visibility impairment in any mandatory Class I area. In determining the significance of the impairment, factors such as geographic extent, intensity, duration and frequency and time of impairment must be considered. The FLM must concur on any EPA BART exemption approval.

The visibility review of new sources is performed through the new source review process. If a State has been delegated the PSD program, the visibility regulation will constitute a very slight extension of that program. The review process is as follows:

1.  A potential applicant must apply for a new source/PSD permit. Baseline visibility monitoring data may need to accompany the application.

2.  The State/EPA notifies the FLM of receipt of application and provides the FLM with pertinent information including the source's assessment of visibility impacts on nearby Class I areas.

3.  The FLM has the opportunity to demonstrate that the source will cause an adverse impact on the visibility of his Class I area. If the State agrees, the permit is not issued. If the State disagrees, the State must provide its reasons in writing and then may issue the permit. In the case of a disagreement, the FLM is afforded the opportunity to consult with the Governor.

4.  A variance procedure allowing violations of the Class I increments involves visibility. The applicant is provided an opportunity to demonstrate that visibility will not be impaired even though Class I increments are predicted to be violated. If the FLM and the State agree, the permit may be issued. However, Class II increments must not be violated. If the Governor recommends the variance and the FLM does not concur, the matter is decided by the President.

5.  The same factors for determining adversity as described for existing sources apply to the new source review. BACT must be met by the source. The opportunity for a public hearing must be provided on all new source reviews.

The potential impacts of the visibility portion of the new source review process may be classified in three types. First, siting restrictions may be implicit. However, the 50 mile radius buffer zones around Class I areas as suggested by the utility industry have no factual basis. This is a misrepresentation of the practical implementation of the visibility requirements. Second, controls which represent a greater degree of control than would have been defined as BACT, i.e., less emphasis on economics, may be required. Third, a source may be allowed to construct even though violations of the Class I increments will occur.

EPA Region VIII examples

Through the PSD process, Region VIII has performed visibility analysis for several coal fired power plants, coal mines, and a few oil shale facilities. Methods used for analysis have included both the use of contractor prepared visibility prediction computer models and hand calculations utilizing computer generated air quality concentrations.

Two guidance documents have been made available recently by EPA concerning the procedures to be used for calculating visibility impacts. The documents are entitled "Workbook for Estimating Visibility Impairment" and "User's Manual for the Plume Visibility Model". These documents are available from the EPA Office of Air Quality Planning and Standards in North Carolina and the National Technical Information Service. An example calculation for the Colony oil shale visibility assessment is provided as figure 1 of this paper. EPA Region VIII visibility analyses have been of greater and lesser complexity than the referenced example. The analyses have been performed on a case-by-case basis considering such factors as the source type and emissions and proximity to the Class I area.

The criteria to be used as yardsticks against which to assess the adverse nature of source's impact have not been accurately defined. These are case-by-case decisions to be made by the FLM. Visibility impairment indices considered include change of contrast ($\Delta$C), visual range(VR) reduction (%), plume coloration (blue-red ratio), and plume perceptibility ($\Delta$E). Quantitative threshold levels of visibility perception considered in recent Region VIII analyses are approximately:

$\Delta$C = 0.03 - 0.05
VR reduction = 5 to 10 percent
blue-red ratio = 0.90 - 0.95
$\Delta$E = 5 to 10

It must be emphasized that these are approximate numbers and do not have official EPA endorsement as guidance.

Conclusion

Visibility regulations will impose additional procedural, and perhaps substantive, requirements on new energy facilities. A few existing sources may need to install retrofit control technology. The visibility provisions allow variances to the Class I increments and may allow a source to construct which otherwise would have been denied.

Visibility regulations will protect Class I areas from future degradation and provide the mechanism to mitigate existing impairment.

The protection of visibility in Class I areas will provide continued tourism and recreational benefits. The balancing of energy production goals and economic considerations with the desired protection of integral vistas will allow the States to provide for a proper balance between energy development and visibility protection goals.

Figure 1

Visibility Analysis

Colony

o Charlson Equation

$$VR = \frac{constant}{mass\ concentration}$$

o $b_{scat}/SO_4$ conc.

$$0.05\ to\ 0.1\ X\ 10^{-4}\ m^{-1}\ (\mu g/m^3)^{-1}$$

o USFS Analysis

- Compared to power plants
- 50 mile viewing distance
- Maintain clarity within

o USFS Point Source Report

- Vistas identified in and out
- Used EPA calculated $\Delta C$

o GLC vs. Plume

o $\Delta C = aRe^{-bR}dB$

a = factor for Co
R = observer to target distance
b = background $b_{scat}$
dB = ($b_{scat}/SO_4$ conc.) (predicted $SO_4$)

o Lost Lakes to Sleepy Cat Peak

$$\Delta C = (0.7)(40\ km)\ \left[e^{(-.01)(40)}\right]\ X$$

$$\left[(0.072)\ (1.0\ X\ 2\ X\ 0.01)\right]$$

$\boxed{SO_2\ \mu g/m^3}$  $\boxed{SO_2 - SO_4}$  $\boxed{1\%\ per\ hr.\ X\ plume\ travel}$

$\Delta C = 0.027$

o "Yardstick"

.03 "perceptible
.10 "day and night"

18

# VISIBILITY OF A NO$_2$ PLUME IN NORTHERN ARIZONA

**Neil S. Berman**
Arizona State University

## Introduction

The brown color of urban atmospheres is often partially attributed to nitrogen dioxide (NO$_2$) in the air. Robinson (1) has discussed the effects of both particulates and NO$_2$ with examples typical of concentration levels in Los Angeles. Nitrogen dioxide absorbs light strongly over the blue-green area of the visible spectrum. Particulates or aerosols scatter and absorb light leading to the reduction in visibility compared to the atmosphere without these particles.

When both aerosols and NO$_2$ are present, the intensity of light from an object is attenuated by scattering and absorption.

$$I = I_o \ e \ -(\sigma + kc)x \ , \tag{1}$$

where $I_o$ is the source intensity, $\sigma$ is the extinction coefficient containing the effects of scattering and absorption by aerosols, k is the absorption coefficient of NO$_2$, c is the concentration of NO$_2$, and x is the path length or distance from the object. This equation assumes that the effects are constant over the distance from the observer to the object. The effect of NO$_2$ alone can be seen by taking the ratio of the intensity in Equation (1) to the intensity without NO$_2$. When visibility is limited and x is relatively small, the concentration of NO$_2$ necessary to influence coloration of visible objects is much higher than when the path length can be very large. Robinson suggests that target discoloration would be noticeable when c = 0.25 ppm and x = 2 miles. In Northern Arizona the average visibility was found by Roberts et al. (2) to be 70 miles. Thus a similar discoloration would require only 0.007 ppm of NO$_2$ in the air. Such levels of NO$_2$ (13 $\mu$g/m$^3$) are not exceeded very often in non-urban areas of Arizona based on State of Arizona reports (3). Although the highest concentrations of NO$_2$ are probably accompanied by high aerosol levels and consequently low visibility, it is of interest to examine a typical power plant plume and the ambient air for effects of $NO_2$ on visibility.

In addition to the reduction in intensity of illumination, the modification of the luminance of the horizon sky by absorption due to $NO_2$ is important. Robinson (1) shows that the scattered luminance from the horizon, B, is related to the luminance without $NO_2$ absorption, $B_o$, by

$$B = B_o b/(kc + b),$$
(2)

where b, the scattering coefficient, is a function of path length or visibility. Typically b is taken to be 3.9 divided by the visibility in miles. Again the visibility is a function of aerosol concentration and changes in the color of the horizon sky require less $NO_2$ when the visibility is large.

Concentration Data

The maximum $NO_2$ concentration at selected sites in Arizona are given in Table 1.

TABLE 1

Maximum $NO_2$ Concentration at Selected Sites

ppm. 24 hour average

| Site | 1979 | 1978 | 1977 |
|------|------|------|------|
| Grand Canyon | 0.02 | 0.02 | 0.03 |
| Joseph City | 0.03 | 0.06 | 0.02 |
| Phoenix | 0.13 | 0.07 | 0.13 |
| Phoenix (1 hr Ave) | 0.32 | 0.54 | 0.42 |

These data are taken from State of Arizona reports (3). In Phoenix, nitrogen oxide emissions from motor vehicles are significant. These emissions coupled with night ground based temperature inversions lead to high concentrations of all primary pollutants in the winter months. In the summer, atmospheric mixing during the longer day results in low concentrations. Typical daily one hour average NO and $NO_2$ concentrations taken at the central Maricopa County station are shown in Figure 1 for days in January and July.

Another $NO_2$ source is illustrated in Figure 2. The data were taken by Aerovironment Inc., at a site on the west side of Phoenix near the Salt River. The highest $NO_2$ levels are recorded after the ozone begins to decline and reactions that are the reverse of those in the presence of light occur. Only the early morning residuals would seem important to the visibility and color of the atmosphere. Similar conditions would prevail in the non-urban areas of the state even though sources are not local. In the absence of other data the factor of 3 relationship between the one hour average and the 24 hour average for Phoenix will be assumed for the entire state.

A point source emission of 750g/s $NO_2$ at a height of 150 meters will be considered to compare the effect of $NO_2$ on visibility compared with the background. Typical Northern Arizona meteorological conditions of wind speed, 4 m/s; stability, class B; and inversion height, 1500 m will be used along with the Gausian plume formulas (4). Concentrations of $NO_2$

as a function of distance from the source are shown in Figure 3. The maximum concentration at ground level is 0.74 ppm.

Results

The relative intensity of illumination, Equation (1), for a concentration of 0.74 ppm as a function of wavelength for a path length of two miles is shown in Figure 4. The intensity reduction over the entire band from 4000 to 7000 angstroms can be found by integrating the curve in Figure 4 and subtracting from the total area. When the integration is done for a two mile path length and plotted as a function of concentration, Figure 5 is obtained. Many factors go into determining when the reduction can actually be perceived by an observer so no definitive prediction can be made. If 25% reduction is used as the criteria, then 0.2 ppm is needed for this two mile path.

A similar determination of reduction in Luminance for the horizon sky can be made when the visibility is fixed. Equation (2) is integrated when b is fixed. Equation (2) is integrated when b is obtained from the visibility and k is a function of wavelength. The results are shown in Figure 6 for visibilities of 70, 7, and 3.5 miles. The 25% reduction criteria gives concentrations of 0.05 ppm for 70 mile visibility, 0.3 ppm for 7 mile visibility and 0.6 ppm for 3.5 mile visibility.

The calculations show that the maximum background concentrations of $NO_2$ would affect visibility only slightly in Northern Arizona and only when the visibility was greater than 70 miles. The highest Phoenix concentrations combined with much lower visibility than in Northern Arizona would be noticeable. Table 2 gives the concentration of $NO_2$ which would match the reduction in visibility (horizon sky luminance) as particulates.

TABLE 2

| Visibility, Miles | $NO_2$ Concentration to have the same effect as particulates ppm |
|:---:|:---:|
| 1 | 8.30 |
| 5 | 1.66 |
| 10 | 0.83 |
| 20 | 0.41 |
| 70 | 0.12 |
| 100 | 0.08 |

In Table 2 the visibility determines b in Equation (2), and k was taken as constant at 5500 $A^o$, the center of the visible range. This calculation again shows that $NO_2$ background is not as important as particulates in Northern Arizona. With particulate levels corresponding to a 30 mile visibility, the Phoenix concentration of $NO_2$ would color the sky brown on the horizon.

The effect of $NO_2$ in the plume would certainly be noticed by an observer in the plume looking back at the source. Generally the particulates in the plume would have an even greater effect than $NO_2$. It is interesting also to determine visibility changes looking at right angles through the plume. Table 3 shows the concentration of $NO_2$ necessary to obtain 25% reduction in intensity looking through a plume of the given width.

TABLE 3

| Plume Width km. | $NO_2$ Concentration for visible color, ppm |
|---|---|
| 0.1 | 2.2 |
| 0.5 | 0.44 |
| 1.0 | 0.22 |
| 2.0 | 0.11 |

These concentrations and widths are compared with the example plume in Table 4. The intensity reduction at 4000 A$^O$ is used for comparison. A reduction of at least 50% would be necessary to have a noticeable effect.

TABLE 4

| x, km. | Intensity Reduction | $NO_2$ Concentration at $\sigma_y$ ppm |
|---|---|---|
| 1 | 36% | 0.45 |
| 2 | 35% | 0.22 |
| 3 | 28% | 0.12 |
| 4 | 22% | 0.07 |
| 5.5 | 13% | 0.03 |

The $NO_2$ in the plume would not be observable perpendicular to the plume. Also the concentrations reach the maximum background only 5.5 km. from the source.

Conclusions

Current nitrogen dioxide effects on visibility in Northern Arizona do not appear to be significant. For the maximum short time ambient $NO_2$ concentration of 0.1 ppm, discoloration of the horizon sky or of white objects would be noticeable only if the aerosol level is low enough to give visibilities greater than 70 miles. A typical power plant plume from a single stack would not show discoloration looking at right angles. Only observations on the plume line would be affected by the $NO_2$ in the plume. Combinations of stacks could add a factor of ten to the widths in the example and some meteorological conditions could also lead to higher concentrations. Since the calculations do show effects of $NO_2$ without such increases, visibility changes would certainly be observed. A factor of ten increase in the plume width would be observed as a brown plume from the source to approximately 5 km. away.

References

1.  Robinson, M. "Effect on the Physical Properties of the Atmosphere," Chapter 11 in Stern (editor), Air Pollution, Volume 1, Academic Press, New York, 1968.

2.  Roberts, E.M., J.L. Gordon and R.E. Kary, "Visibility Measurements in the Painted Desert by Photographic Photometry," Dames and Moore report.

3.  "Air Quality Data for Arizona," State of Arizona Department of Health Services Reports 1974-1979.

4.  Turner, D.B., "Workbook of Atmospheric Dispersion Estimates," Environmental Protection Agency AP-26, Office of Air Programs, Research Triangle Park, North Carolina, 1970.

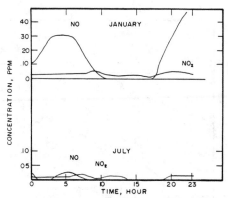

FIGURE 1. TYPICAL NO AND NO₂ CONCENTRATIONS PLOTTED AS A FUNCTION OF TIME OF DAY PHOENIX, ARIZONA

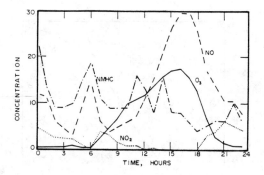

FIGURE 2  MEASURED CONCENTRATIONS OF NON-METHANE HYDROCARBONS (NMHC), NITROGEN OXIDES (NO AND NO₂) AND OZONE (O₃) IN PHOENIX ON JUNE 27, 1976. THE CONCENTRATION UNITS ARE PPHM EXCEPT FOR HYDROCARBONS WHERE THE UNITS ARE PARTS PER TEN MILLION.

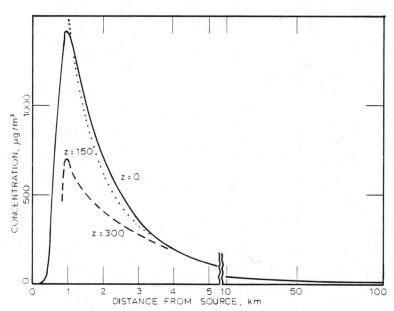

FIGURE 3  NO₂ CONCENTRATIONS AS A FUNCTION OF DISTANCE FROM SOURCE, X, AT GROUND LEVEL, Z=0, AND TWO HEIGHTS.

23

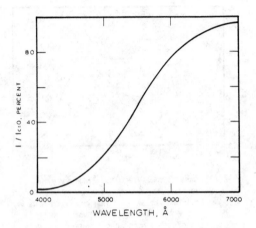

FIGURE 4 INTENSITY OF ILLUMINATION, I, COMPARED TO
INTENSITY WITH NO₂ CONCENTRATION OF ZERO.

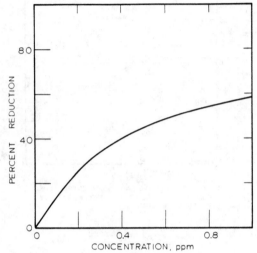

FIGURE 5. REDUCTION IN INTENSITY OF ILLUMINATION
INTEGRATED OVER VISIBLE WAVELENGTHS FOR
PATH LENGTH OF TWO MILES.

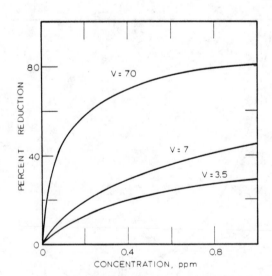

FIGURE 6 LUMINANCE OF THE HORIZON SKY REDUCTION
INTEGRATED OVER VISIBLE WAVELENGTHS FOR
DIFFERENT VISIBILITIES ( MILES )

# A SULFATE VISIBILITY RELATIONSHIP
# DERIVED FROM FIELD DATA IN NEW YORK STATE

**John M. Stansfield**
Brown and Root, Inc.

**James B. Homolya**
U.S. Environmental Protection Agency

Introduction

During the fall of 1978, an EPA sponsored research project was conducted in the Albany, New York, area to determine the impact of a
large point source on ambient sulfate levels.  This project, known as
SIPSES (Study of the Impact of a Primary Sulfate Emission Source), has
been previously reported upon.[1]

During SIPSES a network of ambient air quality monitoring stations was
established in the vicinity of an oil-fired power plant that burned a
high vanadium residual fuel oil.  At two of these stations both ambient
sulfate and nephelometer visibility data were collected for 28 consecutive days.  These data indicated that when the ambient sulfate levels
increased, there was a decrease in nephelometer visibility.

This report is a continuation of the data analysis associated with
SIPSES.  Specifically this report is a somewhat more detailed examination of the visibility/sulfate relationship noted during SIPSES.  Also
presented in this report is an examination of sulfate data collected by
the New York State Department of Environmental Conservation (NYSDEC).
The NYSDEC has maintained a high volume sampler network for many years.
At some of these stations the filters are routinely processed for water
soluable sulfates.  A 14 year period of data collected at Troy, New
York, was made available for this study by the NYSDEC.  Nearby is a
first order weather station, staffed by the National Weather Service, at
the Albany Airport.  Subjective visibility data were available for this
same period of record.

SIPSES Results

During the SIPSES experiment, nephelometer visibility measurements were taken at two locations. These locations, Base and Site 5 are shown on Figure 1. In addition to these data, surface weather data from the Albany airport were obtained.

The nephelometer data were recorded as $b_{scat}$, or the scattering coefficient of the airborne material in the atmosphere. While the nephelometer is a good monitor for airborne particulates, it is relatively insensitive to relative humidity. To convert from scattering coefficient, or $b_{scat}$, to a local visibility ($L_{vd}$) in miles, the following expression was used:

$$L_{vd} = 4.7/b_{scat}$$

Moreover, the nephelometer measures the visibility at what is essentially a single point. Thus, the term local visual distance or $L_{vd}$ is appropriate for nephelometer data.

Visibility measurements taken at the Albany airport are subjective estimations by a human observer. These measurements represent the prevailing visibility and as such, are an area measurement. That is, the subjective visibility measurements represent the visibility over an area, where the nephelometer measurements represent visibility at a single point within an area. Subjective visibility measurements are considered to be more sensitive to relative humidity than the nephelometer.

Two methods were used to compute a representative 24 hour visibility. One method was to take the 24 hourly observations and express them as an average. This was the method used for the nephelometer-sulfate data. The second method was to take the hours 0700, 1000, 1300, and 1600 LST and express them as an average to represent a day. This latter method was used for the sulfate comparison as there is some question about the compatibility of daytime subjective visibility measurements vs. nighttime measurements.

Figure 2 is a comparison of the various visibility measurements during SIPSES. As can be seen, the two nephelometer measurements compare favorably with each other. The two methods for computing a representative 24 hour subjective visibility also compare favorably with each other. However, there is less agreement when the nephelometer data are compared to the airport subjective data. This is probably due to several factors. For example, there is a sizeable distance between the airport and the two nephelometer stations. Also, the area in the immediate vicinity of the airport does not have the industry that is located along the Hudson River. Moreover, the two techniques are not exactly compatible, i.e., area measurements vs. point measurements.

In addition to the hourly subjective visibility data available during SIPSES, relative humidity data were also available. Reiss and Eversole[5] have examined daytime subjective visibility measurements taken at Newburg, New York, during the period 1948-1970. Newburg is located on the Hudson River about 80 miles south of Albany. Their data suggest a crude semilogarithmic relationship between subjective visibility (log) and relative humidity (linear).

Table 1 shows the correlation between daytime subjective visibility measurements and relative humidity for data collected during SIPSES. These measurements are for the four daytime hours and do not include any hours for which precipitation was recorded. Less than half of the variation in the subjective could be explained using a parsimonious relative humidity model. Moreover, no great improvement was obtained when semilogarithmic transformations were made. This lack of a good correlation between visibility and relative humidity has been previously reported on by Paterson[4] in data collected at Sydney, New South Wales, Australia. A possible explanation for the low correlation at Albany may be due to the fact that during SIPSES, subjective visibilities were frequently greater than 10 miles. Thus the data tended to be clustered with a lack of representation in the low visibility range.

As part of the follow-up data analyses for SIPSES, scatter diagrams of the nephelometer visibility vs. sulfate were made. Two of these diagrams for the Base site are shown in Figures 3 and 4. While a linear relationship, $Y = a + bX$, could be fit to the data, there appeared to be an apparent curvature in the scatter pattern when plotted on a linear scale. When the same data are plotted on a log-log scale, the result is to remove the curvature at the ends of the scatter diagram. An improved fit is the result.

At this point, it was decided to perform linear regression analyses for the two monitoring sites using different transformations of the nephelometer visibility and sulfate data. The results are summarized in Table 2. For the Base site it was found that an improvement in the fit could be obtained. The fit improved from 51.80% to 89.06%. That is, the proportion of the total variation about the mean of the dependent variable explained by the regression was increased by 37.26%.

However, the improvement is less marked for the data collected at Site 5. For that data an improvement of only 13.68% could be obtained. Moreover, there is a much lower fit to the data at Site 5. At this location, only about half of the variation in the 24 hour average nephelometer visibility could be explained by the various transformations used.

Site 5 was in a rural area with woods in the immediate vicinity and no significant sources in the immediate area. This site could receive industrial particulates from evaluated plumes diffusing downward.

Long Term Results

In examining the long term subjective visibility and sulfate data, it was found that a significant amount of data was collected on precipitation days. That is, a day when a trace or more of moisture (rain and/or snow) was observed at the Albany airport. In fact, over half (56.8%) of the filters were collected on such days as indicated in Table 3.

TABLE 3

NUMBER OF FILTERS

Troy, New York

| | |
|---|---|
| Precipitation Days | 428 |
| Non-Precipitation Days | 325 |
| Total | 753 |

There appears to be a somewhat higher sulfate level during precipitation days. This is suggested by the geometric means shown in Table 4. The cause for the somewhat higher sulfate levels during precipptation is a subject beyond the scope of this paper. Two of several possible explanations are the scavenging effects of precipitation and the possibility that the related meteorological conditions favor $SO_2$ $-SO_4^=$ sources located in one area and not another. Consider a cold front extending southward from a low pressure area near the St. Lawrence River. This could result in a southernly flow from the Atlantic seaboard, northward to Albany. This would provide moisture for rainfall as well as the advection of $SO_2$ $-SO_4^=$ from that source region.

Over the years, there has been a variation in the amount of sulfate that has been measured at Troy. This can be seen in Table 4 and in Figure 5. Figure 5 indicates that an apparent cyclic pattern may exist. This pattern shows a decrease in the late sixty's, an increase in the early seventy's, followed by another decrease. No firm explanation can be given at this time for this pattern. It is not known if this is the result of shifting climatic conditions, changes in emission characteristics, or some other factor.

It was decided to divide the data into two categories: precipitation and non-precipitation days. This resulted in an improvement in the fit of a parsimonious regression model. This can be seen in the following three examples:

TABLE 5

EFFECT OF PRECIPITATION ON REGRESSION

Troy, New York
1966-1979

| Model | $r^2$(%) All Days | Non-Precipitation |
|---|---|---|
| $Y = a + bX$ | 20.67 | 40.51 |
| $1/Y = a + bX$ | 2.78 | 45.56 |
| $lnY = a + blnX$ | 21.72 | 38.35 |

Since rain (or snow) can of itself easily reduce visibility, this result is not suprising. The precipitation cases obviously contain many variables and are beyond the scope of this study. Therefore, attention will be directed towards the non-precipitation day data.

Regression analyses were performed on the non-precipitation data to determine what improvement, if any, could be obtained through the use of transformations. The results of these regressions are shown in Table 6. An improvement of some 7.31% could be obtained when a semi-logrithmetic relationship was used for the full 14 year data base.

However, this result is data base dependent. Figure 6 shows the value of $r^2$ on a yearly basis for two of the transformations studied. In one year, 1970, only an insignificant relationship was found to exist between subjective visibility and sulfate. In another year, 1976, an $r^2$ of 88.23 could be obtained using a reciprocal subjective visibility relationship. Yet in 1974, the same reciprocal relationship obtained an $r^2$ of only 2.31%.

These results indicate that for the data base used, that no one transformation was always superior to any other transformation. Moreover, the correlation between subjective visibility varied from one year to the next.

Lab Analysis Procedures

During SIPSES, ambient sulfate concentrations were measured at several locations surrounding the power plant. These data were collected on a daily basis using General Metal Works high volume samplers, Model GMWL 200H.

These samplers were run on a midnight to midnight basis. General Metal Works GMW-810 glass fiber filters were used. On the morning after exposure, each filter was collected, folded, and placed in an individual glassine envelope. Each high volume sampler had a Dickson pressure recorder and each pressure chart was also placed in the manila envelope. All high volume samplers received a multipoint calibration before and after the field program.

The filters were sent to the laboratory for batch processing. After obtaining final weights, they were processed for water soluable sulfates. Two inch strips were digested in 50 ml of distilled water for one hour. The solution was diluted to 250 ml and an 50 ml alquot was analyzed for sulfate ion concentration using turbidimetric techniques. All filters were corrected for blanks, individual running time, and flow rates.

The State of New York Department of Health (NYSDH) used two different methods to process the NYSDEC filters for sulfates. Up until July, 1971, they used the turbidimetric technique. Beginning with the July, 1971, filters they used the methylthymol blue (MTB) method. In the turbidimetric technique a solution of barium chloride and hydrochloric acid is mixed with the water soluable sulfate extract. This results in the precipitation of barium sulfate. The resulting turbid solution is read at 420 nm. The result is compared against a standard curve to determine $SO_4^=$/ml.

In the automated methylthymol blue (MTB) method, an equimol solution of barium chloride and methylthymol blue indicator is mixed with the sulfate extract, precipitating barium slufate. The remaining barium ion is complexed by the methylthymol blue, and the amount of free MTB is determined at 480 nm. The sulfate concentration in the water extract is determined by comparison of the sample's absorbance to a standard curve.

The NYSDH indicates that a statistical analysis of 228 pairs of samples analyzed by both the turbidimetric and MTB methods showed a correlation coefficient of 0.984. Known sulfate solutions were analyzed and the standard deviations were:


TABLE 7

COMPARISON OF NYSDEC METHODS

| Level | Standard Deviation | |
| | Turbidimetric | MTB |
| --- | --- | --- |
| 5.0 µg $SO_4^=$/ml | $\pm$ 0.4 | $\pm$ 0.2 |
| 25.0 | $\pm$ 0.5 | $\pm$ 0.6 |

Until January, 1980, MSA glass fiber filters were used by the NYSDEC. After that date, Schleicher & Schuell filters were used.[2]


Conclusions

While it is possible to develop a parsimonious regression equation relating sulfate to visibility, the degree of fit likely to be obtained is data base dependent. Nephelometer visibility data and sulfate data collected concurrently over a 28 day period have demonstrated at least at one site that a good degree of fit can be obtained through the use of a suitable transformation. Both of the following two transformations gave significant results:

$$1/Y = a + bX \qquad (r^2 = 89.06\%)$$

and

$$\ln Y = a + bX \qquad\qquad (r^2 = 81.99\%)$$

However, this may not be a universal conclusion. During this same period of time, another site, similarily equipted, gave good results, but not near as good. This indicates that a universal parsimonious equation may not be possible.

During this same field study, daytime subjective visibility and relative humidity were examined. These data indicate that while a regression equation can be made to fit these data, the degree of fit, as expressed by $r^2$, is about the same as that for nephelometer visibility and sulfate for the untransformed data.

When a long period of airport subjective visibility and surface data were compared, it was found that considerable improvement could be obtained when the analysis was restricted to days when there was no precipitation.

It was found that the correlation between subjective visibility and sulfate varied from one year to the next, In fact, in one year there was no relationship for all practical purposes. Yet in other years, there was a definite relationship between the two parameters.

The correlation between subjective visibility and sulfate could be improved somewhat through the use of transformations. A semilogrithmetic transformation resulted in the most improvement when the 14 year period of record was considered. However, for any given year, there was some variation as to the best transformation to use.


Acknowledgement

The authors wish to thank Mr. Bill Delaware of the New York State Department of Environmental Conservation for supplying the sulfate data collected by that agency.

References

1.  Boldt, K. R., C. P. Chang, E. J. Kaplin, J. M. Stansfield and B. R. Wuebber, May, 1980, Impact of a Primary Sulfate Emission Source on Air Quality, EPA-600/2-80-109, U. S. Environmental Protection Agency, ESRL, RTP, N. 27711.

2.  Delacqua, Barbara and Sue Koblantz, September, 1980, Personal Communication, New York State Department of Health Albany, New York.

3.  MRI, 1980, Integrating Nephelometer, Meteorology Research, Inc. Altadena, California 91001.

4.  Paterson, M. P., March, 1973, Visibility, Humidity, and Smoke in Sydney, Atmospheric Environment, Vol. 7, p 281-290.

5.  Reiss, Nathan M. and Rae Ann Eversole, 1978, Rectifiction of Prevailing Visibility Statistics, Atmospheric Environment, Vol. 12, p. 945-950.

TABLE 1

REGRESSION SUMMARY

Daytime Visibility vs Relative Humidity
Albany Airport During SIPSES
Sept. 18 – Oct. 15, 1978

| Model | $R^2$ (%) | $F_o$ | Rank |
|-------|-----------|-------|------|
| Y = a + bX | 43.16 | 78.20 | 2 |
| Y = a + b$ln$X | 44.46 | 82.44 | 1 |
| $ln$Y = a + bX | 41.88 | 74.22 | 3 |

Y = Hourly Subjective Visibility (Miles)

X = Hourly Relative Humidity (%)

Computed on 105 observations where no precipitation was
recorded.  Daytime hours were 0700, 1000, 1300, and 1600 LST.


TABLE 2

SUMMARY OF
NEPHELOMETRY/SULFATE REGRESSION RESULTS ($r^2$)
DURING SIPSES

| Model | Base | Site 5 |
|-------|------|--------|
| Y = a + bX | 51.80 | 41.04 |
| Y = a + bX$^2$ | 35.16 | 29.72 |
| Y = a + b/X | 78.75 | 54.72 |
| 1/Y = a + bX | 89.06 | 52.14 |
| Y = a + b$ln$X | 71.15 | 52.24 |
| $ln$Y = a + bX | 77.55 | 47.83 |
| $ln$Y = a + b$ln$X | 81.99 | 51.53 |

TABLE 6

Visibility/Sulfate
Regression Model Summary
Non-Precipitation Days
Troy, New York
1966 – 1979

| Model | $R^2$ (%) | $MS_E$ | $F_o$ | Rank |
|---|---|---|---|---|
| Y = a + bX | 40.51 | 92.46039 | 219.96 | 3 |
| Y = a + b/X | 9.33 | 140.92405 | 33.23 | 6 |
| 1/Y = a + bX | 45.56 | 0.02964 | 270.32 | 2 |
| Y = a + b$ln$X | 39.31 | 94.32381 | 209.23 | 4 |
| $ln$Y = a + bX | 47.82 | 0.28441 | 295.97 | 1 |
| $ln$Y = a + b$ln$X | 38.35 | 0.33602 | 200.89 | 5 |

$MS_E$ = Mean Square of the Error

$F_o$ = Test Statistic (F-test)

X = 24 Hr. $SO_2^=$ ($\mu g/m^3$)

Y = Average Daytime Visibility (Miles)

TABLE 4

Sulfate Geometric Mean, $\mu g/m^3$
Troy, New York

| Year | Precipitation Days | Non-Precipitation Days | All Filters |
|---|---|---|---|
| 1966 | 11.8 | 8.5 | 10.3 |
| 1967 | 8.8 | 8.1 | 8.5 |
| 1968 | 7.8 | 9.3 | 8.4 |
| 1969 | 8.5 | 6.5 | 7.7 |
| 1970 | 7.0 | 7.1 | 7.1 |
| 1971 | 7.4 | 8.4 | 7.8 |
| 1972 | 10.9 | 10.1 | 10.4 |
| 1973 | 11.0 | 10.8 | 11.0 |
| 1974 | 12.8 | 7.7 | 10.2 |
| 1975 | 10.7 | 10.6 | 10.7 |
| 1976 | 9.3 | 9.0 | 9.2 |
| 1977 | 7.9 | 8.0 | 7.9 |
| 1978 | 7.1 | 6.5 | 6.8 |
| 1979 | 8.2 | 8.2 | 8.2 |
| 1966 – 1979 | 9.0 | 8.4 | 8.8 |

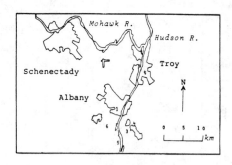

Figure 1

Albany, Schenectady, Troy Metropolitan Area

1, 3, 4, 5, 6, b = SIPSES Stations

t = Troy Station

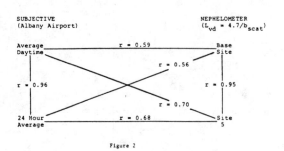

Figure 2

**VISIBILITY CORRELATIONS
DURING SIPSES**

NOTE:  Includes all weather conditions
       Nephelometer data are 24 hr. averages

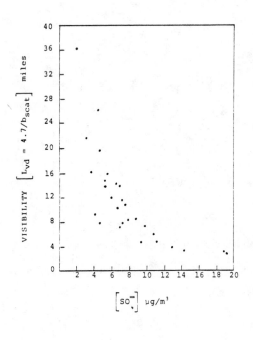

Figure 3

NEPHELOMETER VISIBILITY vs SULFATE
BASE SITE
SIPSES

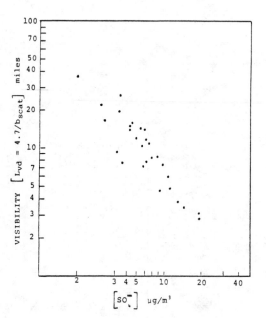

Figure 4

NEPHELOMETER VISIBILITY vs SULFATE
BASE SITE
SIPSES

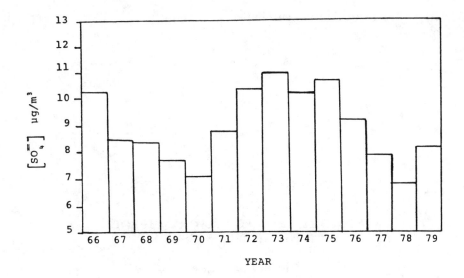

Figure 5

SULFATE
ANNUAL GEOMETRIC MEAN
TROY, NEW YORK

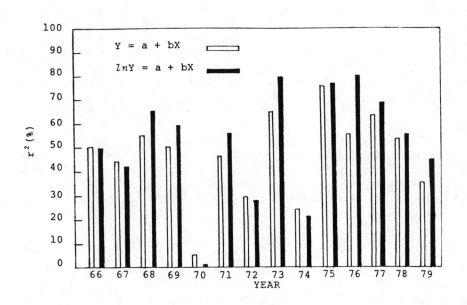

Figure 6

SUBJECTIVE VISIBILITY vs SULFATE
REGRESSION CORRELATION
TROY, NEW YORK
NON-PRECIPITATION DAYS

35

## STATUTORY AND REGULATORY ASPECTS
## OF THE VISIBILITY ISSUE

**Michael L. Teague**
Hunton and Williams

Good afternoon, I am Mike Teague, an attorney with Hunton & Williams, which is counsel to the Utility Air Regulatory Group. I have worked closely with the Visibility Protection Committee of this ad hoc group of electric companies since December 1977 when the Committee was formed to participate in EPA's visibility rulemaking under the Clean Air Act. I am here today to discuss the statutory and regulatory aspects of the visibility issue.

Legislative Summary

Two portions of the Air Act are relevant to this discussion. One is entitled "Visibility Protection for Federal Class I Areas," and appears in section 169A of the Act. The other is contained in the provisions for the prevention of significant deterioration (PSD) -- § 165(d), in particular.

The Visibility Protection Program

The visibility protection program has four components, some of which are already in place. First, EPA listed those mandatory class I federal areas to be protected under this program on November 30, 1979. The list contains 156 of the 158 possible areas, many of which are here in the southwest. Second, EPA submitted a report to Congress on visibility impairment in the class I areas in November of 1979. While this report is to contain the "present" scientific knowledge on visibility, it is already outdated. Third, EPA published visibility regulations on December 2, 1980. Fourth, the states are to use those regulations to revise their implementation plans to contain emission limitations,

schedules of compliance and other measures to make reasonable progress
toward meeting the goal of the visibility protection program.

Let's take a closer look at that goal. Congress declared:

> as a national goal the prevention of any future, and
> the remedying of any existing, impairment of visibility
> in mandatory class I Federal areas which impairment results
> from manmade air pollution.

Progress toward meeting this goal was to be made by the states' requiring
best available retrofit technology, or BART, for certain existing
stationary sources and developing long-term strategies of 10 to 15
years. More on the content of these requirements later.

## The PSD Program

The PSD program also deals with visibility. Under this program, a
facility proposed for areas attaining the ambient standards is subject
to a preconstruction analysis of, among other things, the ambient air
quality and visibility both at the proposed site and in the area
potentially affected by the proposed facility. That analysis will be the
basis for (1) demonstrating compliance with the applicable increments
and (2) assessing the impact of the facility's emissions on the
visibility in class I federal areas.

The procedures for resolving disagreements as to the effects of a
proposed facility on visibility are complex and will involve the applicant, the state, and the affected federal land managers. If disagreements
cannot be resolved, the relevant Governor, and perhaps the President,
could ultimately decide on the terms and conditions of the PSD permit,
if one is granted.

The recently signed visibility regulations also contain provisions for
the PSD visibility analyses.

## EPA's Visibility Regulations

Let's turn now to those final regulations. They take a phased approach.
Control of regional haze has been postponed until the technology for
dealing with it is developed. I will touch on the adequacy of the
technology for the first phase of visibility protection, which focuses
on plume effects, in a moment.

Moving to their substantive requirements, the new visibility regulations
apply to 36 states -- 21 of the 24 states west of the Mississippi River,
including Hawaii and Alaska. These states have nine months to revise
their implementation plans to include BART requirements for certain
existing sources of visibility impairment and long-term strategies to
ensure reasonable progress toward the national goal for visibility
protection. I will discuss both in turn, but let me emphasize that the
states are charged with making the substantive visibility decisions.
EPA must, however, review the revised implementation plans and promulgate
its own requirements if a state submits an inadequate plan.

## BART Analyses

BART analyses will be required for those stationary sources to which
visibility impairment in the mandatory class I areas or in vistas

integral to them has been reasonably attributed if they:

1. are in any of 26 enumerated categories;

2. were in existence on August 7, 1977;

3. did not go into operation before August 7, 1962;

4. have the potential to emit at least 250 tons per year of any pollutant; and

5. have not been exempted from BART requirements by EPA.

Focusing on some of the key terms that I have just mentioned, we see that EPA has defined:

1. "Visibility impairment" to mean any humanly perceptible change in visibility from that which would have existed under natural conditions.

2. "Natural conditions" to include naturally occurring phenomena that reduce visibility.

3. "Integral vistas" to mean a view perceived from within a mandatory class I federal area of a specific landmark or panorama located outside the boundary of the mandatory class I federal area. Integral vistas will be identified according to criteria developed by the federal land managers after notice and opportunity for public comment. When disagreements arise on the validity of the identification, EPA has required the states' Governor to give the federal land manager an opportunity to be heard. The electric industry and others have commented that they see no statutory basis for the integral vista concept, which appears to be an unconstitutional infringement on the states' sovereignty over its land.

4. "Reasonably attributable" to mean attributable by visual observation or any other technique that the state deems appropriate.

What is BART? EPA has defined BART to be an emission limitation which is to be based on the degree of reduction achievable through the application of the best system of continuous emission reduction for each pollutant which is emitted by an existing stationary facility. The emission limitation must be established, on a case-by-case basis, taking into consideration:

1. the technology available;

2. the costs of compliance with alternative emission limitations;

3. the energy and non-air quality environmental impacts of compliance;

38

4. any pollution control equipment in use or in existence at the source;

5. the remaining useful life of the source; and

6. the degree of improvement in visibility which may reasonably be anticipated to result from the use of such technology.

BART for large power plants must be conducted pursuant to "Guidelines for Determining BART for Coal-Fired Power Plants and Other Existing Stationary Facilities," EPA publication no. 450/3-80-009b, which is incorporated by reference into the regulations. Since this massive book was just made available, I have not had a chance to digest it. The preamble to the visibility regulations, however, suggests that the states should assess the available technology capable of removing the most pollution to see if an improvement in visibility can be obtained. If not, the analysis ends. If some improvement can be obtained, the states are to investigate the spectrum of controls, from no change to full control, to determine which retrofit technology is the "best" -- which one produced the most favorable balance of the factors that must be considered.

In addition to emission limitations, the revised state implementation plans must establish procedures to ensure the proper operation and maintenance of control equipment. When confronted with this issue in the rulemaking for revised new source performance standards for power plants, the Agency essentially passed the buck. It will be interesting to see what EPA does in the yet unpublished BART Guidelines.

As I mentioned earlier, BART should be stated in terms of an emission limitation. The Agency has, however, provided the states with the option of imposing a design, equipment, work practice, or other operational standard when measurement technology for a source is technologically or economically infeasible.

Let's take a look at the specifics of a BART assessment. The states are required to develop source-receptor relationships -- relate incremental reductions in emissions of various pollutants to changes in the psycho-physical reaction in some representative segment of the population. To do this, the states must be able to characterize emissions, their atmospheric transformations, the optical properties of those emissions and their secondary pollutants, their location, and of course, what all this means in terms of a human visual response. Such a characterization should be made in terms of a frequency distribution and must be preceded by an understanding of meteorological and natural visibility conditions often over large geographic areas. This frequency distribution may have to be formulated for distances from the source of emissions and for terrain for which no dispersion coefficients have been developed.

These difficult technical tasks must be performed with little guidance from the Agency, which has not incorporated by reference into the regulations visibility monitoring or modeling guidelines. The BART Guidelines may provide some help on these tasks -- the proposed version of this document did not.

The electric industry and others have filed comments with EPA, stating that the visibility protection rulemaking is premature because the techniques necessary to implement the visibility protection program do not exist. The Agency recognizes these deficiencies by stating in the

39

preamble of the final regulations that further evaluation of visibility monitoring and modeling must precede their routine use in regulatory programs. While the use of these techniques is left to the states' discretion, EPA does not explain what they can use or what the Agency will use to review revised implementation plans. While in some cases the attribution of visibility impairment to specific facilities may be possible, explaining with precision the ability of emission reduction to improve visibility is another matter.

The Act explicitly recognizes that available emission control technology may only "reduce" the visibility impairment caused by the facility. This fact will influence the amount of control that can be required since its ability to improve visibility is a critical factor in establishing BART. The societal value associated with small changes in the intensity or size of a visible plume may not justify spending vast sums to obtain such changes.

A state could conclude, following a BART analysis, that the plant under scrutiny already has the best available retrofit technology and that no change in emission limitations is required. At the same time, more stringent emission limitations could be imposed. Such limitations must be obtained as expeditiously as practicable, but within five years of the approval of a revised implementation plan.

The final visibility rules contemplate that some pollutants will escape BART requirements during the first round of implementation plan revisions. Emission of such pollutants will be reanalyzed for BART as new technologies for them become available. I do not believe that Congress anticipated more than one BART assessment per facility.

As I mentioned earlier, EPA can exempt facilities from BART analyses. An exemption will become effective if:

1. The source gives prior written notice of its application for the exemption to affected federal land managers.

2. The source's application demonstrates that it will not emit any air pollutant which may be reasonably anticipated to contribute to a significant impairment of visibility in any mandatory class I federal area and in vistas integral to them. Decisions as to the significance of an impairment will be made on a case-by-case basis and will focus on the enjoyment of the visitor's visual experience of the class I areas and take into account the intensity, extent, duration, frequency and time of the visibility impairment and its relationship with natural conditions that reduce visibility.

While the Act does not provide for state concurrence, EPA argues that it is necessary since the Agency does not want to preempt the states' authority to impose restrictions more stringent than EPA's.

The most curious aspect of the exemption process is that it is scheduled to follow the BART analyses. State resources expended on such analyses could be wasted on visibility effects eventually found to be insignificant.

## Long-Term Strategies

The second major visibility protection component required of the revised
state plans is a long-term strategy. This strategy must lay out a
mechanism for dealing with each example of existing visibility impairment
in the mandatory class I federal areas and vistas integral to them. The
plan must include a strategy for those facilities which contribute to
local and interstate visibility impairment. With respect to these
strategies, the state must:

1.  Explain why its strategy is inadequate.

2.  Coordinate with the plans of the federal land
    managers.

3.  Provide for periodic reports which discuss progress
    and problems.

4.  Review the impacts of new sources in accordance
    with the PSD regulations, the ruling on new source
    reviews for nonattainment areas, and "any other
    binding guidance provided by the Agency" on
    visibility matters.

5.  Consider such tactics as "additional emission
    limitations," mitigation of construction
    activities, source retirement, and smoke management
    techniques.

In developing their long-term strategies, the states must take into
account (1) the effect of new sources and (2) the costs of compliance,
the time necessary for compliance, the energy and non-air quality
environmental impacts of compliance, and the remaining useful life of
any affected existing source and equipment therein.

As to new sources, there appear to be two types of analyses. The first
occurs during the PSD review for a facility proposed for a clean area.
In this analysis, the facility's predicted impacts on visibility in any
class I federal area will be compared against an adversity standard. In
short, will the impact interfere with a visitor's visual experience in
the class I area? When a state rejects the attempts of a federal land
manager to show that an adverse impact exists, the state must publish its
reasons. When the federal land manager convinces the states that an
adverse impact will result, no PSD permit will be issued.

The second type of new source analysis will study the impacts of PSD
facilities on the integral vistas of any mandatory class I federal area
and the impacts of facilities proposed for nonattainment areas on any
mandatory class I federal area including its integral vistas. The
substantive test to be applied appears to be the adverse impact test
that I have just discussed and the vague notion of consistency with
reasonable progress toward meeting the national visibility goal.

These requirements are accompanied by little explanation of the nature
and scope of the regulatory program. What, for example, is an adverse
impact; how does one demonstrate that an adverse impact exists; what is
an integral vista; what is reasonable progress; what is an approvable
implementation plan? The visibility rules suggest that the states will

be left to work out these details and that EPA will defer to their judgment.

I might add that there is no express statutory authorization for the regulation of the visibility effects of facilities proposed for nonattainment areas. In addition, the PSD program is to be the exclusive source of regulation for the visibility effects of those facilities whose construction commenced after August 7, 1977 in clean areas. The final visibility regulations ignore this exclusivity.

## Visibility Monitoring

I would like to address the visibility monitoring requirements in the visibility regulations. They permit the state to require monitoring of visibility in any federal class I area near proposed new stationary sources for such purposes and by such means as the State deems necessary and appropriate. This regulation is related to a general requirement that the state plans include a strategy for evaluating visibility in any mandatory class I federal area. Two comments are in order. First, I do not see that the states have the jurisdiction over federal lands necessary to run a monitoring program, or to require private individuals to monitor, in such areas. Second, EPA has not explained how meaningful data can be collected or how it should be used. The final regulations merely refer to guidance "as is provided by the Agency," rather than making direct assistance available to the states now.

## The Practicability of the Visibility Regulations

EPA provides the states with nine months to revise their implementation plans. That schedule appears to be impracticable because of all the steps that must be taken in that period. What are those steps?

1. The states have one month to set up lines of communication with the federal land managers.

2. The federal land managers have two additional months to certify that visibility impairment exists and to identify integral vistas.

3. The state must then determine if it can reasonably attribute the impairment to a BART source.

4. If the attribution can be made, the state must make a BART assessment. If not, the state has to develop a long-term strategy to reduce the visibility impairment.

5. Once a draft plan revision exists, the state must give the federal land managers two months to review it.

6. The state must then hold a public hearing and make a tentative decision.

7. If a facility owner is dissatisfied with a BART decision for his facility, he can apply for an exemption which involves:

a.  application preparation;

b.  obtaining concurrence from the state;

c.  EPA's taking up to three months to announce
    a public hearing;

d.  conduct of a public hearing;

e.  time for EPA to make a final decision; and

f.  time for the federal land manager to decide
    whether he concurs with EPA's judgment.

Just looking at the time available for notice for various activities in
this schedule, eight of the nine months could be consumed.  That would
leave one month for data acquisition, analysis, and documentation on
the very complex environmental, technical, economic, energy, and social
decisions.

But is this scheduling problem real.  EPA has stated that its preliminary
analyses, performed in conjunction with certain federal land managers,
could not identify any existing sources that will need to install additional controls.  The Agency states that few, if any, sources will be
subject to BART requirements and that many of the basic elements of the
long-term strategies are already in place.  If this is the case, I can
only wonder why the regulations on existing facilities were promulgated
at this time.

Conclusion

I believe that there is general agreement that visibility in our national
parks is important.  I believe there is less agreement on what the
visibility is in those parks, what it would be without anthropogenic
activity in and around them, which anthropogenic activities affect the
national parks, and what the social cost would be to produce an optimal
level of visibility.  Congress, EPA, and the federal land managers have
devoted much of their time and energy to these issues, as have several
segments of the public in an effort to assist the government in
resolving visibility issues.  Even so, information that has surfaced
in the last two months from EPA's VISTTA program and revised regulatory
analysis suggests that most of those resources have been poorly spent
and that the visibility protection program will have little, if any,
effect.

# WATER AVAILABILITY AND REQUIREMENTS
# FOR WESTERN OIL SHALE DEVELOPMENT

**G. A. Miller**
U.S. Geological Survey

Introduction

No oil-shale industry exists today in the area, but information is available from which to draw a few general conclusions as to both water supply and needs for such an industry.

This discussion is limited to the physical occurrence of water supplies. Most of the specifics concern the Piceance Creek Basin in northwestern Colorado, which contains about 3/4 of the rich oil shales of the Green River Formation. For purposes of this paper, the physical occurrence of water is assumed to be the ultimate constraint on its potential for use by an oil-shale industry. Other constraints (e.g., legal, economic, political, social) must operate under this constraint and are not considered in detail herein.

Water Supply, Water Use, and Water Management

There are a variety of recognized on-site needs for water at oil shale mines and plants. Estimated water needs, the supplies available, potential environmental effects related to their use, and potential constraints on water availability all dictate prudent management of the water resource.

### Projected Water Use by an Oil-Shale Industry

Water uses for oil shale, as recognized today, may be categorized into industry uses at an oil-shale mine and plant, and ancillary uses by the supporting population. In this article, the use by industry is assumed to be consumptive; that for ancillary purposes is only for intake water. For most municipal systems, only a fraction of intake water is consumptively used, the remainder being returned to the local hydrologic system.

Use by industry. Water use at an oil-shale mine/plant site may include water for dust control, mining, moisture removed by mine ventilation, drilling, process steam, power generation, cooling and moistening retorted shale, reclaiming and revegetating retorted shale and other disturbed lands, stack scrubbers, and other minor uses including domestic water. The estimates of water use, based to date largely on projections rather than on operational data, range from less than 1 to more than 6 barrels of water per barrel of shale oil (BW/BO) produced.[1-8] Most estimates are in the range of 1-4 BW/BO. Thus, a 400,000-BPD (barrels per day) industry may use as little as 20,000 ac-ft/yr or as much as 80,000 ac-ft/yr.

The wide range in projected use is in part related to different methods of oil-shale mining, extraction, processing, and reclamation. In part it reflects the uncertainties of projecting full-scale uses in a new industry based on extrapolation from only small-scale experiences and from applicable specific uses by other industries.

Ancillary uses. Use by the increased population that will accompany large-scale development of oil shale is necessarily based on a projection of population growth. Published estimates suggest that a 50,000-BPD mine/plant may employ about 3,000 people.[1-8] Multiplier factors to convert workforce to total population vary, but many are in the range of three to five. This numerical factor may decrease as the size of industry increases. A factor of four would result in an estimated population of about 100,000 people for a 400,000-BPD industry. The figure may be substantially less if much of this production is from surface mines.

Estimates of water use per capita vary over wide ranges (perhaps 50-200 gpd/person covers most). Assuming 100 gpd per person as the required water system intake, then 100,000 people would need about 11,000 ac-ft/yr. The present population centers are located on the Colorado River and White River, and thus could be expected to obtain some or most of their additional water from these streams. These estimates do not take into account any large-scale growth in other industries or power generation in the area, and thus could be considered as minimum requirements.

Potential Water Supplies for an Oil-Shale Industry

Sources of water that may be used include local supplies of ground water and surface water, nearby supplies from the White and Colorado Rivers, and water imported from other areas. The following discussion is limited to local supplies which probably will be utilized first and which potentially could supply the needs of a sizable oil-shale industry. Several published reports discuss the potential for water supplies from local sources.[1-11; 14]

Ground water. Ground water in the Piceance Creek Basin occurs in two broad types of aquifer systems: (1) a near-surface aquifer system that

45

recharges and discharges within the basin, and (2) a series of deep aquifers that apparently both recharges and discharges outside the basin.

The geologic units in the near-surface system include the alluvial deposits along stream courses, the Uinta Formation, and the underlying oil-shale bearing Green River Formation. This aquifer system, in the northern part of the basin within the drainages of Piceance and Yellow Creeks, has been estimated to contain from 2.5 to 25 million acre-feet of ground water in storage.[10] Additional ground water occurs in the southern part of the basin. The quality of most of this water is such that it can be used by an oil-shale industry. Dewatering of oil-shale mines will be necessary in much of the basin, thus making a portion of this stored water available for use at the mine site. Off-site supply wells could furnish additional water. Dewatering rates for mines have been estimated to range from several hundred gpm to about 20,000 gpm. At Federal Lease Tract C-b, about 1000-1500 gpm of ground water was being produced during shaft sinking in late 1980. At Federal Tract C-a, dewatering rates for a small-scale experimental mining operation were typically in the 1500-2000 gpm range during 1978-80. Much of this water at Tract C-a was returned to the aquifer through injection wells and thus is available for future use. This rate of dewatering, about 3,000 ac-ft/yr, probably would be adequate to support a mine/plant producing about 15,000 to 60,000 BPD of shale oil. Extrapolating further, if one million ac-ft of the estimated 2.5-25 million ac-ft stored in this aquifer system in the Piceance basin were available for oil shale uses, it would theoretically support a 400,000-BPD industry for a period ranging from about 10-15 years to perhaps 50 years.

The deep aquifer system consists of several stratigraphic units beneath the Green River Formation. The lack of drill hole and test data on these units constrains any discussion of their water-bearing character-istics. Some extrapolation can be made from known geologic data and by inference from other areas where their water-bearing characteristics are better known. Several deep aquifers may be a potential source of ground water for an oil-shale industry: sandstones of both the Glen Canyon Sandstone and the Entrada Sandstone, the Weber Sandstone, and the Leadville Limestone. These four units, which lie several thousand feet below the base of the Green River Formation, aggregate a thickness of many hundred feet, and potentially could furnish significant amounts of ground water. Conservatively estimated, their ground water storage capacity may be many millions of acre-feet. No data are available on their water-bearing characteristics in the Piceance Creek Basin. However, these rocks are used as aquifers in several other areas where they lie at shallow depths, and where well yields typically are in the 10s-100s gpm range from each unit. Definitive data on water quality are not available for the Piceance Creek Basin, but their water is of usable quality in many other areas.

Preliminary data suggest that several million acre-feet of stored ground water exists beneath the oil shale area in the Uinta Basin, Utah.[6, 11, 13]

Surface water. Almost all the local runoff leaves the Piceance Creek Basin in four streams: Piceance and Yellow Creeks, which flow northward into the White River, Parachute and Roan Creeks, which flow south into

46

the Colorado River. The average total annual outflow in these streams is about 70,000 acre-feet.[12] Most of this flow is measured downstream from the present irrigated agricultural areas in the basin. The water is of such quality that it is suitable for many uses at an oil-shale plant.

Thus, much of this water could be utilized for an oil-shale industry. This locally derived streamflow could potentially supply most or all the water needs for an industry of about 400,000 BPD to 1½ million BPD.

The major surface water supplies are in the Colorado and White Rivers; the average annual flow in those two streams near the oil shale deposits is about 2.8 and .5 million acre-feet, respectively, sufficient for a very large oil-shale industry.[12]

### Potential Constraints on the Use of Water

Potential constraints on water use by an oil-shale industry are made up of a broad spectrum of legal, institutional, and economic factors. Included are international treaties and agreements with Mexico, the Colorado River Compact between upstream and downstream States, salinity goals or limits on the Colorado River system, State and Federal water quality standards and rules, and State laws pertaining to water rights. These potential constraints are simply mentioned here; the degree to which they operate will greatly affect the management of water supplies for an oil-shale industry.

### Potential Environmental Effects, Water Management Opportunities

Potential future effects of an oil-shale industry on the hydrologic system can be described in a qualitative way. The water management practices that are used, along with oil-shale extraction methods and land reclamation techniques, will greatly influence the magnitude and timing of these effects. The water rights system will impose certain management practices that can greatly affect streams, ground water, and water quality.

In brief, potential general effects of an oil shale industry on the hydrologic system and some applicable management methods are as follows:

1.  Mine dewatering and use of local water supplies will cause flows to diminish in many local streams, springs, and seeps, and thus in the Colorado and White Rivers. Because of probable offsetting effects, the reduction in flow may be less than consumption by industry. These effects include an increased capacity for local ground water recharge and reduced evapotranspiration use because of lowered ground water levels. Conjunctive use of surface water supplies and ground water in storage can minimize effects on streamflow.

2.  Ground water levels in many of the mining areas will decline in response to mine dewatering. Such declines can be dramatic.[10] However, declines can be minimized early on by reinjecting excess water into the same aquifer during the start-up phase of mining, thus making the water available for later use.

3. Changes will occur in the water quality of many streams, springs, and seeps. Such changes can either diminish or enhance the water quality, depending mainly on water-management practices. A few specific examples:

   A. The natural flows from Piceance Creek and Yellow Creek contribute about 1/5 to 1/4 of the salt load in the White River, but only about 1/20 of the total flow in the river. These salts are in large part derived from brackish to saline ground water inflows in the lower reaches of the two streams. Diverting these flows for utilization by an oil-shale industry would slightly diminish the flow in the White River below Piceance Creek but would significantly reduce the river's salt load.[8, 12]

   B. The concentration of dissolved solids (TDS) typically varies seasonally in both the White and Colorado Rivers. During the fall-winter baseflow months, the concentration of TDS commonly is 1.5 to 2 times higher than during the spring-summer snowmelt months. Maximizing diversion from these streams during periods of the more saline flows, perhaps to off-stream reservoirs or to ground-water recharge sites near the oil-shale plants, would tend to improve the overall water quality downstream. The effects could be measured by mass loading analysis.

   C. Water quality and the quantity of runoff and ground-water recharge in mined-out areas will be dramatically influenced by the extraction methods used and by the land reclamation techniques. Surface retorting of shale and subsequent surface disposal of the retorted shale is most likely to directly affect the quantity and quality of surface water, whereas in situ methods have the greatest potential to affect ground water. Reclamation practices for mined areas can significantly affect local runoff, recharge, evapotranspiration, and water quality. Sound water-management practices need to be applied to land reclamation because the effects achieved will be permanent.

Summary

The oil shale areas of Western Colorado have adequate water resources to support a very large oil-shale industry. Local runoff in Piceance Creek Basin and the ground water in storage beneath the basin constitute a resource that appears adequate for a large industry. The water-management practices used will greatly affect the water requirements by such an industry, as well as both short-term and long-term hydrologic and environmental effects.

# REFERENCES

1.  U. S. Dept. of the Interior, Final environmental statement for the prototype oil shale leasing program: 6 v., USDI, 1973.

2.  Rio Blanco Oil Shale Project, Detailed development plan, tract C-a: Gulf Oil Corp., Standard Oil Co. (Indiana), 3 v., 1976.

3.  Rio Blanco Oil Shale Project, Revised detailed development plan, tract C-a: Gulf Oil Corp., Standard Oil Co. (Indiana), 3 v., 1977.

4.  C-b Shale Project, Detailed development plan and related materials: Ashland Oil, Inc., Shell Oil Co., 2 v., 1976.

5.  C-b Shale Oil Project, Modification to detailed development plan: Ashland Oil, Inc., and Occidental Oil Shale, Inc., 1980.

6.  White River Shale Project, Detailed development plan, federal leases U-a and U-b; 2 v., White R. Shale Proj., 1976.

7.  U. S. Dept. of the Interior, Final environmental impact statement, proposed development of oil shale resources by Colony Development Operation in Colorado: USDI, 1977.

8.  U. S. Dept. of the Interior, Final environmental statement, proposed Superior Oil Company land exchange and oil shale resource development: USDI, Bureau of Land Management, 1980.

9.  U. S. Bureau of Reclamation, Alternative sources of water for prototype oil shale development, Colorado and Utah: USBR, Salt Lake City, 1974, 115 p.

10. J. B. Weeks, and others, Simulated effects of oil-shale development on the hydrology of Piceance basin, Colorado: USGS Prof. Paper 908, 1974.

11. D. Price, and T. Arnow, Summary appraisals of the nation's ground-water resources--Upper Colorado region: USGS Prof. Paper 813-C, 1974, 38 p.

12. U. S. Geological Survey, Water resources data for Colorado, v. 2, 3, Water Year 1979: USGS, 1980.

13. W. F. Holmes, U. S. Geological Survey, Written Communication, 1980.

14. U. S. Dept. of the Interior, Water management in oil shale mining: U. S. Bureau of Mines, prepared by Golder Associates, 1977.

15. S. G. Robson and G. J. Saulnier, Jr., Hydrogeochemistry and simulated solute transport, Piceance basin, northwest Colorado: USGS Open-File Report 80-72, 1980, 89 p.

## AIR QUALITY IMPACT OF OIL SHALE DEVELOPMENT

**Terry L. Thoem**
U.S. Environmental Protection Agency

Introduction

The potential oil shale industry may be characterized as the only example of a 60-year pregnancy. The promise and the potential of development of the vast oil shale resources has been discussed by, written about, planned for, and evaluated by countless individuals and organizations since and even before the famous 1918 National Geographic article. It is perhaps fitting that the special energy issue of National Geographic to be released early in 1981 will devote a substantial amount of space to oil shale. During the next year it is anticipated that we will see the birth of a few oil shale projects. This birth has been a painful ordeal to most of us. EPA's contribution to these labor pains varies depending upon who is assessing the pain. My presentation today should clarify the extent of influence that EPA's regulations, policies, and philosophies have upon the development of oil shale.

It must be clearly understood that EPA has never been an anti-oil shale advocate. We have, since we were created in 1970, advocated some limited development of oil shale technologies. We still maintain that philosophy The Prototype Oil Shale Leasing Program was and remains to be a good framework for the industry, government, and the public to evaluate the positive and negative aspects of the development of oil shale. Advantages which accrue to this approach are several. Industry can iron out technological flaws at an early stage and can better design commercial facilities. Industry and the financial community can better define the capital and operating costs associated with oil shale production. Industry, environmental regulators, and environmental organizations can more accurately quantify waste streams and their impacts on the environment.

Recent major pronouncements and decisions by government and industry constitute a potential threat to this "limited development approach"...

o   DOE is evaluating a 5 million barrel per day open pit operation in Piceance Basin.
o   DOI has decided to pursue additional leasing.
o   An Exxon planning study suggests that a goal of 8 million barrel per day of shale oil by 2010 is achievable.

Many of us have worked long and hard to provide a setting for the orderly development of oil shale in Colorado and Utah.  A delicate balance between oil shale development and  environmental protection is necessary. In my opinion suggestions that 5 or 8 million BPD are possible will fade away under an analysis of its associated costs.  Consider for a moment the air quality impacts of an 8 million barrel per day industry.

   Emission densities comparison

. PM        Oil shale   2X      NY NJ Conn. AQCR
            Oil shale   15X     Denver
            Oil shale   2 1/2X  NY-DC coast

. $NO_x$    Oil shale   2 1/2X  NY NJ Conn AQCR
            Oil shale   15X     Denver
            Oil shale   5 1/2X  NY  NJ Conn AQCR

. $SO_2$    Oil shale   =       NY NJ  Conn AQCR
            Oil shale   8X      Denver
            Oil shale   =       NY - DC coast

An additional perspective may be gained by comparing the particulate emissions from an 8 million BPD industry to all coal fired power plants projected for construction in the U.S. between 1980 and 2000.  They are essentially equal.

Further perspective on the air quality impacts may be gained by a discussion of the visibility impacts which would be associated with an 8 million BPD industry.  Existing visibility in Piceance Basin has been measured at better than 180 miles (300 km).  An 8 million BPD case analysis shows that visibility would be reduced to less than 12 miles (20 km) throughout the entire Piceance Basin.

Development of Colorado's and Utah's oil shale resource is a necessary part of the Nation's effort to reduce oil imports.  Energy Conservation measures, utilization of solar energy, utilization of the Nation's vast coal resources, and continued development of oil and gas resources are also parts of the necessary solution.  The key is a "determination to develop shale correctly".

The development must be "correct" for Colorado in terms of the rate of development, the eventual size of the industry, and the mix of technologies used by the industry.  Balance is needed.  Oil shale development should occur in an orderly fashion.  EPA Region VIII has provided specific guidance on its posture toward oil shale development in its Energy Policy Statement issued as a draft in Fall of 1979 and finalized in April 1980.  Items addressed in the Statement which relate directly to oil shale development include 1) expedited permit processing,

51

2) grandfathering, 3) rate of development, 4) "Better than Bact" controls, 5) EPA input to DOI on additional leasing, 6) planning and implementing a coordinated research effort and 7) information and communication efforts.

## Air Quality Impacts

The Clean Air Act Amendments of 1977 (PL 95-95)(Figure 1) provide the statutory basis for EPA's air quality regulatory development. The principal applicable portions of the Act include 1) Prevention of Significant Deterioration (PSD), 2) visibility, 3) State implementation plans, 4) New Source Performance Standards and 5) National emission standards for hazardous air pollutants. The States of Colorado, Utah and Wyoming have similar air pollution control legislative and regulatory authority.

PSD . . .

There are two major components of the PSD permit application review process. First, the applicant must demonstrate that the facility will employ Best Available Control Technology (BACT). Second, the controlled emissions must not cause ambient concentrations in excess of applicable PSD increments. Increments have been defined (Figure 2) for all areas of the country. (PSD Class I, II and III). Increments for National Parks and Wilderness areas allow only a slight amount of degradation while most other areas can accommodate a reasonable amount of industrial growth. There has been litigation regarding the PSD concept since 1972. The recent Alabama Power vs EPA decision changes some of EPA's Prevention of Significant Deterioration of Air Quality (PSD) regulations. Final regulations to satisfy the court ruling were published on August 7, 1980. Changes include 1) the need for a BACT assessment for all pollutants regulated under the Act (Figure 3) except where controlled emissions will be below de minimis levels (Figure 4), 2) the requirement for an analyses of baseline air quality concentrations for all pollutants regulated under the Act, and 3) guidance on modelling, phased construction, and stack heights. The analyses for baseline pollutant concentrations may include one year of continuous monitoring. Alternatives would include either a showing that existing data (for NAAQS pollutants) collected at other locations were representative or modelling existing emissions (for non-NAAQS pollutants) in the area to estimate concentrations. Modelling guidance contained the opinion that air quality models are on the frontier of science but must be used. It should be noted that the opinion was rendered that if a PSD increment was initially judged not be be violated but was later found to be violated either through monitoring data or improvements in modelling techniques, the State Implementation Plan could/should be revised to require additional control.

The size of an oil shale industry may be limited by air quality Class I increments at the Flat Tops Wilderness Area and/or the Class II increments in areas where facilities cluster. This concept has caused significant activity in the past year by both industry and government. Preliminary analyses indicate that development of 200,000 to 400,000 BPD could consume the increment. (Figures 5 and 6) Several oil shale developers considered rushing in to get their share of the increment. This has not occurred. EPA has recognized that a Regional Complex

52

Terrain Model applicable for oil shale country must be developed.  A
"planning accuracy" modelling effort has recently been supplied in draft
form under contract to EPA.  Results of this study were unfortunately
inconclusive due to the assumptions and methodology employed.  The
design of a two year field data gathering effort followed by model
development or refinement has been initiated.  It is our goal to have a
useable Regional model for use in permitting decisions by 1982.  Joint
funding between government and industry is being sought.

The review of PSD permit applications is establishing a better emissions
data base than previously available.  To date PSD permits have been
issued for five oil shale projects.  Figure 7 shows permitted emission
rates on a per barrel of oil produced for four of the permits.  The fifth
permit was for a research-type operation.  Given the concern over
consumption of the PSD increments it is safe to conclude that these
permitted limits should provide an upper benchmark for future applications.  In order to provide a comparison between an oil shale facility
and a conventional fossil fuel plant, Figure 8  compares the Colony
emissions rate to a conventional 1000 MWe coal fired power plant meeting
New Source Performance Standards (NSPS).  These facilities are comparable
on an energy basis of coal input and oil putput.

Visibility ...

EPA is involved in litigation regarding visibility regulations.  Regulations were to have been promulgated by August 1979.  Since they were not,
FOE and EDF sought court relief.  We are under court order to propose
regulations by May 18, 1980 and to promulgate by November 18, 1980.
Regulations were  proposed   on May 15, 1980 (Federal Register of May 22,
1980).  A public hearing was held on July 2, 1980 in Salt Lake City.
Based upon limited visibility analyses performed via PSD reviews on oil
shale projects to date, visibility protection could provide additional
constraints, beyond the Class I PSD increments, to the size of an
industry.  These constraints could materialize based upon single source
impacts (plume blight) or cumulative impacts (regional haze).  The
reality of this additional constraint depends to a large extent upon
policy decisions made by EPA on Congressional intent of the visibility
provisions (the in/out issue) and case-by-case decisions by the Federal
Land Manager on the adversity of impacts in addition to the visibility
goals established for the area.

State Implementation Plans ...

States have the primary regulatory responsibility for controlling air
pollution sources within the State.  Colorado has established an
emission standard of 0.3 pounds of $SO_2$ per barrel of oil produced.
Three of the active oil shale companies have petitioned the Colorado
Air Pollution Control Commission to change this standard.  Colorado has
also designated certain areas of the State as Category I with increments
the same as EPA Class I.

New Source Performance Standards ...

NSPS for oil shale facilities have not yet been established.  EPA has
taken several actions designed to provide guidance in lieu of NSPS.
On January 18, 1980, EPA established a management structure designed to
be responsive to the President's synthetic fuels and Energy Mobilization
Board initiatives.  Two policy groups, the Alternate Fuels Group and

the Permits Coordination Group, will operate under the direction of the EMB Task Force. The Alternate Fuels Group will develop the Agency's regulatory and research strategy for synthetic fuels industries. Working groups within the AFG have been established for oil shale, direct coal liquefaction, indirect coal liquefaction and coal gasification, and ethanol from biomass. The AFG will coordinate preparation of environmental guidance for these emerging technologies for use by industry planners and permitting officials. The AFG will also recommend and oversee preparation and promulgation of control technology based standards for synthetic fuels facilities. The preparation of Pollution Control Guidance Documents (PCGD) will provide the first component of the basis for these standards. The first draft of the oil shale PCGD is scheduled for early 1981. Industry, government and the public will be asked to review and provide comment on the draft PCGD. Final publication is slated for Fall 1981. The second policy group, Permits Coordination Group, will provide national management of permitting mechanisms in order to respond effectively to demands for expediting permit decisions. The PCG will develop a system for tracking permits, establish project decision schedules for permit review, and assure that potential environmental and permitting problems are recognized early.

Another area of EPA action in oil shale is the implementation of an oil shale environmental research program.

The major thrust and accomplishments over the past year have included: the development of the document "Environmental Perspective on the Emerging Oil Shale Industry" which presents general information revelant to oil shale pollution problems and their control as they are viewed today; the initiation of work on a pollution control technology guidance document which will discuss the applicability, performance and costs of pollution control alternatives available for the oil shale industry; and the completion of several field research studies.

The first document became available in draft form in the Summer of 1979. Following extensive review by EPA, other government agencies, the oil shale industry, and the environmental community, numerous changes were made. This document titled "Environmental Perspective on the Emerging Oil Shale Industry" is scheduled for release in December 1980. This report, prepared by the EPA Oil Shale Research Group, will convey EPA's understanding and perspective of environmental aspects of oil shale development by providing a summation of available information on oil shale resources; a summary of major air, water, solid waste, health, and other environmental aspects; analysis of potentially applicable pollution control technology; a guide for the sampling, analysis, and monitoring of emissions, effluents, and solid wastes from oil shale processes; suggestions for environmental goals, and a summary of major retorting processes, emissions and effluents.

Other air research efforts include 1) assessment of appropriate $H_2S$ and other sulfur compounds control technology, 2) characterization of various oil shale technologies air emissions, 3) assessment of abandoned retort fugitive $SO_2$ emissions, 4) ambient visibility monitoring, and 5) evaluation of trace element concentrations in oil shale.

## Conclusion

Mining and conversion of oil shale will degrade air quality. The question is the magnitude and the significance of the degradation. Key air quality questions such as the following exist:

1. What are the concentrations of various sulfur species in retort off gas streams and what is the best approach to control of these compounds?

2. What will be the air quality and visibility impacts on the Flat Tops Wilderness Area (nearest Class I area)?

3. What are the expected trace element concentrations in air, water, and solid waste residual streams?

4. Is conventional pollution control technology directly applicable to oil shale gas streams? Is it effective?

5. What are the air quality impacts from the expected population growth associated with the development of an oil shale industry?

These questions may sound familiar to some of you. They should be, since they are the same ones posed in our EPA presentation at the 11th, 12th, and 13th Annual Oil Shale Symposiums. The fact that they remain unanswered points out to us that the environment is not much closer to being ready for oil shale than it was a year ago.

Answers to the above questions (and perhaps other questions not yet posed) will in part determine the ability of individual plants and of an oil shale industry to be compatible with the desired environment for oil shale country.

Answers to some of the above questions may be partially answered by theoretical research work and limited-scope field investigations in the absence of any oil shale facilities. Answers to the remaining questions will necessarily be developed through rigourous testing programs and data analyses performed on facilities representative of commercial size. Reliable answers will be obtained only through a cooperative coordinated effort among government and industry.

Development of oil shale appears to be closer to reality than ever before. EPA has for years advocated that some small scale development occur so that some of the unanswered environmental questions could be answered. This remains to be our philosophy toward the development of oil shale. There are too many uncertainties associated with its development to go full speed ahead. Despite the reasonably good efforts by industry to date there have still been water discharges in excess of NPDES permit limits, ignition of processed/raw shale mixtures, and surprises in the sulfur species content of gas streams.

If oil shale development proves to be economically attractive and technologically feasible, EPA advocates that development occur in an orderly phased manner. Development up to some as yet undetermined size can and must occur in a manner compatible with environmental standards and objectives.

FIGURE 1

EPA LEGISLATIVE MANDATES

| | |
|---|---|
| CLEAN AIR ACT AMENDMENTS OF 1977 | PL 95-95 |
| CLEAN WATER ACT AMENDMENTS OF 1977 | PL 95-217 |
| SAFE DRINKING WATER ACT OF 1974 | PL 93-523 |
| RESOURCE CONSERVATION & RECOVERY ACT OF 1976 | PL 94-580 |
| TOXIC SUBSTANCES CONTROL ACT OF 1976 | PL 94-469 |
| NOISE CONTROL ACT OF 1972 | PL 92-574 |

FIGURE 3

POLLUTANTS REGULATED UNDER THE
CLEAN AIR ACT AMENDMENTS OF 1977

| NAAQS | NCN-NAAQS |
|---|---|
| PARTICULATE MATTER | BE |
| $SO_2$ | HG |
| $NO_x$ | ASBESTOS |
| CO | VINYL CHLORIDE |
| NHHC | F |
| $O_3$ | $H_2SO_4$ MIST |
| $P_B$ | TOTAL REDUCED SULFUR |
| | BENZENE? |

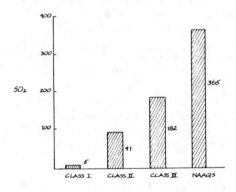

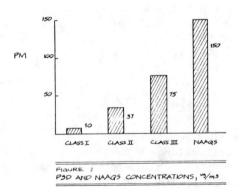

FIGURE 2
PSD AND NAAQS CONCENTRATIONS, ᵁᵍ/m3

FIGURE 4

POLLUTANT DE MINIMIS LEVELS (CONTROLLED)

| POLLUTANT | GUIDELINE FOR BACT | | GUIDELINE FOR DETAILED IMPACT ANALYSIS | |
|---|---|---|---|---|
| CO | 100 TON PER YEAR | | 500 µG/M³ | 8 HOUR |
| $NO_2$ | 10 | " | 1 | " ANNUAL |
| TSP | 10 | " | 5 | " 24 HOUR |
| $SO_2$ | 10 | " | 5 | " 24 HOUR |
| $O_3$ (VOC) | 10 | | - | - |
| $P_B$ | 1 | | 0.03 | 3 MONTH |
| HG | 0.2 | | 0.1 | 24 HOUR |
| BE | 0.004 | | 0.005 | 24 HOUR |
| ASBESTOS | 1 | | 1 | 1 HOUR |
| F | 0.02 | | 0.01 | 24 HOUR |
| $H_2SO_4$ | 1 | | 1 | 24 HOUR |
| VINYL CHLORIDE | 1 | | 1 | MAX |
| TRS | | | | |
| $H_2S$ | 1 | | 0.5 | 1 HOUR |
| METHYL MERCAPTAN | 1 | | 0.5 | 1 HOUR |
| DIMETHYL SULFIDE | | | | 1 HOUR |
| DIMETHYL DISULFIDE | | | | 1 HOUR |
| $CS_2$ | 10 | | 200 | 1 HOUR |
| $CO_S$ | 10 | | 200 | 1 HOUR |

SEPTEMBER 5, 1979 FEDERAL REGISTER 40 CFR 51 AND 52

FIGURE 5

EPA MODELLING RESULTS
24 HOUR CONCENTRATIONS, IN $\mu G/M^3$

PROTOTYPE LEASE EVALUATIONS - 1976

o   200,000 BPD

|  | TSP | SO$_2$ |
|---|---|---|
| DINOSAUR NATIONAL MONUMENT | ~18 | 5 |
| COLORADO NATIONAL MONUMENT | < 3 | < 1 |
| WHITE RIVER NATIONAL FOREST | ~ 8 | < 2 |
| ASHLEY NATIONAL FOREST | ~ 5 | ~ 1 |

PSD PERMITS - 1979

o   IMPACTS ON FLAT TOPS - 62,000 BPD

|  | TSP | SO$_2$ |
|---|---|---|
| C-A | < 0.1 | 0.3 |
| C-B | 0.5 | < 0.1 |
| COLONY/UNION | 1.4 | 1.1 |
| TOTAL | ~ 2.0 | 1.5 |

$$\frac{5.0}{1.5} \times 62,000 \text{ PBD} = \boxed{207,000 \text{ BPD}}$$

FIGURE 6

EPA MODELLING RESULTS - 1980

ASSUMPTIONS

o   BOX MODEL - 40 KM WIDE, VARIABLE HEIGHT
o   400,000 BPD
o   "COLONY TYPE" EMISSIONS
o   NO CHEMICAL TRANSFORMATION OR DEPOSITION
o   NO DISPERSION OUT OF BOX
o   HOMOGENEOUS MIXING

RESULTS, 24 HOUR CONCENTRATIONS

| METEOROLIGICAL CASE<br>MIXING HEIGHT AND WIND SPEED | TSP | SO$_2$ |
|---|---|---|
| 2500 M, 6 M/SEC | 0.4 $\mu G/M^3$ | 0.5 $\mu G/M^3$ |
| 1000 M, 5 M/SEC | 1.3 | 1.6 |
| 500 M, 5/M/SEC | 2.6 | 3.2 |
| 500 M, 3 M/SEC | 4.4 | 5.4 |
| 400 M, 4 M/SEC | 4.1 | 5.0 |

NOTES

MEAN ANNUAL METEOROLOGICAL CONDITIONS

A.M. MIXING HEIGHT          400 M
P.M. MIXING HEIGHT          2500 M
A.M. WIND SPEED             4 M/SEC
P.M. WIND SPEED             6 M/SEC

FIGURE 7

PSD PERMIT EMISSION RATES

POUNDS PER BARREL OF OIL

| POLLUTANT | COLONY | UNION | C-B | C-A |
|---|---|---|---|---|
| SO$_2$ | 0.164 | 0.237 | 0.160 | 0.668 |
| NO$_x$ | 0.903 | 0.291 | - | - |
| HC | 0.158 | 0.125 | - | - |
| PM | 0.134 | 0.102 | - | - |
| CO | 0.036 | 0.172 | - | - |

FIGURE 8

OIL SHALE VS POWER PLANT COMPARISON

| | TONS PER YEAR | |
| | 50,000 BPD | 1000 MWE |
| POLLUTANT | COLONY | POWER PLANT |
|---|---|---|
| SO$_2$ | 1239 | 13,790 |
| PM | 1008 | 1,062 |
| NO$_x$ | 6817 | 17,714 |

# ESTIMATION AND MITIGATION OF SOCIOECONOMIC
# IMPACTS OF WESTERN OIL SHALE DEVELOPMENT

**Dee R. Wernette**
Argonne National Laboratory

Introduction

Rapid, large-scale energy developments occuring in rural areas frequently result in negative social and economic impacts on the host communities.[1] Many of these impacts are caused by the rapid in-migration of large numbers of workers with their dependents, which can overwhelm the local social, housing, public service, and economic infrastructures. These impacts have been identified with coal and other related energy developments in the Western United States,[1,2,3,7] and have been suggested as likely consequences of the rapid development of a large oil shale industry in the area.[5,6]

Two general approaches may be used in reaction to the threat of these negative social and economic impacts. One approach, labeled here as impact avoidance, entails planning the location, timing, and size or scale of the facility so as to avoid or minimize negative impacts. By siting small scale facilities in rural areas, for example, and larger facilities in urban areas, negative socioeconomic impacts can potentially be reduced or avoided. The second approach may be labeled impact mitigation. Even when large facilities are sited in rural areas, some measures may be taken to reduce the negative impacts. Such measures could reduce the numbers of in-migrants, hasten and/or facilitate the development of the necessary new support systems, or some combination of both. One could, for example, encourage workers (through financial or other incentives) to not bring their families to the development site. Or, one could facilitate the construction of needed new public service facilities (police and fire stations, water and sewage treatment systems, etc.) in advance of the population influx through loans or grants to impacted communities.

The purpose of this paper is two-fold: to identify the likely levels of negative socioeconomic impacts of the development of oil shale facilities in the West; and to show the potential value of avoidance and mitigation measures in reducing these impacts. Other factors, such as environmental and public health impacts, or economic or technical considerations, are beyond the scope of this analysis, although they would also likely affect the siting and development of an oil shale industry. The following discussion is concerned solely with estimating and avoiding and/or mitigating the socioeconomic impacts from Western oil shale development.

Methods and Measures

The findings discussed here were produced by a computer model (SEAM - the Social and Economic Assessment Model) which can simulate the siting of various energy facilities in any of over 3,000 counties in the continental United States. A detailed description of the model is not feasible here.[4] A general understanding of the model's procedure is necessary, however, for accurate interpretation of the findings presented below.

Projecting socioeconomic impacts with SEAM consists of three steps. First, the age- and sex-specific populations of the impact and adjacent counties are projected to the year 2000, using the cohort survival method. These projections are based on the best available fertility, mortality, and migration data. By comparing national and county-level age- and sex-specific basic and secondary labor force participation rates, and adding the estimated numbers of available commuters from other counties, SEAM projects the annual number of basic and secondary workers available for work. The annual number of basic workers required to construct and op-erate various types of energy facilities is part of the model's data base. In the second step this data is combined with secondary employment multi-pliers, to project the annual numbers of additional secondary workers associated with the energy development. By comparing the annual numbers of locally available and required basic and secondary workers, SEAM com-putes the annual numbers of in-migrant workers and households. The final section of SEAM computes the additional costs of providing 11 different public services and facilities for the construction peak and operation populations in the impact county. As will be noted below, these results may be produced for two or more simultaneous energy developments in one or more counties.

Western Oil shale deposits are found in a number of counties in Colorado, Utah, and Wyoming. The deposits which have attracted the most attention and interest to date, however, are located in the Piceance basin, in the Colorado counties of Rio Blanco, Garfield, and to a lesser extent Mesa. Consequently, simulated oil shale facilities are sited only in these three counties in this study. This is done to show the potential socio-economic impacts of the initial shale oil development projects, since these are likely to occur in this area. The developments discussed show construction of commercial scale oil shale facilities beginning in 1983. This is the earliest likely date for the construction of such facilities, and is probably somewhat optimistic. 1983 was chosen to facilitate comparison of our results with the findings of the Office of Technology Assessments (OTA) study.[5] Whether these or other levels of production capacity will be developed by 1990 is unclear. The findings presented here indicate the likely impacts of such development, but neither advocate nor assume that such development will occur.

A variety of demographic and fiscal impact measures are provided by SEAM. Of these, the average annual growth rate during the peak construction years will be used because it has the closest empirical link to negative impacts. Specifically, county-level annual growth rates at or above 7% will be interpreted as indicative of significant local fiscal, economic, and social impacts.[7]

Findings

Table 1 presents the labor requirements for the construction and operation of three commercial-scale oil shale facilities which might be built within the next decade. These labor requirement figures reflect best engineering estimates, since such facilities have not yet been constructed and tested. The Paraho/Tosco facility has the shortest construction period, but also has the greatest labor demand at construction peak (4750 in year 4). The 57,000 bbl/day Occidental facility ranks second in the length of its construction period, and has the smallest labor demand at construction peak (2875). One concern in the discussion which follows will be with estimating the maximum likely level of oil shale development which could be constructed in the shortest time period, with average annual population growth rates at or below the 7% level. Therefore, our simulations will include only these two types of developments.

Table 2 presents the projected impacts of these three oil shale facilities sited in each of these three counties. As noted, one impact avoidance technique consists of siting only appropriately sized facility in the variously sized counties. Rio Blanco has the smallest population; Mesa has the largest; and Garfield in intermediate in size. It is not surprising, therefore, that simulated impacts, measured here in average annual population growth rates, are greatest in the smallest county, and smallest in the largest county. The simulated impacts also vary between facility types, with Paraho/Tosco associated with the greatest simulated impacts, and Occidental 57,000 bbl/day (hereafter Occi-57) with the fewest. Impact avoidance through matching county and energy facility sizes can clearly reduce the projected levels of socioeconomic impacts. If one were to use this type of impact control measure only, one would site the Paraho/Tosco facility only in Mesa county, and site Occi-57 facilities in Garfield and Rio Blanco counties. For each of these facility-county combinations the projected average growth rates are below the 7% level. The scenarios discussed below vary from these combinations only in siting the Paraho/Tosco facility in Garfield county. As will be seen, this is made possible through the simulated mitigation measures under study.

Tables 3 and 4 show the effects on projected annual growth rates of the simulation of two additional impact avoidance measures: the presence or absence of simultaneous developments in adjacent counties (Table 3), and the timing of more than one facility constructed within the same county (Table 4). Simultaneous developments in adjacent counties could reduce the numbers of commuters available to work on a construction project, with resulting increased inmigration and higher projected population growth rates. As we see from Table 3, however, the projected effects of this are very small. This is due primarily to the relatively small projected numbers of commuters from adjacent counties. One would expect these effects to be greater in a more densely populated area, with a larger potential commuter pool. The differences in projected impacts shown in Table 4, in contrast, are much greater. The timing of the construction of more than one facility within a county affects the projected level of socioeconomic impacts: the closer the construction starting dates, the greater are the potential impacts. This means that concern for socio-economic impacts, among other things, could limit the rate of growth of an oil shale industry sited within this restricted geographical area. This is because the more rapid such growth, the greater is the projected socioeconomic impacts, all else equal. This factor, like the facility-county siting combinations discussed above, may be termed impact avoidance, since it is affected by the macro-level planning for the development of an oil shale industry. Clearly such impact avoidance measures could be important in affecting the levels of socioeconomic impacts of the construction of an oil shale industry in this area.

In contrast to the measures discussed thus far, impact mitigation measures take the siting and timing of a facility as given. The mitigation measures discussed here attempt to reduce the impact of such development through either (A) reducing the numbers of inmigrant workers associated with the development; or (B) reducing the number of dependents accompanying inmigrant workers. The projected number of inmigrant workers required can be changed in one of two ways. One way is to increase the number of potential basic and secondary commuter workers available to work in the impact county through such means as providing free commuting transportation, or even paying workers for part or all of their commuting time. Since no studies on the changing willingness to commute as affected by financial compensation are known to the author, simulation of this measure is arbitrary. For our purposes it is estimated that 90% of the available workers in adjacent counties, and 66% of the available workers in counties once removed, could be induced to commute to the impact county, given adequate incentives. These workers are in counties linked by major roads to the three counties under study. The effects of this simulated impact mitigation measure are shown in column four of Table 5. As can be seen by comparing the values in this column with these in column one (No Mitigation Measures), encouraging increased commuting from adjacent areas can significantly reduce the projected average annual population growth rates associated with oil shale development in this area.

A second means of reducing the number of immigrant workers is through reducing the number of secondary workers associated with each basic worker. Feeding and housing workers en masse in barracks or other company operated living units, with reduced retail goods and services available, is one way of accomplishing this. This mitigation measure is simulated in the figures in column two, Table 5, by setting the secondary employment multiplier at .5, i.e. one additional secondary job created for every two additional basic jobs created. Like the changed commuter willingness levels discussed above, this value is arbitrarily set and should not be taken as an accurate measure. This level set reflects the author's impressions from having worked in a remotely-sited Canadian-gold mine. Clearly a more accurate and up-to-date measure of the lowest feasible level of this multiplier is desireable. Comparing columns one and two in Table 5 shows that reducing the secondary employment multiplier to this level can reduce projected annual growth rates, especially in smaller counties such as Rio Blanco and Garfield.

The third impact mitigation measure simulated in the Table 5 findings is changing the household size for inmigrant workers. One could reduce this in one of two ways: hire only single inmigrant workers; or pay inmigrant workers extra or leave their spouses and/or dependents elsewhere. The difficulty and costs, financial and others, entailed in either of these methods have not been studied. Lacking such data, the household size for inmigrant basic and secondary workers was set at 1.33, i.e. only one in three inmigrant workers would bring one dependent with him/her, and none would bring more than one dependent. When set at this level, the changed household size factor has a clear impact on the projected annual average growth rates. This can be seen by comparing the figures in columns one and three.

Each mitigation measure reduces the projected annual growth rate, as is shown in Table 5. The reductions, and thus the potential value of these measures, are greatest in the smaller, otherwise more heavily impacted counties (Rio Blanco and Garfield). Clearly the effects of these measures are not purely additive, as may be seen in the last column in Table 5, in which the three mitigation measures are combined. The effects of reducing the average household size of inmigrant workers will be much less, for example, if one has already reduced the projected numbers of such workers by increasing the projected number of available commuter workers. The combination of all three mitigation measures reduces the projected average annual growth rates in all three counties below the 7% level. These impact measures do not reflect the simultaneous developments in the adjacent counties, however. The projected impacts with this simultaneous development are shown in Scenario 2 of Table 6. Since the projected growth rates for each of these counties are at or below the 7% level, this suggests that with the use of the impact avoidance and mitigation measures discussed here, approximately 250,000 bbl/day of oil shale capacity could be constructed in these three counties by 1990, with projected over-all levels of socioeconomic impacts which are considerably more manageable than otherwise. Indeed, Scenario 3 indicates that over 300,000 bbl/day capacity could be constructed in this time period in this area without unmanageable socioeconomic impacts. This is, however, the likely upper limit for oil shale construction in this time period and geographical area, without additional measures or costs.

Discussion

Socioeconomic impact considerations are only one factor affecting the siting and timing of the construction of an oil shale industry. Economic, environmental, and land ownership or access considerations are among the other factors affecting this. Some of these factors, such as the economic, will be affected by socioeconomic considerations. The economic attractiveness of oil shale development would almost certainly be decreased by the higher labor costs resulting from the mitigation measures discussed here. The costs of strict control of environmental pollutants could likewise leave relatively few funds for socioeconomic impact mitigation. The viability, timing, and location of an oil shale industry will likely be affected by these and related factors. The issues and options discussed here are thus only one part of this larger analysis of a potential oil shale industry.

A number of caveats are appropriate in interpreting the findings presented above. The first concerns the interpretation of the impact mitigation measures and their projected effects. As was noted, the values set for the employment multiplier, household size, and commuter willingness levels were not based on empirical findings, and can only be considered hypothetical until replaced or substantiated by studies. A second caveat concerns the degree of congruence between the construction skills required and those available in the local labor force. Data on the skill requirements for these technologies are not yet available. This could obviously affect the projections presented here. For these and other reasons these findings should be viewed as tentative rather than certain, and approximate rather than completely accurate. In addition, other impact mitigation measures not examined here could also lessen the socioeconomic impacts of oil shale development. These shortcomings suggest the need for more research in this area. For only with such research findings can the development of American's energy resources be planned in such a way as to minimize the negative socioeconomic impacts on the people and communities involved. Therein lies the true value of the estimation and mitigation of the socioeconomic impacts.

Acknowledgements

The assistance and support of the following individuals are gratefully acknowledged:  David South, for developing the labor requirements for these oil shale technologies, and for general advise on the feasibility and timing of oil shale developments;  Mark Bragen, for his yoeman efforts in developing, refining, and facilitating the use of SEAM;  Marla Kozelka and Sharon Ryan for their patient typing and retyping of this manuscript;  and Richard Winter and Danilo J. Santini, for their comments, criticisms, and suggestions on the first draft. Any errors and shortcomings which remain are in spite of their efforts, and are the sole responsibility of the author.

References

1.  For a good introduction to this boomtown literature, see Charles F. Cortese and Bernie Jones, Boom Towns:  A Social Impact Model with Propositions and Bibliography, Denver: Social Change Systems, Inc., 1976 and Wm. Frenderburg, Social Science Perspectives on the Energy Boomtown:  A Literature Survey, New Haven, Conn.:  Yale University Institute for Social Policy Studies, 1976.
2.  Jos. Davenport III and J. Davenport, eds., The Boomtown: Problems and Promises in the Energy Vortex, Laramie, Wyoming, 1980
3.  Steve H. Murodock and F. Larry Leistritz, Energy Development in the Western United States, New York, N.Y.:  Praeger Publishers, 1979
4.  David W. South and Mark Bragen, Argonne National Laboratory, unpublished report, 1980.

5.  Office of Technology Assessment, An Assessment of Oil Shale Technologies, Washington, D.C.:  U.S. Government Priting Office, 1980.
6.  Peter D. Miller, "Stability, Diversity, and Equity:  A Comparison of Coal, Oil Shale, and Synfuels" in Unseld, Chas. T., et. al., eds., Sociopolitical Effects of Energy Use and Policy, Washington, D.C.:  National Academy of Sciences, 1979.
7.  Danilo J. Santini, L. Gardner Shaw, and Edward Tanzman, "Fiscal Impacts of Energy Facilities on County Governments", Proceedings of the Pittsburgh Modeling and Simulation Conference, Pittsburgh, PA, 1980

Table 1  Annual Basic Worker Labor Requirements for the Construction and Operation of Three Shale Oil Production Facilities

| Year | Occidental Modified In-Situ, 100,000 bbl/day | Occidental Modified In-Situ, 57,000 bbl/day | Paraho/Tosco Surface Retort 96,000 bbl/day |
|---|---|---|---|
| 1 | 0 C* | 0 C* | 250 C* |
| 2 | 225 C | 225 C | 1000 CO |
| 3 | 350 C | 350 C | 4000 CO |
| 4 | 1125 C | 1125 C | 4750 CO |
| 5 | 1600 C | 1600 C | 3500 CO |
| 6 | 2500 C | 2500 C | 3550 CO |
| 7 | 3375 C | 2875 C | 2350 CO |
| 8 | 3500 C | 2000 C | 2125 O |
| 9 | 3000 C | 1625 O | 2125 O |
| 10 | 2700 C | | |
| 11 | 2600 O | | |
| 12 | 2600 O | | |
| Total Person/years for Construction | 18,375 | 10,675 | 13,250** |
| Person/years of construction per bbl/day capacity | .18 | .19 | .14** |

* "C" indicates that these basic workers are employed in the facility construction. "O" indicates workers operating the plant. "CO" indicates some workers are constructing parts of the facility, while others are operating it at the same time.

** This figure is based on only the construction employment in those years identified by "CO."

Source:  David W. South and Mark Bragen, Argonne National Laboratory unpublished report, 1980.

Table 2  Projected Mean Annual Population Growth Rates During the 7 to 10 Year Construction Periods of Three Oil Shale Facilities in Three Colorado Counties*

| Colorado | Occidental Modified In-Situ 100,000 bbl/day | Occidental Modified In-Situ 57,000 bbl/day | Paraho/Tosco Surface Retort 96,000 bbl/day |
|---|---|---|---|
| Garfield | 7.86% | 7.14% | 12.9% |
| Mesa | 2.48 | 2.2 | 5.04 |
| Rio Blanco | 17.05 | 16.34 | 30.86 |

* These growth rates are compounded annually, i.e. the inmigrants of one year are counted as part of the base population in computing the growth rate for the next year. Each simulated development was projected to begin in 1983. These results reflect the projected construction-induced growth, assuming that only the development in question is occuring in that and adjacent counties. The results of simultaneous developments in one or more adjacent counties are shown in Tables 3 and 5 which follow.

Table 3  Projected Mean Annual Growth Rates Over the 7 to 8 Year Construction Periods from Simulated Oil Shale Development with and without Simultaneous Developments in Adjacent Counties

| County, Development Type and Size, and Date of Construction Start | Without Simultaneous Development | With Simultaneous Development |
|---|---|---|
| Garfield, Colorado Paraho/Tosco 96,000 bbl/day Facility, Construction Start: 1983 | 12.9% | 13.14% |
| Mesa, Colorado, Paraho/Tosco 96,000 bbl/day Facility, Construction Start: 1983 | 5.05 | 5.87 |
| Rio Blanco, Colorado, Occidental Modified In-Situ, 57,000 bbl/day Facility, Construction Start: 1983 | 16.34 | 16.74 |

64

Table 4  Projected Mean Annual Growth Rates Over the 7 to 14 Year Construction Periods from the Simulated Siting of Two Oil Shale Developments with Various Phasings of Construction Schedules

| County and Development Type and Size | Phasing of Construction Schedules | | |
|---|---|---|---|
| | One Development Begins in 1983, the Other Begins in 1990 | One Development Begins in 1983, the Other Begins in 1987 | Both Developments Begin Construction in 1983 |
| Garfield, Colorado Paraho/Tosco 96,000 bbl/day | 10.68% | 13.08% | 22.06% |
| Mesa, Colorado Paraho/Tosco 96,000 bbl/day | 4.38 | 5.65 | 9.65 |
| Rio Blanco, Colorado Occidental Modified In-Situ | 14.2 | 17.06 | 29.88 |

Table 6  Projected Annual Growth Rates from the Simulated Sitings of Simultaneous Oil Shale Developments in Three Colorado Counties

| Counties | Scenarios | | |
|---|---|---|---|
| | 1 | 2 | 3 |
| Garfield, Colorado | 13.14% | 6.27% | 6.27% |
| Mesa, Colorado | 5.87 | 3.06 | 3.62 |
| Rio Blanco, Colorado | 16.74 | 7.1 | 7.1 |

Scenario 1: Paraho/Tosco Facilities in both Garfield and Mesa Counties, and one Occidental 57,000 bbl/day In-Situ Facility in Rio Blanco county, all construction starts in 1983, no mitigation measures used.

Scenario 2: Same as Scenario 1, but all 3 mitigation measures used.

Scenario 3: Same as Scenario 2, with an additional Occidental 57,000 bbl/day In-Situ Facility sited in Mesa County, construction starting in 1983.

Table 5  Projected Annual Growth Rates from the Simulated Siting of Oil Shale Developments in Three Colorado Counties, with the Simulation of Three Possible Impact Mitigation Measures

| County, Development Type and Size, and Date of Construction Start | Presence and Type of Mitigation Measure Simulated | | | | |
|---|---|---|---|---|---|
| | No Mitigation Measures | Changed[1] Secondary Employment Multiplier | Changed[2] Household Size Factor | Changed[3] Commuter Willingness Levels | Combined All 3 Mitigation Measures |
| Garfield, Colorado Paraho/Tosco 96,000 bbl/day, 1983 | 12.9% | 8.28% | 9.78% | 7.17% | 5.86% |
| Mesa, Colorado Paraho/Tosco 96,000 bbl/day, 1983 | 5.05 | 3.9 | 2.42 | 3.16 | 2.79 |
| Rio Blanco, Colorado Occidental Modified In-Situ, 57,000 bbl/day, 1983 | 16.34 | 11.74 | 4.7 | 5.45 | 4.04 |

[1] This simulates the providing of secondary services to the workers by the construction firm, which reduces numbers of secondary workers. This multiplier was changed from 1.8 to .5.

[2] This simulates the construction company encouraging the workers to leave spouses and dependents at other locations through financial and other incentives. This multiplier was changed to 1.33 from 1.9 for basic workers, and to 1.33 from 2.46 for secondary workers.

[3] This simulates the construction company providing free transportation for commuting workers within the region and additional incentives to encouraging commuting, as necessary. This parameter was changed to 90% of available workers in adjacent counties and 66.6% of available workers in counties once removed, from approximately 26% and 3.9% respectively.

65

# OIL RECOVERY AND RECYCLE WATER TREATMENT
# FOR IN SITU OIL SAND PRODUCTION

**J. Nagendran**
Alberta Environment

**S. E. Hrudey**
University of Alberta

## Introduction

The Alberta Oil Sands represent one of the world's largest hydrocarbon reserves. They consist of 4 major deposits (Figure 1) with total in-place reserves estimated at $950 \times 10^9$ barrels (B) of bitumen.

Of this total reserve only $74 \times 10^9$ B are available with less than 50 m of overburden, the current economic cutoff for existing surface mining technology. The majority of the reserve, $741 \times 10^9$ B, is below 150 m and will require the application of *in-situ* recovery technologies to allow exploitation.

The available and/or proposed processes for recovery of these hydrocarbons are summarized in Figure 2.

Currently, there are two full scale surface mining plants (Suncor and Syncrude) in operation with several more in various stages of planning and approval. *In-situ* processes have not progressed to full scale operation although a recent survey[1] reported 16 active pilot projects involving more than 266 test wells. One *in-situ* project proposal by Esso Resources Canada Ltd.[2] for the Cold Lake deposit has progressed to the final approval stage. This project would employ the cyclic steam stimulation ("Huff and Puff") process to produce 160,000 B/d ($25.4 \times 10^3$ m$^3$/d) of bitumen which would be upgraded to 141,000 B/d of synthetic crude ($22.1 \times 10^3$ m$^3$/d). This *in-situ* oil sand recovery process provided the basis for this investigation.

# In-Situ Oil Sand Production by Steam Stimulation

Oil Sand Bitumen Characteristics.    The terms bitumen, tar and/or heavy oil are often used interchangeably to refer to hydrocarbons with a gravity < 12° API.  The chemical properties of bitumen from the Cold Lake deposit are summarized in Table 1.[3]

Process Description.    The characteristics and applicability of the steam stimulation process for heavy oil or bitumen recovery have been reviewed by Farouq-Ali[4,5], Nicholls and Luhning[1] and Humphreys[6].

The basic prerequisites of an in-situ technology for oil sand production are the need:

- to reduce the viscosity of the bitumen to provide mobility in the reservoir
- to establish communication within the reservoir between the source of the driving force and the production wells
- to provide a drive mechanism for oil mobilization.

The Esso Resources cyclic steam stimulation process proposal calls for the drilling of a well pattern on a 48 ha grid[2].  These wells will be used to inject steam to the bitumen deposit for a 4-8 week period.  This will be followed by a short "soak-in" period during which heat transfer and reservoir pressure equalization can occur.  The resultant increase in reservoir bitumen temperature will dramatically reduce bitumen viscosity, as illustrated in Figure 3.

The heated bitumen may then be produced back at the same well under the combined driving forces of solution gas drive, gravity forces and reservoir pressurization due to the steam injection.

The production phase will normally continue for 3 to 6 months until production drops below economic limits.  The steam injection cycle may then be repeated.  It is anticipated that individual wells will tolerate several cycles to provide an ultimate economic life of 6 to 8 years.  The total proposal calls for 8300 wells to be drilled over the 25 year life of the project.

The overall water balance for the production and bitumen upgrading process is summarized in Table 2.

It is apparent that, despite a high level of water recycle in the process, a large fresh water makeup (93,000 m$^3$/d) is projected.  Fresh water availability may become a limiting factor to full development of bitumen extraction facilities in Alberta.  For example, application of the current objective of keeping water withdrawal from the Athabasca River (for the large Athabasca deposit) at below 10% of annual minimum flow would limit total water demand to 905,000 m$^3$/d.  Currently, the two operating surface mining operations combined with one plant at the final approval stage would account for a surface water demand of 270,000 m$^3$/d.  The remaining water availability would provide for only about six in-situ plants without any allowance for other major water users.

Furthermore, it has been estimated[4] that as much as 60% of the injected heat may be produced back at the surface.  Hence, energy as well as environmental considerations favour maximum reuse of the produced water which accompanies the bitumen to the surface.

The process proposed for handling water produced from recovery wells at the Esso Resources plant is summarized in Figure 4. This demonstrates the intention to renovate the produced water to become boiler feed water for subsequent steam generation.

The wastewater renovation scheme is challenged by a highly stable oil-in-water emulsion which occurs with produced water from the cyclic steam stimulation process.

Process Evaluation

## Electrolytic Flotation

Even though electrolytic flotation has not been used previously for treating recycle water from *in-situ* oil sand production plants, it has been used successfully in treating many other types of emulsified wastewater.[7] However, some of these emulsions were likely less stable than the recycle water encountered in this study. The characteristics of the process as they relate to this treatment application are discussed elsewhere.[7]

## Wastewater Characteristics

The range of wastewater parameters encountered in raw, recycle water is summarized in Table 3.

Except for the large fluctuation in oil concentration the other parameters exhibited limited variability. In particular, pH remained rather constant from sample to sample, suggesting a well buffered system. The electrolytic flotation process evaluation was based on the oil removal capability of this process as measured by a modified partition-gravimetric method[8] as described by Nagendran and Hrudey.[7]

## Treatment Performance

The bench scale experimental apparatus consisted of a direct current power supply, automatic timer, electrode grid, cylindrical reaction vessel, heating coil and insulation as illustrated in Figure 5.

Flotation and Potential Field Effect Alone. To test the destabilizing influence of an electrical potential field without concurrent dissolution of the anode, the electrode grids were made from platinum.

Initial testing with platinum electrodes encountered a sudden drop of power supply current towards zero. Examining the electrodes after each such occurrence it was evident that a black precipitate was coating the anodes of the platinum electrodes. This precipitate was attributed to the presence of sulphur containing organic compounds in solution. This interference has been reported by Mikheeva et al.[9] We overcame the precipitate problem by alternating the polarity of the electrodes thereby creating a self cleansing effect on the alternating anodes. An automatic timer was used to alter the polarity once every 10 minutes. This enabled continuous use of the platinum electrodes. Unfortunately, oil removal with this mode of operation was not effective. While there was adequate bubble (hydrogen and oxygen) production, no significant separation and flotation was taking place, suggesting that the emulsion was not being adequately destabilized.

Flotation, Electrostatic Coagulation and Electrode Coagulant.  In order to improve the treatment performance the use of a more electropositive metal for the electrode such as aluminum or iron was implemented.  Under these circumstances, up to 98% oil removal (from 2890 mg/L to 46 mg/L) was measured.  However, the rate of anode dissolution was high enough to require electrode replacement after two to three hours of operation as indicated by the highly corroded condition of electrodes as illustrated in Figure 6.

The metal ion contribution to solution from these sacrificial electrodes was calculated for current densities of 200 mA/cm$^2$, and was found to be up to 783 mg/L of ferric ions ($Fe^{+++}$) over a 1.5 hour operating period. The high rate of metal dissolution and the resulting need to frequently replace electrodes, makes the use of electrode coagulation impractical for a continuous, fullscale treatment process.

Flotation, Electrostatic Coagulation and External Coagulant.  The stable platinum electrodes were used again and different external coagulants were tried.  These included, alum, calcium hydroxide, ferric sulphate, humic acid, a frothing agent (Cetyl trimethyl ammonium bromide), sulphuric acid and ferric chloride.  Ferric chloride was found to be the most suitable coagulating agent.  In order to compare the treatment achieved with electrolytic flotation using an external coagulant to that of the treatment achieved with chemical coagulation alone, replicate chemical coagulation tests were performed.  It was found that the treatment achieved with 80 mg/L of ferric ion and electrolytic flotation was not significantly different (2 sided "t" test; $P > 0.05$) from the treatment achieved with 125 mg/L of ferric ion with strictly chemical coagulation. The overall superior performance of the electrolytic flotation process with external coagulant over chemical coagulation alone is shown in Figure 7.

In order to avoid sample and experimental variation from day to day, electrolytic flotation with ferric chloride treatment and chemical coagulation tests were performed simultaneously.  The results are summarized in Table 4.

Electrolytic flotation process samples were collected at 15 minute and 30 minute intervals.  The chemical coagulation test was stirred for 30 minutes and allowed to settle for 4 hours.  A one way analysis of variance indicated a highly significant difference ($P < 0.001$) among the three mean final oil concentrations.  However, a Duncan's multiple range test showed that the electrolytic treatment of samples for 15 minute and 30 minute residence times did not differ significantly.

Based on these observations, the best process conditions obtained for the electrolytic flotation process were a current density of 132 mA/cm$^2$, a residence time of 15 minutes and a ferric ion dosage of 125 mg/L.  These process conditions were obtained at an operating temperature of 55°C and achieved a final oil concentration falling consistently below 50 mg/L. Inspection of Figure 7, however, indicates a diminishing return for coagulant dosage beyond 100 mg/L.  Hence, the highest oil removal conditions are not necessarily the most cost effective.  Cost analysis, as developed in the next section, indicated that a dosage of 100 mg $Fe^{+++}$/L was the most cost effective dosage among the observations of this study. This coagulant dosage, with the other process conditions as described above, produced an average final oil concentration below 60 mg/L.  Such a treatment level is considered to be adequate as an initial step in a sequential treatment scheme for upgrading produced water to boiler feed grade.

## Treatment Costs

Full evaluation of the potential of the electrolytic flotation process requires consideration of treatment costs. A cost analysis was based on a rate of \$90 per 160 kg drum of ferric chloride and a daily production rate of 141,000 B of synthetic crude with a produced water recycle rate[2] of $38 \times 10^3$ m$^3$/d. On this basis the chemical cost for the electrolytic flotation process (at 100 mg/L ferric ion concentration) was estimated at approximately 4.43¢/B of synthetic crude produced. Adding the cost of power estimated at 0.22¢/B (for 15 minutes of 1 Amp and 7 volts) makes the total treatment cost 4.65¢/B synthetic crude produced.

The hydrogen production rate during electrolysis was evaluated to determine if it could offset some of the power and chemical costs. However, the volume of hydrogen produced was only 0.4 L/ampere/hour, hence the benefit was negligible.

The chemical cost of treating the wastewater by chemical coagulation alone to a level of treatment equivalent to the electrolytic flotation process would cost approximately 6.33¢/B. In addition a much longer total residence time was required (1 to 4 hours vs. 15 mins. for the electrolytic flotation process) which would increase the capital cost of a chemical coagulation reactor.

Biesinger et al.[10] reported treatment costs for various industrial applications of dissolved air flotation. Costs varied widely by application ranging from a low value of \$8.00 per 1000 m$^3$ to a high of \$129 per 1000 m$^3$ for chemical manufacturing industries. Two oil refinery cases were cited with costs of \$24 and \$32 per 1000 m$^3$. An inflation factor of at least 100% should be applied to these figures to update them to 1980 \$.

The cost of the electrolytic flotation process using 100 mg Fe$^{+++}$/L and current densities sufficient to achieve a treatment efficiency of nearly 90% to 99+% would be \$172 per 1000 m$^3$.

Dissolved air flotation costs cannot be directly related to the electrolytic flotation process costs estimated by this study due to differences in wastewater characteristics. However, such a comparison does provide a gross indicator of the relative magnitude of wastewater treatment costs by the electrolytic flotation process with external chemical addition. It is clear that chemical costs provide the largest cost component and that improved economics for this process will require more effective and/ or less expensive chemical aids.

## Conclusions

Full exploitation of the immense hydrocarbon reserves of Alberta's Oil Sands deposits will require the use of emerging *in-situ* recovery technologies. Fresh water supply may become a limiting factor to the implementation of such technology. Consequently, a need exists to optimize process water treatment and reuse. The highly emulsified oil content of produced water for the cyclic steam stimulation process provides an example of the treatment challenge.

Electrolytic flotation offers some promise for the treatment of production recycle water from *in-situ* oil sand recovery processes. Evaluation of the process in this context indicated the following:

1. Electrolytic flotation using inert platinum electrodes without external coagulant addition was ineffective;

2. Electrolytic flotation using iron or aluminum electrodes was effective but involved an unacceptably high rate of anode corrosion;
3. Electrolytic flotation using platinum electrodes in combination with ferric chloride was significantly more effective than treatment with ferric chloride alone.
4. Electrolytic flotation with 100 mg $Fe^{+++}$/L was able to achieve consistent oil removal to less than 60 mg/L (corresponding to 90-99+% removal).
5. The cost of electrolytic flotation with ferric chloride addition would be high relative to reported costs for the use of dissolved air flotation in other industrial applications.
6. The major component of process cost was the chemical coagulant cost and thus improved economics for the process would require less expensive and/or more effective chemical additives.

Acknowledgements

This study was funded by Petro-Canada under the technical direction of Mr. W.J. Rowley. Esso Resources Canada Ltd. provided the samples of produced water from their pilot steam stimulation plant at Cold Lake. The project was also supported by the education leave provided to one author (J.N.) by Alberta Environment.

References

1. Nicholls, J.H. and R.W. Luhning. "Heavy oil sand *in-situ* pilot plants in Alberta (past and present)", The Oil Sands of Canada-Venezuela 1977, C.I.M. Spec. Vol. 17, 527, (1977).

2. ESSO Cold Lake Project Final Environmental Impact Assessment, Vol. 1, Esso Resources Canada Limited, Calgary, 1979.

3. Strausz, O.P. "The chemistry of oil sand bitumen.", The Oil Sands of of Canada-Venezuela 1977, C.I.M. Spec. Vol. 17. 146, (1977).

4. Farouq-Ali, S.M. "Current status of steam injection as a heavy oil recovery method.", J. Can. Petrol. Technol. 13 (1), 54, (1974).

5. Farouq-Ali, S.M. "Current status of in-situ recovery from the tar sands of Alberta", J. Can. Petrol. Technol. 14 (1), 51, (1975).

6. Humphreys, R.D. "An overview assessment of in-situ development in the Athabasca deposit", Alberta Oil Sands Environmental Research Program, Manage. Rep. PM-1, (1979).

7. Nagendran, J. and S.E. Hrudey. "Electrolytic flotation of in-situ oil sand production recycle water", Environ. Eng. Tech. Rep. 80-1, U. of Alberta, (1980).

8. APHA. Standard Methods for the Examination of Water and Wastewater. 14th Ed. American Public Health Association-American Waterworks Association-Water Pollution Control Federation, New York, 1975.

9. Mikheeva, E.P., Kagasov, V.M. and Mamakov, A.A. "An electroflotation method of purifying coking industry effluent from oils", Electrochem. Ind. Proc. Bio., 6, 40, (1974).

10. Biesinger, M.G., Vining, T.S. and Shell, G.L. "Industrial experience with dissolved air flotation", Proc. Ind. Wastes Conf., Purdue Univ., 29, 290, (1974).

TABLE 1  Composition of Cold Lake Bitumen (after Strausz[3])

| ELEMENTAL COMPOSITION (%) | | GROSS CHEMICAL COMPOSITION (%) | |
|---|---|---|---|
| Carbon | 83.93 | Asphaltene | 15.3 |
| Hydrogen | 10.46 | Deasphalted Oil | 84. |
| Nitrogen | 0.23 | • acids | 15.2 |
| Oxygen | 0.94 | • bases | 6.38 |
| Sulphur | 4.70 | • neutral N compounds | 1.15 |
| | | • saturates | 21.3 |
| Hydrogen/Carbon | 1.49 | • monoaromatics | 8.3 |
| Molecular weight | 490 | • diaromatics | 3.6 |
| | | • polyaromatics & non-defined polar compounds | 24.35 |
| | | • thiophenoaromatic from mono- and diaromatic fractions | 0.8 |

TABLE 3  Typical Produced Water Characteristics from the Steam Stimulation *In-Situ* Process

| PARAMETER | CONCENTRATION RANGE |
|---|---|
| Oil (centrifuged sample) | 479 - 5692 mg/L |
| pH | 7.0 - 7.2 |
| Alkalinity (as $CaCO_3$ equivalent) | 240 - 360 mg/L |
| Electrical Conductivity | 5750 - 6900 micromhos/cm |

TABLE 4  Simultaneous Electrolytic Flotation and Strict Chemical Coagulation (with 125 mg/L $Fe^{+++}$ dosage)

| | ELECTROLYTIC FLOTATION | | CHEMICAL COAGULATION |
|---|---|---|---|
| Reaction Time | 15 min | 30 min | 30 min* |
| Mean Effluent Oil (mg/L) | 45.6 | 43.2 | 79.6 |
| Standard Error | ±1.6 | ±1.2 | ±2.6 |

*30 minutes of mixing followed by 4 hours of settling

TABLE 2  Water Balance for Esso Resources Cold Lake Project (after ESSO[2])

| BALANCING FUEL | COAL | GAS |
|---|---|---|
| | $10^3$ m³/d | |
| **WATER REQUIREMENTS** | | |
| Injection Steam | 63.6 | 63.6 |
| Process Water and Steam | 19.7 | 19.7 |
| Cooling Tower | 25.8 | 25.8 |
| Water Treatment Backwash | 17.6 | 15.5 |
| Utility Water | 12.7 | 12.7 |
| Potable Water | 3.8 | 3.8 |
| TOTAL: | 143.2 | 141.2 |
| **WATER CONSUMPTION** | | |
| Consumed in Operations | | |
| • Displacing Bitumen | 25.4 | 25.4 |
| • Cooling Tower Evaporation | 20.0 | 20.0 |
| • Hydrogen Synthesis | 2.4 | 2.4 |
| • Coke Gasification | 5.1 | 5.1 |
| | 52.9 | 52.9 |
| Water Disposal | | |
| • Deep Wells: | | |
| • Water Treatment Backwash | 7.6 | 7.6 |
| • Sour Water | 2.4 | 2.4 |
| • Desalter Brine | 2.4 | 2.4 |
| • Chemical Wastes | 0.8 | 0.8 |
| | 13.2 | 13.2 |
| • Surface: | | |
| • Water Treatment Backwash | 10.0 | 7.9 |
| • Cooling Tower Blowdown | 3.3 | 3.3 |
| • Boiler Blowdown | 2.7 | 2.7 |
| • Utility Water Waste | 10.7 | 10.7 |
| | 26.7 | 24.6 |
| **RECYCLED** | | |
| • Produced Water | 38.2 | 38.2 |
| • Sour Water | 4.4 | 4.4 |
| • Sanitary Wastes | 3.8 | 3.8 |
| • Oily Wastes | 4.0 | 4.0 |
| TOTAL: | 50.4 | 50.4 |
| | 143.2 | 141.1 |
| **WATER SUPPLY** | | |
| Recycled Water | 50.4 | 50.4 |
| Fresh Water Make-up | 92.8 | 90.7 |
| TOTAL: | 143.2 | 141.1 |

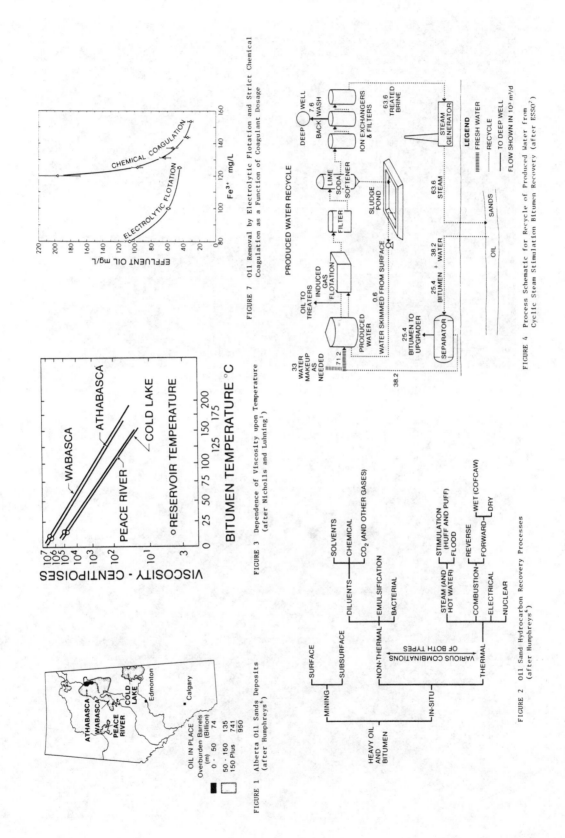

FIGURE 7  Oil Removal by Electrolytic Flotation and Strict Chemical
Coagulation as a Function of Coagulant Dosage

FIGURE 3  Dependence of Viscosity upon Temperature
(after Nicholls and Luhning[1])

FIGURE 4  Process Schematic for Recycle of Produced Water from
Cyclic Steam Stimulation Bitumen Recovery (after ESSO[2])

FIGURE 1  Alberta Oil Sands Deposits
(after Humphreys[6])

FIGURE 2  Oil Sand Hydrocarbon Recovery Processes
(after Humphreys[6])

73

FIGURE 5   Bench Scale Treatment Apparatus

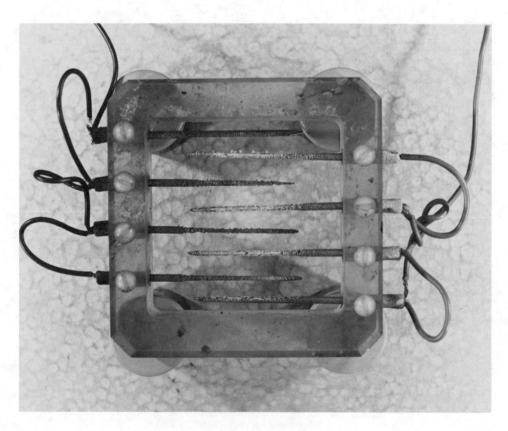

FIGURE 6   Corroded Electrodes

## MATERIALS: EXCESSES, SHORTAGES OR DISRUPTIONS
## INTRODUCTORY REMARKS

**Edmund J. Rolinski**
University of Dayton

There are two aspects to the problem of materials. The first is in the area of critical technology which deals with kinds and classes of commodities which the technology of the U.S. is in the forefront and where there are restrictions to due to the Export Control Act. The second is in the area where shortages and long lead time are required for materials and where stockpiling is necessary.

The fact that materials and the steady supply of critical and strategic metals and minerals ever becomes an important issue is based upon the role of the U.S. and its relations with the rest of the world. Because of this, there are two sets of interrelated problem areas and can be summarized as political issues and technology issues. Political issues involve critical supplies, stockpiles, priorities, price and energy-environmental implications (See Figure 1).

The U.S. reliance on certain metals such as columbium, titanium, chromium, manganese, tantalum, cobalt, plantinum group metals, tin, nickel, cadmium, mercury, zinc and tungsten are dangerously high. This situation is somewhat the same for Japan but is in sharp contrast to the USSR. The first paper in this session will address the current problems in non-fuel materials policy and international implications.

The second paper will address the environmental consequences of a coal conversion industry in Western Kentucky. This paper is concerned with the large amounts of high sulfur coal utilized in their effects on air and water quality, in addition to land use, coal and water resources as well as various socio-economic issues. Although the materials issues are not considered, the use of coal is highly significant as an economic factor.

When asked about materials and their importance as raw materials in the economic picture one can easily see the importance of scrap and recycl- able materials. However, impact can be readily visible if new and emerging materials technology and materials processing can be implemented to make production costs and improved parts production possible. The area of high temperature materials utilization is a natural outgrowth for higher efficiency process by virtue of the second law of thermody- namics (See Figure 2).

Two significant developments occurred this year which will impact the total picture of materials supply. The first is that Crucible Specialty Metals Division of Colt Industries, Inc., announced commercial production of a new cobalt-free steel used to make cutting tools for machining highly alloyed materials in space and aircraft. In past years, cobalt bearing tool steels have accounted for about 3% of total U.S. cobalt demand. The second is that United States Steel Corporation announced the development of a new process called cyclite that make use of water based lubricants and does away with the use of tin in the manufacture of steel cans. If this process is implemented, the new process would reduce tin consumption and would allow higher and cheaper steel cans, meanwhile allowing greater ease of recycling to be possible.

Resource conservation and air pollution control presents a case study on the compatibility issue of cleaning the environment before it becomes a problem. Since the cost of raw materials has increased drastically over the past few years, the case study involving organic solvent emissions releases from a gasket airing oven shows how engineering design can become an economical method for recovering raw materials. The next paper on the environmental considerations in siting resource recovery facilities in New York City presents three possible choices which are mass burning, coarse refuse derived fuel and refined refuse derived fuel. Site specific and process related issues such as air quality, odor and ash disposal are governing considerations. These choices and their possible impact on materials shortages can have a great impact when resource recovery methods are implemented.

The final invited paper in this session is concerned with the disruption of a steady supply of JP-4 as required by the Department of Defense. As a result of continued requirements for jet fuel the Air Force has launched on an alternative plan to utilize shale oil for the production of JP-4 fuel. As an informed customer, the Department of Defense program on shale oil production and processing to jet fuel is reviewed.

POLITICAL ISSUES

● CRITICAL SUPPLIES
    • STOCKPILES
    • PRIORITIES
● SUPPLY AND DEMAND
● ECONOMIC FACTORS
● ENVIRONMENTAL FACTORS
● FUEL-ENERGY REQUIREMENTS
● FUEL-ENVIRONMENTAL IMPLICATIONS

## POLITICAL ISSUES AND MATERIALS IMPLICATIONS
Figure 1

TECHNOLOGY ISSUES
● NEW AND EMERGING MATERIALS TECHNOLOGY
● MATERIALS PROCESSING
    • PARTS PRODUCTION
● MATERIALS SUBSTITUTION
● DISPOSABLE MATERIALS
    • RECYCLE DESIGN
    • RESOURCE RECOVERY
● HIGH TEMPERATURE PROCESSES
● MATERIALS AND ECONOMICS

## TECHNICAL ISSUES AND MATERIALS IMPLICATIONS
Figure 2

# CURRENT PROBLEMS IN NONFUEL MATERIALS POLICY: PRIORITIES, TRADE-OFFS, INSTITUTIONAL CONSTRAINTS, AND COUNTERMEASURES TO COPE WITH SHORTAGES IN SUPPLY AND DISRUPTION IN ESSENTIAL PRODUCTION SERVICES

**Michael J. Deutch**
Consulting Engineer

## Introduction

In two World Wars, in Korea, and to a lesser degree in Vietnam, materials supply problems ranked high in the preoccupations of the U.S. and its Allies, even though our resources of minerals and materials are abundant, and our mining, refining and fabricating industry ranks high among the largest and the best. No country is endowed with all the materials necessary in war and peace, and rich deposits are, more often than not, quite distant from refineries and fabricating plants. Thus in times of international tensions, the perception of shortages in non-fuel minerals and materials (because of their pervasive and essential role throughout the economy) causes prohibitive price swings, threatens the disruption of production schedules, and inevitably influences our defense posture, our foreign policy, as well as domestic employment and productivity.

The serious problems that arise in the so-called Materials Cycle, have been well defined over a decade, in the Reports of the National Commissions on Materials Policy; on Supplies and Shortages; in the comprehensive studies of the National Academy of Sciences; its committees; and in more recent inquiries. Unfortunately, these studies have not as yet resulted in the formulation and implementation of a cohesive policy on non-fuel minerals and materials.

With the international outlook becoming increasingly cloudy, defense preparedness, transportation, and energy projects are now visibly affected by erratic prices and uncertainties in supply of imported materials essential to our national defense, and to civilian productivity and services. Even when the economy and employment do improve, problems with materials supply may hinder the recovery.

This Symposium will discuss considerations of national security, the problems in industry, and the need for objective cost/benefit and feasibility analysis for trade-offs in policy formulation. Since the Legislative Process is necessarily accommodative, until supply problems are perceived as pressing, and conflicting views of industry, of the environmentalists and of labor are reconciled with other priority preoccupations in both Branches of the Government, it is unrealistic to assume that the recently enacted Materials Policy Bill will be truly implemented (beyond a few more grants for academic research).

1. National Security

A growing number of local confrontations in the spreading arc of international instability that now extends from North Africa to the Persian Gulf threaten the vital interests of the U.S.A. so seriously that our defense posture in both tactical and strategic weapons and our Industrial Preparedness Base must be strengthened as soon as possible. To do so it is necessary to insure and expedite the supplies of critical materials and components needed for new weapons systems as soon as the plans call for the production stage for our own defense requirements, as well as those of our Allies. In so doing we will encounter obstacles that are not due exclusively to the magnitude of the requirements, or insufficient appropriations for defense in the past.

> First, Changing Requirements: Every confrontation requires a different technology. In this day and age aerospace, communications, automation, guidance systems, etc., military and civilian technology may all compete for the supply of exotic or special materials, minerals or organic compounds not previously produced or stockpiled.

> Second, our Strategic Stockpile was conceived after World War II, under Congressional and budgetary strictures that were well justified and motivated at the times of amendment or oversight review, when policy objectives and the strategic posture of the U.S. were different.

> • It is not certain that is still appropriately linked to the changing requirements of the Industrial Preparedness Posture in terms of what will be needed in the next emergency (rather than in "prior wars").

> Thirdly, our materials problems relate to three areas of foreign policy and international security which extend our potential vulnerability over far distant horizons: Our security arrangements with NATO, Japan and some countries in Africa and Asia require that we provide them with the logistic support necessary to their participation in joint defense planning on a vast geographical scale: Figure 1, entitled "Potential Choke Points in the Sea Routes and Ports of Trans-shipments of Petroleum and Non-Fuel Minerals" shows how far flung our supplies are. Figure 2, entitled "Percent of Total Supply of Selected Metals and Oil to West Germany" illustrates the dependency of a single NATO member on supplies of metals and petroleum over worldwide lanes of trade.

Discontent and turmoil in the countries of the so-called North/South Confrontation is increasing instability along many routes, and even at the original sources of supply. Unfortunately, our international monetary and trade negotiations, and our posture in the International Lending Agencies have not yet achieved a stable system of fair and secure access to the unprospected and un-mined resources of raw materials in the less developed or poorer nations of the world.

## 2. Effects on the Domestic Economy

Even in the assumption that producers of critical materials "can not succeed in emulating OPEC" (as we have been repeatedly assured in 1973-76), materials costs are likely to rise substantially over the years, causing political concern and monetary problems in 5 areas of domestic economic policy, i.e., balance of payments, employment, and loss of productivity that will exacerbate conventional preoccupations in industry and trade, as well as cyclical fluctuations in supply and demand, labor productivity, and cost of living. In a more detailed study not as yet available for publication, this author had the privilege to correlate and analyze the impact of interruptions of supply of energy and non-fuel materials on essential civilian activities and services, as well as the economy and the balance of payments. There is a clear correlation between interruption of supply of essential materials and dislocation of industry, transportation and trade.

## 3. Institutional Problems in Industry

The U.S. materials industry has also structural problems of its own, that are not due exclusively to our energy predicament, the high cost of environmental litigation or compliance, "Washington indifference," or uncertainty in imports of materials from abroad: Some are due to changing times, others to increasing capital intensiveness; higher labor and transportation costs, changes in end-products and areas of use, and adverse trends in the demand from the best historical markets for materials, i.e., the automobile, construction and housing industries, themselves severely hit in periods of inflation, high interest rates and unemployment.

In short, the materials industry is facing two unprecedented and unforeseen perils: A rapidly changing financial and economic environment (such as insufficient capital formation, cost-push inflation in the midst of a protracted recession, etc.). In addition, the materials industry faces a perilous fall-out from the ominous clouds on the international military, political and monetary horizon.

Past experience and tradition in the materials industry may not always provide solutions in these difficult times, particularly since management is simultaneously faced with: adverse public attitudes, with a host of regulatory and environmental constraints; growing suspicion of multinational companies' motives; the "No Growth Syndrome"; the "Laissez Faire Philosophy"; pressing demands from the poorer countries of the World for a "New Economic Order" and from Academe for a Technological Fix (irrespective of cost, institutional and attitudinal constraints) to implement conservation, substitution and recycling of materials.

After the heated political and economic debate on energy, it is harder to assess objectively the possible effects of sporadic interruption of materials imports. Past experience tells us that localized shortages of certain materials will bring about dislocations of industry and essential services, causing sporadic unemployment and more cost-push inflation.

4. Excesses in Environmental Regulation, or Shortfall
   in Applied Environmental R&D?

In the interface between Energy, the Environment and Materials, U.S. policy formulation has been somewhat hesitant, and not always well informed on the true costs (in time and money) of compliance with the voluminous body of changing environmental standards and control, or of the irretrievable delays in the completion of urgent new facilities and conversion projects needed to remedy our deficiencies in indigenous production of energy and non-fuel materials, be it because of uncertainty as to legislative intent, jurisdictional conflicts in both branches of Government, or between Federal and State policies, attitudes, intervention, or indecision on siting, size, type or operation of new facilities. Whatever the merits of each case, more often than not Time handed down a Death Sentence....

There may have been, in 1967-70, some justification to promulgate as many environmental strategies as could spur R&D on new pollution abatement techniques (with costs passed on to the consumer, in an era when the credo of "Limits to Growth" was in fashion). Now new perils require new priorities: It is a fact that the materials industry did not have in 1973-74 sufficient liquidity to cover the capital costs of compliance with environmental controls, and in addition, make additional investment in exploration and in modernization of refining and smelting of lower-grade domestic reserves. In periods of political and monetary uncertainty and high interest rates, it is the capital intensive, long turn-over industries that are unable to tap capital markets.

In the 1980's we will have to reconcile the conflict between Environmental Regulation and the national imperative to insure timely construction and prompt operation (or conversion) of critical materials, synfuels and other costly, long lead-time projects, already delayed by unavoidable R&D or technological difficulties in selection. A prompt decision to proceed or to abandon must be made at a certain point in time, with sober analytical appraisal of the remaining uncertainties, but without further hesitation. I am confident that EPA's control strategies will come to grips with cost/benefit, feasibility and environmental trade-offs in urgent critical materials projects.

A Symposium on Materials cannot ignore the perception (widespread in the western states) that the need of environmental protection have not been reconciled with the needs for energy and materials. For example,

- The Clean Air Act bars from certain areas new facilities that do not meet Standards for Prevention of Significant Deterioration of Air Quality. These standards are based on computerized Models that contain elements of uncertainty, which may have to be reduced through the use of actual air quality measurements and more regard for level of actual risk of unacceptable air pollution, and thus provide maximum flexibility in balancing environmental, economic and strategic materials values where possible under the law.

- The requirement to gear pollution abatement in new facilities to the level of Best Available Control Technology (BACT) is still challenged in industrial and engineering circles because of the lack of explicit definition of processes and equipment (e.g., scrubbers) that are actually available in the real world of engineering and procurement, rather than on paper.

Financial institutions avoid projects where there is uncertainty as to the specifications availability and cost of equipment needed to comply with the pollution abatement standard considered by EPA as meeting BACT. Unfortunately, EPA's research efforts and management have been perceived over the years to be mainly supportive of EPA's regulatory role, rather than focusing on the design, development and engineer pollution abatement hardware (or standardize equipment) which could reliably comply with EPA's control strategies.

The acceptance and implementation of environmental standards in urgent energy and materials projects is predicated on the perception that the standards are based on computer simulation of validated models whose precision can be assessed, and which were derived from unbiased inputs, from typical, rather then worst case, annual meteorology.  In short, the control strategies and standards must be determined with a high degree of objectivity and implemented with some flexibility, so that the Government is free to adjust regulation to special circumstances, higher priorities or emergencies. The August 1980 revision of EPA regulations on Air Pollution is a step in the right direction, and it will probably be followed by further improvements under the capable leadership of Douglas Costle.  It is important that the dialogue between industry and Environmental Protection centers more on problem solving and less on litigation.  The dialogue cannot become constructive as long as it remains adversary and couched in legalistic terms.


5.  <u>Excesses in Regulatory Limits to Exploration</u>

The critical problem of access to materials and minerals essential to our national security and economy is not limited to those that do not occur in the U.S.A. and must be imported to the full extent of our needs (e.g., tin, industrial diamonds or beryllium):  There are also a number of conventional minerals which we now produce from high cost, lower yield deposits than those mined out in the past, and it is our reliance on imports from the undeveloped, low cost deposits abroad that has helped meet U.S. needs while continuous technological improvement by our mining and refining industry increases production at acceptable costs.

Still, the political and social turmoil in the developing countries makes it imperative for our mining and materials industry to continue exploration and development of new indigenous sources of supply.  This requires large speculative investments (the element of risk that is always high prior to discovery) and vast L.T. capital investments to develop the mines and refinery installations after discovery.

The map of U.S. land areas with favorable geological indications to justify exploration is now reduced by a number of Federal and State land management restrictions and outright withdrawal from exploration, be it for environmental or other reasons.  Some are historical, others may be quite valid, or inspired by the laudable intent to avoid speculative windfalls to land speculators or lease owners that do not intend to contribute to exploration, but only to benefit from discovery on surrounding lands.

Whatever the case may be, the map of the United States with overlays representing the wide variety of regulatory restrictions to mineral exploration looks to an institutional investor as if it was closed by a forbidding curtain. The efforts of the U.S. mining industries to discover new deposits of needed materials will remain constrained until more flexibility and more cost/benefit trade-offs are introduced in the administration of the vast and complex set of prohibitions.

## 6. Excesses in Demand?

Two criticisms are voiced abroad: The first is that ownership and development of foreign deposits by U.S. corporations is not sufficiently profitable for the host country. This criticism grows louder whenever there is a gap in supply, and is usually seized upon as an argument for better terms of trade, or a larger contribution to alleviate budgetary or employment difficulties in the host country.

The second criticism, voiced by industrialized nations that are more dependent on imported materials is that U.S. per capita consumption of energy and of non-fuel minerals is excessive. This should be taken with a grain of salt: per capita materials usage is always greater in highly industrialized countries, and is a function of the level and nature of manufacturing and construction activity in a given country. Without denying that there may have been an insufficient conservation effort in the past when materials were plentiful and cheap, there is now a pressing need for an aggressive effort in substitution, recycling and new technology, U.S. demand for materials is not excessive: it has not been stimulated by pricing and other government policies as in the case of energy.

- International comparison of per capita use and assumptions of optimum demand based on theoretical estimation of elasticity are often invalid and lead to a "hinged" approach to policy formulation, since assumptions on price/demand elasticity can be carried to extreme conclusions: A high coefficient may lead to the conclusion that higher prices will achieve conservation by reducing demand so sharply that we may substitute labor and capital for conventional fuels, as well as for non-fuel materials in a "soft path" to prosperity. However, the reverse assumption is an elasticity coefficient so low it would lead us to conclude that our demand for materials is irreducible and must be satisfied at any cost. In reality, price/demand elasticity varies not only in time and circumstance, but also depends on the essentiality of a given material, the perception of the marketplace, the overhang of stocks, the outlook for the weather, and the short term swings in prices of freight and futures.

## 7. Problems with Policy Formulation

Our ability to face the challenges of resource scarcity and to provide peaceful solutions to the economic problems of the world, depends on the economic strength, the national vitality and the quality of leadership in the U.S.

A strong materials industry that understands the issues and priorities of the Nation, and is given the means to achieve desirable growth (in harmony with defense and other national priorities) is needed to develop new sources of materials, through exploration, new technology, conservation or substitution. This is easier said than done under adverse international conditions, by an industry that is not particularly profitable, with problems that are over-shadowed by other difficulties in high priority segments of the economy.

To survive and to achieve some balanced growth in these difficult years, the materials industry must solve a number of institutional problems, some of which require and deserve assistance of the Government (e.g., massive new investments, new international trade arrangements and protection of invest-ments abroad, new technology and modernization).

83

This symposium may not be the right forum to catalogue the institutional problems of, or the improvements needed in, the materials industry. However, since the general area of technological innovation ranks so high in recent legislative proposals on materials, one pitfall should be mentioned here: It concerns not only "roles and missions" in R&D on materials, but also the right mix in allocation of funds between advanced research and applied technology:

- With the exception of defense, aerospace and, to only some extent, energy, R&D on materials is supported primarily at the basic research end of the spectrum, and to a smaller extent in the final steps, as part of specific hardware demonstration programs. In the middle ground of applied research or exploratory development where the reduction to practice of new fundamental knowledge is first tested as to probability of scientific premises, support of R&D is most lacking. This is a critical deficiency in the institutional arrangement throughout our overall economy which may affect its ability to quickly and effectively cope with change.

In the real world, demonstration and scale-up projects rarely receive adequate attention or support until the need is upon us at which time "crash" programs often are initiated. By their very nature there is a tendency to limit the risk, and restrict the goals of such programs to reasonably assured probabilities. If the "middle ground" R&D has not been done prior to the full maturation of the need, the solutions available from the "crash" programs will be limited and may be inadequate. Yet, there is little support for such R&D which is neither basic nor tied to the specific requirements of a hardware or process demonstration program. There is little government/industry/university infrastructure to define and accomplish it. In addressing the obstacles to overall national economic strength as well as narrow defense in materials needs, we must consider ways of correcting this deficiency.

*     *     *

How then do we reconcile our interdependence in the World Economy with the need to reduce excessive linkage of our economy and national security to distant and politically unstable sources of supply? There are three ways to do so:

The first is to increase, as much as possible, exploration and production of needed materials in the U.S.A. and other areas of safe access, and to step up recycling and substitution of materials. This first path requires capital incentives, advances in the application of technology, and improved equipment.

The second way is to limit non-essential or wasteful uses of scarce materials, reduce exports insofar as possible, and seek alternate--even if more costly--sources of supply from safer foreign sources, encouraging the consuming industries to make necessary adjustments.

The third is to improve our strategic stockpile in content and size, seeking higher grade materials in more readily usable forms. Critical materials on which we are excessively import-dependent should be, whenever possible, stored close to points of use. This will be costly, and we may have to dispose of stockpiled items that are not now of high defense priority (whatever the book loss to the Treasury, and the effect of such disposal on commodity prices).

After the acrimonious debates on energy, it seems even harder to sound today a second warning, namely, that events abroad may result in sporadic interruption of supply and localized shortages of certain materials, causing dislocations, increasing unemployment, and cost-push inflation: In the past, (e.g. 1974-1976) the extent of our energy predicament, the costs, lags and difficulties in the implementation of various alternatives were sometimes downplayed.

In facing our materials problems, public opinion has reason to doubt that we do have the planning capabilities necessary to assess indications of adverse change, to adapt our policies and procedures rapidly enough to circumscribe the damage. Much remains to be done:

In the international arena, where Government/industry cooperation could be closer, there is need for increased international exchanges of materials technology, and increased cooperation with other industrialized nations to step up geophysical exploration and development of new reserves, and in the financing of mining development under mutually acceptable trade arrangements with the host countries. Last but not least, a joint effort by the International Lending Agencies is needed to remove a number of obstacles and tensions that now becloud the productivity and future of mines located in the developing nations.

While the U.S. materials industry may not be able to counteract, by itself, the unpredictable hazards abroad, mining and manufacturing management should increase their efforts to:

- Monitor all potential shortages of essential materials and their consequences on users, design, prototypes and production lags.

- Explore the feasibility of cooperative R&D programs (in industry-wide beneficial partnership with the Federal Government) to develop new technology, as recommended in every legislative initiative on Materials Policy. This may well prove to be the most promising, non-controversial area of Government/industry cooperation: Industrial innovation is especially risky in the basic materials industry, because of the magnitude of the investment needed to modernize mining, processing, refining, and conservation modes in the fabrication and use of materials. Thus, high margins of profit (or assistance in financing) will be needed for an adequate ROI in projects subjected to price controls.

- Government should apply in different industries a variety of appropriate means to foster innovation to meet different constraints rather than adopt "across the board" measures. Policies such as accelerated depreciation methods, or low-interest financing could be selectively applied to those segments of the materials industries (or to specific priority projects wherein the required capital constitutes the greatest obstacle to innovation. In order to make such incentives politically acceptable, it is essential that the public become aware that it is in their own best interest to support innovation. A continuing effort by industry is needed at this juncture to demonstrate that innovation is essential to the attainment of the Nation's goals of national defense, high employment, and an improved quality of life. Without public recognition of the necessity of innovation, P.L. 479 will lack effectiveness.

I would like to voice a caveat: Our vulnerability to interruption of supply of imported non-fuel minerals does not necessarily follow the percentage of imports to total U.S. consumption of a given material. In a national emergency the security of a highly industrialized country depends not only on its military strength, but also on the health of its industry, the integrity of its currency and the sophistication of its technology. Some segments of industry, and some services are more essential than others. The surge capability of the Industrial Preparedness Base and the ability to promptly mesh the most essential segments of the Mobilization Base into a set of feasible priorities throughout the economy are a prime concern that goes beyond the Materials Policy.

In January 1980, U.S. consumption reached $2,500 billion. Of this, $225 billion were in the form of Domestic Processed Non-Fuel Materials, and $20 billion of Imported Processed Materials. These 9% of our economy used $25B of U.S. raw materials, $1B of old scrap, and $5B of imported non-fuel minerals; employed 5 million workers (excluding stone, clay, glass and fuel); another 23 million workers used these processed materials in manufacturing, transportation, utilities and construction. In short, the non-fuel materials industry is essential, and its growth may have been held back by excessive regulation, or even benign neglect in planning -- but it does not constitute our entire mobilization base: There are other segments of the economy, smaller or less pervasive than materials use, which assume particular importance in a national emergency and, in normal times, account for much employment, tax revenue, and consumer goods. In bad times, such industries as the automotive, construction, and railroad industries compound the problems of the materials and minerals industry, and are as anxious to secure relief from burdensome regulations as well as the understanding of the U.S. Government.

Materials problems in non-critical areas of the economy are properly left to resolution by competitive innovation in the free market, which has functioned so well in normal times. When there is a serious threat to our National Security, we need a comprehensive and consistent set of policy choices, that will expedite innovation and modernization in priority sectors of the economy, presently handicapped by capital shortages, and inability to comply with the growing number of regulations and controls.

Government policies dealing with a variety of materials problems require coordination, consistency in regulatory objectives and realistic goals for compliance in a realistic time frame. When regulations appear to shift without warning, compliance suffers.

Existing $SO_2$ control strategies are perceived as inflexible, with no practical variances for urgent fuel conversion or critical minerals mining or refining projects. I submit that the standards could well recognize the existence of a margin of uncertainty in the attainment of EPA goals. For example,

- In geographically non-controversial cases involving industrial plants that are not particularly injurious to the environment, the decision may be considered in the center of the margin of uncertainty.

- In non-vital, low-priority facilities, EPA can take the best estimate (of BACT or PSD) available.

- In vital, high-priority projects, the decision should be taken in between the two objectives.

## Conclusions

• Increasing dependence of the U.S.A. and its NATO Allies on imports of non-fuel minerals is likely to be caused by regional confrontations, acts of terrorism or crowding in the sea-lanes or ports of trans-shipment, rather than due to scarcity of resources, or emulation of OPEC.

• Short term disruption of supply can increase further the lead-time in defense preparedness and procurement schedules. Our vulnerability is greater in materials and minerals that have special properties of applications in advanced technologies (military and civilian), in materials for which we have no indigenous deposits or which have not been sorely needed in the past, and for those where alternative sources are distant or unfriendly, and where substitute materials are difficult to develop.

• Short term disruption of non-fuel mineral imports may also be damaging to our civilian economy in delaying an upswing, as well as in a protracted world recession. Some nations are more fearful of surpluses and lower prices for raw materials when the world economy recovers than they are apprehensive of shortages now. This view is already detectable in discussions with the LDC's on ocean nodule exploitation in international meetings on the Law of the Sea.

• Much goodwill, interdisciplinary know-how and judicial temperament will be needed to reconcile the need for increased exploration, mining and modernization of processing smelting and refinery with the existing environmental and land management regulations that paralyzing litigation will not improve. The problems in the materials/environment interface are not insoluble, unless erroneous, adversary perceptions preempt an area where engineering problem solving, within the framework of available technology, rigorous cost/benefit analysis and scheduling would do much good.

• Special financial incentives and procedural assistance by the Government are necessary and justified in the case of marginal prospects for new sources of critical materials, particularly when costly exploration and developing technology is needed.

• Whether the country's vulnerability in the supply of certain critical materials results from a chronic Resource War, from benign neglect or institutional rigidities, I am encouraged by the serious attention given to the problem in the last six months by the Undersecretary of Defense for Research and Engineering and the Defense Science Board for the first time in years. I am confident that at both ends of Pennsylvania Avenue concrete action will be taken in January 1981, in cooperation with industry and defense contractors — not limited to appropriations for inconclusive studies.

Major Ocean Choke Points

1. La Perouse Straits
2. Tsugaru Straits
3. Tushima Straits
4. Formosa Straits
5. Malacca Straits
6. Lombok Straits
8. Hormuz Straits
9. Bab el Mandeb
10. Madagascar Straits
11. Cape of Good Hope
12. Gibraltar Straits
13. Straits of Bonifacio
14. Straits of Messina
15. Bizerte, Cagliari (La Gallita)
16. Bosporus
17. Suez Canal
18. Straits of Aqaba
19. Panama Canal
20. Cape Horn
21. Denmark Straits
22. Iceland - Faeros Gap
23. Faeroes - Shetlands and Shetlands - Orkneys Gap
24. Baltic Approaches
25. Atlantic Ocean off Western Africa

POTENTIAL CHOKE POINTS IN THE SEA ROUTES AND PORTS OF TRANS-SHIPMENTS OF
PETROLEUM AND NON-FUEL MINERALS

Figure 1

| Sea Route | Iron Ore | Copper | Lead | Zinc | Bauxite | Crude Oil |
|---|---|---|---|---|---|---|
| The Mediterranean | 0.1 | — | — | --- | 16.5 | 48.0 |
| The Eastern Atlantic, including the North Sea (for suppliers in West Africa and Scandinavia) | | | | | | |
| Africa | 18.7 | 33.0 | 4.6 | 0.8 | 18.7 | — |
| Sweden and Norway | 31.3 | — | — | --- | — | 6.4 |
| Across the Indian Ocean and the Eastern Atlantic (for suppliers located east of the Malacca Strait, such as Australia) | | | | | | |
| Australia | 2.8 | --- | 1.5 | 0.2 | 20.2 | — |
| Asia | 1.2 | — | 0.9 | — | — | --- |
| Across the Atlantic (from countries in North and Latin America) | | | | | | |
| North America | 4.6 | 11.0 | 26.4 | 24.8 | — | — |
| Latin America | 16.7 | 32.5 | 7.8 | 4.0 | 4.8 | 2.2 |

PERCENT OF TOTAL SUPPLY OF SELECTED MATERIALS TO A NATO COUNTRY (WEST GERMANY)
CARRIED OVER DIFFERENT SEA ROUTES

Figure 2

88

Weapon System Families

Fighter Bombers (Jet + Prop)
Fighter/Interceptors
Assault Helicopters
Transport Helicopters
Transports (Jet + Prop)
Bombers (Jet + Prop)
Main Battle Tanks
Light Tanks
Armored Personnel Carriers
Heavy Artillery
Medium Artillery
Light Artillery

SP Medium/Heavy Artillery
Medium/Heavy Mortars
Light Mortars
Antiaircraft Guns
Surface-Air Missiles
Air-Surface Missiles
Air-Air Missiles
Surface-Surface Missiles
Antitank Missiles
Machine Guns
Recoilless Weapons
Antitank Guns

SP Light Artillery

Not included end items:  RPV's, CBW, Special Weapons

SHORTAGES OF CRITICAL MATERIALS IN U.S. ARMY
STRATEGIC END-ITEMS 1979-85

Figure 3

- Precision Miniature Bearings
- Jewel Bearings
- Specialty Metals

- Precision Components and Timing Devices
- Aircraft Clocks

- Aerial Refueling Hose
- Chaff
- Fine Wire Mesh Filters (Aircraft)
- 500 Gal Fabric Drums
- Hydraulic Drive Motors (Sonar Systems)
- Rubber Footwear
- Ultra-fine Cobalt Powder

- High Purity Silicon
- Cathode Ray Tubes (B&W)
- Thermal Batteries
- Large High Quality Forgings

- Acrylic Sheet
- UDMH
- Vacuum Tubes
- TWT Tubes
- Tubeaxial Fan (MK87 Gun System)

- Industrial Fasteners
- Low Noise Bearings (Subs)
- Aerial and X-ray Film
- Rayon

Not included bottlenecks:  Energy, lack of inland transportation, upgrading facilities

SHORTAGES OF CRITICAL MATERIALS IN U.S. ARMY
STRATEGIC END-ITEMS 1979-85

ENDANGERED SPECIES

Figure 4

## ENVIRONMENTAL CONSEQUENCES OF A COAL CONVERSION INDUSTRY IN WESTERN KENTUCKY

**Harry G. Enoch**
Kentucky Department of Energy

Introduction

The U.S. must develop new energy sources to replace the dwindling
supplies of domestic oil and natural gas and to decrease our
dependence on imported oil.  Synthetic fuel from coal can play a
major role in this process and, at the same time, provide a cleaner
and more environmentally acceptable use for coal than has been possible in the past.  Coal is the nation's most plentiful fossil fuel--
abundant resources are available in Appalachia, the Midwest, and the
Northern Great Plains (Fig. 1).  The richest coal deposits are
in the western U.S., but there are problems with synfuels development
in those areas--due to the commitment of water resources, difficulties
in reclamation, lack of an existing population base, distance from
markets, and the proximity to national landmarks.  Industrial development has long been a problem in Appalachia, chiefly due to the
rough terrain.  Perhaps the most favorable area for synfuels development is along the major waterways in the Eastern Interior Coal
Basin (Fig. 2).  In addition to the coal and water resources,
the region has a well-developed transportation network, gently rolling
terrain, a large labor force, and existing social structure and services.

The Synfuels Industry in Western Kentucky

Four commercial-scale synfuel facilities have been proposed in the
western Kentucky coal fields to be built in the 1980 decade (Fig.
3).  These plants can provide many benefits for the state in the
form of new jobs, increased revenue for state and local governments,

increased business activity, and, of course, increased production of scarce liquid and gaseous fuels. If we are to realize these benefits, it is important that we recognize the potential undesirable effects of this development on the local communities and the environment and take the necessary steps to prevent or minimize them.

The earliest synfuel development in this region will be the Solvent Refined Coal (SRC I) demonstration plant, which will begin construction in 1981. This project is a cooperative effort between the International Coal Refining Company (a joint venture between Air Products and Wheelabrator-Frye), the U.S. Department of Energy, and the Kentucky Department of Energy. Following successful demonstration of the technology the plant will be expanded to commercial size.

W. R. Grace has plans for a facility converting coal to methanol to gasoline via indirect liquefaction. The project is currently in the preliminary design phase and is funded by U.S. DOE.

Texas Eastern and Texas Gas Transmission Co.'s have proposed to build an indirect coal liquefaction plant based on the commercial model in operation in South Africa (SASOL II). The design, feasibility study, and environmental assessment will be carried out with joint funding by the companies and U.S. DOE.

A project based on the H-Coal process is planned by Ashland Oil and Airco. The H-Coal pilot plant in Catlettsburg, Kentucky, began operations in 1980 and a commercial plant is being designed with funding by U.S. DOE.

A summary of commercial plans and schedules is listed in Fig. 4. If built as proposed, these four plants will consume about 40 million tons of coal a year and produce the equivalent of 260,000 barrels of oil each day in liquid fuels, gases, and chemical feedstocks. In addition to the synfuel plants, there are new coal-fired power plants planned by the utilities in the region (see Fig. 3). These plants, which will be under construction between 1981 and 1990, will add 6,180 MW of generating capacity and will use about 20 million tons of coal a year.

## Synfuels Industry Assessment

In July 1980, Governor John Y. Brown appointed a "Synfuels Coordination Task Force" to guide the state in planning and preparing for the synfuels development. As part of this effort the Kentucky DOE has begun a comprehensive assessment of the proposed synfuels industry. This work is to provide an information base for other state agencies, local governments, and the public. Although it will not be in the same detail as the individual Environmental Impact Statements (EIS), it will be complementary. The EIS's are largely directed toward localized effects of a single facility, while our assessment will be more regional in scope and will consider the overall effects of all the proposed industrial and related developments.

The industry's design data and timetables are being used whenever possible. In the absence of industry data, we are working with the best estimates we can make at Kentucky DOE. What follows are some of the implications of the synfuels industry based on our work thus far. A more detailed and documented version of this preliminary assessment will be available soon.

## Socio-Economic Effects

This section will deal with the labor requirements and population increases rather than social and economic effects per se--these will be detailed elsewhere, but are directly related to population changes. The socio-economic impacts of the synfuels industry may be more significant than the environmental impacts and of more concern to local residents.

Synfuel plants will be among the largest construction efforts in the country in the 1980's. Due to their size the plants in western Kentucky may have to compete nationwide for skilled labor. Possible shortages are already foreseeable for certain crafts (e.g. pipefitters and electricians). Thus, a large in-migrating work force appears inevitable. The peak construction force for most commercial synfuel projects has been estimated to be 5-6,000--an exception is the SASOL plant, which is estimated to be 15,000. As seen in Fig. 5, maximum construction activity will occur in late 1985 (based on today's most optimistic schedules), when three projects reach a peak employment of about 25,000. For discussion purposes I have assumed that all of this labor will come in from outside the region--this allows the local labor pool to be fully absorbed by the utility construction effort (actually, the two industries will directly compete for this pool since both have similar skill requirements). Eight coal-fired units will be under construction in late 1985, requiring about 2-3,000 workers.

The social problems associated with construction are due to the rapid buildup of a large population, followed by a rapid decrease. Applying standard multipliers for secondary employment and families, the peak population increase due to synfuels in 1985-86 is estimated to be 93,000. The affected communities may require considerable assistance in order to mitigate the adverse effects of this large influx. The nearest communities will not have to absorb all the growth--it is expected that the work force will be located within a 60-90 minute commuting radius.

In the operating phase of the industry, the synfuel plant employment, estimated to be 5,000, should not cause a large population increase. Assuming that 1/2 to 2/3 of the operating labor (largely unskilled) can be drawn from the local unemployed, the resulting new 1990-1992 population is estimated to be only 10-15,000 for the region. The new coal mining work force required will be larger, 7-10,000, and may have a greater impact, since it comes in counties which have smaller populations (Fig. 6).

## Coal Resources

A recently completed survey[1] showed that w. Kentucky has 39 billion tons of coal resources (Fig. 7). The record production from this field was 56 million tons in 1975. The production had slipped to 43 million tons by 1979 as the market for high-sulfur coal softened. It is difficult to estimate what proportion of the 40 million tons needed for synfuel plants will be produced in Kentucky. One basis for such an estiamte is the coal used by the w. Kentucky power plants: 90% of their coal comes from the Eastern Interior Basin and 90% of that comes from w. Kentucky. Thus 80-90% of the coal for synfuels could come from w. Kentucky. It is also uncertain what proportions will come from surface mining--the current split in w. Kentucky is 54%

surface and 46% underground. However, basing future predictions on current trends can result in large inaccuracies.

The synfuels plants will strengthen the markets for w. Kentucky coal, calling for as much as a doubling of current production. The synfuel plants will use about 0.1% of the resource each year. The planned power plants in the region will use another 8-10 million tons of high-sulfur coal each year.

Surface Water Resources

Synfuels plants have large water demands (Fig. 8) -- mostly required for evaporative cooling. In surface water streams, a critical flow must be maintained for transportation, aquatic life, and assimilation of contaminants. The critical flows (the 10-year low flow minus commitments) must be matched against the synfuel plant demands. The Ohio River Basin Commission has a policy for the allocation of up to 10% of the low flow of that river for energy use.[2] The two synfuel plants on the Green River require together 16% of the low flow (21% when the new power plant is added in). This could cause some problems and needs to be further addressed. Alternative cooling technologies (e.g. dry cooling) could reduce the above requirements by 50%. No problems are predicted on the Ohio River. Previous studies have found the Ohio River has sufficient water for a greatly expanded energy industry.[3,4] By comparison with synfuel plants and power plants, the water needs for mining and for the new population are small.

Regional Air Quality

We are carrying out a detailed regional air quality modeling study for all the synfuel plants, power plants, and other planned industrial expansions in the w. Kentucky region. The results of this study are not yet available. Such a study is needed to determine if the region has enough clean air space to accommodate the synfuels plants and new power plants, if future growth will be compromised, if more emission controls are needed, and whether alternative sites should be considered.

The region is currently experiencing problems with total suspended particulates (TSP) and photochemical oxidants ($PhO_x$). The cities of Henderson and Owensboro and Vanderburgh Co., Indiana, are in nonattainment of the air quality standards for TSP; the same areas have had problems recently with $PhO_x$ (measured as ozone). The synfuel plants will cause only a small increase in TSP but the 25% increase in hydrocarbon emissions and the 130% increase in nitrogen oxide emissions may be significant (Fig. 9) -- both contribute to $PhO_x$ formation. Both are a problem largely of urban areas, where automobiles are the predominant source. Fortunately, their control is prescribed in the Clean Air Act and will soon be much less of a problem if the Congress can avoid giving further extensions on deadlines for achieving auto emission standards.

The sulfur and nitrogen oxide emissions of synfuel plants and new power plants will be quite an increase over 1976 levels. The standards are not likely to be endangered, however, as the present concentrations of these two pollutants are very low in w. Kentucky, in spite of the large amount of high-sulfur coal used there.[5] The resulting increases in sulfate levels will be undesirable in view of the present visibility problems they are causing. Furthermore, sulfur and nitrogen emissions are increasingly coming under attack, whether warranted or not, as contributing to interstate pollution and acid precipitation.

The increased mining activity will add to the levels of TSP, although quantitation is difficult. In the mining region, Muhlenberg Co. is not meeting the TSP air quality standards and the problem has been partially attributed to present surface mining operations. The new population growth will contribute to air pollution mainly through the higher use of automobiles and will impend on an already borderline problem.

The possibility of adverse health effects from hazardous air pollutants or local air standard violations should be identified during the EIS and permitting processes. Kentucky is actively promoting synfuels development, but only with the best available controls needed to comply with all existing state and federal environmental laws, regulations, and standards.

Water Quality

Synfuel plants and most other new industries in Kentucky should have little effect on water quality. This is due in part to the Clean Water Act of 1972, which mandated "fishable, swimmable waters" by 1983 and "zero discharge" by 1985. The synfuel plants will be designed for zero discharge (no process effluents to streams). The solid and hazardous waste disposal programs recently implemented under the Resource Conservation and Recovery Act will do a good job of protecting groundwater from future industrial contamination. The greatest potential for contamination will be from accidental spills of synfuels products during transportation.

Mining is another potential source of water pollution from sediment and acid drainage. Both of these problems can be controlled, and it will be required under the new strip mining regulations. Probably the greatest source of water pollution from the synfuels development will be from the increase in urban runoff and improperly treated sewage resulting from the increased population.

Land Use

Synfuel facilities will require 1-2,000 acres of land (Fig. 10), which should be relatively level and near a major waterway. In w. Kentucky this will generally be found to be agricultural land, and each of the four sites involves a significant acreage of farmland. The amount of farmland taken countywide will be rather insignificant and may be more than offset by future increases in productivity. Solid waste disposal will require additional acreage, but will not involve farmland, if possible. Synfuel plants will produce approximately 3,000 tons/day of solid waste or about 500 acre-feet/year. At a depth of 50 feet, the four plants would require 56 acres per year. This land could be returned to productive use.

The mining of additional coal for the synfuel plants will result in significant land disturbance. An estimate for 40 millions tons of coal with 50% from surface mining is 4,000 acres per year (based on 1800 tons of coal/acre-foot, 45 inch seams, and 80% recovery and adding 10% for additional disturbances: haul roads, prep plants, etc.). Some of the coal is under agricultural lands. The new strip mining laws will not allow farmland to be disturbed unless it can be returned to equal quality for its original use.

The population growth may contribute to some urban sprawl and this will require land. The estimates for the maximum additional urban areas needed in 1990-92 are: 2,400-3,600 acres resulting from plant operation and 4,800-9,500 acres resulting from coal mining (based on 2,682 persons/square mile in Kentucky urban-suburban areas[7]). A summary of the land use for a 260,000 barrels of oil equivalent/day coal conversion industry is shown below:

| | |
|---|---|
| 4 Plants | 6,000 acres |
| Waste disposal | 56 acres/year |
| Coal mining | 4,000 acres/year |
| Urban areas | 7-13,000 acres |

The land needs may seem high, but need to be compared against an alternative source of energy to gain perspective. For the production of liquid fuels, one could consider a renewable source: ethanol from corn, which is also being promoted in Kentucky. To produce the equivalent amount of energy would require almost 300 commercial size ethanol plants, which would use more land than the coal conversion plants (assuming 20 million gallon/year plants and 50-100 acres/plant):

295 Plants   15-30,000 acres

This would result in one plant in every 132 square miles of Kentucky or one plant every 12 miles throughout the state. In contrast to the coal resource which could be produced in several counties of w. Kentucky, the area needed for corn production would be staggering (assuming 2.5 gallons/bushel and 100 bushels/acre):

Corn production   24,000,000 acres/year

This acreage is not available in Kentucky--the entire state is only 25.5 million acres. In fact, the 2.4 billion bushels of corn required is about 40% of the U.S. corn production.

Synthetic fuels from coal and ethanol from biomass may both make a contribution to U.S. energy production in the future. It should be recognized, however, that a biomass industry with a significant energy output will have significant social and environmental consequences, some of which will be greater and some less than for coal conversion. Analyses, like the one above, continue to show us that all of man's activities (and energy production in particular) have effects on his social and physical environment. The choice of future energy sources, including conservation, must be tempered by a careful study of the trade-offs involved.

## Acknowledgements

In preparing this article I have drawn liberally from the work of others involved in the "Synfuels Industry Assessment", including Dr. James Jones and David Holmes of the Kentucky DOE and Dr. Mickey Wilhelm of the University of Louisville and have had advice and suggestions from many others. Responsibility for the data and conclusions above, of course, rests entirely with the author.

## References

1.  Kentucky Geological Survey, Western Kentucky Coal Resources, March 1980.

2.  Ohio River Basin Commission, Water Resources Assessment of Emerging Coal Technologies, September 1977 (draft).

3.   A.D. Shepherd, A Spatial Analysis Method of Assessing Water Supply and Demand Applied to Energy Development in the Ohio River Basin (Oak Ridge National Laboratory), August 1979.

4.   U. S. DOE, Synthetic Fuels and the Environment: An Environmental and Regulatory Impacts Analysis, June 1980.

5.   H. G. Enoch and J. E. Jones, "Environmental Aspects of the Developing Synthetic Fuels Industry in Kentucky" in Energy Technology VII (Government Institutes, Inc.), June 1980.

6.   Council on Environmental Quality, Environmental Quality-1979, December 1979.

7.   Kentucky Development Cabinet, Kentucky's Settlement Patterns, December 1974.

8.   Coal Age, p. 71, May 1980.

9.   Hittman Associates, Inc., Baseline Data. Environmental Assessment of a Large Coal Conversion Complex. Volume 1, p. 32, May 1975.

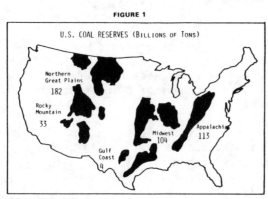

FIGURE 1

U.S. COAL RESERVES (BILLIONS OF TONS)

**Adapted from Reference 8.**

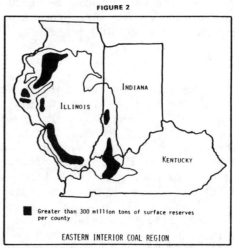

FIGURE 2

EASTERN INTERIOR COAL REGION

**Adapted from Reference 9.**

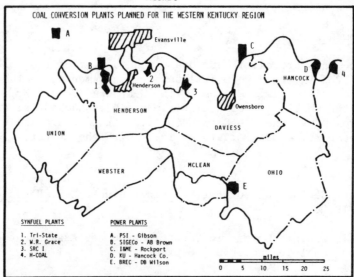

FIGURE 3

COAL CONVERSION PLANTS PLANNED FOR THE WESTERN KENTUCKY REGION

**FIGURE 4**

COMMERCIAL SIZE SYNFUEL PLANTS PLANNED FOR WESTERN KENTUCKY

| | SRC I | W.R. GRACE | H-COAL | TRI-STATE |
|---|---|---|---|---|
| COUNTY | DAVIESS | HENDERSON | BRECKINRIDGE | HENDERSON |
| PROCESS | DIRECT LIQUEFACTION | COAL-METHANOL-GASOLINE | DIRECT LIQUEFACTION | SASOL/FISCHER-TROPSCH |
| COAL USE (T/D) | 30,000 | 29,000 | 22,500 | 30,000 |
| PRODUCTS | BOILER FUEL COKE NAPHTHA | GASOLINE LPG | SYN-CRUDE BOILER FUEL | GASOLINE FUEL OIL SNG CHEMICALS |
| BOE/DAY* | 100,000 | 60,000 | 50,000 | 50,000 |
| START-UP | 1990# | 1987 | 1987-88 | 1987 |

*BARRELS OF OIL EQUIVALENTS/DAY

#6000 TPD DEMONSTRATION PLANT WILL START UP IN 1984

**FIGURE 5**

SYNFUEL PLANT CONSTRUCTION LABOR REQUIREMENT

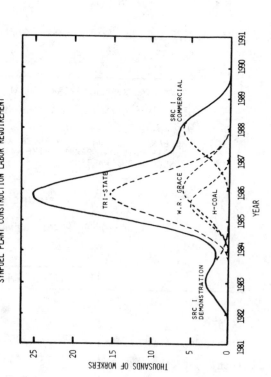

**FIGURE 6**

POPULATION CHANGES FOR THE SYNFUELS DEVELOPMENT: OPERATIONAL PHASE

| COUNTY | 1990 POPULATION PROJECTIONS* | 1990 POPULATION INCREASES ASSOCIATED WITH SYNFUELS DEVELOPMENT |
|---|---|---|
| HENDERSON | 43,686 | |
| DAVIESS | 89,672 | |
| HANCOCK | 7,928 | 10-15,000 FOR PLANT OPERATION |
| BRECKINRIDGE | 17,593 | |
| VANDERBURGH, IND. | 168,772# | |
| UNION | 20,376 | |
| WEBSTER | 17,548 | |
| MUHLENBURG | 34,182 | 20-40,000 FOR MINING |
| OHIO | 25,950 | |
| MCLEAN | 12,963 | |
| BUTLER | 11,410 | |
| GRAYSON | 23,058 | |

*KENTUCKY DEPT. OF COMMERCE, KENTUCKY DESKBOOK OF ECONOMIC STATISTICS, 1979-80

#1970 POPULATION

**FIGURE 7**

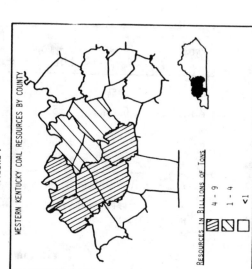

WESTERN KENTUCKY COAL RESOURCES BY COUNTY

RESOURCES IN BILLIONS OF TONS

4 - 9

1 - 4

<1

Adapted from Reference 1.

**FIGURE 8**

EFFECT OF SYNFUEL PLANT WATER CONSUMPTION ON CRITICAL
SURFACE WATER FLOW

| Project | River | 7D-10Y Low-Flow* (Million gallons/day) | Synfuel Plant Consumption | Cumulative % of Low Flow# |
|---|---|---|---|---|
| SRC I | Green | 255 | 17 | 7 (8) |
| W.R. Grace | Green | 255 | 23 | 16 (21) |
| H-COAL | Ohio | 9,300 | 12 | 0.1 (0.1) |
| Tri-State | Ohio | 9,700 | 23 | 0.8 (1.4) |

*7 day - 10 year low flow

#Values in parenthesis include new power plant
water consumption

**FIGURE 9**

EFFECT OF SYNFUEL PLANTS ON AIR EMISSIONS IN THE
EVANSVILLE-OWENSBORO-HENDERSON AIR QUALITY CONTROL REGION#

| Emission Sources | Thousands of Tons/Year | | | | |
|---|---|---|---|---|---|
| | $SO_2$ | TSP | $NO_2$ | CO | HC |
| 1976 -Point Emissions* | 263 | 34 | 42 | 88 | 8 |
| -Area Emissions* | 3 | 16 | 15 | 12 | 21 |
| -Total* | 266 | 50 | 57 | 100 | 29 |
| 1976-1992 New Power Plants | 267 | 20 | 180 | 11 | 3 |
| Four Proposed Synfuel Plants | 50 | 4 | 74 | 15 | 7 |

TSP, TOTAL SUSPENDED PARTICULATE; HC, HYDROCARBONS

*EPA, 1976 National Emissions Report, August 1979

#Includes Kentucky counties: Daviess, Hancock, Henderson,
McLean, Ohio, Union, Webster and Indiana counties:
Dubois, Gibson, Perry, Pike, Posey, Spencer,
Vanderburgh, Warrick

**FIGURE 10**

SYNFUEL PLANT LAND USE BY COUNTY

| | TRI-STATE | W.R. GRACE | SRC I | H-COAL |
|---|---|---|---|---|
| Plant Site (acres) | 2,000 | 1,500 | 1,500 | 1,000 |
| County | Henderson | | Daviess | Breckinridge |
| County Area (acres) | 277,295 | | 295,577 | 354,766 |
| County Cropland (acres) | 167,784 | | 173,443 | 167,864 |
| % of Area Used | 1.3 | | 0.5 | 0.3 |
| % of Cropland Used (max) | 2.1 | | 0.9 | 0.6 |

98

# RESOURCE CONSERVATION AND AIR POLLUTION CONTROL: A CASE STUDY OF COMPATIBILITY

Anthony J. Buonicore, Joseph Bilotti and Pankaj R. Desai
York Services Corporation

INTRODUCTION

Resource conservation is of paramount importance today particularly due to spiralling inflation and raw material costs, the non-availability of plentiful, inexpensive energy supplies, and increasingly stringent environmental regulations. Such a situation necessitates that more attention be given to material and energy recovery to ensure continued economic and successful plant operation. The problem seems to be whether industrial growth can be met without significantly harming the environment. There is no question that the environment must be protected and the quality of life improved. At the same time, however, economic stability must also be maintained. Recent developments provide persuasive evidence that these two objectives, in all their many ramifications, will rank high on the nation's list of priorities and be prime determinants of domestic and foreign policy for many years to come.

The conflict between industrial growth and pollution control is an extremely complex and emotional issue. However, a number of areas do exist where the goals of each are mutually compatible. When intelligently implemented, resource conservation can reduce the environmental impact of process operations and enhance the reliability of future material supplies. One such case study involves solvent emissions from a number of gasket curing ovens located at a major manufacturing facility in the northeastern United States.

PROCESS DESCRIPTION

The curing operation is one part of an overall process to produce ring-shaped gaskets (or wafers) for use as friction surfaces in the clutches of automatic transmissions. The asbestos gaskets, after forming, are saturated in a warm bath containing a phenolic resin with a solvent

99

(principally ethanol) carrier. The saturated gaskets then pass through a set of rollers to squeeze out excess solution and into three natural gas-fired ovens for curing at approximately 300-350°F. In the curing process, the solvent carrier is vaporized and exhausted to the atmosphere. The resin dries and sets in the gasket making it hard and brittle. A solvent balance on the plant indicated that emissions from these curing ovens exceeded the state air pollution control regulations for allowable hydrocarbon emissions.

EMISSION TEST DATA

In order to confirm the hydrocarbon emission levels indicated by the solvent balance, an emission test program was conducted on the exhaust stacks from the three ovens. There were two zones in each oven, both exhausting to separate stacks. Each oven was classified as a single source by the state regulatory agency. Emission data for each zone is summarized in Table 1. The total hydrocarbon emission level for each oven (operating 18 hours per day) was determined to be:

Oven # 95 - 1158 lb/day
Oven #96 - 1181 lb/day
Oven #101 - 932 lb/day

The allowable hydrocarbon emission level per oven (source) under existing air pollution control regulations was 800 pounds per day. This level was exceeded in all three ovens. Hence it was necessary to evaluate alternatives to bring the units into compliance.

CONTROL STRATEGY

An investigation into potential process modifications to reduce or eliminate solvent emission levels indicated that such modifications were not feasible for these specific ovens. However, an effort could be made to minimize exhaust flow rates and, in effect, reduce the size and cost of any air pollution control equipment required.

A detailed engineering feasibility study was then performed to evaluate various alternative control techniques. The methods investigated include:

(1) Absorption
    a) Wet Scrubbing with wastewater treatment system
    b) Wet Scrubbing with no wastewater treatment facilities
(2) Incineration
    a) Thermal Incineration
    b) Thermal Incineration with Air Preheating
    c) Thermal Incineration with Waste Heat Boiler
    d) Catalytic Incineration
    e) Catalytic Incineration with Air Preheating
    f) Catalytic Incineration with Waste Heat Boiler
(3) Adsorption (with solvent recovery)

The absorption route was eliminated because it resulted in a water pollution problem and the necessity for an expensive system for wastewater treatment.

The various incineration routes were ruled out due to auxiliary fuel requirements and the incompatibility of recovered heat energy with existing and projected plant steam loads.

Adsorption by activated carbon and recovery of reusable solvent presented the most attractive approach to the client. The savings associated with directly reusing the solvent in the curing process would amount to $210,000 annually at a conservative recovery rate of 90% and $1.50 per gallon solvent cost.

## THE BUBBLE CONCEPT

At the time the engineering analysis was being completed, the EPA formally proposed as policy the "bubble" concept introduced under the PSD provisions of the Clean Air Act amendments of 1977. The "bubble" policy reflected a change in the traditional regulatory approach. Rather than focusing on individual emission sources within a plant (such as individual stacks, vents, or other emission points), under the "bubble" policy, the plant would be viewed as if it were contained within an imaginary bubble with a single stack. Using this concept, individual plants could offer their own mix of air pollution controls to comply with the requirements of State Implementation Plans.

The "bubble" policy was an important part of a larger incentive system designed to encourage businesses to reduce emissions and to use those reductions to meet their own financial objectives. It was intended to be particularly useful to industries with many process operations emitting similar pollutants but for which the marginal costs of control were different. If it was found to be less expensive to tighten control of a particular pollutant at one point in the system and relax controls at another, this would be possible under the "bubble" policy as long as the total emissions from the plant did not exceed the sum of the current emission limits on the individual sources. Properly applied, this alternative approach will promote greater economic efficiency and increased technological innovation.

For the particular case in question, under the "bubble" concept, the three curing ovens could be considered a single source. This would introduce new alternatives for consideration. For example, it might be more economical to control a single oven with pollution control equipment capable of achieving greater than the required efficiency and leave the remaining ovens uncontrolled.

## APPLICATION OF BUBBLE STRATEGY

The deciding factor on whether the "bubble" route would be more advantageous rested with the economics, i.e., cost advantage to the plant. However, prior to any financial analysis, it was first necessary to evaluate whether the "bubble" could be applied at this particular plant. Careful evaluation indicated that the situation did meet all of the restrictions that applied to the "bubble" concept, including:

1. Emissions must be quantifiable and trades among them must be even. Each emission point must have a specific emission limit and that limit tied to enforceable testing techniques.

2. Pollutants of the same type must be traded, in this case, hydrocarbons for hydrocarbons.

3. Alternatives must insure that air quality standards would be met.

4. Development of the "bubble" plan must not delay enforcement of federal and state requirements.

5.  The state implementation plan for the control of hydrocarbons had to have been approved by the EPA.

With adsorption as the selected control alternative, the possibility of using a "bubble" approach was investigated for several alternatives as presented in Figures 1 through 3. These alternatives with the resultant hydrocarbon emission rates are summarized in Table 3.

Any of the three alternatives could meet the allowable total emissions of 2,400 lb/day. However, only Alternative 3 could be considered a true "bubble" solution. In both Case 1 and Case 2, the emissions from each oven would be less than 800 lb/day and, hence, the individual ovens would each have been brought into compliance. Case 3 would result in emissions of greater than 800 lb/day from two of the three sources (ovens), but with a total for all three less than the allowable 2,400 lb/day, hence a true "bubble" solution.

If only capital investment and annual operation and maintenance costs were considered, the "bubble" approach (controlling one oven only) with a relatively small, highly efficient activated carbon system would have been the best alternative. However, for these particular curing ovens and in view of the quantity and current market value of the recoverable solvent, Alternative 2 (partial control on each oven - which limits emissions to less than 800 lb/day/oven, and thus bring each individual source into compliance) is the optimum control route. Table 4 provides a summary of the uniform annual cost for all three alternatives. In another situation, the economics could very well have been in favor of applying for a "bubble" permit.

CONCLUSION

The solvent recovery case study presented is a specific example of an application where neither air pollution nor plant profitability goals had to be compromised. The solvent recovery activated carbon adsorption system will reduce solvent emissions to acceptable levels and at the same time provide a return on investment in the form of recovered solvent.

ACKNOWLEDGEMENT

The authors are grateful to Mr. Roger Kniskern for collection of the emissions test data and Mr. Edward Whitlock for engineering assistance on the evaluation.

TABLE 3

TOTAL EMISSIONS ANTICIPATED FOR ALTERNATIVE CONTROL OPTIONS

| Alternative | Description | Resultant Emissions* |
|---|---|---|
| 1 | Adsorption controls on both zones, all three ovens | 327 lbs/day |
| 2 | Controls on heaviest emitting zones from each oven | 1,072 lbs/day |
| 3 | Controls on only heaviest emitter. No controls on other ovens. | 2,208 lbs/day |

* Applying conservative equipment hydrocarbon removal efficiency of 90%.

102

Table I
Process Parameter
(at Test Points)

| Parameter | Oven #95 Zone 1 | Zone 2 | Oven #96 Zone 1 | Zone 2 | Oven #101 Zone 1 | Zone 2 |
|---|---|---|---|---|---|---|
| Temperature, °F | 205 | 211.3 | 215.8 | 213 | 209 | 211 |
| Stack Pressure, in. Hg.abs | 29.89 | 30.06 | 30.42 | 30.45 | 29.99 | 30.0 |
| Moisture, % by volume | 2.4 | 2.1 | 1.9 | 2.0 | 1.9 | 2.1 |
| Molecular weight of dry gas | 28.86 | 28.86 | 28.86 | 28.86 | 28.86 | 28.86 |
| Molecular weight of mixture | 28.60 | 28.63 | 28.66 | 28.64 | 28.65 | 28.63 |
| Flow rate - ACFM | 8103 | 3127 | 4798 | 8733 | 2545 | 5827 |
|        - DSCFM | 6294 | 2426 | 3754 | 6882 | 1982 | 4516 |
| Total Hydrocarbons (max. in zone) (as $C_6H_{14}$ ppmv | 1265 | 352 | 695 | 973 | 2458 | 567.1 |
|      lb/hr) | 53.88 | 10.48 | 17.62 | 48.01 | 33.91 | 17.86 |
| Total Particulates avg. grain/SCFD | 0.007 | 0.045 | 0.004 | 0.012 | 0.00164 | 0.0039 |

Total Emissions

| | Oven #95 | Oven #96 | Oven #101 |
|---|---|---|---|
| Hydrocarbons, lb/hr | 64.36 | 65.63 | 51.77 |
| Particulates, gr/SCFD | 0.0176 | 0.0092 | 0.0032 |

Table 2
Economic Analysis of Pollution Control Alternatives for Curing Ovens

| Air Pollution Control Method | Total Capital [6] Costs | Yearly Operating [8] and Maintenance Cost | Uniform Annual [1] Cost |
|---|---|---|---|
| Wet Scrubber | | | |
|   w/treatment system | 3,140,632 [4] | 158,611 [5] | 714,553 |
|   w/o treatment system | 91,432 | 30,148 | 46,330 |
| Incineration | | | |
| Thermal | | | |
|   w/no heat recovery | 199,286 | 1,703,471 [3] | 1,738,745 |
| Thermal | | | |
|   w/Air Pre-heater | 349,367 | 587,216 | 649,054 |
| Thermal | | | |
|   w/Waste Heat Boiler | 500,366 | 1,712,564 | 663,889 |
| Catalytic | | | |
|   w/no heat recovery | 570,091 | 911,685 | 1,012,582 |
| Catalytic | | | |
|   w/Air Pre-heater | 635,975 | 168,191 | 280,749 |
| Catalytic | | | |
|   w/Waste Heat Boiler | 871,175 | 924,897 | 414,881 |
| Adsorption | | | |
|   w/solvent recovery system | 876,641 [2] | 116,851 | 62,002 [7] |

1. Uniform Annual Cost based on 10 year service life of equipment and a 12% cost of money.
2. This cost includes the solvent recovery system producing a solvent with a water content of 5-6%.
3. Based on burning either #2 fuel oil or natural gas.
4. This cost includes waste water treatment facility.
5. Operating and maintenance costs for the waste water treatment system is an estimate based on experience for similar systems.
6. All costs at December 1979 level.
7. This cost was calculated by subtracting the $210,000 per year savings on solvent from the uniform annual cost of $270,562.
8. Electricity costs estimated using $0.04/kw-hr, operating 8760 hrs/yr.

TABLE 4
## Uniform Annual Cost Summary

| | Ovens 95,96 and 101-Zones 1 and 2 Emissions | Ovens 95,96 and 101 Heaviest Zone Emission | Oven #96 Zones 1 and 2 Emissions |
|---|---|---|---|
| **Estimated Equipment Cost ($, March 1980)** | | | |
| Charcoal Adsorption System (including activated carbon, condenser, valves, Air/Air Exchanger, Solvent Recovery System) | 220,000 | 184,000 | 160,000 |
| HEAF Filter | 140,000 | 120,000 | 106,000 |
| I.D. Fan | 10,000 | 6,000 | 4,000 |
| Duct work | 35,330 | 21,130 | 13,400 |
| Guyed stack (50', 7 gage C.S.) | 4,790 | 4,660 | 4,100 |
| Tax and Freight | 32,810 | 26,860 | 23,000 |
| Installation | 221,460 | 181,330 | 155,250 |
| Subtotal | 664,390 | 543,980 | 465,750 |
| Captured Cooling Water System (including cooling towers, pumps, motors, fans and installation) | 64,160 | 45,150 | 33,360 |
| Engineering | 72,860 | 58,910 | 49,960 |
| Subtotal | 801,410 | 648,040 | 549,570 |
| Contingency | 80,140 | 64,800 | 54,960 |
| Total Capital Cost | 881,550 | 712,840 | 604,530 |
| **Operating and Maintenance Cost (per year)** | | | |
| Steam | 75,006 | 40,588 | 22,830 |
| Electricity | 32,662 | 18,351 | 12,484 |
| Activated Carbon (based on replacement life of 5 yrs) | 7,083 | 4,281 | 3,148 |
| Maintenance (based on 7 hrs maintenance per week) | 2,100 | 2,100 | 2,100 |
| Total O&M Costs ($/yr) | 116,851 | 65,320 | 40,562 |
| Annualized Capital Cost (10 yrs, 12% money; CRF = 0.1770) | 156,034 | 126,173 | 107,002 |
| Credits | (210,000) | (157,000) | (75,800) |
| Uniform Annual Cost, $/yr | 62,885 | 34,493 | 71,764 |

## ALTERNATIVE SYSTEM 1

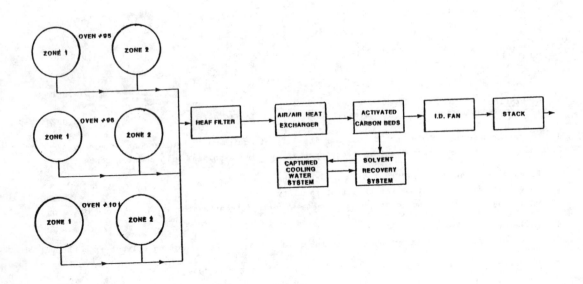

Figure 1

Adsorption Control on Both Zones, All Three Ovens

# ALTERNATIVE SYSTEM 2

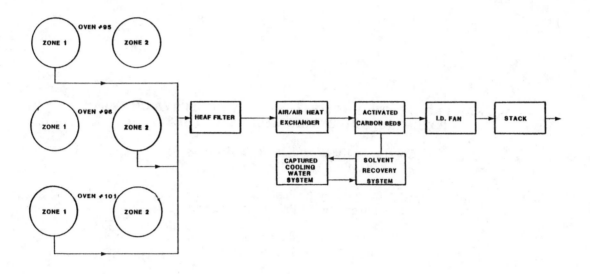

Figure 2
Controls on Heaviest Emitting Zones From Each Oven

# ALTERNATIVE SYSTEM 3

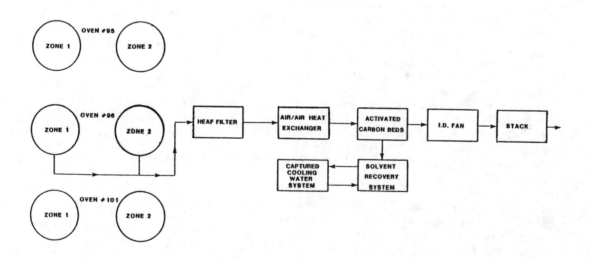

Figure 3
Controls on Only Heaviest Emitter

# NEW PROCESS TECHNOLOGY: ENERGY CONSERVATION POTENTIAL AND ENVIRONMENTAL IMPACTS

**Charles L. Kusik** and **James I. Stevens**
Arthur D. Little, Inc.

**Harry E. Bostian**
U.S. Environmental Protection Agency

**Herbert S. Skovronek**
Allied Chemical Corporation

Introduction

The United States, with under six percent of the population of the world, consumes a disproportionate amount of its materials, namely, forty-two percent of the world's aluminum production, thirty-three percent of the copper production, and twenty-eight percent of the iron and steel as shown in Table 1. Since converting raw materials to semi-finished products is energy-intensive, it is not surprising to also see that the United States also consumes a large percentage of the world's fossil fuels. Table 2 shows that about thirty-seven percent of the nation's energy requirements is consumed in the industrial sector where it is used primarily for chemical processing, steam raising, drying, space cooling and heating, preheating, and miscellaneous other purposes. Table 3 shows the distribution of energy in the industrial sector. It is seen that five industrial sectors, headed by the primary metals industry, account for about three-quarters of the energy used in the manufacturing industries.

Significant reductions in energy consumption have been achieved in many industrial sectors by better "housekeeping" (i.e., shutting off standby furnaces, better temperature control, elimination of steam and heat leaks, etc.) and by greater emphasis on optimization of energy usage. However, energy costs have been rising rapidly, much more rapidly than plant equipment costs as shown in Figure 1. As a result, further changes in the pattern of energy consumption can be expected by the introduction of new industrial processes in order to take advantage of a more readily available raw material or less costly fuel.

The energy conservation potential and potential environmental consequences of 55 new industrial processes were analyzed in a study[1] completed in 1979 by Arthur D. Little, Inc., under the auspices of the Industrial Environmental Research Laboratory (Ci) of the U.S. Environmental Protection Agency (EPA). These 55 new industrial processes fell largely into the five industry sectors accounting for a major portion of U.S. industrial energy use (see Table 3). This paper summarizes some of the energy and environmental findings of new process technology. Details on the study's methodology process analyses and environmental aspects can be found in 21 reports[1].

Even a cursory review of the technical journals shows that there are an enormous number of potential new processes that could be considered within the 13 industry sectors studied. Given the scope of our study, we focused on identifying changes in the primary production processes which could be implemented within the next two decades.

Since most processes in the heavy industrial chemicals and metal industries take anywhere from five to twenty years from initial inception to commercialization, it is clear that few if any of the processes being considered for implementation have been developed in the United States as a result of a concern with high cost energy. "Energy-conserving technology" when it is being applied now in the United States has been developed largely abroad, mainly in Europe or Japan. Examples that can be cited are collection of carbon monoxide gas from basic oxygen furnaces (developments by Japanese and French companies), suspension preheaters in the cement industry (German and Japanese developments), and flash smelting in the copper and nickel industries (a Finnish and Canadian development).

Energy conservation is defined broadly here to include conservation of energy or energy form (gas, oil, coal) by a process or feedstock change. Natural gas and oil were considered to have the highest energy form value followed in descending order by electric power and coal. Thus, a switch from gas to electric power would be considered energy conservation because electric power could be generated from coal existing in abundant reserves in the United States in comparison to natural gas.

Some of the "groundrules" for this study were, for example, that we were interested primarily in the battery limits of the actual manufacturing processes; such aspects as boiler fuel, steam raising, electric generation, etc., were usually excluded. The quality of the final product from the manufacturing process remained the same or consisted of a directly substitutable material. The substitution of less critical fuel or feedstocks (e.g., coal for oil or natural gas) was considered to be "energy conserving" even if there was no reduction in the energy per unit of product, as long as process modifications or substitutions were required. Waste heat use within existing processes was recognized for its energy conserving nature but was (usually) excluded if the manufacturing process remained essentially unchanged.

Finally, technological changes were eliminated from further consideration within the study if the only changes that could be envisioned were energy conservation as a result of: better policing or "housekeeping;" better waste heat utilization; fuel switching in steam raising; or power generation or production of synthetic fuels. The focus here is on new processing techniques in the first stage of the production sequence involving the treatment of raw materials to produce a semi-finished or intermediate product such as metal ingots and monomers.

In undertaking this study, selected State air emission regulations along with the Federal government's stationary source performance standards and effluent limitation guidelines were surveyed to establish the most probable limits of air and water emissions, and obtain a perspective of the types of pollution control systems to be considered. At the State regulatory level, there are a large number of different regulations for airborne emissions. Nevertheless, we found that approximately the same type of air pollution control systems would be required, regardless of what State or Federal regulations were to be met.

For water effluents we chose the EPA Best Available Technology Economically Achievable (BAT) Guidelines (1983) as the effluent limitations that would have to be met for both currently practiced and alternative processes considered. The rationale for this choice was that any plant employing the technologies evaluated in these reports should install wastewater treatment systems capable of meeting BAT standards, although at the time of construction the New Source Performance Standards might be applicable. Because regulations for the handling and disposal of solid waste are either non-specific or non-existent, we chose various types of controlled landfill disposal methods, where our judgment suggested potential adverse environmental impacts might occur from uncontrolled disposal. Further documentation describing the methodology used is contained in fifteen industry-oriented volumes of reports[1]. Processes selected for study are shown in Table 5.

Approach

For the processes analyzed, important aspects of our approach included:

- Establishing a baseline technology against which the process changes could be assessed. Normally this baseline was a technology currently practiced in a major portion of the industry to make a given product. In comparing the baseline and alternative processes, a deliberate attempt was made to start with the same or similar raw materials and produce the same or similar end products. For example, ammonia can be made using natural gas as a feedstock in the conventional technology; this process forms the baseline against which the coal-based process was assessed.

- Determining energy usage in baseline and alternative technologies for both production and pollution control, using common conversion factors shown in Table 4. In this paper, energy use is to be interpreted as direct energy requirements for the process (e.g., fuels, coke, steam, electricity, carbon used for electrodes, etc.). It does not include "embodied" energy for materials used in the production process.

In addition, we developed estimated capital and operating costs for both production and pollution control for each of the processes studied with details presented in reference 1.

In order to provide a perspective of the future (1989) situation, we assumed that (a) current or anticipated environmental problems could be overcome with existing, economically viable technology; (b) the new process would totally replace the existing process by 1989 for incremental production; and (c) that production would continue to grow at its current rate in each industry. Using these arbitrary bases, a preliminary comparison of predicted energy use, cost, and environmental impact could be made. Of course, the real world situation will be considerably more complex. The new process may not be the only one used for (new) production in 1989. On the other hand, the current process may not continue to account for the same (base) production volume as plants become obsolete and are replaced. Although attempts have been made to incorporate the most stringent but realistic pollution control technology, other, as yet undefined, environmental needs and new control capabilities will certainly also occur. Relative costs for fuel, equipment, money, etc., will certainly also change in an undefinable manner.

Estimates of national energy conservation potential were then made by assuming that the new energy conserving technology would be applied to all new plant construction over a 15-year span of time. Growth rates used to calculate production increases assumed here over 15 years are shown in Table 5 along with the technologies examined in each industry sector.

Results

Using the growth projections shown in Table 5, discharges in 1989 were calculated assuming the indicated processes were to account for 100% of all new projected production installed from 1974 to 1989. Table 5 shows estimated national changes in energy use and controlled pollutant discharges from the base case process. Changes up to $\pm$ 10 million lbs. (or 10 million Btu) or less are indicated by half-shaded diamond. Increases of 10 million to 100 million lbs. are shown by a single upward facing open triangle, while decreases of 10 million to 100 million lbs. by a single downward facing shaded triangle. Changes of 100 million lbs. or more are shown by two triangles, with open triangles representing increases and shaded triangles representing decreases. For example, Table 5 shows that if refractory metal cathodes are used, this new aluminum technology could reduce energy use by more than $100 \times 10^{12}$ Btu in 1989. Likewise, controlled $SO_x$, and $NO_x$, would each decrease by more than $100 \times 10^6$ lbs. and particulates by $10$ to $100 \times 10^6$ lbs./yr., due to reduction in emissions at the power generating station. Similarly, solid residue discharges at aluminum smelters would change by under $10 \times 10^6$ lbs. if this new technology were implemented as indicated by the half shaded diamond. These summary tables should be used carefully because of underlying assumptions as described in reference 1; for example, 1) solid residue discharges from electric power plants were not included in this study, 2) nitric acid ion exchange process for production of alumina from domestic clays shows large $SO_x$ emissions because we assume coal will be used as a boiler fuel while the Bayer process calculations are based on natural gas.

To provide some perspective on the impact of pollutant discharges, Table 6 shows national emissions of $NO_x$, $SO_x$, and particulates. In addition, solid residue discharges from the utility sector alone are estimated to be 2.5 million tons of sludge per year. As in other estimates made in this study, we assume discharges are proportional to production for new plants installed between 1974 and 1989. Different processes give different levels of discharges resulting in a range of potential emissions for new plants installed from 1974 to 1989. Summing the range of estimates in the 13 industries studied gives the results shown at the bottom of Table 6 with details presented in each of the pollutant-specific summary reports.

Since preparation of our original estimates for pollution control costs, there has been an increasing impetus toward the development of more specific regulations concerning emissions to the environment. The increasing pressures toward further reduction of sulfur and nitrogen oxide emissions in order to alleviate the occurrence of acid rain is but one example of how increasing environmental regulations will require larger capital investments and contribute to increased operating costs as the nation relies upon the environmentally less desirable energy sources such as coal. The recently proposed regulations for hazardous waste disposal under the Resource Conservation and Recovery Act, the potential emission limits under the National Emission Standards for Hazardous Air Pollutants, the forthcoming effluent regulations on priority pollutants, and the prevention of significant degradation and visibility regulations are all likely to contribute to requiring greater efforts for environmental control in these industries than was considered in the original study. In Table 7 we have recorded our estimates of pollution control costs as a percentage of production costs for selected energy conserving process options. Since pollution control costs have usually increased at a greater rate than other costs, we have indicated in Table 7 those processes where increasing percentages of production costs may be attributable to pollution control. We do not believe that there are likely to be any dramatic breakthroughs in pollution control technologies that will drastically lower costs. Consequently, presuming that regulations do not exceed best available pollution control technologies, the costs of pollution control for energy conserving manufacturing options are not likely to present any additional problems not now faced such as the competition for capital and the costs incurred in a nonproductive (pollution control) system.

Conclusion

Examination of Table 5 indicates there is a strong correlation between conversion to coal and heavier feedstocks and increases in $SO_x$ emissions as shown by: hydrocracking of heavy bottoms; use of coal in cement kilns; use of coal in production of ammonia; and direct combustion of asphalt in petroleum refining.

Some new technologies indicate a decrease in controlled discharges of one pollutant while increasing another such as for $SO_x$-$NO_x$ emissions in: glassmaking; natural gas fired furnace to electric melting; some of the new $SO_x$ control technology in nitric acid manufacture.

Estimating weighting factors for the various pollutants is needed to determine whether such processes provide environmental benefits. In this way one can then make judgments on whether it is environmentally beneficial to accept an increase in one pollutant while decreasing another.

Finally, there is a strong correlation between energy (Btu) saving and decreases in controlled discharges of $NO_x$, $SO_x$, and particulates as seen in Table 4 for: new aluminum technology; BOF off-gas recovery in steelmaking; dry quenching of coke as a replacement for wet quenching; cleanup of wet process phosphoric acid to make detergent grade product rather than use of the electric furnace route; de-inking of newsprint; and preheating in general as evidenced by the examples in glassmaking and cement technology.

Clearly there would be a national environmental benefit if such processes are installed which simutaneously reduce $SO_x$, $NO_x$, and particulate discharges. Programs to demonstrate such new technologies should be fostered in order to alleviate industry's concerns about process feasibility and thus provide for a potentially more rapid implementation of such new technology.

Table 8 shows energy saving potential when production increases are multiplied by energy saving potential (Btu-ton) for selected new process technologies. Examination of Table 8 shows that energy conservation in terms of reducing specific energy consumption or critical fuels (e.g., oil and natural gas) by new process technology can potentially have a large and significant impact. Implementing new energy conserving technology can conserve up to 8 to 9 quads per year in a 15 year time frame. This amounts to a saving of over fifty percent of energy used if conventional technology were used in future plant expansions.

## Acknowledgments

Most of the information used herein was developed under the auspices of the EPA for which we gratefully acknowledge support granted under EPA contract No. 68-03-2198. A large number of industry experts worked on this assignment at Arthur D. Little, Inc. The authors are especially indebted to the contributions of the following principal investigators: R.W. Hyde,(aluminum); Dr. P. Huska,(cement); R. Shamel,(chloralkali); Dr. D.W. Lee,(glass); Dr. M. Mounier,(steel); F. Iannazzi,(pulp and paper); S. Dale,(olefins); R. Stickles,(petroleum refining); W. Keary, (phosphoric acid); J. Sherff,(ammonia and fertilizers); Dr. D. Shooter, (textiles); R. Nadkarni (copper).

## References

Arthur D. Little, Inc., "Environmental Considerations of Selected Energy Conserving Manufacturing Process Options,"

Industry Reports:

| | |
|---|---|
| Volume I -- | Industry Summary Report (EPA-600/7-76-034a) |
| Volume II -- | Industry Priority Report (EPA-600/7-76-034b) |
| Volume III -- | Iron and Steel Industry (EPA-600/7-7-76-034c) |
| Volume IV -- | Petroleum Refining Industry (EPA-600/7-76-034d) |
| Volume V -- | Pulp and Paper Industry (EPA-600/7-76-034e) |
| Volume VI -- | Olefins Industry (EPA-600/7-76-034f) |
| Volume VII -- | Ammonia Industry (EPA-600/7-76-034g) |
| Volume VIII -- | Alumina/Aluminum Industry (EPA-600/7-76-034h) |
| Volume IX -- | Textiles Industry (EPA-600/7-76-034i) |
| Volume X -- | Cement Industry (EPA-600/7-76-034j) |
| Volume XI -- | Glass Industry (EPA-600/7-76-034k) |
| Volume XII -- | Chlor-Alkali Industry (EPA-600/76-0341) |
| Volume XIII -- | Phosphorus/Phosphoric Acid Industry (EPA-600/7-76-034m) |
| Volume XIV -- | Copper Industry (EPA-600/7-76-034n) |
| Volume XV -- | Fertilizer Industry (EPA-600/7-76-034o) |

Pollutant Specific Reports

| | |
|---|---|
| Volume XVI -- | Sulfur Oxides |
| Volume XVII -- | Nitrogen Oxides |
| Volume XVIII -- | Particulates |
| Volume XIX -- | Solid Residues |
| Volume XX -- | Organics and Toxics |
| Volume XXI -- | Study Overview and Pollution Projections |

## Disclaimer

The report on which this paper is based was reviewed by the Industrial Environmental Research Laboratory, U.S. Environmental Protection Agency, and approved for publication. Approval does not signify that the contents necessarily reflect the views and policies of the U.S. Environmental Protection Agency, nor does mention of trade names or commercial products constitute endorsement or recommendation for use.

TABLE 1

U.S. CONSUMPTION OF MATERIALS
AS A PERCENT OF WORLD TOTAL

Selected Metallic

| | | | |
|---|---|---|---|
| Aluminum | 42% | Cobalt | 32% |
| Molybdenum | 40% | Platinum | 31% |
| Nickel | 38% | Iron | 28% |
| Copper | 33% | Zinc | 26% |

Fossil Based

| | |
|---|---|
| Natural Gas | 63% |
| Coal | 44% |
| Petroleum | 33% |

Source: MIT "limits to Growth".

TABLE 2

DISTRIBUTION OF ENERGY CONSUMPTION BY END USE SECTOR

| Sector | Purchased Fuels Plus Electricity $10^{12}$ Btu | % |
|---|---|---|
| Industrial including Manufacturing and Non-Manufacturing | 28,791 | 36.9 |
| Residential/Commercial | 29,526 | 37.8 |
| Transportation | 19,777 | 25.3 |
| Electrical Generation | - | - |
| Total | 78,094 | 100.0 |

Source: DOE

TABLE 3

DISTRIBUTION OF ENERGY CONSUMPTION WITHIN MANUFACTURING

| Five Energy Intensive Manufacturing Industries | Purchased Fuels and Electricity* (%) |
|---|---|
| (1) Primary Metals Industry | 25 |
| (2) Chemicals & Allied Products | 18 |
| (3) Petroleum & Coal Products | 16 |
| (4) Paper & Allied Products | 8 |
| (5) Stone, Clay & Glass Products | 8 |
| Total of Five | 75 |
| Other Manufacturing | 25 |
| Total Manufacturing | 100 |

*Purchased electricity valued at an approximate fossil fuel equivalence of 10,500 Btu/kWh.

Source: DOE. Based largely on purchased fuels except for primary metals industries and petroleum and coal products where captive consumption of energy from byproducts is included.

TABLE 4

ENERGY CONVERSION FACTORS

| | |
|---|---|
| Electric Power | 10,500 Btu/kWh |
| Fossil Fuel Equivalent: | |
| Coal | 26.2 Million Btu/Ton |
| Distillate Fuel Oil | 5.826 Million Btu/Bbl |
| Residual Fuel Oil | 6.286 Million Btu/Bbl |
| Natural Gas | 1,036 Btu/Cu.Ft. |

**TABLE 5: NATIONAL INCREASE OR DECREASE IN ANNUAL ENERGY USE
AND CONTROLLED DISCHARGES IN 1989 COMPARED TO BASE YEAR OF 1974***

| Commodity (Growth Rate) | Process | $10^6$ BTU — Energy | lbs. — SO$_x$ | NO$_x$ | Particulates | Solid Residue* |
|---|---|---|---|---|---|---|
| **Petroleum** (1.5% per yr.) | **Base Case: East Coast Refinery** | | | | | |
| | • Direct Combination of Asphalt in Process Heaters and Boilers | ◆ | △ | △ | △ | ◆ |
| | • Flexicoking | △ | △ | ◆ | ◆ | △ |
| | **Base Case: Gulf Coast Refinery** | | | | | |
| | • On-site Electric Power by Combustion of Vacuum Bottoms | ◆ | ◆ | ◆ | ◆ | ◆ |
| | **Base Case: West Coast Refinery** | | | | | |
| | • Hydrocracking of Heavy Bottoms | △ | △ | ◆ | ◆ | ◆ |
| | • High Purity Hydrogen via Partial Oxidation of Asphalt | ◆ | ◆ | ◆ | ◆ | ◆ |
| **Cement** (2% per yr.) | **Base Case: Long Dry Kiln** | | | | | |
| | • Suspension Preheater | ▼ | ◆ | ▼ | ◆ | ◆ |
| | • Flash Calciner | ▼ | ◆ | ▼ | ◆ | ◆ |
| | • Fluidized Bed | ▼ | ▼ | ▼ | ◆ | ◆ |
| | • Coal as Fuel Instead of Gas in Long Dry Kiln | ◆ | △△ | ◆ | ◆ | N.A. |
| **Olefins** (8% per yr.) | **Base Case: Ethane-Propane Process** | | | | | |
| | • Naphtha Process | △△ | △ | ▼ | ◆ | ◆ |
| | • Gas—Oil Process | △△ | △△ | ▼ | ◆ | ◆ |
| **Ammonia** (6% per yr.) | **Base Case: Ammonia via Natural Gas** | | | | | |
| | • Ammonia via Coal Gasification | ▼ | △ | △△ | ◆ | ◆ |
| | • Ammonia via Heavy Fuel Oil | ▼ | △ | △ | ◆ | ◆ |
| **Alumina** (6% per yr.) | **Base Case: Bayer Process** | | | | | |
| | • Hydrochloric Acid Ion Exchange | △△ | ▼ | △△ | ▼ | △ |
| | • Nitric Acid Ion Exchange | △△ | △△ | ◆ | N.A. | △ |
| | • Clay Chlorination (Toth) | △△ | △△ | ▼ | N.A. | ◆ |
| **Aluminum** (6% per yr.) | **Base Case: Hall—Heroult (Current Practice, C.P.)** | | | | | |
| | • Hall—Heroult (New) | ▼▼ | ▼▼ | ▼▼ | ▼▼ | ◆ |
| | • Alcoa Chloride | ▼▼ | ▼▼ | ▼▼ | N.A. | ◆ |
| | • Refractory Hard Metal Cathode | ▼▼ | ▼▼ | ▼▼ | ▼▼ | ◆ |
| | **Base Case: Bayer with Hall—Heroult (C.P.)** | | | | | |
| | • Clay Chlorination (Toth Alumina) and Alcoa Chloride | ▼ | ▼▼ | ▼▼ | N.A. | ◆ |
| **Kraft Pulp** (5% per yr.) | **Base Case: Kraft Pulping** | | | | | |
| | • Alkaline Oxygen Pulping | ▼ | ▼▼ | ▼ | ◆ | N.A. |
| | • Rapson Effluent-Free Kraft Pulping | ▼ | △ | ▼ | ◆ | N.A. |
| **Newsprint Pulp** (2.5% per yr.) | **Base Case: Refiner Mechanical Pulp (RMP)** | | | | | |
| | • Thermo—Mechanical Pulp (TMP) | ◆ | ◆ | ◆ | ▼ | N.A. |
| | • De-inking of Old News for Newsprint Manufacture | ▼ | ▼ | ▼ | ▼ | N.A. |
| **Glass** (2.5% per yr.) | **Base Case: Regenerative Furnace** | | | | | |
| | • Coal Gasification | ◆ | ◆ | △ | ◆ | ◆ |
| | • Direct Coal Firing | ◆ | △ | △ | N.A. | ◆ |
| | • Coal—Fired Hot Gas Generation | ◆ | ◆ | △ | N.A. | ◆ |
| | • Electric Melting | △ | △△ | ▼ | ◆ | ◆ |
| | • Batch Preheat with Natural Gas Firing | ▼ | ◆ | ▼ | ◆ | ◆ |

Key:

△△  Increase of 100 million or more compared to base case technology
△   Increase of 10 to 100 million compared to base case technology
◆   Increase or decrease of under 10 million compared to base case technology
▼   Decrease between 10 and 100 million compared to base case technology
▼▼  Decrease of 100 million or more compared to base case technology
N.A.  Not Available
C.P.  Current Practiced Process

113

TABLE 5 (continued)

| Commodity (Growth Rate) | Process | 10⁶ BTU | lbs. | | | |
|---|---|---|---|---|---|---|
| | | Energy | SOₓ | NOₓ | Particulates | Solid Residue* |
| **Copper** (3.5% per yr.) | **Base Case: Conventional Smelting** | | | | | |
| | • Outokumpu Flash Smelting | ▼ | ▼▼ | ◆ | ◆ | ◆ |
| | • Noranda | ▼ | ▼▼ | ◆ | ◆ | ◆ |
| | • Mitsubishi | ▼ | ▼▼ | ◆ | ◆ | ◆ |
| | • Arbiter | △ | ▼▼ | △ | ◆ | ◆ |
| **Chlorine, NaOH** (5% per yr.) | **Base Case: Graphite—anode Diaphragm Cell** | | | | | |
| | • Dimensionally Stable Anodes | ▼ | ▼ | ▼ | ◆ | ◆ |
| | • Expandable DSA | ▼ | ▼ | ▼ | ◆ | |
| | • Polymer Modified Asbestos | ▼ | ▼ | ▼ | ◆ | ◆ |
| | • Polymer Membrane | ▼ | ▼ | ▼ | ◆ | ◆ |
| | • Ion Exchange Membrane | ▼ | ▼ | ▼ | ◆ | N.A. |
| | • Mercury Cell | ▼ | △ | △ | ◆ | |
| **Steel** (2.5% per yr.) | **Base Case: No Off—Gas Recovery** | | | | | |
| | • Off—Gas Recovery | △ | ◆ | ▼▼ | ▼ | ◆ |
| **Blast Furnace Hot Metal** (2.5% per yr.) | **Base Case: Blast Furnace** | | | | | |
| | • Blast Furnace with External Desulfurization | ▼ | ◆ | ◆ | △ | ◆ |
| **Coke** (2.5% per yr.) | **Base Case: Wet Quenching of Coke** | | | | | |
| | • Dry Quenching of Coke | ▼ | ▼ | ▼ | ◆ | N.A. |
| **Steel** (integrated) (2.5% per yr.) | **Base Case: Steelmaking Coke Oven, Blast Furnace BOP Route** | | | | | |
| | • Direct Reduction EAF Route | △△ | △△ | △△ | ▼▼ | △ |
| **Phosphoric Acid** (detergent grade) (2.5% per yr.) | **Base Case: Electric Furnace** | | | | | |
| | • Chemical Cleanup of Wet-Process Acid | ◆ | ▼ | ▼ | ◆ | ◆ |
| | • Solvent Extraction of Wet-Process Acid | ◆ | ▼ | ▼ | ◆ | ◆ |
| **Nitric Acid** (4% per yr.) | **Base Case: No NOₓ Control** | | | | | |
| | • Catalytic Reduction | ◆ | ◆ | ▼▼ | ◆ | ◆ |
| | • Molecular Sieve | ◆ | △ | ▼ | ◆ | ◆ |
| | • Grand Paroisse | ◆ | ◆ | ▼ | ◆ | ◆ |
| | • CDL/Vitok | ◆ | △△ | ▼ | ◆ | ◆ |
| | • Masar | △ | △△ | ▼▼ | △ | ◆ |
| **Fertilizers** (Mixed) (4% per yr.) | **Base Case: Natural Gas** | | | | | |
| | • Converting from Natural Gas to Fuel Oil | ◆ | ◆ | N.A. | ◆ | ◆ |
| | • Installing Scrubbers | ◆ | ◆ | N.A. | ◆ | N.A. |
| **Textiles** (2.2% per yr.) | **Base Case: Knit Fabric** | | | | | |
| | • Conventional Aqueous | | | | | |
| | • Advanced Aqueous | ◆ | ◆ | ▼ | ◆ | ◆ |
| | • Solvent Processing | ◆ | ◆ | ◆ | ◆ | ◆ |
| | **Base Case: Woven Fabric** | | | | | |
| | • Conventional Aqueous | | | | | |
| | • Advanced Aqueous | ▼ | ▼ | ◆ | ◆ | ◆ |

*Excludes boiler and power plant solid residue discharges

TABLE 6

NATIONWIDE EMISSION ESTIMATES

| | $10^6$ lb/Year | | | |
|---|---|---|---|---|
| | $SO_x$ | $NO_x$ | Particulates | Solid Residue |
| Stationary Fuel Combustion | | | | |
| –Electric Utilities | 42,000 | 13,000 | 6,800 | 5,000 |
| –Other | 11,400 | 12,800 | 7,200 | NA* |
| Industrial Processes | 12,600 | 1,400 | 21,200 | NA |
| Transportation | 1,600 | 21,200 | 2,600 | NA |
| Solid Waste | 200 | 400 | 1,200 | NA |
| Miscellaneous | 200 | 400 | 1,600 | NA |
| TOTAL in 1974 | 68,200 | 50,000 | 40,000 | 250,000 |
| Range in increase in emissions in 1989 over that in 1974 for 13 industries studied including electric power generation for $SO_x$, $NO_x$, and particulate discharges. | 2,167 to 7,919 | 2,111 to 3,730 | 450 to 1,285 | 32.5 to 91.1 |

*Not available.

Source: National Air Quality and Emission Trend Report, 1975, U.S. EPA, Research Triangle Park, N.C., NTIS PB-263922; EPRI Journal, p. 33, May 1978; and Arthur D. Little, Inc. estimates.

115

TABLE 7

ESTIMATED COSTS OF POLLUTION CONTROL

|  | Pollution Control Costs – As % of Production Costs | Expected Changes |
|---|---|---|
| **Alumina** |  |  |
| Nitric Acid Leaching | 7.7 | ↑↑ |
| Hydrochloric Acid Leaching | 1.5 | ↑ |
| Toth Alumina Process | 5.7 | ↑ |
| Alcoa Chloride Process | 1.5 | -- |
| Titanium Diboride Cathodes | 5.0 |  |
| Toth Alumina with Alcoa Chloride | 3.5 | ↑ |
| **Ammonia** |  |  |
| Ammonia via Coal Gasification | 5.9 | ↑↑ |
| Ammonia via Heavy Oil Gasification | 2.3 | ↑ |
| **Cement** |  |  |
| Fluidized-Bed Cement Process | 4.7 | -- |
| Suspension Preheater | 3.2 | -- |
| Conversion to Coal from Oil and Natural Gas | 4.6 | ↑ |
| **Chloroalkali** |  |  |
| Modern Mercury Cell | 0.5 | -- |
| **Fertilizers** |  |  |
| Nitric Acid with: |  |  |
| - Catalytic Reduction of NOx | 9.8 | -- |
| - Molecular Sieves for NOx | 6.7 | -- |
| - Grande Paroisse for NO2 | 4.0 | -- |
| - CDL/Vitok for NOx | 2.7 | -- |
| - Masar for NOx | 3.3 | -- |

TABLE 7 (Continued)

ESTIMATED COSTS OF POLLUTION CONTROL

|  | Pollution Control Costs – As % of Production Costs | Expected Changes |
|---|---|---|
| **Glass** |  |  |
| Direct Coal Firing | 13.5 | ↑↑ |
| Electric Melting | 3.5 | ↑ |
| **Iron & Steel** |  |  |
| External Desulfurization of Blast Furnace Hot Metal | 9.7 | ↑ |
| Direct Reduction – Electric Furnace | 5.4 | ↑ |
| **Paper and Allied Products** |  |  |
| Alkaline Oxygen Pulping | 13.3 | ↑ |
| Rapson Effluent – Free Kraft Process | 2.3 | ↑ |
| Deinking of Waste News as a Substitute for Mechanical Pulping | 6.9 | ↑ |
| Thermomechanical Pulping | 2.8 | ↑ |
| **Petroleum Refining** |  |  |
| Direct Combustion of Asphalt in Heaters/Boilers | 52.4 | ↑ |
| Asphalt Conversion by Hydro-cracking | 13.9 | ↑ |
| Asphalt Conversion by Flexi-coking | 26.2 | ↑ |
| Internal Power Generation Using Asphalt | 15.0 | ↑ |
| Hydrogen Generation by Partial Oxidation | 1.4 | ↑ |
| **Primary Copper** |  |  |
| Conventional Smelter | 15.6 | ↑ |
| Outokumpu Flash Smelting | 12.8 | ↑ |
| Noranda Process | 10.3 | ↑ |
| Mitsubishi Process | 14.8 | ↑ |
| Arbiter Process | 5.2 | ↑↑ |

-- virtually no change
↑ minor change
↑↑ significant change

TABLE 8

POTENTIAL IF MOST ENERGY CONSERVING TECHNOLOGY IS EMPLOYED
IN ALL NEW PLANT EXPANSIONS OVER THE NEXT 15 YEARS

| Commodity | Process | Energy Use-Conventional Process* (Quads/yr) | Conservation Potential of New Process (Quads/yr) | Million bbl/day | Main Contribution to Energy Conservation |
|---|---|---|---|---|---|
| Aluminum | Refractory metal cathodes | 2.99 | 0.51 | | Electric energy saving |
| Ammonia | Coal gasification | 0.82 | 0.82 | | Critical feedstock conservation |
| Cement | Suspension preheating | -0- | 0.16 | | Critical feedstock conservation |
| | Coal firing (additional conservation) | 0.605 | 0.21 | | |
| Chloralkali | New membrane technology | 0.84 | 0.127 | | Electric energy saving |
| Fertilizers (HNO$_3$) | New NO$_x$ control technology | 0.042 | 0.015 | | Energy saving |
| Glass | Batch preheating | 0.32 | 0.06 | | Critical fuel conservation |
| | Additional saving with coal options | - | 0.26 | | |
| Coke | Dry quenching | - | 0.09 | | Energy saving |
| Iron | External desulfurization | 1.45 | 0.04 | | Saving of low sulfur met. coal |
| Steel | CO collection from BOP | 0.03 | 0.10 | | Energy saving |
| Olefins | Alternative feedstock | 2.79 | 2.79 | | Critical feedstock conservation |
| Petroleum | Bottoms upgrading | 4.00 | 2.90 | | Critical feedstock conservation |
| Pulp | Rapson | 0.24 | 0.18 | | Energy saving |
| | Recycled newsprint | 0.08 | 0.05 | | Energy saving |
| Phosphoric Acid (Det. grade) | Chemical cleanup | 0.18 | 0.17 | | Energy saving |
| (Wet acid) | Strong acid process | 0.20(a) | 0.17(a) | | Energy saving |
| Copper | Flash smelting (coal firing) | 0.066 | 0.066 | | Energy saving |
| Textiles (knit) | Solvent | 0.012 | 0.009 | | Energy saving |
| Textiles (woven) | Adv. aqueous | 0.25 | 0.14 | | Energy saving |
| TOTAL | | 14.92 | 8.87 | | |

*Basis: Projected incremental energy use resulting from industry expansion over 15 years.
(a) Includes heating value of sulfur.

117

## ENVIRONMENTAL, OPERATIONAL AND ECONOMIC ASPECTS
## OF THIRTEEN SELECTED ENERGY TECHNOLOGIES

**Larry Hoffman**
The Hoffman-Muntner Corporation

This paper is an overview of our report for EPA entitled, "Environment-al, Operational, and Economic Aspects of Thirteen Selected Technologies" (EPA-600/7-80-173). The intended purpose of the report is to present an assessment of possible options that potentially could provide increase efficiency and/or utiliza our more abundant fossil fuel resources (e.g., coal, oil shale) for generation of electricity and other energy needs. For more details, the reader is referred to the EPA publication.

In the case of <u>conventional coal-fired steam-electric power plants</u>, current efficiencies range from approximately 31 to 38 percent. The prospect for the foreseeable future is that newer plants will have efficiency values below 40 percent. It is unlikely that truly oper-ational efficiency values in excess of 40 percent from conventional plants will be realized within the foreseeable future. In the absence of pollution control measures, coal fired steam-electric plants would provide very substantial undesirable environmental impacts. However, the current state-of-the-art of environmental control and resulting control measures are capable of substantially mitigating currently identified undesirable pollution and other environmental effects. Continuing environmental control activities are expected to provide the means of control for the near-term any potential overall undesirable effects resulting from increasing use of coal to fire steam-electric plants.

<u>Diesels</u> have been commercially utilized in excess of 80 years. They are used extensively to power moderate size stationary electric generators for a variety of services. Even though the output of a large diesel generator is small compared to the output of a typical utility fossil-fuel steam-electric generator, the attainable efficiency is generally as

great. Recently, concern has developed relating to the potential carcinogenic aspects of diesel exhaust. Future utilization of stationary diesel generators may well depend on diesel emission control standards. The cost of diesel derived electric energy is somewhat higher than that from a conventional steam-electric plant. This is due to the relatively high operating cost (per kwh electric energy) of a diesel generator installation. DOD experience indicates diesel derived electric energy is at least twice as expensive as that purchased from an electric utility. Even so, for selected applications, diesel generators are very appropriate.

Current fluidized-bed combustion efforts are largely in the research, development, and demonstration stages. Some manufacturers have recently begun to advertise the availability of atmospheric commercial/industrial scale units. The attainable boiler efficiency is limited by the same general loss components as for a conventional boiler. Boiler efficiency values equal to those attainable by conventional boilers will depend on the ability to achieve substantially complete carbon burn-up. The environmental aspects of a fluidized-bed boiler are similar to that of an equivalent capacity conventional boiler with flue gas desulfurization (FGD) burning the same coal. A major difference is the relatively low $NO_x$ emission and the amount and nature of the spent bed material as compared to the effluent from a FGD system. For fluidized-bed combustion with the same $SO_x$ removal, almost three times as much limestone is required. Spent bed material from a fluidized-bed boiler contains appreciable CaO (i.e., quicklime) that may present handling and disposal problems. Hopefully, commercial uses will be found for the spent bed material. In the near term, fluidized-bed boilers are projected to compete with industrial/ commercial scale conventional boilers with $SO_x$ emission control. Such units when developed would permit coal to be burned more conveniently at such locations as schools, hospitals, shopping centers, office buildings, small industrial parks, etc.

There are many gas turbine-steam combined-cycle power plants currently in operation which achieve overall efficiencies around 40 percent. However, these systems currently rely upon gas or oil the price and future availability of which have become of serious concern. Therefore, there is major emphasis on making today's turbines run more efficiently on these scarce fuels and to develop improved turbines that will operate efficiently on the synthetic fuels that will one day replace oil and natural gas. In addition to improved efficiency, such combined-cycle power plants utilizing gas-turbine and steam-turbine technology have a number of other key features which could make them particularly appealing to the utility industry. Besides very fast start-up capabilities, these features include relatively low capital investment per kilowatt of electric generation, relatively low operating costs, and the capability for use as a base-load or peaking power plant. Another potentially promising aspect of the combined-cycle power plant is its projected ability to use low-energy gas from coal. The environmental implications of this are significant. Since such low-Btu gas can be clean burning, much of the environmental control problems and expense associated with conventional coal-fired steam generating plants would be avoided. A variation of the combined gas turbine and steam-turbine system features the direct combustion of coal in a pressurized fluidized-bed (PFB). Although internal particulate control is still required, the PFB offers the potential for direct combustion of high-sulfur coal without stack gas cleanup while achieving an overall coal pile-to-bus bar plant efficiency of approximately 40 percent.

The low/medium-Btu gasification of coal is essentially an existing
technology. In fact, gas manufactured from coal was first produced in
the eighteenth century. Currently, low/medium-Btu coal gasifiers are in
use in Europe, South Africa, and to a very limited extent, in the United
States. Coal can be gasified by any of several processes: synthesis,
pyrolysis, hydrogasification. In synthesis, coal or char is reacted
with steam and oxygen or air and produces the heat for a reaction that
produces a mixture of hydrogen and carbon monoxide. In pyrolysis, coal
is heated in a starved air atmosphere. In the process, some gas and
liquids result, the major product being a coke residue. In hydrogasification, coal, coke, or char is reacted with hydrogen to form methane.
Pipeline gas is produced by upgrading a medium-Btu gas. Environmental
problems common to coal associated energy generating systems will generally also apply to coal gasification facilities. Additional adverse
environmental aspects of proven and pilot plant processes are difficult
to assess because of the very limited data available from such operations. The conversion efficiency as based on total energy input, is
somewhat process and site specific and is estimated to be in the 70 to
80 percent range including raw gas cleanup. The value without gas
cleanup (i.e., raw hot gas output) is estimated to be as high as 90+
percent when the sensible heat of the gas is included. Since this is
basically a developed technology, over the foreseeable future, efficiencies are not expected to improve significantly. The cost is currently estimated at $2.50 to $4.00 per million Btu.

The chemically active fluid bed (CAFB) process uses a shallow fluidized-
bed of lime or lime-like material to produce a clean, hot gaseous fuel
from high sulfur feedstock (e.g., residual oil). Solid fuel feedstocks
such as coal are also feasible. The initial CAFB pilot unit (2.39 Mw)
was developed by the Esso Research Centre in Abingdon, England. A 10
Mw demonstration plant has subsequently been constructed by Foster
Wheeler at the La Palma Power Station (Central Power and Light Company)
in San Benito, Texas. EPA is sponsoring the demonstration of this technology. Environmental data are very limited. Principal environmental
concerns relate to the size of the particles in the product gas stream,
the vanadium (bound in a mixture of oxides) emission level, and the disposal of spent, sulfided limestone. The solid waste disposal problem
appears to be the major environmental concern. Since all activities are
R&D, no actual full scale performance data are available. In this
regard, the total gasification efficiency is estimated to be approximately 87 percent. Similarly, economic values are also projections.
EPA estimates that a retrofit CAFB plant to fuel a 500 Mwe plant would
cost $]72 per kw of installed capacity; the operating cost is estimated
at 2-3 mills per kwh (]977 dollars).

Coal liquefaction provides the means to produce liquid fuels from coal.
In indirect liquefaction, the coal is gasified to make a synthesis gas
and then passed over a catalyst to produce alcohols (methanol) or par-
affinic hydrocarbons. In direct liquefaction the coal is liquefied
without a gasification intermediate step. Specific processes are generally directed toward converting coal to liquid fuels with minimal production of gases and organic solid residues. The liquid products that
are produced vary with the type of process and the rank of coal that is
utilized. Research and development of coal liquefaction has been underway for many years. The first practical uses of coal-derived liquid
fuels were about 1790 when the fuels were used for experimental light-
ing, heating, and cooking. During World War II, Germany produced liquid
fuels from coal in industrial amounts (45 million bbl/year). Since
then, coal liquefaction plants have been constructed in a number of

120

countries but currently only South Africa is producing liquids from coal. Commercial demonstration of coal liquefaction has never been accomplished in the United States. Current U. S. activities are limited to research and development and pilot plant programs. Environmental problems common to fossil energy facilities will also apply to coal liquefaction facilities. Liquefaction processes present some unique problems such as the need for the characterization of materials with carcinogenic effects, characterization and treatment of wastes, fugitive emissions, and effluents and the disposal of sludges and solid wastes. These problems are generally common to all liquefaction processes, however, since no large scale plants are in operation in the U. S., the only available data on emissions and effluents are estimated from pilot plant operations and cannot be completely quantified for a commercial operation. Projected efficiencies for coal liquefaction facilities are in the 55 to 70 percent range. Accurate values for coal conversion efficiencies are difficult to estimate and thus an exact value cannot be given until commercial demonstration takes place. Estimated costs for indirect coal liquefaction plants are in the $7-10 per million Btu range (1980 dollars). Generally, the estimated cost for direct coal liquefaction plants is less than for indirect liquefaction.

High-Btu gasification of coal can be accomplished by any of several processes: synthesis, pyrolysis, or hydrogasification. In synthesis, coal or char is reacted with steam and oxygen and produces the heat for a reaction that produces a mixture of hydrogen and carbon monoxide. In pyrolysis, coal is heated in a starved air atmosphere. In the process, some gas and liquids result, the major product being a coke residue. In hydrogasification, coal, coke, or char is reacted with hydrogen to form methane. To produce a pipeline quality gas, medium-Btu gas (e.g., from hydrogasification) is cleaned and further treated. This further treatment could include a shift conversion to obtain the proper carbon monoxide to hydrogen ratio followed by a second purification process, followed by a methanation process. To an extent, environmental concerns common to coal-fired boiler facilities will also generally apply to coal gasification facilities. Additional unique adverse environmental impacts are difficult to estimate. No commercial plants are in operation anywhere in the world and assessments must be based on limited information from pilot plants. In addition, information from a pilot plant may not be representative of a commercial operation. Projected overall energy efficiencies for coal gasification have been estimated to be approximately 75 percent. The estimated at gate costs of high-Btu gas produced by a gasification plant are $4 to $6 per million Btu (1977 dollars).

Oil shale resources can be processed either by conventional mining followed by surface processing or by in situ (in place) processing. In situ processing can be accomplished by either true or modified in situ methods to extract oil from shale. Oil shale resources in the United States are estimated to exceed two trillion barrels of petroleum and of the total, 25 to 35 percent is presently projected as commercial. Shale oil has been produced commercially at various time intervals in eleven countries since the initiation of shale oil operations in France in 1838. In Canada and the Eastern United States, a very small industry was operating around 1860, but disappeared when petroleum became plentiful. Currently, the only commercial production is in Russia (Estonia) and China with a combined production of approximately 150,000 barrels per day. The conventional process (conventional mining and surface retorting) to produce a crude is composed of four major steps: mining the shale; crushing it to the proper size for the retort vessel; retorting the shale to release the oil; and refining the oil to a high-quality product. True in situ processes involve fracturing the shale bed via vertical well bores to create permeability without mining or removal of material followed by underground retorting. Retorting can also be done via well bores utilizing natural permeability where it may exist. The modified in situ process involves mining or removing by other means (such as leaching or underreaming) up to 40 percent of the shale (i.e., in the retorting section) in order to increase the void volume and allow rubblization before retorting. In modified in situ, the mined shale can be surface retorted. Considerable environmental questions are associated with oil shale processing and until these uncertainties as well as the demonstration of an economically acceptable commercial scale viable technology are resolved, future development of a viable oil shale industry is uncertain.

The fuel cell, by converting chemical energy directly to electricity, can efficiently use fuels without an intermediate mechanical step. Fuel cell power plants offer many attractive characteristics such as modular construction, low environmental emissions, high efficiency and rapid response to load demand fluctuations. Because of the modular construction, fuel cells are easily transported and installation times and costs reduced. The fuel cell concept itself is not new: such cells have already provided power for moon landings and, between 197] and 1973, provided electric power to 50 apartment houses, commercial establishments, and small industrial buildings. What is new is an effort to capitalize on the fuel cell's inherent flexibility, safety, and efficiency by putting together a generator system that can use a variety of fuels to meet today's utility-scale power need economically. Environmental considerations like low water requirements, limited emissions, and quiet operation help make fuel cell plants an attractive power option. Whereas fossil fuel and nuclear plants require large quantities of water for cooling, fuel cells generate less heat and can be air-cooled by low speed fans. Because fuel cells can use a variety of hydrocarbon fuels, they share with conventional generating processes the environmental problems currently associated with extracting and processing fossil fuels. The required hydrogen for the fuel cell power section can be derived by gasifying coal. In such a case the coal gasifier would be an integral part of the fuel cell power plant. The Energy Conversion Alternatives Study (ECAS) team estimated an overall efficiency of 50 percent for its conceptual molten carbonate fuel cell power plant. Although still in the prototype stage, the fuel cell offers a means to produce electricity efficiently on both small and large scales. These systems could be used to complement existing facilities or supply new generating capacity where environmental considerations restrict conventional combustion plants.

Magnetohydrodynamics (MHD) is a potential energy alternative in which electricity is generated directly from thermal energy, thus eliminating the conversion step of thermal to mechanical energy encountered in conventional steam-electric generators. However, due to the nature of the process, it would be inefficient to apply MHD by itself to the large scale generation of electricity. Therefore, its eventual implementation is being planned around combining MHD with a conventional steam plant to make use of the waste heat from the MHD generator. The efficiency of such a combined MHD/steam plant is predicted to be about 50 percent as compared to 38 percent projected for conventional coal-fired power plants with flue gas desulfurization (FGD) systems. Unlike rotating machines, much of this increase in efficiency is attributed to the fact that all the rigid structures in MHD generators are stationary, thus permitting operation at elevated temperatures approaching $5000^{\circ}$F. These temperatures are much higher than even the most advanced contemporary plants, resulting in much higher efficiencies through the entire thermal cycle than are attainable in such conventional plants. Although much work remains before the widespread application of the magnetohydrodynamic energy conversion process to electric utility power generation, there is experimental evidence that MHD can significantly improve overall power plant efficiencies. Another promising aspect of this technology is the ability to remove, during the process, pollutants such as $SO_x$, $NO_x$, and particulates generated in the combustion of coal, thereby eliminating the need for external flue gas scrubbing to meet environmental standards.

# ENVIRONMENTAL CONSIDERATIONS AND COMPARISONS OF ALTERNATE GENERATION SYSTEMS

**Kennard F. Kosky** and **Jackson B. Sosebee**
Environmental Science and Engineering, Inc.

## Introduction

To respond to the decline in availability of fuel oil and to the national policy of reducing United States dependence on foreign oil imports, alternative means of generating electricity must be found. The emphasis on developing alternative generation systems has focused on improving thermal efficiency, conventional and innovative use of coal, recovering energy from discarded resources, and displacing generation through energy storage systems. Alternative energy systems are not, however, without impacts to the environment. The evaluation of environmental impacts and compliance with regulations is a necessary part of constructing and operating new systems which has become apparent in this last decade. However, information on the environmental effects of new systems has not developed to a degree similar to that of conventional generation. Nevertheless, insight can be gained through a comparison of alternative energy systems and conventional generation. This paper is such a comparison of four alternative generation systems:

1. Coal/oil mixtures,
2. Resource recovery,
3. Slow speed diesels with heat recovery, and
4. Compressed air energy storage.

There are three main areas--air, water, and solid waste--where environmental comparisons are made. Air pollution impacts will be evaluated on the basis of emissions. While comparisons can also be made on air quality impacts, in-stack effluent characteristics do not allow appropriate comparisons to be made. The impacts to the water will be evaluated only in terms of quantitative water consumption and discharges. Environmental impacts associated with impingement and entrainment are site-specific in nature and cannot be quantified. Impacts related to specific water quality standards are similarly site specific. These site-specific impacts will be treated qualitatively. The potential for solid waste impacts is dependent upon both site-specific and waste characteristics. As a consequence, this third area will be reviewed from a qualitative standpoint.

Coal/Oil Mixtures

## System Description

An alternative fuel for existing oil-fired generating facilities is a coal/oil mixture. In this system, coal and oil are mixed and simultaneously fired in the existing units. Figure 1 presents system characteristics of coal/oil mixture (COM) firing for a 400-MW unit. The coal/oil mixture is based upon a 50 percent coal and 50 percent oil mixture by weight. Using coal of 12,000 Btus per pound and oil with 18,600 Btus per pound (high heating values), a fuel weight of 235,320 lbs/hr will supply 3600 million Btu/hr. The COM fuel is approximately 40 percent coal on a Btu basis. On a 400-MW unit, coal use will displace 5,430 barrels of oil per day, or 1,600,000 barrels of oil per year.

The COM firing can be accomplished by installation of on-site systems. Plant capacity for a 400-MW facility requires a 12,000 barrels per day facility. Coal unloading, storage and reclaims systems are required. Pulverizers rated at 10.5 tons per hour are needed to reduce the size of the coal. Special storage of the COM at 125°F with continual mixing is necessary to ensure the mixture will not solidify. Steam atomization on the burner guns is necessary to ensure proper firing of the COM.

## Environmental Considerations and Comparisons

Figure 2 shows the flow diagram of a COM facility for an existing oil-fired unit. Coal handling, storage, reclaim, coal pulverizing, coal and oil mixing and COM storage must be installed. Appropriate consideration is necessary for air pollution control equipment. In addition, ash (and potential scrubber sludge) disposal is necessary.

Additional air emissions from a COM facility will be in the form of fugitive particulate emissions. A majority of fugitive emissions will occur from coal storage areas. Bottom ash and economizer and precipitator ash disposal has potential for fugitive emissions. Coal pulverization and coal storage are generally controlled by the use of bag filters. Stack emissions have a potential for being significantly greater than with oil firing. The generally higher percent sulfur in coal coupled with reduction in Btu content per unit weight increases the pounds of sulfur dioxide emissions per Btu. In addition, the increased ash content in coal will create particulate emissions from COM firing that can be an order of magnitude higher than oil firing. As a consequence, stack emission will, as a minimum, be controlled by electrostatic precipitation or bag filtration. Sulfur dioxide scrubbing is also a possible requirement.

Water discharges from rainfall run-off on coal and ash areas may occur. Groundwater contamination by leachate formation is also likely. Depending upon the site area, these impacts can be significant. The level of significance is dependent upon the source of the coal, handling/storage practices, and site-specific geohydrological conditions.

Technology can be adapted from present-day coal plants. Coal and ash handling procedures have been developed to adequately mitigate these potential environmental impacts. The availability of space for air pollution control equipment must be evaluated. Proper operation of coal handling facilities and use of dust suppressants can mitigate fugitive emissions. Careful placement of coal storage, ash and scrubber sludge areas can significantly reduce the potential for groundwater contamination. Lining of storage and disposal areas, along with flow segregation from run-off, can significantly mitigate impacts to surface waters.

Presented as Figure 3 is an environmental comparison of COM firing and conventional generation. Particulate, sulfur dioxide and nitrogen oxides can be less or essentially equivalent to that of oil-firing. Available particulate control equipment will significantly reduce COM firing to a level at least one-third lower than oil firing. Estimated sulfur dioxide emissions are slightly above oil-firing, based on equivalent fuels (% S). Application of sulfur dioxide scrubbing or use of low sulfur coal will reduce the sulfur dioxide emissions from COM firing from 50 to 90 percent. Nitrogen oxides will be approximately equivalent.

The potential for fugitive emissions from coal handling ash disposal can present significant environmental impacts if not adequately controlled. Fugitive emissions could cause impacts on air quality that exceed standards or create a nuisance.

Generation of 174 tons per day of ash is a significant new impact for an already existing oil unit. Violations of water quality standards could occur. If sulfur dioxide scrubbing is required, solid waste generation could double or triple, necessitating even more careful evaluation. The leachate potential, however, depends on type of coal, site location, and operating procedures.

Resource Recovery

## System Description

Municipalities have long been faced with the problem of inefficient disposal of solid waste. With the advent of greater awareness of resource recovery, the desire to implement new technologies in the disposal of solid waste has increased. Such technologies can recover a significant amount of energy in the form of electricity.

Figure 4 presents system characteristics of a resource recovery process designed by Black Clawson, Fiberclaim, Inc. Refuse and trash amounting to 940,000 tons per year will be delivered to the facility through computerized logging of each disposal truck.

The refuse is transferred to a wet pulping system called a Hydropulper. Each pulper will be 20 feet in diameter, constructed of reinforced concrete with a steel lining, and powered by a 1200-hp motor. When the waste has been pulped to a size sufficiently small enough to pass through a 1-inch diameter opening, it is pumped to liquid cyclones. Oversized wastes are ejected out of the pulper. This slurry is pumped to liquid cyclones where lighter combustible fraction is separated from the heavier, inorganic fraction. The organic portion of the slurry is pumped to a two-stage dewatering process. Here the solid content of the slurry is increased from 3 percent to approximately 50 percent, which is sufficient for combustion. The excess water from the slurry is recycled into the process.

The pulp fuel, which is approximately 55.9 percent moisture, 4,219 Btus per pound, is conveyed pneumatically to a boiler storage area. The fuel is similar to bark or baggasse, for which steam generators have been in service for many years. Approximately 514,800 tons of fuel per year will be produced from 940,000 tons of refuse and trash.

Four identical steam generating bark boilers with a nominal capacity of 765,000 pounds of steam per hour will be coupled with two 28.5-MW electric generators. The boiler will be equipped with a high efficiency electrostatic precipitator. Cooling water for the condensers will consist of a 6-cell mechanical draft cross-flow cooling tower.

Organic material and oversized waste are separated through various processes for ferrous metal, non-ferrous metal, and glass. Approximately 65,500 tons of ferrous metal, 34,400 tons of glass, 7,000 tons of aluminum, and 4,700 tons of other non-ferrous metal are recovered each year.

## Environmental Considerations and Comparisons

Presented in Figure 5 is a flow diagram of the resource recovery process. The major environmental considerations are air emissions from steam generators, odors emitting from handling solid waste, cooling tower evaporation and drift, solid waste leachate, and impingement and entrainment.

The majority of air emissions will result from 4 pulped fuel boilers. Combustion of pulped fuel will result in emissions of particulate matter and sulfur dioxide. Particulate emissions can be controlled by high efficiency precipitation. Sulfur dioxide emissions are sufficiently low that no control is required. Incoming garbage odors and process odors must be eliminated before venting to the environment. Odor emissions can be mitigated by venting air to the forced draft fans of the boilers, thus providing control by combustion. In the event the boilers are unable to handle all the air, activated carbon filters could be used before exhausting this air outside the plant.

As shown on the figure, there will be no discharge to surface waters due to the recycling and use of water in the process. Surface and ground water can be affected by runoff and leachate from fly-ash areas. Leachate can be controlled by lining and runoff by segregation of run-off flows. Impingement can be mitigated by proper design of intake structure (less than 0.5 ft/sec) while entrainment can be mitigated through proper location of the structure.

Figure 6 presents the environmental comparisons of a resource recovery facility and conventional coal-fired generation. Air emissions comparisons are slightly greater for particulate yet reduced for sulfur dioxide. The water consumption is significantly less with resource recovery because of the reuse of water. There is no anticipated water discharge. The volume of incoming solid waste will be significantly reduced. However, boiler and precipitator ash and those materials that cannot be used in the process will create a volume of waste greater on a per megawatt basis than conventional generation. The two processes are equivalent in terms of solid material storage. The odors from a resource recovery facility must be minimized.

Slow Speed Diesels with Heat Recovery

### System Description

Increasing the thermal efficiency of generating facilities can reduce the demand for foreign oil. The use of diesel generating units (an existing technology) with heat recovery, can be effective in improving thermal efficiency. The diesel generating units are of the large, slow-speed, two-cycle type presently used for power generation in Europe, Africa and Asia. Slow speed diesel engines, an established European technology, are those rotating in the range of 90-150 revolutions per minute (RPM) in contrast to the locomotive diesel engine which operates at 900 RPMs. The slow speed diesel has an advantage over currently used, medium speed (275-520 RPM) diesel engines because the slow speed engines can burn poor quality and synthetic fuels. Manufacturers that offer slow speed engines are Burmeister and Waine AIS, Denmark; Grandi Motori Trieste Spa (Fiat), Italy; Maschienefabrik Augsburg-Nurnberg (M.A.N.), West Germany; and Sulzer Brothers Limited, Switzerland.

The major use of slow-speed diesels is in the marine industry. These were developed primarily because of the mechanical efficiencies gained by direct driving huge ship propellers. Engines (see Figure 7 for characteristics) are available in varying numbers of cylinders from 4-12, ranging from 450-1,050 millimeters (18-41 inches) in diameter and with strokes from 900-2,180 millimeters (35-86 inches). Engine power outputs from 3-40 megawatts are available, depending upon the manufacturer, engine series, and number of cylinders. These engines are turbo-charged and have high net thermal efficiencies without exhaust heat utilization. With exhaust heat utilization systems, often referred to as bottoming cycle, approximately 7-8 percent of the engine generator nameplate capacity can be additionally obtained. Heat rates ranging from 7,600 to 7,830 Btus per KWh can be reasonably expected.

Heavy residual oils can be used in these engines with very little effect on performance. The future liquid fuels derived from coal, oil shale, or tar sands are expected to be acceptable in such engines, provided that the pilot injection is used to initiate combustions of those having low octane ratings, such as SRC2. Recent experiments with coal/oil mixtures indicate that such fuels can be burned. However, problems of cylinder liner and injector wear make use of such abrasive fuels as yet impractical.

## Environmental Considerations and Comparisons

Presented as Figure 8 is a flow diagram of a slow speed diesel engine with heat recovery. Major environmental considerations are the impacts of air emissions, evaporation and drift effects, and impacts of cooling tower make-up and blow-down.

Air emissions, to a great degree, can be controlled by residual fuel oil specifications. Engines also can be detuned to reduce $NO_x$ emissions. Effects on water bodies can be mitigated through the same use as conventional steam electric facilities.

Figure 9 presents an environmental comparison of a slow speed diesel with heat recovery to both a base load steam electric power plant and a gas turbine. Although each of the sources of fuel is different, the comparison shows some interesting results.

Emissions from gas turbines using distillate oil are significantly lower than slow speed diesels or base load coal-fired steam electric generation. However, the slow speed diesel plant, which is a base load facility, unlike a gas turbine, has less emissions per megawatt than a coal facility due to the greater efficiencies and more complete combustion. Nitrogen oxides ($NO_x$) are formed from the high temperature in the cylinders which fixes the nitrogen in the fuel and in the combustion of air. Greater emissions of $NO_x$ are formed by the generally more complete combusion and efficiency of the slow speed diesel. These engines can meet the proposed standards of $NO_x$ control by ignition retard.

Water consumption and discharge is significantly less with a slow speed diesel facility primarily due to the greater efficiency and reduced need for heat rejection. The slow speed diesel will not have any solid waste generation.

## Compressed Air Energy Storage

### System Description

Figure 10 presents the system characteristics of a compressed air energy storage (CAES) facility. Compressed air energy storage facility is not a pure energy storage system. Heat must be added from either burning oil or natural gas in order to increase the specific work of the compressed air to an efficient level. In a conventional gas turbine generator, three-quarters of the oil or gas is used to compress the air or is lost in exhaust heat; only one-quarter of the fuel energy is converted to electrical energy. By using stored compressed air in place of the compressor portion of the gas turbine generator, only one-third as much oil or natural gas is required to generate a given amount of electricity. In the CAES system, a synchronous motor generator replaces the generator used in the conventional gas turbine system. The motor generator is connected by clutches to either the compressor or turbine parts of the CAES. This arrangement enables the motor generator to use base load generating capacity to drive the compressor during off peak periods and generate electricity during peak load periods.

The CAES system has all the components located on the surface with the exception of the compressed air storage facilities. Subsurface facilities consist of air and water shafts and underground storage cabins or reservoir. The air and water shafts provide the connection between the surface components and the air storage cavern as this system is water compensated. The water shaft is connected to a surface reservoir and hydraulically compensates the air storage cavern as air is stored or released maintaining the cavern at a constant pressure. Maintaing constant pressure in the cavern minimizes inflow of surrounding groundwater and/or loss of stored air and reduces stresses induced by cyclical pressure fluctuations.

## Environmental Considerations and Comparisons

Figure 11 shows a flow diagram of a compressed air energy storage system. Major emissions are from the combustion of fuel oil or gas. Because heat is rejected from the compressor portion of the system, a cooling system is required that necessitates consideration of associated environmental impacts. With a closed-cycle system (cooling tower) make-up and blow down can result in impingement and entrainment of aquatic organisms. Geologic impacts, consisting of groundwater system disruption, are possible. In addition, disposal of bedrock material must be evaluated.

Figure 12 presents the environmental comparison of compressed air energy storage and a convential gas turbine facility. Air emissions for particulate and sulfur dioxide are approximately one-third that of a gas turbine. Since combustion occurs under similar conditions with both turbine facilities, $NO_x$ emissions are approximately equivalent. Water consumption is slightly greater for the CAES system to allow for make-up of the water reservoir. To fill the reservoir, a significant initial quantity of water is needed. For operation, a significant amount of the water is consumed, primarily to control $NO_x$ emissions. Water discharges are minimized by discharge of blow down back into the reservoir. Geological impacts, which can be significant, are mitigated through the water compensated system.

## Conclusions

Alternative energy systems evaluated can greatly reduce the need for importing foreign oil. However, environmental impacts will occur. Most of the systems can be established with minimum harm to the environment by using existing and advanced mitigative techniques. The impacts comparisons of these systems can be summarized as follows: (1) Coal/oil mixtures--the utilization of this system can actually improve air emissions yet has the potential for increasing groundwater contamination. Environmental impacts are considered equivalent or slightly greater than conventional generation. (2) Resource recovery--this alternative system can solve solid waste problems while recovering valuable resources. Impacts are not unlike incinerators, traditionally used a decade ago to reduce solid waste. Odors and leachates must be evaluated appropriately to ensure minimum impacts. Compared to conventional generation resource recovery is considered to have less environmental impacts. (3) Slow speed diesel with heat recovery--this form of generation is the most efficient for using fossil fuel. Air emissions are comparable to that of coal generation. The use of alternative fuels make this alternative attractive in the future. Impacts are considered less than conventional generation. (4) Compressed air energy storage--the impacts will be less than gas turbines, however, the emissions are displaced. It does allow the use of coal base load generation during off peak hours and reduces the need for oil during peaking periods. From an environmental standpoint, impacts are considered equivalent to or slightly greater than conventional generation.

## Bibliography

K.F. Kosky, P.E. and D.H. Kohlhepp, P.E. "A Resource Recovery Facility: Energy--Environmental Interactions," Energy and the Environment Proceedings of the Sixth National Conference. p. 487.

Preliminary Reservoir Engineering Report on Compressed Air Storage Facilities for the CAES System; Environmental Science and Engineering, Inc.; 1979.

Prevention of Significant Deterioration Analysis for the Proposed Diesel Generating Plant; Environmental Science and Engineering, Inc.; 1979.

PSD/BACT for Coal/Oil Mixture Test at Florida Power and Light Sanford Unit 4; Environmental Science and Engineering, Inc.; 1979.

- 50/50 COAL/OIL MIXTURE BY WEIGHT

- COAL: 12,000 BTU/LB, 13 PERCENT ASH, 2.5 PERCENT SULFUR

- OIL: 18,600 BTU/LB, LOW ASH, 1.7 PERCENT SULFUR

- COAL: 40% ON A BTU BASIS, DISPLACES 5430 BARRELS OIL PER DAY, 1,600,000 BARRELS PER YEAR

- COM PLANT CAPACITY OF 12,000 BBL/DAY

- FOUR PULVERIZERS RATED AT 10.5 TONS/HOUR

- STORAGE TANK TEMPERATURE OF 125°F

- STEAM ATOMIZATION SYSTEM ON BURNER GUNS

Figure 1 COAL/OIL MIXTURE FIRING SYSTEM
CHARACTERISTICS

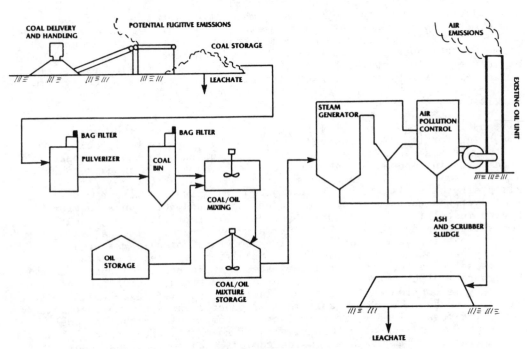

Figure 2 FLOW DIAGRAM OF COAL/OIL (COM)
MIXTURE FIRING AT EXISTING OIL
FIRING POWER PLANT

| ENVIRONMENTAL FACTOR | COM FIRING | OIL FIRING |
|---|---|---|
| AIR EMISSIONS (POINT SOURCE) | | |
|   — PARTICULATE | 360 LB/HR | 1224 LB/HR |
|   — SULFUR DIOXIDE | 9883 LB/HR | 9363 LB/HR |
|   — NITROGEN OXIDES | 2627 LB/HR | 2629 LB/HR |
| FUGITIVE EMISSIONS | 45 TONS/YEAR | NONE |
| SOLID WASTE GENERATION | 174 TONS/DAY | NONE |

Figure 3 ENVIRONMENTAL COMPARISON OF COM
FIRING AND CONVENTIONAL GENERATION

- **REDUCE 940,000 TONS/YEAR OF SOLID WASTE**
  **— VOLUME REDUCTION OF GREATER THAN 97%**

- **GENERATE 77 MW ELECTRICITY**

- **RECOVER ANNUALLY 65,500 TONS FERROUS METAL;**
  **37,400 TONS OF GLASS; 7,000 TONS OF ALUMINUM**
  **AND 4,700 TONS NON-FERROUS METAL**

- **DISPLACE OVER 1 MILLION BARRELS OF OIL**

- **100 GPM OF BOILER AND COOLING TOWER**
  **BLOW DOWN EVAPORATED WITH FUEL**

- **PULPED FUEL: 4050 — 4350 BTU/LB,**
  **0.14 — 0.2% SULFUR, MOISTURE APPROX.**
  **50% AND 10.5 — 12.1% ASH**

**Figure 4    RESOURCE RECOVERY SYSTEM**
**CHARACTERISTICS**

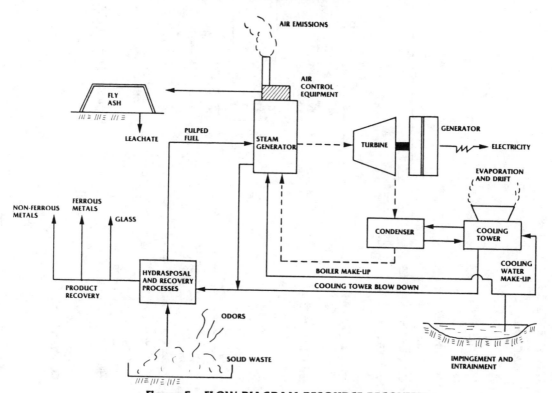

**Figure 5    FLOW DIAGRAM RESOURCE RECOVERY**

| ENVIRONMENTAL FACTOR | RESOURCE RECOVERY PER MW | STEAM ELECTRIC PLANT (COAL FIRED) PER MW |
|---|---|---|
| **AIR EMISSIONS** | | |
| — PARTICULATE | 7 TONS/YEAR | 4 TONS/YEAR |
| — SULFUR DIOXIDE | 28 TONS/YEAR | 43 TONS/YEAR |
| **WATER CONSUMPTION** | 22,800 GALLONS/DAY | 116,000 GALLONS/DAY |
| **WATER DISCHARGES** | 0 GALLONS/DAY | 1400 GALLONS/DAY |
| **SOLID WASTE GENERATION** | 1560 TONS/YEAR | 454 TONS/YEAR |
| **ODORS** | HIGH POTENTIAL | LOW POTENTIAL |

**Figure 6 ENVIRONMENTAL COMPARISON OF RESOURCE**
**RECOVERY AND CONVENTIONAL GENERATION**

131

- EUROPEAN DESIGN TWO AND FOUR STROKE ENGINES

- CAPABLE OF GENERATING 40 MW PER MACHINE

- TYPICAL SPEEDS FROM 90 TO 170 RPM

- BORE AND STROKE RANGES 18 TO 41 INCHES AND 35 TO 86 INCHES, RESPECTIVELY

- ENGINE POWER OUTPUT FROM 725 TO 3500 KW PER CYLINDER

- CAPABLE OF USING RESIDUAL AND COAL DERIVED FUELS

- HIGH EXHAUST TEMPERATURES ALLOW HEAT RECOVERY CYCLE

- HEAT RATES RANGE FROM 7600 TO 7830 BTU/KWH WITH BOTTOMING CYCLE

**Figure 7  SLOW SPEED DIESEL WITH HEAT RECOVERY SYSTEM CHARACTERISTICS**

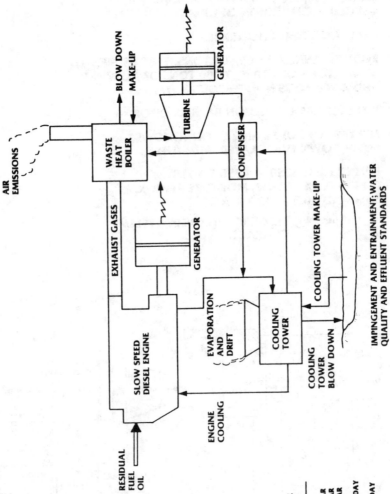

**Figure 8  FLOW DIAGRAM OF SLOW SPEED DIESEL WITH HEAT RECOVERY**

| ENVIRONMENTAL FACTOR | SLOW SPEED DIESEL PER MW | STEAM ELECTRIC PLANT PER MW | GAS TURBINE PER MW |
|---|---|---|---|
| AIR EMISSIONS | | | |
| — PARTICULATE | 3 TONS/YEAR | 4 TONS/YEAR | 0.7 TONS/YEAR |
| — SULFUR DIOXIDE | 33 TONS/YEAR | 43 TONS/YEAR | 5.3 TONS/YEAR |
| — NITROGEN OXIDES | 92 TONS/YEAR | 25 TONS/YEAR | 4.2 TONS/YEAR |
| WATER CONSUMPTION | 91,000 GALLONS/DAY | 116,000 GALLONS/DAY | 337 GALLONS/DAY |
| WATER DISCHARGES | 1100 GALLONS/DAY | 1400 GALLONS/DAY | 67 GALLONS/DAY |
| SOLID WASTE GENERATION | NONE | 454 TONS/YEAR | NONE |
| HEAT RATE | 7600 BTU/KW | 9500 BTU/KW | 11,139 BTU/KW |
| FUEL | RESIDUAL OIL | COAL | DISTILLATE OIL |

**Figure 9  ENVIRONMENTAL COMPARISONS OF SLOW SPEED DIESEL AND CONVENTIONAL GENERATION**

- 220 MW FACILITY
- COMPRESSION OF AIR USING OFF PEAK BASE-LOAD POWER
- AIR STORAGE VOLUME 4,000,000 FT³ @ 1000 PSI
- HEAT RATE OF TURBINE ≤ 5000 BTU/KWH
- CAPABLE OF 10 HOURS OF OPERATION PER DAY

Figure 10  COMPRESSED AIR ENERGY STORAGE SYSTEM CHARACTERISTICS

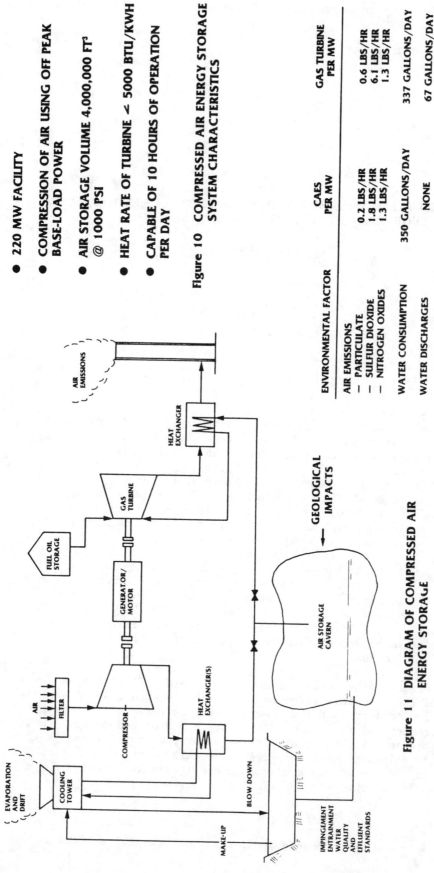

Figure 11  DIAGRAM OF COMPRESSED AIR ENERGY STORAGE

| ENVIRONMENTAL FACTOR | CAES PER MW | GAS TURBINE PER MW |
|---|---|---|
| AIR EMISSIONS | | |
| — PARTICULATE | 0.2 LBS/HR | 0.6 LBS/HR |
| — SULFUR DIOXIDE | 1.8 LBS/HR | 6.1 LBS/HR |
| — NITROGEN OXIDES | 1.3 LBS/HR | 1.3 LBS/HR |
| WATER CONSUMPTION | 350 GALLONS/DAY | 337 GALLONS/DAY |
| WATER DISCHARGES | NONE | 67 GALLONS/DAY |
| GEOLOGICAL IMPACTS | 673 CUBIC YARDS | NONE |

Figure 12  ENVIRONMENTAL COMPARISON OF COMPRESSED AIR ENERGY STORAGE (CAES) AND CONVENTIONAL GENERATION

# ENVIRONMENTAL ASSESSMENT AND SITE-SELECTION METHODOLOGY FOR SELECTED ENERGY STORAGE SYSTEMS

**Louis J. DiMento**
NUS Corporation

## Introduction

The storage of electrical energy during off-peak power demand periods and using this energy during peak periods may be a practical way of decreasing the need to construct new generating plants. Two types of systems which are technically feasible but have not yet been developed in this country are underground pumped hydro (UPH) and compressed air energy storage (CAES). The former involves the provision of hydroelectric power by allowing water to flow into a subsurface reservoir and the latter requires storing air at high pressures in underground storage volumes and releasing it later to provide power.

Although neither of these types of facilities has been built in the United States, they are likely to be developed in the future because of their beneficial impact on fuel consumption and new power plant construction. For this reason, a site selection methodology which includes environmental as well as technical factors has been developed with the objective of balancing cost factors with environmental concerns in the site selection process.

This paper describes the site selection methodology, including a description of the generic environmental impact assessment which was devised in order to determine the environmental criteria. The remainder of this introduction provides a brief description of the energy storage systems. This is followed by a description of the environmental assessment and site selection methodologies. The third section summarizes the results, using technical and socioeconomic criteria as examples. The conclusions are presented in the final section.

### Underground Pumped Hydro

UPH operates on the same basic principle as conventional surface-sited pumped storage--the production of power by turbines driven by falling water. In the case of UPH, head (difference in elevation) is created by providing two reservoirs--one at the surface and one underground. The surface reservoir can be a natural water body or an artificially constructed pond. The underground reservoir can be an existing or specifically excavated cavern in rock. Conventional hydroelectric pumping/generating equipment is located in an underground powerhouse adjacent to the lower reservoir.

As noted above, the system stores power during off-peak periods, and releases it during peak periods. During off-peak periods, power is obtained from the grid through a switchyard. The UPH pumps are driven by the motor-generator operating as a motor to pump water from the underground reservoir to the surface reservoir. During peak periods, water is conveyed downward through a penstock (pipe) and through the turbine located in the powerhouse near the lower reservoir. The turbine drives the motor-generator in the generator mode, thereby supplying power to the grid. A simplified diagram of a UPH facility is provided in Figure 1.

### Compressed Air Energy Storage

The CAES system utilizes the same principle of using off-peak power from the grid to store energy and then releasing this energy to provide power during peak periods. However, in the CAES system, air under high pressure is stored underground rather than water.

Figure 2 illustrates the major components of the CAES system. During off-peak periods the turbines are disengaged and the compressors are driven by the motor-generator operating as a motor using grid electricity. Air is compressed, cooled, and pumped into the underground storage reservoir. During peak-load periods, the air is released from storage, mixed with fuel in a combustion chamber, fired, and expanded through a series of turbines. The turbines drive the motor-generator in a generator mode, thereby supplying peaking power to the grid. During power generation, the compressor is disengaged and the entire output from the turbines is used to drive the generator.

The intercoolers and aftercooler remove the heat generated during compression. This is done to prevent excessive degradation of the storage media and to minimize its size and cost. Air is released from the cavern during peak periods to run the turbines, and the combustors are utilized to heat the air. Heating is necessary in order to efficiently run the turbines. In some systems waste heat from the exhaust gases generated by this process is circulated to a recuperator to heat the air released from the cavern. During this stage of operations, the motor-generator functions as a generator and is turned by the turbine. A compensating reservoir can be used to keep the cavern air pressure constant and thus decrease the required volume of the cavern. As the air is released, the liquid from the reservoir fills the cavern. The switchyard provides switching capability between the transmission line and the motor-generator. The cooling towers are needed to dissipate the heat generated during the compression process.

Three types of CAES facilities are being examined by utility companies in this country: CAES in hard rock, CAES in salt caverns, and CAES in confined aquifers. In the first type, a cavern is excavated from rock in the same manner as for UPH. For the second, water is pumped into a salt dome to dissolve the cavern from the salt, and the resulting brine is generally disposed of in underground salt aquifers. CAES in a confined aquifer requires pumping the air into the aquifer to displace the groundwater. Figure 3 presents underground views of the three facilities.

Methodology

## Generic Environmental Impact Assessment

Energy storage facilities can have significant impacts on a number of environmental categories. Those categories which are most likely to be affected include socioeconomics, geology, hydrology, air quality, noise, meteorology (fogging), aquatic ecology, and terrestrial ecology. Public safety is another issue to be considered when siting these facilities. The different types of energy storage facilities will vary in their potential impacts on each of these environmental categories. However, the general approach used for the impact assessment is the same for all systems and for each category. The basic approach which has been used is to determine the potential magnitude, duration, extent, and variability of the impacts associated with each envrionmental component and to summarize them in tables. General definitions of these terms are as follows:

Magnitude may be low, moderate, or high, and is a professional judgement of the potential size of the impact. The table specifies the greatest magnitude that may occur at any location. Duration refers to the length of time the impact will take place. The duration is considered short-term if the impact only occurs during the construction period; it is con·· sidered long-term if it may last longer than the construction period but for a finite time. This would generally be for the length of the operating period. Otherwise it is permanent. The three categories for the extent of the impact are onsite, referring to only the site; site area, which includes up to a few miles from the site; and site region, which covers a larger area such as a county or river basin. The variability of the impact describes whether the magnitude, duration, or extent of the impact can vary substantially depending on the location of the site. As shown below, these terms are important in developing the criteria for the site selection methodology. A more specific meaning of these terms depends on the type of impact under consideration and on the area where the sites are located.

This terminology is used in the generic assessment for all the environmental categories noted above. Figure 4 presents the results for the socioeconomic impacts of UPH in hard rock as an example. The socioeconomic impacts for the other systems are very similar to those for UPH. As the figure indicates, these impacts include the effects on economics, demographics, infrastructure, land use, aesthetics, cultural resources, and social welfare. The magnitude, duration, extent, and variability of these impacts, which are presented in the figure, are used to establish the criteria in the site selection methodology.

The analyses for the other environmental categories are conducted in a similar manner. However, in contrast to the socioeconomic impacts, some of the other categories vary greatly depending upon the system. For example, the magnitude of the air quality impact due to the fugitive dust resulting from the disposal of excavated rock can be high for UPH but is nonexistent for CAES in dissolved salt. The results of the

136

environmental analysis for any type of energy storage system will be a
series of tables summarizing the impacts for each environmental category
noted above. These results are then used to develop the environmental
criteria as discussed below.

## Site Selection Methodology

The site selection methodology for the four energy storage projects consists of five steps which are outlined in Figure 5. This methodology is an adaptation of methods developed by NUS, Acres American, and
Fenix and Scisson for previous energy projects, including a study of
hard rock energy storage performed for the Potomac Electric Power Company (1) and a salt dome siting study performed for Middle South
Services (2).

The first step is to define the study region based on projected energy
demand considerations, including the utility's service area and the load
center within that area. The load centers are those geographic portions
of the service area with the highest projected demand for electricity.
It may be decided, for example, that an energy facility should be located
within a certain distance of these centers to minimize costs of transmitting power to them. The availability of competing alternative energy
sources within this area may also affect the definition of the study
region. Certain sections of the service area may be very suitable for
mine-to-mouth coal plants, conventional hydropower, dispersed solar
power, and other forms of energy generation. Such sections would be excluded from the study area if their suitability for alternative sources
is indisputable. Figure 6 presents a hypothetical service area,
the load center and conventional energy sections within it, and the
study area.

The second step in the methodology is to determine the potential siting
area--those portions of the study area which have the appropriate
geological characteristics. This step, which can be conducted concurrently with Step 1, involves an examination of the literature describing the geological history of the region to delineate the areas
possessing the general rock properties for hard rock facilities, salt
features for dissolved salt facilities, or porous media for confined
aquifer facilities. The shaded area in Figure 7 represents the
potential siting area which is the intersection of the suitable geologic
region with the study area.

The third step is to select the energy storage method and to define the
plant characteristics. This includes a determination of the type of
plant, the power and energy storage capacities desired, and the approximate area required for surface facilities. Other factors which would be
included are the depth of potential storage areas, and the water and
fuel requirements.

The next step is to eliminate all areas from the study area which are
technically, socially, or environmentally unsuitable for the energy
storage facility. Areas which are technically unsuitable include those
near formation contacts and faults, those possessing very rugged terrain
or volcanic effects, or areas without nearby water. An area would be
excluded for social reasons if it is a population center, national forest,
national, state or local park or monument, or a recreational center. It
would also be excluded if it would severely conflict with local plans or
zoning laws. Environmentally unsuitable areas include important wetlands,

137

critical habitats for endangered species, and conservation and wilderness areas. The candidate siting area is what remains after these exclusions have been made. This step is indicated schematically in Figure 8.

Finally, sites in the candidate siting area are selected for detailed evaluation. The number of sites selected depends on the size of the potential siting region and the time and resources available for applying the site-selection criteria. A review of these criteria can help in determining which sites are selected. For example, sites which are very near to transmission lines and roads (two technical criteria) might be included in the selection. Another factor which would influence the selection would be current ownership. For example, it may be more difficult to negotiate for a site consisting of parcels with many different owners than for one which is owned by a single entity. After the sites have been selected, they would be rated by using the methodology discussed below.

The site-selection criteria fall into ten categories. These consist of a technical category and the nine environmental categories noted above. Each category is assigned a total weight of 100 units. The 100 units are divided among the separate impacts in that category based on judgements of their importance relative to one another. For example, in the socioeconomic category, the cultural impact (included in Figure 4) of the loss of historic resources was weighed 7 units, while the infrastructure impacts (in Figure 4) were assigned a weight of 20 units.

Each impact is ranked on a scale of zero to 10 for a specific site. The rank is zero if the site characteristics are such that impact would be severe, and 10 if there would be minimal impact. For example, the historic resources ranking would be zero if there were an historically significant resource on the site and 10 if there is none. A specific site would be ranked somewhere between zero and 10 if its characteristics are such that the impact would be between the two extremes. A similar weighting and ranking technique is applied for the technical category included in the site evaluation. The weights in this case are based on the percentage of cost attributable to each technical factor, and the ranking is based on the site characteristics which affect that technical factor.

The weight and rank of each item are multiplied to provide a numerical measure of the quality of the site with respect to that item. For instance, in the example cited above where the historical resource criterion was assigned a weight of 7 units, if a particular site were located in an area where there would be a minimum impact on historic resources, the rank is 10, and the numerical measure for that impact is 7 x 10 = 70. The maximum rating a site can receive for each category is 1,000.

In order to assess the importance of one category relative to another, a weight must be assigned to each of the ten categories. These weights can be assigned on a scale of zero to 1 such that the sum of the ten weights is 1. In this way, the maximum rating that a site can receive when all categories are included would still be 1000. The actual weights which are chosen would be affected by local conditions and attitudes, existing legislation, and other factors. They could be obtained informally or by such formal techniques as the Delphi method. Figure 9 presents the site rating methodology in mathematical terms.

Results

The environmental and technical criteria for all four energy storage systems have been developed to be used directly in the site selection process. This section presents an outline of the criteria for the technical and socioeconomic categories as an example of the overall results.

## Technical Criteria

The technical criteria for CAES in salt caverns are based on the site-selection methodology developed by Fenix and Scisson for comparing 47 salt domes in Louisiana and Mississippi (3). In this methodology, the costs of all the items required to develop and operate a CAES facility in a salt dome are estimated. Fourteen site characteristics are then identified and their contributions to each of the costs are estimated. For each of the site characteristics, the sum of its contributions to cost is then obtained. The weight assigned to a particular characteristic is equal to its percentage of the total costs attributed to all site characteristics. These technical characteristics and the weights assigned to each are presented in Figure 10. All the costs except for transmission and site access are constants based on engineering estimates for a typical facility. The transmission and site access costs depend on the best and worst sites under consideration, and, therefore, the weights of factors whose costs depend on them are variables.

As noted above, the ranking for each technical factor is a function of the characteristics a site may possess with respect to that factor. For example, the first factor, depth to salt, receives a ranking of 10 if the depth is less than 2000 feet, 7 if it is between 2000 and 3000 feet, and 0 if it is below 3000 feet. The other technical criteria are ranked similarly and then an overall technical rating is obtained by multiplying the weights and ranks.

## Environmental Criteria

The weightings and rankings for the environmental criteria are based on the environmental impact assessment described in the previous section. A significant impact which is potentially large and highly variable will receive a rather high weight. Sites are ranked according to their characteristics as shown in the example below. Variability is important in determining the weight of a potential impact because, if there is no variation among sites, they all receive the same ranking. The following example of the socioeconomic category demonstrates the site selection methodology for all the environemntal categories. The socioeconomic criteria are the same for all four types of energy storage systems, although this is not the case for the other environmental categories. The impacts and their weightings are summarized in Figure 11. The other environmental impacts are weighted similarly.

The ranking for the environmental criteria are developed in a manner similar to those for the technical criteria. For example, in the socio-economics category, infrastructure impacts result from the requirements of the facility itself and from the requirements of the population that inmigrates to work on the project. The best location would be one where all the required labor force is available locally so that no inmigration would be necessary, and one in which all public services are operating below capacity. Such a location would be given a ranking of 10. A site in an area with a small labor force and with most public services operating near capacity would be given a ranking of 0. The rankings will vary between these extremes depending on the number of public services operating near capacity and the size of the labor force available locally. The rankings for the other socioeconomic impacts as well as for the impacts included in the other environmental categories are developed in a similar manner.

To complete the site selection methodology the weightings of the ten categories relative to one another are assigned as discussed in the previous section. It is then a straightforward matter to rate any collection of sites.

Conclusions

Energy storage facilities offer a feasible alternative to new power plant construction and have the advantage that less fuel is required. The major differences in the environmental impacts between these facilities and conventional power plants stem from the underground storage requirements. For example, the potential ground water impacts of excavation operations and brine disposal are unique to this type of power facility. Also, for UPH and CAES in hard rock, the large construction force required to excavate the caverns may have substantial socioeconomic impacts.

The site selection methodology described in this paper is an additive approach which includes invironmental and technical criteria. It has the advantages that it can be easily applied by a utility company to evaluate specific sites and that it requires that environmental factors be explicitly included in the process. At the same time, it is flexible because it allows for variation in the category weights to reflect differing values and concerns.

Since the methodology is additive in nature, it has the disadvantage that it does not adequately quantify site characteristics which individually are not significant but in combination would be serious. A multiplicative approach would be more suitable if such situations are likely. However, in general, the impacts of the energy storage facilities tend to be independent, and the additive approach is adequate.

Although the methodology is conceptually straightforward, it can be rather tedious to apply, and, therefore, its application would be simplified if it were automated. Automation would have the additional advantage that the category weights can easily be modified to perform sensitivity analysis. Also, if the criteria are examined in order of decreasing significance, some sites may be eliminated before all the criteria have been examined. The software can be written to test when this is the case and to signal that these sites need not be examined for the remaining criteria.

1.  Acres American, Inc., <u>Preliminary Design Study of Underground Pumped Hydro and Compressed Air Energy Storage in Hard Rock</u>, Columbia, MD, 1979
2.  Fenix & Scisson, Inc., <u>Summary of Research on Forty-Seven Selected Salt Domes in Louisiana and Mississippi</u>, Volume I, Tulsa, Oklahoma, undated.
3.  Ibid.

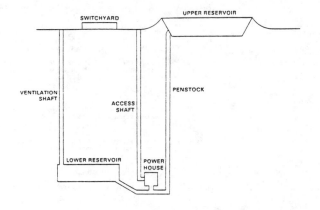

SIMPLIFIED VIEW OF A UPH FACILITY
Figure 1

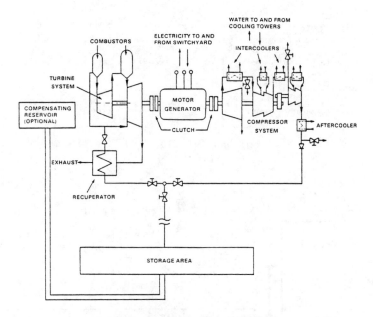

MAJOR COMPONENTS OF THE CAES SYSTEM
Figure 2

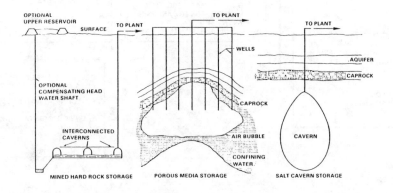

SIMPLIFIED UNDERGROUND VIEWS OF THE THREE TYPES
OF CAES STORAGE AREAS
Figure 3

141

| | MAGNITUDE | DURATION | EXTENT | VARIABILITY |
|---|---|---|---|---|
| **ECONOMIC IMPACTS** | | | | |
| EMPLOYMENT | | | | |
|     CONSTRUCTION | MODERATE | SHORT TERM | SITE REGION | LOW |
|     OPERATION | LOW | LONG TERM | SITE REGION | LOW |
| EARNINGS | | | | |
|     CONSTRUCTION | MODERATE | SHORT TERM | SITE REGION | MODERATE |
|     OPERATION | LOW | LONG TERM | SITE REGION | MODERATE |
| PROPERTY VALUES | HIGH | PERMANENT | SITE AREA | HIGH |
| **DEMOGRAPHIC IMPACTS** | | | | |
| INMIGRATION | | | | |
|     CONSTRUCTION | MODERATE | SHORT TERM | SITE REGION | HIGH |
|     OPERATION | LOW | LONG TERM | SITE REGION | LOW |
| **INFRASTRUCTURE IMPACTS** | | | | |
| CONSTRUCTION | MODERATE | SHORT TERM | SITE REGION | HIGH |
| OPERATION | LOW | LONG TERM | SITE REGION | LOW |
| **LAND USE IMPACTS** | | | | |
| CHANGE IN EXISTING LAND USE | HIGH | PERMANENT | ONSITE | HIGH |
| CONFLICT WITH PRESENT | | | | |
|     SURROUNDING LAND USE | HIGH | PERMANENT | SITE AREA | HIGH |
| CHANGE IN EXISTING AND | | | | |
|     PLANNED ZONING | HIGH | PERMANENT | SITE AREA | HIGH |
| **AESTHETIC IMPACTS** | | | | |
| DECREASE IN VISUAL SCENIC | | | | |
|     QUALITY | HIGH | PERMANENT | SITE AREA | HIGH |
| LOSS OF AMENITIES | HIGH | PERMANENT | SITE AREA | HIGH |
| **CULTURAL IMPACTS** | | | | |
| LOSS OF HISTORIC SITES | HIGH | PERMANENT | ONSITE | HIGH |
| DECREASE IN VALUE OF | | | | |
|     NEARBY HISTORIC SITES | MODERATE | PERMANENT | SITE AREA | HIGH |
| LOSS OF RECREATIONAL AREAS | HIGH | PERMANENT | ONSITE | HIGH |
| DECREASE IN VALUE OF | | | | |
|     NEARBY RECREATIONAL AREAS | MODERATE | PERMANENT | SITE AREA | HIGH |
| **SOCIAL WELFARE IMPACTS** | | | | |
| DISPLACEMENT OF FAMILIES | | | | |
|     LIVING ON THE SITE | HIGH | PERMANENT | ONSITE | HIGH |
| LOSS OF COMMUNITY CHARACTER | HIGH | LONG TERM | SITE AREA | HIGH |

POTENTIAL SOCIOECONOMIC IMPACTS OF UPH IN HARD ROCK

Figure 4

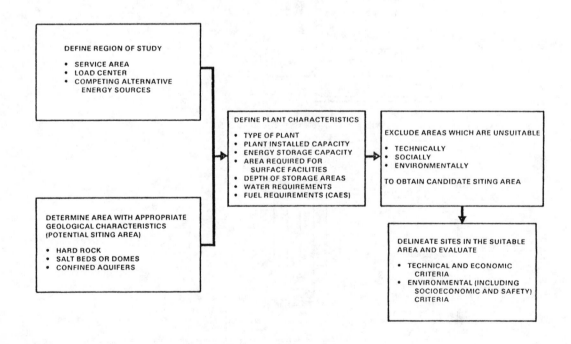

SITE SELECTION METHODOLOGY FOR ENERGY
STORAGE PROJECTS

Figure 5

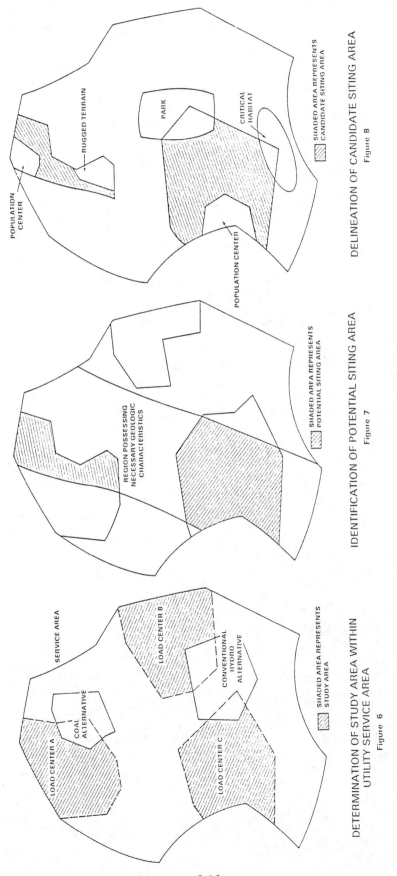

POPULATION
CENTER

RUGGED TERRAIN

PARK

CRITICAL
HABITAT

POPULATION CENTER

SHADED AREA REPRESENTS
CANDIDATE SITING AREA

DELINEATION OF CANDIDATE SITING AREA
Figure 8

REGION POSSESSING
NECESSARY GEOLOGIC
CHARACTERISTICS

SHADED AREA REPRESENTS
POTENTIAL SITING AREA

IDENTIFICATION OF POTENTIAL SITING AREA
Figure 7

SERVICE AREA

LOAD CENTER A

COAL
ALTERNATIVE

LOAD CENTER B

CONVENTIONAL
HYDRO
ALTERNATIVE

LOAD CENTER C

SHADED AREA REPRESENTS
STUDY AREA

DETERMINATION OF STUDY AREA WITHIN
UTILITY SERVICE AREA
Figure 6

$$\text{SITE RATING} = \sum_{i=1}^{10} W_i \times (\sum_j w_{ij} \, r_{ij})$$

WHERE

$i$ = THE CATEGORY

$\sum_{i=1}^{10}$ = THE SUM OVER ALL TEN CATEGORIES

$W_i$ = THE WEIGHT FOR THE $i$th CATEGORY

$w_{ij}$ = THE WEIGHT FOR THE $j$th ELEMENT OF THE $i$th CATEGORY

$r_{ij}$ = THE RANK FOR THE $j$th ELEMENT OF THE $i$th CATEGORY

$\sum_j$ = THE SUM OVER ALL THE ELEMENTS WITHIN A CATEGORY

### METHOD FOR DETERMINING SITE RATING
Figure 9

| FACTOR | SIGNIFICANCE |
|---|---|
| DISTANCE TO PLANT ACREAGE | $\frac{1.72}{\text{sum}} \times 100$ |
| DISTANCE FROM EXISTING RAILROAD LINE OR NAVIGABLE WATERWAYS | $\frac{.302}{\text{sum}} \times 100$ |
| A. RAILROADS | |
| B. NAVIGABLE WATERWAYS | |
| DISTANCE TO LEACHING | $\frac{4.394}{\text{sum}} \times 100$ |
| A. SUBSURFACE WATER | |
| B. SURFACE WATER | |

### CAES IN DISSOLVED SALT TECHNICAL CRITERIA
Figure 10a

| FACTOR | SIGNIFICANCE |
|---|---|
| DISTANCE TO BRINE DISPOSAL | $\frac{4.394}{\text{sum}} \times 100$ |
| LAND USAGE OF SITE | $\frac{5.084 = .752 \times (W_T - L_T)}{\text{sum}} \times 100$ |
| SEISMIC SURVEY REQUIRED | $\frac{.66}{\text{sum}} \times 10$ |
| FUTURE PLANT EXPANSION POSSIBLE | $\frac{3.466 + .752 \times (W_{SA} - L_{SA}) + .564 (W_T - L_T)}{\text{sum}} \times 100$ |

### CAES IN DISSOLVED SALT TECHNICAL CRITERIA
(Continued)
Figure 10b

| FACTOR | SIGNIFICANCE |
|---|---|
| DEPTH TO SALT | $\frac{5.344}{\text{sum}} \times 100$ |
| SURFACE OF LAND | $\frac{14.17 - .752 \times (W_{SA} - L_{SA})}{\text{sum}} \times 100$ |
| CURRENT USE OF DOME | $\frac{14.672 + .752 \times (W_T - L_T)}{\text{sum}} \times 100$ |
| FLOOD HAZARD | $\frac{7.644 + .752 \times (W_{SA} - L_{SA})}{\text{sum}} \times 100$ |
| SEISMIC RISK | $\frac{4.804 + .752 \times (W_T - L_T)}{\text{sum}} \times 100$ |

### CAES IN DISSOLVED SALT TECHNICAL CRITERIA
(Continued)
Figure 10c

| FACTOR | SIGNIFICANCE |
|---|---|
| TRANSMISSION | $\frac{.664 + .752 \times (W_T - L_T)}{\text{sum}} \times 100$ |
| SITE ACCESS | $\frac{.0008 + .752 \times (W_{SA} - L_{SA})}{\text{sum}} \times 100$ |

### CAES IN DISSOLVED SALT TECHNICAL CRITERIA
(Concluded)
Figure 10d

| ENVIRONMENTAL IMPACT | SIGNIFICANCE | |
|---|---|---|
| ECONOMIC IMPACTS | 10 | |
| PROPERTY VALUES | | 10 |
| DEMOGRAPHIC IMPACTS | 5 | |
| INFRASTRUCTURE IMPACTS | 20 | |
| LAND-USE IMPACTS | 18 | |
| CHANGE IN EXISTING LAND USE | | 3 |
| CONFLICT WITH PRESENT SURROUNDING LAND USE | | 8 |
| CONFLICT WITH EXISTING AND PLANNED ZONING | | 7 |
| AESTHETIC IMPACTS | 15 | |
| DECREASE IN VISUAL SCENIC QUALITY | | 7 |
| LOSS OF AMENITIES | | 8 |
| CULTURAL IMPACTS | 16 | |
| LOSS OF HISTORIC RESOURCE | | 7 |
| DECREASE IN VALUE OF NEARBY HISTORIC SITES | | 1 |
| LOSS OF RECREATIONAL AREAS | | 7 |
| DECREASE IN VALUE OF NEARBY RECREATIONAL AREAS | | 1 |
| SOCIAL IMPACTS | 16 | |
| DISPLACEMENT OF FAMILIES LIVING ON THE SITE | | 8 |
| LOSS OF COMMUNITY CHARACTER | | 8 |

## UPH IN HARD ROCK AND CAES IN HARD ROCK, DISSOLVED SALT AND CONFINED AQUIFER SOCIOECONOMIC CRITERIA

Figure 11

145

# AN ENVIRONMENTAL ASSESSMENT OF COAL-USING PRIME MOVERS FOR RESIDENTIAL/COMMERCIAL TOTAL ENERGY SYSTEMS *

**M. L. Jain, J. C. Bratis** and **T. J. Marciniak**
Argonne National Laboratory

## INTRODUCTION

Environmental protection and deployment of new technologies and systems for conservation of scarce fuels have become of paramount, though sometimes conflicting, concern. The total energy system is a conservation concept based on meeting the heating, cooling, and electricity demands of a community by on-site generation of electricity and by use of the reject heat from the prime movers for heating and cooling. In contrast, conventional energy systems are based on buying electricity from a utility and on burning oil or natural gas on-site for heating.

Total energy systems based on scarce fuels have long been used in the industrial sector. However, total energy systems based on coal and coal-derived fuels have not yet found a niche in the marketplace. The widespread implementation of coal-based total energy systems hinges on several factors besides technical and economic considerations. For instance, questions related to fuel availability, increased local emission of pollutants, and compliance with environmental regulations must be resolved. Since the conservation and economic aspects of coal-using prime movers for residential/commercial total energy systems were reported in a previous paper,[1] the present work is primarily concerned with the environmental aspects.

*Work performed under the auspices of the U.S. Department of Energy, Office of Fossil Energy as part of the Total Energy Technology Alternative Studies (TETAS) project.

ENVIRONMENTAL REGULATIONS AFFECTING COMBUSTION OF FOSSIL FUELS

Public concern over the quality of the environment during the late 1960s and the 1970s resulted in pressure for a governmental policy to clean up the environment through legislation and appropriation of funds. The National Environmental Policy Act of 1969, which became effective on January 1, 1970, established a national environmental policy and the Council of Environmental Quality to assist and advise the President on environmental and ecological matters. In late 1970, Congress created the Environmental Protection Agency (EPA) for the purpose of standard-setting, research, and monitoring and enforcement of environmental laws. The EPA has developed several regulations to comply with the environmental laws enacted by Congress during the past decade. A summary of key federal laws and corresponding EPA regulations concerning combustion of fossil fuels is given in Table 1. An understanding of these laws and regulations is essential before making any environmental assessment.

The Clean Air Act of 1970, as amended in 1977, is intended to attain and maintain air quality sufficient to protect public health and welfare. To meet the objectives of this act, the EPA established National Ambient Air Quality Standards (NAAQS), setting maximum allowable limits for ambient concentrations of carbon monoxide (CO), sulfur dioxide ($SO_2$), total suspended particulates (TSP), photochemical oxidants ($O_x$), hydrocarbons (HC), and nitrogen oxides ($NO_x$). To achieve the NAAQS, the EPA set New Source Performance Standards (NSPS) for stationary fuel combustion devices and New Source Review (NSR) regulations for various sources. Presently, the NSPS limit only $SO_x$, $NO_x$, and TSP emissions from several stationary sources. The NSR is more site-specific and applies only to new power sources or modification of existing power sources that emit 100 tons or more (after controls) of any of the pollutants. The NSR may be based on either the Non-Attainment (NA) or the Prevention of Significant Deterioration (PSD) criterion. The NSPS for fossil fuel-fired boilers and stationary internal combustion engines are summarized in Table 2.

The Clean Water Act of 1972, as amended in 1977, is concerned with the control of toxic substances in waste water. All industries must meet the Best Available Treatment (BAT) requirements of this law by July 1, 1984. The Resource Conservation and Recovery Act of 1976 is concerned with the management of solid wastes, including residue and sludge by-products from air and water pollution control equipment and ash from combustion processes. Details of the environmental regulations affecting stationary combustion devices can be found in some recent studies.[2,3]

The air pollutants emitted from combustion devices are likely to disperse widely in the atmosphere and could adversely affect the health and well-being of human populations. Therefore, air pollution is likely to be a major short-term impediment to implementation of coal-using prime movers in residential/commercial total energy systems. For this reason, the analysis presented here focuses primarily on emissions of air pollutants. Since NSR is a site-specific regulation, only the environmental criteria established by the NSPS need be considered in a comparative environmental assessment of various prime mover technologies.

ESTIMATION OF AIR POLLUTANT EMISSIONS

Several variables influence the impact of the environment due to air pollutant emissions. These variables include the type of fuel and fuel specifications for both the conventional and total energy systems, the utility fuel mix, and the residential/commercial mix in the community. For this analysis, it is assumed that both the conventional and total energy systems use fuels with the same specifications. Five utility fuel mixes, covering a wide spectrum of fuels, and four communities, representing a range of residential-to-commercial ratios, are considered.[4]

The fuels assumed for this study are given in Table 3. A recent study[5] for DOE on industrial cogeneration is also based on these fuels. The utility fuel mixes are presented in Table 4. As this table shows, the amount of the fossil fuels that produce the air pollutants ranges from 0% to 100% for coal, 4% to 82% for oil, and 2% to 5.4% for natural gas. Table 5 gives the energy consumption data for the communities considered. These data are derived from a previous study.[4] Projected residential/commercial fuel mix data are given in Table 6. These data, along with projected national utility fuel mix data for 1990, are used to assess air pollutants on a national level.

## Conventional Systems

To determine the utility and local air pollutant emissions from conventional energy supply systems, a knowledge of emission factors for various stationary combustion devices is desirable. An emission factor is an estimate of the rate at which a pollutant is released to the atmosphere as a result of fuel combustion in a particular device. The EPA has compiled generalized emission factors for various stationary combustion devices.[8] These emission factors were adjusted for the fuels assumed in this analysis and are presented in Table 7. For the emissions associated with generation of electric power by the utility, it is assumed that compliance with the NSPS is achieved with suitable controls. For emissions from the heating devices used locally in residential/commercial applications, no pollution controls are assumed.

The air pollutant emissions were calculated for each community using the emission factors in Table 7 and the fuel consumption data in Table 5. The results for a community with a mix that is 48% residential and 52% commercial are given in Table 8. As the table indicates, maximum pollutants are produced when the utility is using coal only (represented by utility E). The local emissions reflect the quantities of fuel used to meet the heat loads.

## Total Energy Systems

The total energy systems considered here are based on diesel engines, simple gas turbines, oil- and coal-fired steam turbines, current and advanced Stirling engines, and closed and open cycle externally-fired gas turbines (EFGTs). To determine the environmental impact of using a total energy system in any of the four community mixes considered, air pollutant emissions were calculated for each prime mover technology. This required estimates of emission factors for each prime mover, together with estimates of the fuel quantities needed for each community.

148

The emission factors for all the prime movers used in the total energy systems are given in Table 9. Among the key assumptions used in estimating these factors was compliance with NSPS standards for $SO_x$, $NO_x$, and TSP emissions, where applicable. For example, for coal-fired prime movers, it was assumed that 90% sulfur removal is achieved by use of atmospheric fluidized bed combustors (AFBCs) and that 99.5% of the TSP are removed by use of particulate controls. Emissions of $NO_x$ were taken from recent studies and are lower than the NSPS.[10,11] Gas turbines were also assumed to meet the NSPS for all three pollutants. For diesel engines, the proposed NSPS for $NO_x$ was used in calculating the $NO_x$ emission factor, but since no other NSPS are set, the $SO_x$ and TSP emissions were estimated using the EPS-compiled emissions data for oil-fired burners.[8] Likewise, the oil-fired Stirling engines and EFGTs were assumed to have emission characteristics similar to oil-fired burners.

The fuel consumption of each community was estimated in an earlier study.[9] Using those estimates and the prime mover emission factors, the air pollutants were calculated for all four communities. Typical results for a community with a 48% residential/52% commercial mix are given in Table 10.

## ESTIMATION OF SAVINGS IN AIR POLLUTANT EMISSIONS

As Table 8 shows, the air pollutant emissions calculated for conventional systems are influenced by the utility fuel mix. Therefore, the savings in air pollutant emissions by replacing a conventional energy system with a total energy system depend on the utility fuels that are displaced. With present emphasis on reducing or eliminating oil and natural gas use by the utilities, coal use will undoubtedly expand rapidly and, in the future, the utilities will probably rely mostly on coal to produce electricity. Thus, assuming a 100% coal-using utility as a basis for comparison, the percentage emission savings from total energy systems were estimated. The detailed results are available in a recent study.[12]

## DISCUSSION OF RESULTS

### Factors Affecting Environmental Impact of Total Energy Systems

The estimations of major air pollutants and the possible savings presented above are discussed in this section to evaluate their possible impact. Three factors that influence the environmental impact of a total energy system are discussed: the utility fuel mix; the residential/commercial mix of the community; and the fuel type used by the prime movers. Although not discussed, the environmental impact also depends on many other site-specific conditions, such as the local topography and meteorology, which determine the pollutant concentration and dispersion.

Effect of Utility Fuel Mix on Overall Emissions. The fossil fuels combusted by the utilities contribute to air pollutants. In general, a coal-burning utility produces more pollutants than an oil-burning utility. The pollutants produced by the conventional system based on five utility fuel mixes are compared in Figs. 1 and 2 with those produced by total energy systems.

As Fig. 1 indicates, for a 48% residential and 52% commercial community, all of the oil-fueled total energy systems would have a negative impact on air quality in the region if the conventional system were based on utility A, which relies primarily on oil (see Table 4.) However, compared to conventional systems based on utility E, a 100% coal-burning utility, all oil-fueled total energy systems except for the one using diesel engines would have a positive impact on the environment. Finally, compared to conventional systems based on utilities B, C, and D, which use a mixture of fuels, all but the Stirling engine options would produce more pollutants. Similar observations can be made for communities with different residential/commercial mixes.

As Fig. 2 demonstrates, all coal-using total energy systems would produce more emissions than a conventional system based on Utility A. Compared to utilities B, C, and D, both Stirling and EFGT options would produce somewhat lower emissions. Finally, compared to a conventional system based on utility E, all of the coal-using total energy systems would produce considerable savings in air pollutant emissions.

Effect of Residential/Commercial Mix on Overall Emissions. Since the residential/commercial mix determines the relative heating, cooling, and electricity demands, the fuel savings obtained vary for the four communities examined. The manner in which heating demand is met under the conventional and total energy systems also varies for these communities. Thus, the emission savings associated with total energy systems depend on the residential/commercial mix. As Fig. 3 shows, the maximum savings in overall emissions are achieved where the residential/commercial mix is roughly equal. Once again, the advanced Stirling engine option produces maximum savings, followed by the current Stirling and the closed cycle EFGT.

Effect of Fuel Used by Total Energy Systems. Because externally-fired heat engines can use both coal and oil as fuel, the effects of both fuels on savings in air pollutant emissions should be examined. The percentage savings for total energy systems based on Stirling engines, closed and open cycle gas turbines, and steam turbines for the two fuels are compared in Fig. 4. This figure shows that coal-fueled total energy systems would, in general, save more emissions than the oil-fueled systems. The difference is attributable to the method of fuel combustion and pollution control applied. Because the coal is burned in AFBCs, both $SO_x$ and $NO_x$ emissions are less than those produced by a conventional oil-fired heater. The TSP emissions are also lower because the coal-burning systems must use a control device. Furthermore, except for the steam turbine option, all of the total energy systems produce savings -- regardless of the residential/commercial mix -- when compared to a conventional system based on a coal-using utility.

Comparison Based on Major Air Pollutants

In addition to the overall impact of air pollutant emissions from various total energy systems, the environmental impact of individual pollutants is important to discuss, especially since the emission regulations are based on the individual pollutants. The three pollutants addressed by the NSPS are $SO_x$, $NO_x$, and TSP. These are discussed in the following sections.

$SO_x$ Emissions. The savings in $SO_x$ emissions from oil- and coal-fueled total energy systems are compared in Fig. 5. For oil-fueled systems, the $SO_x$ savings range from a low of -80% for steam turbines to a high of 41% for advanced Stirling engines. However, for coal-fueled total energy systems, the $SO_x$ savings vary from -67% for a steam turbine to a high of 46% for advanced Stirlings. For a community with a roughly equal residential/commercial mix, total energy systems based on steam turbines would have a negative impact on the environment because of their lower overall efficiency. In general, compared to a conventional system based on a coal-using utility, total energy systems based on coal-fueled steam turbines and open cycle EFGTs would have a negative environmental impact.

$NO_x$ Emissions. A comparison of $NO_x$ savings from the total energy systems is presented in Fig. 6. As this figure shows, only diesel engines and oil-fired steam turbines can lead to increased $NO_x$ emissions. However, coal-fueled total energy systems produce less $NO_x$ emissions because of lower combustion temperatures in the AFBCs. The net savings would range from a low of 44% for the steam turbine to a high of 81% for the advanced Stirling engine. Oil-fueled EFGTs and Stirling engines also produce significantly lower $NO_x$ emissions. Thus, with coal-fueled total energy systems, a significant reduction in $NO_x$ emissions can be obtained.

TSP Emissions. Particulates are generated in oil-fueled engines due to ash content in the fuel and ejection of unburned carbon. In coal-fueled systems, in addition to ash, limestone particles are carried by the combustion gases. Because of a large percentage of ash content in the coal and limestone particles used in the fluidized bed, cyclones and baghouses must be used to remove the particulates. However, no particulate control is required for the oil-fueled systems.

Savings in TSP emissions are compared in Fig. 7. As the figure indicates, the coal-fueled Stirling engines and external-fired gas turbines would save in TSP emissions. The savings range from 10% for open cycle EFGTs to 47% for the advanced Stirling engines. All of the oil-fueled options lead to increased TSP emissions because no emission control equipment is used. Simple gas turbine-based total energy systems produce minimum negative impacts because of the distillate fuel used. Steam turbine-based total energy systems, which have the lowest efficiency, burn the most fuel and produce the most TSP emissions.

## Overall National Assessment

To assess air pollutant emissions on a national basis, an analysis of pollutants from conventional systems was carried out using the DRI[7] projections for fuel mixes in 1990. The analysis assumed that the heating demand is met by burning coal, oil, and natural gas. The results of this analysis for a 60% residential/40% commercial community are compared in Figs. 8 and 9 with the air pollutant emissions from total energy systems. Percentage savings in emissions were estimated using the average national fuel mix for the utility. The results are given in Table 11.

As Fig. 8 shows, all of the oil-fueled total energy systems except the one based on an advanced Stirling engine would increase overall emissions. $SO_x$ and TSP emissions would increase in all cases. For the coal-fueled total energy systems, Fig. 9 shows that, in comparison to conventional systems, Stirling engines and the closed cycle EFGT would save emissions on an overall basis; however, $SO_x$ emissions would increase in

151

almost all the systems. As Table 11 shows, of all the coal-fueled total energy systems, only those based on Stirling engines can reduce overall emissions (by 14-22%; the closed cycle EFGT has negligible negative impacts on an overall basis. With regard to individual pollutants, coal-based systems would reduce $NO_x$ emissions by 12-57% due to low temperature combustion in the fluidized bed. However, $SO_x$ emissions would increase by 5-15% for Stirlings and by 35% for closed cycle EFGTs.

## CONCLUSION

The analysis presented here leads to several interesting conclusions. Among the key factors determining the net regional environmental impact of replacing a conventional system with a total energy system are the prime mover and fuel used in the latter system and the utility fuel displaced. Coal-burning total energy systems using AFBCs and particulate control devices are better from an environmental point of view than oil-burning total energy systems; moreover, on a regional basis, they can save $SO_x$, $NO_x$, and TSP emissions relative to conventional systems that use electricity supplied by a coal-burning utility. However, on a national basis, the impacts are mixed: based on a projected 1990 fuel mix scenario, $SO_x$ emissions would increase by 5-118% and $NO_x$ emissions would decrease by 12-57%, depending on the prime mover. The TSP emissions would also decrease except with the open cycle EFGT and steam turbine options. Of all the prime movers used in total energy systems, the Stirling engine has the highest overall efficiency and produces the maximum savings in pollutant emissions.

On a local level, total energy systems would increase emissions due to increased fuel use in the community. The local impact would vary, depending on a range of site-specific conditions. For minimal adverse impact, a roughly equal residential/commercial mix is desirable.

## ACKNOWLEDGMENTS

Grateful appreciation is expressed to R.E. Holtz (ANL) for his constructive suggestions and to Messrs. W.W. Bunker, E. Lister, and C.A. Kinney (DOE) for their support and encouragement of the TETAS project.

## REFERENCES

1.  Bratis, J.C., et al., *Residential/Commercial Cogeneration: Fuel and Technology Options*, paper presented at the 15th Intersociety Energy Conversion Engineering Conference, Paper NO. 809244 (Aug. 1980).

2.  Levin, J.E. and D.L. Hazelwood, *Know Key Energy, Environmental Laws*, Energy Systems Guidebook, Power, pp. 107-113 (1980).

3.  *Cogneration: Its Benefits to New England*, Final Report of the Governor's Commission on Cogeneration, The Commonwealth of Massachusetts (1978).

4.  Marciniak, T.J., et al., *An Assessment of Stirling Engine Potential in Total and Integrated Energy Systems*, Argonne National Laboratory Report No. ANL/ES-76 (Feb. 1979).

5. *Cogeneration Technology Alternatives Study*, DOE/NASA/1062-80/4, NASA TM 8/400 (Jan. 1980).

6. Orlando, J.A. and G.H. Lovin, *Technical Consideration for Joint Industry and Utility Cogeneration*, EPRI Cogeneration Workshop, San Antonio, Texas (April 1979).

7. Data Resources Inc., *Quarterly Energy Review* (Autumn 1979).

8. *Compilation of Air Pollution Emission Factors*, U.S. Environmental Protection Agency Publication No. AP-42, pp. 1.1-3 (12/1977), pp. 1.3-2 (4/1977), pp. 1.4-2 (5/1974), pp. 3.3.1-3 (12/1977), pp. 3.3.3-1 (1/1975).

9. Marciniak, T.J., et al., *An Assessment of External Combustion Brayton Cycle Engine Potential in Total and Integrated Energy Systems*, Argonne National Laboratory Report No. ANL/ES-96 (March 1980).

10. Smith, J.W., D.L. Bank, and W.C. Howe, *B&W/EPRI Fluidized Bed Development Program*, 2nd International Coal Utilization Conference and Exhibition, Houston, Vol. 4, pp. 260-284 (Nov. 1979).

11. Stewart, R.D. and J.A. Garcia-Mallol, *Atmospheric Fluidized Bed Fired Industrial Size Steam Generator Design*, 2nd International Coal Utilization Conference and Exhibition, Houston, Vol. 4, pp. 198-215 (Nov. 1979).

12. Bratis, J.C., M.L. Jain, and T.J. Marciniak, Argonne National Laboratory, unpublished information (1980).

Table 1  Federal Environmental Regulations Applicable
to Combustion Devices

| Legal Authorization | Regulations | Affected Facilities |
|---|---|---|
| Clean Air Act of 1970 (amended in 1977) | New source performance standard (existing) | Fossil fuel-fired steam generators (with over 250 x $10^6$ Btu/hr heat input) |
| | New source performance standard (existing) | Electric utility steam generators (with over 250 x $10^6$ Btu/hr heat input) |
| | New source performance standard (existing) | Stationary gas turbines (with 10.1 x $10^6$ Btu/hr heat input) |
| | New source performance standard (proposed) | Stationary internal combustion engines |
| | New source performance standard[a] | Industrial boilers |
| | Prevention of significant deterioration (existing) | Any stationary source emitting more than 100 tons/yr of criteria pollutant |
| | Prevention of significant deterioration (proposed) | Any stationary sources emitting greater than "de minimis" level of any regulated pollutant |
| | Non-attainment area requirements under state implementation plans (existing) | Any stationary source emitting more than 100 tons/yr of any criteria pollutant |
| Clean Water Act of 1972 (amended in 1977) | Effluent guidelines and standards (existing and proposed) | Steam electric power plants |
| Resource Conservation and Recovery Act of 1976 | Disposal of hazardous wastes | New plants requiring on-site disposal of waste materials |

[a]Regulations under development

Table 2  Federal New Source Performance Standards (NSPS) for Air Pollutant Emissions from Fossil
Fuel Combustion Devices

| Application | Fuel | $SO_x$ | $NO_x$ | TSP | Opacity % | Monitoring |
|---|---|---|---|---|---|---|
| | | Lb/$10^6$ Btu Heat Input | | | | |
| Utility steam generators (over 73 MW or 250 x $10^6$ Btu/hr input) | Coal | 1.20[a] | 0.60[b] | 0.03 | 20 | Continuous except TSP |
| | Oil | 0.80[c] | 0.30 | 0.03 | 20 | Continuous except TSP |
| | Natural gas | ----[d] | 0.20 | 0.03 | 20 | Continuous except TSP |
| Steam generators other than utility (over 73 MW or 250 x $10^6$ Btu/hr input) | Coal | 1.20 | 0.70 | 0.10 | 20 | Continuous except TSP |
| | Oil | 0.80 | 0.30 | 0.10 | 20 | Continuous except TSP |
| | Natural gas[e] | ---- | 0.20 | 0.10 | 20 | Continuous except TSP |
| Diesel engines[f] | Oil | ---- | 1.60 | ---- | -- | ---------- |
| Gas turbines | Oil | 0.50 | 0.49 | 0.036 | -- | ---------- |
| | Natural gas | 0.008 | 0.39 | 0.014 | -- | ---------- |

[a]Upper limit after required 90% emission reduction; if original emissions are less than 0.6 lb/$10^6$ Btu, only a 70% reduction is required by the NSPS.

[b]0.5 lb/$10^6$ Btu for sub-bituminous coal, coal-derived fuels, and shale oil.

[c]Upper limit after required 90% emission reduction; no reduction required if original emissions are less than 0.2 lb/$10^6$ Btu.

[d]No standard established.

[e]For low Btu gas produced on-site, use values given for coal.

[f]Proposed.

Table 3  Assumed Fuels for the Assessment

| Specifications | Petroleum-Derived | | Coal |
| | Distillate | Residual | |
|---|---|---|---|
| Sulfur (% wt) | 0.50 | 0.70 | 3.9 |
| Nitrogen (% wt) | 0.06 | 0.25 | 1.0 |
| Hydrogen (% wt) | 12.70 | 10.80 | 5.9 |
| Ash (% wt) | ---[a] | 0.03 | 9.6 |
| Trace elements | low | high | high |
| Heating value (Btu/lb) | 19,450 | 18,200 | 10,810 |
| Density (lb/gal) | 7.13 | 8.1 | na[b] |

[a]Negligible.

[b]Not applicable.

Table 4  Fuel Mix Used by Electric Utilities (%)

| Fuel Type | Utility[a] | | | | |
| | A | B | C | D | E |
|---|---|---|---|---|---|
| Coal | 0 | 32 | 50 | 53.7 | 100 |
| Oil | 82 | 45 | 4 | 6.8 | 0 |
| Natural gas | 0 | 2 | 4 | 5.4 | 0 |
| Nuclear | 18 | 21 | 42 | 23.2 | 0 |
| Hydroelectric | 0 | 0 | 0 | 8.4 | 0 |
| Miscellaneous | 0 | 0 | 0 | 2.5 | 0 |

[a]Data for utilities A, B, C, and E taken from
reference 6.  Utility D represents the projected national utility fuel mix for the year
1990, as given in reference 7.

Table 5  Annual Energy Consumption Breakdown for the Conventional System in Four Communities

| Residential/ Commercial Mix (%) | Non-HVAC Electricity ($10^6$ kWh) | Cooling Electricity[a] ($10^6$ kWh) | Heating ($10^6$ kWh) | | Fuel Input ($10^9$ Btu) | |
| | | | Electricity | Gas[b] | Utility,[c] Off-Site | Community, On-Site |
|---|---|---|---|---|---|---|
| 0/100 | 35.04 | 14.48 | 12.5 | 0 | 705 | 0 |
| 48/52 | 131.40 | 51.87 | 83.2 | 57.87 | 3032 | 197 |
| 60/40 | 87.60 | 38.90 | 43.4 | 57.87 | 1933 | 197 |
| 89/11 | 26.28 | 11.72 | 0 | 50.32 | 432 | 172 |

[a]Coefficient of performance = 3.0.

[b]Heating efficiency = 75%.

[c]Overall efficiency of electricity generation = 30%.

Table 6  Residential/Commercial Energy Use Mix Nationwide[a]

| Source | 1978 | | 1985 | | 1990 | |
| | $10^{15}$ Btu | % Total | $10^{15}$ Btu | % Total | $10^{15}$ Btu | % Total |
|---|---|---|---|---|---|---|
| Coal | 0.2 | 1.05 | 0.3 | 1.4 | 0.3 | 1.3 |
| Oil | 7.0 | 36.65 | 6.8 | 32.1 | 6.9 | 29.6 |
| Natural gas | 7.7 | 40.30 | 8.7 | 41.0 | 9.6 | 41.2 |
| Electricity | 4.2 | 22.00 | 5.4 | 25.5 | 6.4 | 27.5 |
| Solar | 0 | 0 | 0 | 0 | 0.1 | 0.4 |

[a]From Ref. 7.

Table 7  Air Pollutant Emission Factors for Conventional
Energy Systems ($lb/10^6$ Btu)

| Fuel | Utility[a] | | | Residential/Commercial[b] | | |
|---|---|---|---|---|---|---|
| | $SO_x$ | $NO_x$ | TSP | $SO_x$ | $NO_x$ | TSP |
| Coal | 0.69 | 0.60 | 0.03 | 6.9 | 0.14 | 0.90 |
| Oil | 0.20 | 0.30 | 0.03 | 0.75 | 0.12 | 0.017 |
| Natural gas | ---[c] | 0.20 | 0.03 | ---[c] | 0.08 | 0.015 |

[a]Calculated assuming compliance with NSPS.

[b]From Ref. 8.

[c]Not specified.

Table 8 Estimated Air Pollutant Emissions for Conventional Energy
Systems Serving a 48%/52% Residential/Commercial Mix
Community (tons/yr)[a]

| Utility[a] | $SO_x$ | | | $NO_x$ | | | TSP | | | Overall |
|---|---|---|---|---|---|---|---|---|---|---|
| | Utility | Local | Total | Utility | Local | Total | Utility | Local | Total | |
| A | 249 | 0 | 249 | 373 | 8 | 381 | 37.5 | 1.5 | 39 | 669 |
| B | 471 | 0 | 471 | 502 | 8 | 510 | 35.5 | 1.5 | 37 | 1018 |
| C | 536 | 0 | 536 | 492 | 8 | 500 | 25.5 | 1.5 | 27 | 1063 |
| D | 582 | 0 | 582 | 536 | 8 | 544 | 30.0 | 1.5 | 32 | 1158 |
| E | 1046 | 0 | 1046 | 902 | 8 | 910 | 45.5 | 1.5 | 47 | 2003 |

[a]See Table 4 for fuel mix data.

Table 9  Air Pollutant Emission Factors for Total
Energy Systems ($lb/10^6$ Btu)

| System | Oil-Fueled | | | Coal-Fueled | | |
|---|---|---|---|---|---|---|
| | $SO_x$ | $NO_x$ | TSP | $SO_x$ | $NO_x$ | TSP |
| Diesel engine | 0.70 | 1.57 | 0.070 | na[a] | na | na |
| Simple gas turbine | 0.50 | 0.49 | 0.036 | na | na | na |
| Stirling, current | 0.75 | 0.41 | 0.070 | 0.69 | 0.21 | 0.03 |
| Stirling, advanced | 0.75 | 0.41 | 0.070 | 0.69 | 0.21 | 0.03 |
| EFGT, closed cycle | 0.75 | 0.41 | 0.070 | 0.69 | 0.21 | 0.03 |
| EFGT, open cycle | 0.75 | 0.41 | 0.070 | 0.69 | 0.21 | 0.03 |
| Steam turbine | 0.75 | 0.41 | 0.070 | 0.69 | 0.21 | 0.03 |

[a]Not applicable

156

Table 10  Estimated Air Pollutant Emissions from Fox Valley Center and Villages
(48%/52% residential/commercial mix)

| | Fuel Use | | Emissions (ton/yr) | | | |
|---|---|---|---|---|---|---|
| Total Energy Systems | Oil $10^6$ gal/yr | Coal $10^3$ tons/yr | $SO_x$ | $NO_x$ | TSP | Total |
| Diesel engine | 12.25 | na[a] | 677 | 1418 | 63 | 2158 |
| Single gas turbine | 17.49 | na | 606 | 594 | 44 | 1244 |
| First generation current Stirling | 12.37 | na | 672 | 269 | 63 | 1004 |
| First generation advanced Stirling | 11.73 | na | 615 | 246 | 57 | 918 |
| EFGT closed cycle | 13.66 | na | 755 | 302 | 70 | 1127 |
| EFGT open cycle | 14.70 | na | 813 | 325 | 76 | 1214 |
| Steam turbine | 21.57 | na | 1193 | 477 | 111 | 1781 |
| Second generation current Stirling | na | 82.9 | 618 | 188 | 27 | 833 |
| Second generation advanced Stirling | na | 75.9 | 566 | 172 | 25 | 763 |
| EFGT, closed cycle | na | 93.1 | 695 | 211 | 30 | 936 |
| EFGT, open cycle | na | 100.0 | 746 | 227 | 32 | 1005 |
| Steam turbine | na | 146.8 | 1095 | 333 | 48 | 1476 |

[a]Not applicable

Table 11  Percentage Savings in Air Pollutant Emissions for
a 60%/40% Residential/Commercial Mix[a]

| System | $SO_x$ | $NO_x$ | TSP | Overall |
|---|---|---|---|---|
| Diesel engine | -26 | -249 | -91 | -121.0 |
| Gas turbine | -17 | -52 | -36 | -32.0 |
| Stirling, current (oil) | -25 | 34 | -91 | -3.0 |
| Stirling, advanced (oil) | -14 | 40 | -77 | 16.0 |
| EFGT, closed (oil) | -47 | 22 | -127 | -20.0 |
| EFGT, open (oil) | -59 | 16 | -141 | -30.0 |
| Steam turbine (oil) | -161 | -38 | -300 | -115.0 |
| Stirling, current (coal) | -15 | 53 | 18 | 14.5 |
| Stirling, advanced (coal) | -5 | 57 | 27 | 22.0 |
| EFGT, closed (coal) | -35 | 45 | 5 | -0.5 |
| EFGT, open (coal) | -46 | 41 | -5 | -8.0 |
| Steam turbine (coal) | -118 | 12 | -55 | -62.0 |

[a]Results are based on DRI data for 1990 residential/
commercial and utility fuel mix.

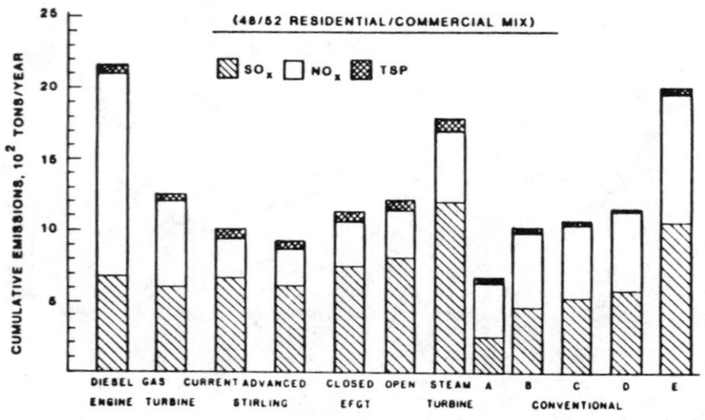

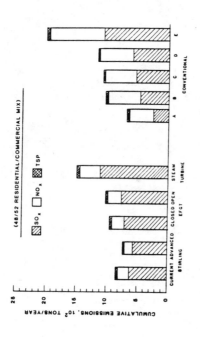

Fig. 1  Comparison of Air Pollutant Emissions from Oil-Fueled Total Energy
        Systems and the Conventional Systems with Various Utility Fuel
        Mixes

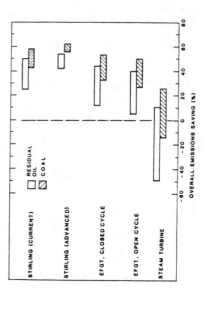

Fig. 2  Comparison of Air Pollutant Emissions from Coal-Fueled Total Energy
        Systems and the Conventional Systems with Various Utility Fuel
        Mixes

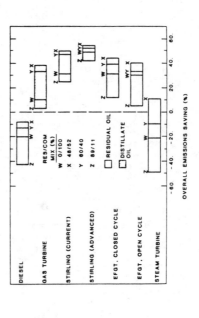

Fig. 3  Effect of Residential/Commercial Mix on Overall Emissions Savings

Fig. 4  Effect of TES Fuel Type on Overall Emissions Savings

158

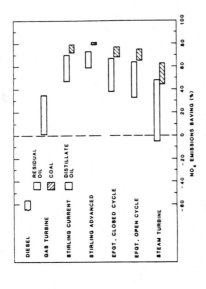

Fig. 6  Comparison of Regional NO$_x$ Emissions Savings

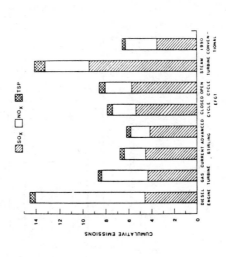

Fig. 8  Comparison of Air Pollutant Emissions from Oil-Fueled Total Energy
Systems with Those from a Conventional System (based on 1990
national fuel mix for utilities and residential/commercial sectors)

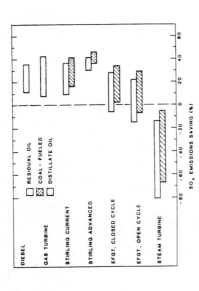

Fig. 5  Comparison of Regional SO$_x$ Emissions Savings

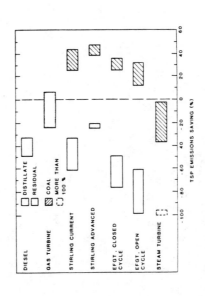

Fig. 7  Comparison of Regional TSP Emission Savings

159

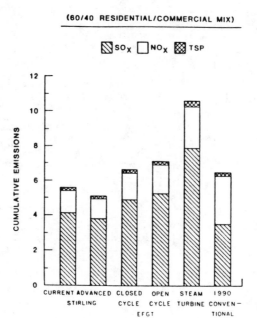

Fig. 9  Comparison of Air Pollutant Emissions from Coal-Fueled Total Energy
Systems with Those from a Conventional System (based on 1990
national fuel mix for utilities and residential/commercial sectors)

ENVIRONMENTAL EFFECTS OF A SUCCESSFUL UNDERGROUND COAL
GASIFICATION TEST IN A DEEP, THIN-SEAM BITUMINOUS COAL

**J. W. Martin** and **A. J. Liberatore**
U.S. Department of Energy

Introduction

In 1979 the United States Department of Energy conducted a successful
6-month test of in situ gasification of bituminous coal. The test was
conducted under the direction of the Morgantown Energy Technology Center
(METC) at the Center's Underground Coal Gasification (UCG) test site
near Pricetown, Wetzel County, West Virginia. Surface lands and mineral
rights for access to the 900-foot deep, 6-foot thick, high ash, and
high sulfur coal seam were provided through a cooperative agreement
with CONOCO, Incorporated.

Figure 1, which illustrates the subsurface locations of the deep wells
used during the Pricetown I field test, will be used to provide a brief
description of the test. Preignition tests were initiated on April 22,
 1979, when the coal seam was pressurized to evaluate in situ pressure
responses and in-seam flow rates. On June 9, 1979, the coal seam was
ignited at the bottom of well P/I-2. Air injected into well P/I-3
percolated through the coal seam to feed the reaction. The reaction
front, which naturally seeks the source of air, progressed to P/I-3
forming a path of coke and partially devolatilized coal between the
two wells.

The formation of this path was necessary to increase the permeability in the coal to permit the high gas flows required during gasification. After the initial path was established, the reaction front was returned to well P/I-2 by injecting into well P/I-2 and producing from P/I-3. The path was then extended across the 57-foot section by injecting into P/I-1 and producing from P/I-2 to complete the preparation of the coal seam for the high flow gasification phase of the test. Gasification, which was initiated on September 23, 1979, with air injection into well P/I-1 and production from well P/I-2, continued until October 5, 1979. The test was shut down on October 19, 1979.

Approximately 850 tons of coal were effected during this small scale test. During the 12-day gasification phase, about 350 tons of coal were gasified at the rate of 25 to 30 tons per day with an average air injection rate of 1.8 million cubic feet per day. A relatively clean combustible gas with a heating value of 127 Btu per cubic foot was produced at an average production rate of 4.2 million cubic feet per day.

More detailed descriptions of the test facility, test plans, and results have been published elsewhere[1,2].

Although Pricetown I was a relatively small scale test, it was considered essential to the technology to conduct an environmental monitoring program. The primary objectives of the program that was established were: (1) to characterize those hydrologic units which might be involved during the test, (2) to establish baseline data on surface streams, near-surface aquifers, and coal steam water quality, (3) to establish baseline data on local air quality and atmospheric conditions, (4) to detect changes in air and water quality on or near the test site during the test, and (5) to evaluate long-term environmental affects resulting from the field test. The environmental monitoring program was conducted by the Department of Geology and Geography and the Department of Civil Engineering at West Virginia University.

Water Quality Monitoring[3]

Conditions at the Pricetown UCG test site are considerably different from the conditions found at other sites where UCG tests have been conducted. In Wyoming, for example, the sites are remote with no major domestic wells located nearby, the coal seams are permeable and shallow, and near-surface aquifers lie adjacent to the targeted gasification zones.

At the Pricetown site, however, there are three domestic supply wells within 1,500 feet and at least 50 domestic water wells within 2 miles. Fortunately, the 900-foot deep Pittsburgh coal seam is relatively impermeable (1 through 2 microdarcies) and the near-surface aquifers are hydrologically isolated from the burn zone. In addition, connate brines, similar in composition to seawater, occur in the coal and several hundred feet of the overburden. These brines have been found to exhibit no movement in the natural state.

In order to assess the near-surface hydrogeologic conditions, a survey of the domestic water wells in a 20-square mile radius surrounding the test site was conducted in 1975[4]. Many of the wells were sampled and analyzed, physical well parameters were determined, and pumping tests were conducted on several wells to estimate aquifer permeabilities. In 1977 the domestic well survey was repeated and all wells previously sampled were again tested for water quality.

The constituents for which the water samples were analyzed are shown in the table. Only a few wells exhibited calcium predominant over sodium and, in general, the principal ground water type for the area is sodium bicarbonate. The domestic well water was found to be of good quality with respect to organic and inorganic constituents, except for relatively high levels of iron and manganese. These metals, how-ever, do not represent a health hazard in the concentrations measured.

Beginning in 1978 the water quality monitoring effort was restricted to wells and streams on and in the close proximity of the test site. The locations of the three surface streams, five shallow water wells, and a deep well (H-1) drilled into the Pittsburgh coal seam are shown in Figure 2. These sites were sampled and the samples analyzed for the constituents listed in the table at monthly intervals. During actual test operations, the sampling frequency was increased to ensure adequate coverage to detect possible environmental problems. Shortly after test shutdown, the sampling frequency was reduced to monthly intervals. This frequency is being maintained in the current long-term water quality monitoring effort which is scheduled to continue at least until October 1981.

The results of the chemical analyses showed that all the water supply wells remained stable throughout the test with the exception of one incident which is described in detail below. The quality of the water in the three streams varied widely as would be expected for small streams under the influence of seasonal changes; however, none of the variations are a result of the UCG test except for the one incident.

Hydrologic Disturbance

On October 5, 1979, a drastic restriction to gas flow from the coal seam occurred. In order to break through the restriction and reestab-lish the high flow gasification regime, coal seam pressures were increased. On October 12, 1979, it was observed that water from well 51 was exhibiting a bad taste and odor. A short time later water was observed to be spouting about 2 feet above the well cap. At this time air pressure on the coal seam was reduced. Within 10 hours, the water spout was pulsing at about 2 inches in height, and 15 hours later water flow had ceased although gas continued to flow from the well cap. On October 13, 1979, well 41 began overflowing and gas bubbles were dis-covered in the stream bed at location D (see Figure 2) and at location G, a small wet-weather spring. By October 14, 1979, water from well 41 was exhibiting an odor and a bad taste.

On October 16, 1979, after the hydrologic disturbance had subsided, the burn zone was again pressurized in an attempt to determine the source of pressurization of the subsurface aquifers. By the morning of October 19, 1979, well 41 was producing very turbid water, and gas bubbles were again observed at location G. At this time pressure was removed from the burn cavity and the field test was shut down.

Several changes associated with the hydrologic disturbance were observed in the stream and shallow ground water quality. First, an increase in phenol levels was recorded for the monitored streams and shallow wells. Figure 3 shows that these levels began increasing about mid-September and the highest levels occurred on October 15, 1979. (For reference, the highest levels were recorded at 0.08 ppm in well 11 and in stream P1.) By November 15, 1979, the concentration of phenol had returned to the baseline levels.

A second change that was observed involved well 51. Normally, the water from this well is predominantly a calcium bicarbonate as shown in Figure 4. However, two samples collected on October 12, 1979, showed that calcium values were reduced below normal and that the water had changed to a concentrated sodium bicarbonate water. By November 1, 1979, samples from well 51 indicated a return to near normal conditions.

Finally, samples taken from well H-4 on October 15, 1979, and October 19, 1979, showed a chloride concentration more than twice the normal values with a possible accompanying increase in sodium levels. By early November the water in this well had also returned to pre-burn baseline levels.

It was not until the post-test evaluation of the wells was completed that a plausible explanation for the hydrologic disturbance was advanced. During the evaluation, a downhole camera located a collar separated from the casing at a depth of 205 feet in well P/I-2. This hole, in all probability, permitted pressurized gas to escape from the coal seam into the near surface sandstone beds where it was transmitted up-dig to the nearby water wells. Fracture zones may have facilitated gas flow to well 51 and permitted the gas bubbling observed at sites D and G in Figure 3. Analysis of the gas collected at well 51 during the disturbance showed a relatively high concentration of carbon monoxide with lesser concentrations of methane and hydrogen. This is a strong indication that gas from the burn zone was probably involved and, in addition, would explain the elevated phenol concentrations in the supply wells and surface streams.

In summary, the gas pressures within the burn zone probably has had very little effect on the shallow ground water body because of the low permeability of the intervening rock layers. Most, if not all, of the manifestations observed appears to be due to pressurized gas escaping from the ruptured well casing at about the 205-foot level in well P/I-2.

Deep Water Quality Changes

Noticeable chemical changes occurred in the deep H-1 well just following completion of the in situ burn test. These changes became apparent at the end of October and in early November 1979. During this period there were significant increasing trends in sulfate, hardness, iron, zinc, and boron concentrations, and a decreasing trend in pH. These changes may have resulted from contaminated deep ground waters or hot gases having migrated from the burn zone to the H-1 well, following fractures in the Pittsburgh coal seam. This well is within 70 feet of the burn zone and a sufficient pressure gradient existed during the burn itself to start the migration. Trends in deep water quality changes are still being examined with respect to chemical sources and burn effects.

## Air Quality Monitoring[5]

The Pricetown I test site is contained in a small, narrow valley with several residences located at the open end of the valley. The hills adjacent to the site are forested and steep and, therefore, wind currents are variable.

The product gas from the UCG process was expected to be a reducing gas composed of hydrogen ($H_2$), carbon monoxide (CO), methane ($CH_4$), carbon dioxide ($CO_2$), hydrogen sulfide ($H_2S$), nitrogen ($N_2$), tar and oil vapors, and water vapor with small amounts of particulate, higher hydrocarbons, ammonia ($NH_3$), and sulfur dioxide ($SO_2$). A full blown gas cleanup system was not within the scope of the test, therefore, the gas stream was passed through a tar knock out vessel to reduce the concentration of tars, oils, and particulates and then incinerated.

The incinerator and stack combination consisted of a 40-foot tall cylindrical vessel with a fire brick lining approximately 5 feet in diameter. Two propane burners with two air blowers were incorporated in the design to provide preheating of the fire brick and to ensure incineration during periods of poor gas quality.

With the entire product gas stream being incinerated, the incinerator became the main source of pollutant with an effluent comprised primarily of $CO_2$, $H_2O$, $SO_2$, and the oxides of nitrogen. Another source of pollutants was the exhaust gases from the electrical generators and the low -pressure, high-volume air compressors.

To monitor the air quality on the site, an air quality monitoring station was installed away from the incinerator toward the open end of the valley. The station was designed to measure three meteorological parameters (wind direction, wind velocity, and height of inversions) and five pollutants [$SO_2$, CO, nitrous oxide (NO), nitrogen dioxide ($NO_2$), and total suspended particles (TSP)].

Unfortunately, many adverse factors resulted in the recording of data which was insufficient for a reasonable analysis of the impact of the incinerator plume on the ambient air. Nearly 50 percent of the time, calm conditions prevailed at the site and the incinerator plume rose directly upward at rapid rate. For an additional 30 percent of the time, the wind direction was away from the monitoring station. For the remaining 20 percent of the time (which included some downtime from instrument malfunction) no discernible correlations between the measured pollutant concentration and the product gas flow rate were observed. At no time, however, did the data collected at the air quality monitoring station indicate a violation of the U.S. Environmental Protection Agency standards for the measured pollutants.

## Summary

The environmental program instituted for the Pricetown I field test proved to be most effective in following a hydrologic disturbance which occurred just prior to test shutdown. The disturbance was manifested by a foul taste and odor, an increased phenol level, and a change in the chemistry of the shallow groundwater. Scheduled sampling and sample analyses showed that the water quality returned to normal baseline levels within a few weeks; however, monitoring is continuing to determine any possible long-term effects.

Since the hydrologic disturbance was traced to a ruptured well casing, it is apparent that an improved design for injection/production well completions is necessary. Well completions must be such that the expansion of the casing resulting from thermal gradients can be accommodated in a controlled manner.

Chemical changes have been recorded in the water from a deep well which is located within approximately 70 feet of the burn cavity. The changes are most likely a result of the migration of pollutants through very small fractures. Long-term changes are still being monitored and, therefore, a complete evaluation of the movement of the pollutant plume is not yet available.

Acknowledgments

A major portion of the work upon which this publication is based was performed pursuant to an Interagency Agreement between the Morgantown Energy Technology Center, the Department of Energy, and the Industrial Environmental Research Laboratory, Environmental Protection Agency.

References

1.  J. W. Martin and A. J. Liberatore, "A Successful Underground Coal Gasification Field Test in a Thin Seam, Swelling Eastern Bituminous Coal," Proceedings of the Sixth Underground Coal Conversion Symposium, Conf. No. 800716, I-8, (1980).

2.  R. E. Zielinski, et al., "Pricetown I Instrumentation and Preliminary Data," Proceedings of the Fifth Underground Coal Conversion Symposium, Conf. No. 790630, 221, (1979).

3.  E. Werner, et al., "Some Environmental Effects of the Pricetown I Underground Coal Gasification Test in West Virginia, U.S.A.," Proceedings of the Sixth Underground Coal Conversion Symposium, Conf. No. 800716, V-20, (1980).

4.  T. L. Sole, et al., Hydrogeologic Assessment of an Underground Coal Gasification Project Site, Grant District, Wetzel Co., West Virginia, U.S. Energy Research and Development Administration, Morgantown Energy Research Center Publication MERC/TPR-76/5, 1976, 50 pages.

5.  Robert N. Eli and David G. Dial, Air Quality Monitoring at the Pricetown I In Situ Gasification Site, West Virginia University, Annual Report for Task No. 21 Under Contract DE-AT21-79MC11284, 1980, 57 pages.

## Table
## Parameters Measured to Monitor
## Water Quality

### Inorganic Parameters

| | | |
|---|---|---|
| specific conductance | suspended solids | turbidity |
| dissolved oxygen | pH | temperature |
| acidity | alkalinity | sodium |
| potassium | calcium | magnesium |
| iron | bicarbonate | sulfate |
| chloride | nitrate | silica |
| phosphorus | alkaline-earth-metal hardness | |

### Trace Elements

| | | |
|---|---|---|
| barium | chromium | copper |
| zinc | cadmium | mercury |
| arsenic | selenium | lead |
| boron | manganese | nickel |

### Organic Constituents

| | | |
|---|---|---|
| phenol | fecal coliform | total organic carbon |
| biological oxygen demand | | |

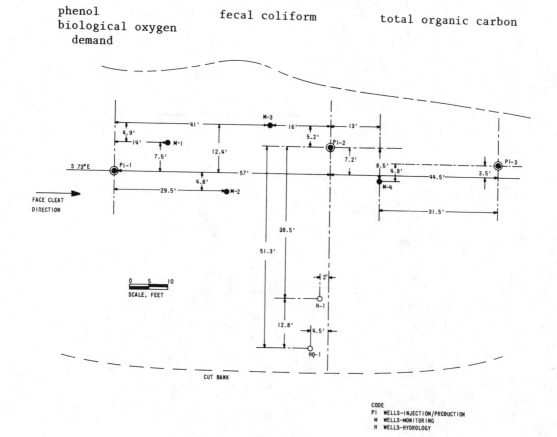

CODE
PI WELLS-INJECTION/PRODUCTION
M WELLS-MONITORING
H WELLS-HYDROLOGY

FIGURE 1.   PRICETOWN I SUBSURFACE WELL PLOT

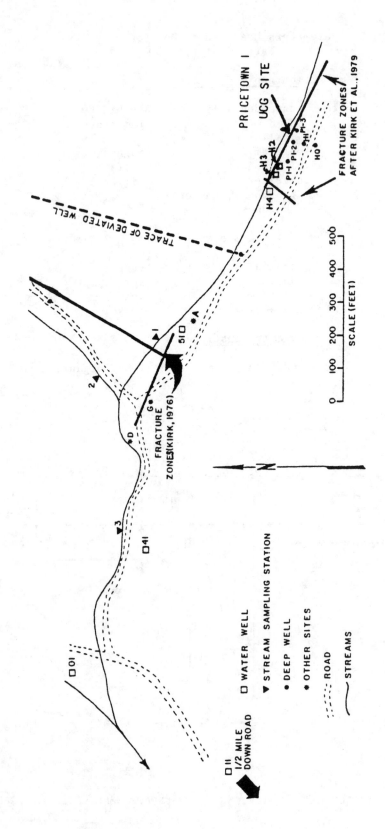

FIGURE 2. LOCATIONS OF OBSERVATION AND SAMPLING POINTS AT
THE PRICETOWN UCG TEST SITE

168

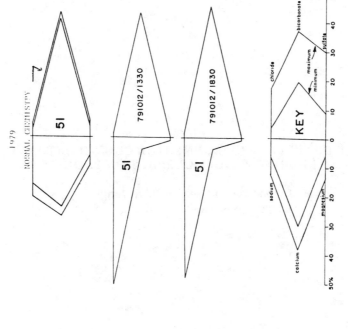

FIGURE 4. MODIFIED STIFF DIAGRAMS OF WATER CHEMISTRY
VARIATIONS IN WELL 51

## A note on space-time diagrams

A convenient way of showing temporal
changes at several locations in one illustration is through the use of the space-
time diagram. It is a two-dimensional
representation of a three-dimensional surface; *i.e.*, a perspective drawing of a
three-dimensional plot. The diagram is
constructed by drawing plots of variations
of a measured variable in space (between
sampling stations) for various times, all
on the same sheet, but offset from each
other diagonally a distance proportional
to the time interval between successive
samples. The result is a plot on which
the distance between stations is scaled on
a horizontal axis, time between observations
is scaled on a diagonal axis, and the value
observed is scaled on a vertical axis:

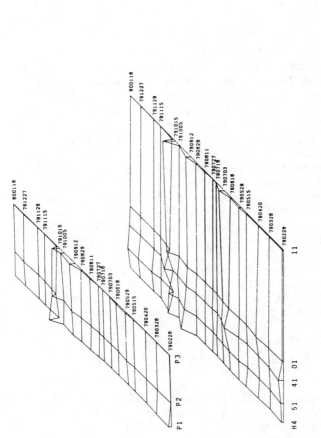

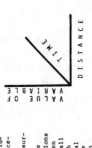

FIGURE 3. SPACE-TIME DIAGRAMS OF PHENOL CONCENTRATIONS
IN STREAM AND WATER SUPPLY WELL SAMPLES

# IMPACT OF DIFFERENT ENVIRONMENTAL CONSTRAINT LEVELS ON AN INDIRECT COAL LIQUEFACTION PROCESS

**P. J. Johnson** and **Suman P. N. Singh**
Oak Ridge National Laboratory

## Introduction

Questions have been raised in many learned quarters regarding the impact on the embryonic synthetic fuel industry of meeting increasingly stringent emission regulations.  Even the very viability of a synthetic fuel industry has been questioned in the light of increasingly stringent pollution control requirements that are anticipated to become the law of the land.  To address the concern(s), a study was performed to evaluate the potential impact of different environmental stringency levels on the economics of producing fuels from coal via the indirect coal liquefaction route.  This assessment was conducted by the Oak Ridge National Laboratory (ORNL) [with a subcontract to Fluor Engineers and Constructors, Inc., Houston (Fluor)] at the request of the U.S. Department of Energy/Assistant Secretary for the Environment-Office of Technology Impacts (DOE/ASEV-OTI).

The study examined the potential impacts of designing indirect coal liquefaction plants to meet increasingly stringent emission control levels that were developed by Fluor in consultation with ORNL and DOE/ASEV-OTI.  Since no emission regulations exist per se for the coal liquefaction industry, the control levels used in the study were developed based on current (1979) federal and/or state regulations for related industries such as coal-fired power plants, coke ovens, and petroleum industries.

It should be emphasized that the environmental control levels developed are NOT current or anticipated emission regulations but are used in the study as a basis for conducting the assessment. However, it is conceivable that some of the control levels could portend the flavor (if not the substance) of emission regulations in the post-1995 time frame.

The results of the impact of the different environmental control levels on the economics of conceptual indirect coal liquefaction plants are summarized in this paper. Detailed results of the study may be obtained from refs. 1 and 2.

Background

The assessment consisted of developing the economic evaluations for conceptual indirect coal liquefaction plants designed to comply with increasingly stringent emission control levels. The control levels examined were divided into four groups, as indicated in Table 1. The conceptual plants were designed as grassroots facilities capable of processing 24,000 tons per stream day (TPSD) [21,767 t/d] of a generic Wyoming subbituminous coal. The analysis of the generic coal is given in Table 2.

Figure 1 is a block flow diagram of the conceptual indirect coal liquefaction process selected for the study. The process basically consists of gasifying the coal in Lurgi Mark IV (dry-ash) pressure gasifiers, converting the syngas (after cleanup and shift) to methanol using the Imperial Chemical Industries (ICI) low-pressure process, converting the methanol to gasoline and other fuels using the fixed-bed Mobil MTG process, and fractionating and alkylating the products to yield a petroleum refinery-like slate of liquid and gaseous hydrocarbon fuels. The conceptual plants were all designed as integrated facilities, with coal and (in three of the four cases) water being the only inputs to the plant. In the fourth case (case C in Table 1), the plant water requirements were met by drilling water wells on the plant site.

In all cases, onsite dedicated facilities were included in the process designs to meet the plants' power, steam, oxygen, wastewater and sour gas treatment requirements. The process designs were modified as needed so as to comply with the different environmental constraint levels given in Table 1.

The choice of the indirect coal liquefaction process was dictated by the need to be consistent with an earlier study conducted by ORNL (and Fluor) for DOE as a part of an ongoing Liquefaction Technology Assessment (LTAS). Details of the LTAS Phase I study may be obtained from refs. 3 and 4. One section of the LTAS Phase I study was used as the design to meet the level C environmental constraint level evaluated in this study.

The details of the emission levels used to design the conceptual plants to meet environmental constraint levels B, C, and D are given in refs. 1 and 3.

Economic evaluations were performed by ORNL to determine the cost of producing products from the conceptual plants based on capital and annual operating costs developed by Fluor and on a pre-established set of economic parameters agreed to by ORNL and DOE. Fluor developed the capital requirements and annual operating costs based on the process designs for the conceptual plants. Fluor estimated the accuracy of the plant capital investment and annual operating costs to be ±30% and ±20%, respectively.  All costs were developed in terms of constant mid-1979 dollars to be consistent with the LTAS Phase I study.  The pre-established set of economic parameters are summarized in Table 3.

The cost of producing gasoline (one of the products from the plants) was determined by ORNL using ORNL developed computer program PRP.[5] Simply stated, PRP estimates the price of producing a product from a plant under a given set of economic parameters using the discounted cash flow (DCF) method.  The calculated product price is the selling price of the product at the plant fence that will just meet all the cost obligations of the plant.  The selling price does not include any additional tax levies on the product, or transportation and marketing charges.

The product price of the other plant products is included in the analysis as a ratio of the selling price of the other products to the selling price of gasoline based on their energy content.  The price ratios used in the present analysis were provided by DOE and are given in Table 4.

Several sensitivity studies were also performed as a part of the economic analysis of the conceptual plants to determine the impact of changing economic parameters on the gasoline selling price.  The sensitivity cases examined are listed in Table 5.

Results

The results of the economic analysis to determine the impact of the different environmental stringency levels on the plant capital investment, annual operating costs, plant energy conversion efficiency, and the gasoline selling price are summarized in Table 6.  The results of the economic sensitivity cases are given in Table 7.

As can be readily seen from Table 6, the impact of designing the plants to meet increasingly stringent environmental control levels is relatively small.  For example, the gasoline selling price from the plant designed to meet current (1979) environmental control levels (level B) is only 12% higher than the gasoline product price from the plant designed with no environmental constraints (level A).  Designing the plant to comply with more stringent than current environmental constraints results in a further 4% increase in the gasoline price (comparing levels B and D).  Other results may be likewise inferred from Table 6.

The results of the economic sensitivity analysis (see Table 7) indicate that step changes in the economic input parameters affects the gasoline selling price to essentially the same extent for all the four environmental constraint levels examined.

Conclusions

Major conclusions that can be readily discerned from this study are that, based on the premises used to develop the assessment, the results indicate that the stringency of the environmental control levels has a relatively small effect on the ultimate gasoline product price, and fears that strict emission control regulations may stifle the synthetic fuel industry appear to be unfounded.

It would appear that the price paid to strictly control the release of emissions from coal conversion plants may well be worth the cost of preventing additional deterioration of the environment.

Acknowledgements

The study was sponsored by the U.S. Department of Energy/Assistant Secretary for the Environment-Office of Technology Impacts under contract W-7405-eng-26 with the Union Carbide Corporation.

References

1. Fluor Engineers and Constructors, Inc., Environmental Control Cost Study, Final Report, ORNL/SUB-80/13838/1 (October 1980).

2. P. J. Johnson et al., The Impact of Environmental Control Costs on an Indirect Coal Liquefaction Process, ORNL-5722 (in preparation).

3. Fluor Engineers and Constructors, Inc., Coal Liquefaction Technology Assessment, Phase I Final Report, ORNL/SUB-80/24707/1 (April 1980).

4. R. M. Wham et al., Liquefaction Technology Assessment-Vol I: Indirect Liquefaction of Coal to Methanol and Gasoline Using Available Technology. ORNL-5664/V.I (in preparation).

5. R. Salmon, A Revised Version of the Discounted Cash Flow Program PRP With Provisions for Escalation, Uniform Reduction of Debt, and Ratio Pricing of Products, ORNL-5723 (in preparation).

6. J. H. Smithson, "Petroleum Product Prices After 1990," memo to C. W. DiBella of DOE, April 26, 1979.

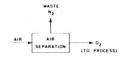

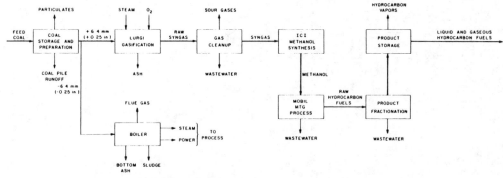

Fig. 1. Block flow diagram of indirect coal liquefaction process evaluated in the study.

Table 1. Emission constraint levels examined in the study[a]

| Level | Description |
|---|---|
| A | No emission control regulations except the control of lethal emissions. |
| B | Current (1979) emission control regulations for related industries (i.e., atmospheric emissions complying with NSPS for related industries, aqueous discharges in accordance with applicable NPDES permits, and solid wastes disposal as nonhazardous wastes). |
| C | Same as level B, except minimum water usage and zero wastewater discharge. |
| D | Projected stringent emission control regulations [i.e., atmospheric emissions complying with more stringent regulations than current (1979) NSPS for related industries (emission levels determined by Fluor and ORNL), zero discharge of priority pollutants in the aqueous emissions, and solid wastes disposed as hazardous wastes in accordance with RCRA guidelines]. |

[a]For detailed emission levels see refs. 1 and 3.

Table 2. Analysis of generic Wyoming Wyodak subbituminous coal used in the study

| Component | wt% |
|---|---|
| Proximate analysis, as-received | |
| Volatile matter | 33.1 |
| Fixed carbon | 33.8 |
| Ash | 5.1 |
| Moisture | 28.0 |
| Total | 100.0 |
| Ultimate analysis, maf basis | |
| Carbon | 74.45 |
| Hydrogen | 5.10 |
| Oxygen | 19.22 |
| Nitrogen | 0.75 |
| Sulfur | 0.45 |
| Chlorine | 0.03 |
| Total | 100.00 |

Higher heating value, as-received: 8509 Btu/lb (19,783 kJ/kg)
Higher heating value, maf basis: 12,720 Btu/lb (29,574 kJ/kg)

Table 3. Economic evaluation guidelines used in the study for the base case evaluations

Discounted cash flow analysis using 100% equity financing and 12% annual after-tax rate of return on equity

Constant mid-1979 dollars

Feed coal price: $1/10^6 Btu ($0.95/GJ)

5-yr plant construction and 25-yr plant operating life

Plant salvage value: 0

Process contingency: 0

Investment tax credit: 10%

Depreciation: Sum-of-the-years' digits for 16 yr

Product price ratios: 1990 price ratio scenario (see Table 4 for values)

Federal income tax rate: 48%

State income tax rate: 3%

Local property tax rate: 2%

Property insurance rate: 0.75%

Annual plant service factors: Year 1 (of operation) = 50%; Year 2 = 75%; Year 3-25 = 93% (established by Fluor)

By-product credits: ammonia at $120/ton ($132.30/t); sulfur at $30/ton ($33.08/t); and crude phenols at 12.5¢/lb (27.6¢/kg).

Table 4. Product price ratios used in the study (based on energy content)

| Component | Price ratios with respect to gasoline[a] | |
|---|---|---|
| | 1979 | 1990 |
| Gasoline | 1.00 | 1.00 |
| SNG | 0.59 | 0.80 |
| C₃ LPG | 0.69 | 0.85 |
| C₄ LPG | 0.76 | 0.87 |
| Crude diesel | 0.43 | 0.62 |
| Fuel oils | | |
| No. 2 | 0.84 | 0.88 |
| No. 6 | 0.62 | 0.73 |

[a]Product prices were given by DOE (ref. 6); price ratios were determined by ORNL as the ratio of the price per unit heating value of the coproduct to the price per unit heating value of gasoline.

Table 5. List of sensitivity cases examined

| Case | Variation |
|------|-----------|
| Feed coal cost | +50% |
| AARR[a] on equity capital | +25% |
| Plant capitalization | 75% Debt[b] + 25% Equity |
| Depreciable capital cost estimate | +25% |
| Annual operating cost estimate | +20% |

[a]Annual after-tax rate of return.

[b]Interest on debt capital is at 9% per annum.


Table 6. Impacts of different environmental constraint
levels on the plant economics and energy efficiency

| | Environmental constraint level | | | |
|---|---|---|---|---|
| | A | B | C | D |
| Plant capital cost, $10^6 | 1963 | 2247 | 2318 | 2332 |
| % change wrt[a] level A | — | 14.5 | 18.1 | 18.8 |
| Annual plant operating cost, $/yr | 315 | 346 | 341 | 352 |
| % change wrt level A | — | 9.8 | 8.3 | 11.8 |
| Calculated plant energy efficiency, % | 63.9 | 61.1 | 61.6 | 60.8 |
| % change wrt level A | — | -4.4 | -3.6 | -4.9 |
| Calculated gasoline selling price,[b,c] $/gal (¢/L) | 1.09 (29) | 1.22 (32) | 1.25 (33) | 1.26 (33) |
| % change wrt level A | — | 12 | 15 | 16 |

[a]wrt = with respect to.

[b]Ex plant gate.

[c]For base case economic premises.


Table 7. Percentage change in the gasoline selling due to
step changes in selected economic parameters

| | Environmental constraint level | | | |
|---|---|---|---|---|
| | A | B | C | D |
| Base case[a] gasoline selling price, $/gal (¢/L) | 1.09 (29) | 1.22 (32) | 1.25 (33) | 1.26 (33) |
| Feed coal cost increased to $1.50/10^6 Btu ($1.42/GJ) | +10 | +9 | +10 | +9 |
| AARR[b] on equity capital changed to 15% | +26 | +27 | +27 | +27 |
| Plant capitalization debt/equity ratio changed to 75%/25%, respectively | -35 | -36 | -36 | -36 |
| Depreciable capital increased by 25% | +17 | +17 | +17 | +17 |
| Annual operating costs increased by 20% | +2 | +2 | +2 | +2 |

[a]See Table 2 for base case economic premises.

[b]Annual after-tax rate of return.

# WATER QUALITY AND SOLID WASTE MANAGEMENT FOR A PROPOSED
# TVA ATMOSPHERIC FLUIDIZED-BED COMBUSTION DEMONSTRATION PLANT

**T.-Y. Julian Chu, D. B. Cox, John W. Shipp, Jr.**
and **L. H. Woosley, Jr.**
Tennessee Valley Authority

## Introduction

Fluidized-bed combustion (FBC) is one of the promising alternative technologies to enable combustion of coal in an environmentally acceptable and economically feasible manner. The Tennessee Valley Authority (TVA) has committed to construct and operate a 20-MW atmospheric fluidized-bed combustion (AFBC) pilot plant to address uncertainties concerning peripheral hardware systems and plant operation which are essential for designing a large-scale AFBC plant. Also, TVA is considering constructing a 200-MW AFBC demonstration plant to develop AFBC technology for full-scale commercial use on the TVA system and possibly by the utility industry. A preliminary draft environmental impact statement is being prepared to aid in the decision-making process. While no firm decision has been made on the schedule of the project as well as the location of the plant, the Shawnee Steam Plant reservation has been identified as the "preferred site."

The Shawnee Steam Plant site is located in McCracken County, Kentucky, on the left bank of the Ohio River at mile 945 (kilometer 1,521) as shown in Figure 1. This paper is to address the preliminary options being considered for water quality and solid waste management for the 200-MW AFBC demonstration plant at the Shawnee site. Detailed discussions include waste characterization, determination of hazardous potential of solid wastes, potential environmental impacts, and possible mitigation measures.

## Waste Characterization

Sources of wastewater and solid wastes associated with FBC power generation include: (1) construction activity--stormwater runoff and concrete batch plant wastes; (2) storage and handling of coal and limestone prior to feeding--discharges can arise from open solids storage piles (runoff and leaching from rainwater) and from such handling steps as coal drying and crushing; (3) the steam cycle--discharges include preoperational pipe cleaning wastes, boiler blowdown, boiler-tube cleaning wastes, cooling water discharge, and effluent from feedwater treatment; (4) miscellaneous wastes--floor drains, sanitary wastes, intake screen backwash, and others; and (5) solid residues--fly ashes and spent sorbents, and sludges from feedwater and wastewater treatment processes. Many of these wastes are similar to those encountered in conventional coal-fired steam plants. Fly ash and spent-bed material from FBC facilities, however, will be characteristically different from fly ash and bottom ash routinely encountered in conventional steam-electric power plants due to the addition of limestone to the boiler and lower combustion temperatures.

## Solid Wastes

At the TVA AFBC demonstration plant, fly ash and spent sorbent will be handled dry. The quantities of solid wastes, expected to be generated over the 35-year operating life of the AFBC facility, are $2.7 \times 10^6$ cubic meters (2,200 acre-feet) for fly ash and $4 \times 10^6$ cubic meters (3,250 acre-feet) for spent sorbent. The amount of solid wastes required for disposal would be substantially reduced when a market is developed for the residues.

Samples of AFBC fly ash and spent sorbent were obtained from an AFBC Technology Test Unit at Oak Ridge National Laboratory. The Test Unit was operated under the conditions similar to the 200-MW AFBC demonstration plant. These samples were analyzed for a variety of chemical constituents, and the results are shown in Table 1. The baghouse and cyclone fly ashes and spent sorbent contained relatively higher concentrations of calcium and sulfur than those in fly ash and bottom ash from conventional coal-fired power plants, obviously resulting from calcium sulfate and unreacted calcium oxide and calcium carbonate.

Samples of spent sorbent and a combined fly ash (baghouse and cyclone) mixture were subjected to four leaching tests including the EPA and ASTM extraction procedures.[1][2] The results of the inorganic analysis of the laboratory leachates (Tables 2 and 3) indicated that neither the combined fly ash nor the spent sorbent could be classified as hazardous as defined by EPA. The concentrations of radioactivities in the fly ash and spent sorbent were less than the hazardous criteria (5 pCi/g of Radium-226) that EPA is currently considering.[3] A highly exothermic reaction was found to result when water was contacted with the spent sorbent.

## Wastewater Effluents

Potential wastewater streams and their estimated quantities discharged from the AFBC demonstration plant are shown in Figure 2. A closed-cycle spray pond with ground water as the source of makeup water is being considered as the preferred heat rejection system based on economic and environmental considerations. Runoffs from fly ash and spent sorbent disposal areas are discussed below. Chemical characteristics of other wastewater discharges that are similar to TVA coal-fired power plant wastewaters were described elsewhere.[4][5]

Solid waste runoffs were simulated in the laboratory. Table 4 indicates that concentrations of dissolved heavy metals in runoffs were generally low as a result of the high pH (varying from 11.8 to 12.3) of the runoff. Small amounts of volatile and semivolatile organics were detected in runoffs (Tables 5 and 6), but the concentrations of organic priority pollutants were below the Kentucky water quality standards for aquatic life protection.[6] The concentrations of Radium-226 and Radium-228 in runoffs were less than the National Primary Interim Drinking Water Standard of 5.0 pCi/l for combined radium.[7]

Potential Environmental Impacts

## Solid Waste Disposal

Fly ash and spent sorbent from the facility, if not utilized, would be handled and disposed of separately in an 81-hectare (200-acre) area adjacent to the plant. Initially, only a small portion of the area required for storage will be prepared. Appropriate mitigation measures will alleviate the possibility of ground water contamination from leachate. Since the spent sorbent will react with water exothermically, there is a potential for fog formation at the site during the winter.

## Wastewater Discharge

The Ohio River from the Shawnee Steam Plant to the confluence with the Mississippi River is currently classified by the Kentucky Division of Water Quality (KDWQ) for all uses except public water supply and food processing industries. KDWQ Surface Water Standards[5] specify that the zone where wastes mix with the receiving stream shall be free from pollutants which are in excess of 0.44 of the 96-hour median lethal concentrations (96-hour $LC_{50}$) for a representative indigenous aquatic organism. Furthermore, the standards specify that for protection of aquatic life the allowable instream concentrations of accumulative toxic substances shall not exceed either the given numerical criteria or 0.01 of the 96-hour $LC_{50}$'s for a representative indigenous aquatic organism for substances for which there are no numerical criteria. The criteria for nonaccumulative toxic substances such as copper are 0.1 of the 96-hour $LC_{50}$. These numerical criteria and 96-hour $LC_{50}$'s for the most sensitive indigenous fish are listed in Table 7.

The preferred method of discharging the combined treated AFBC wastewater stream to the Ohio River is via the existing Shawnee Steam Plant ash pond effluent ditch, which flows into the Shawnee Steam Plant condenser cooling water discharge channel about 335 meters (1,100 feet) upstream from its confluence with the Ohio River (Figure 1).

Potential impacts on water quality were evaluated by projecting the expected "worst case" concentrations of selected constituents at the mouth of the condenser cooling water channel and by estimating mixing zones in the Ohio River required to meet the appropriate standards. Several "worst case" scenarios were evaluated since the concentrations of some constituents are occasionally higher in the Ohio River than in the AFBC waste stream. Table 8 shows the projected "worst case" concentrations of the constituents of interest in the existing steam plant ash pond effluent, AFBC waste stream, and combined AFBC-ash pond discharge. This projection assumes complete mixing of the AFBC waste stream with the ash pond effluent. Also contained in Table 8 are the projected "worst case" concentrations of these constituents at the mouth of the condenser cooling water discharge channel. These projections were made using a simple one-dimensional mathematical dispersion model.

Based on these projections, the "worst case" concentrations of copper, iron, lead, mercury, nickel, total residual chlorine, unionized ammonia, and zinc at the mouth of the condenser cooling water discharge channel exceed the KDWQ Surface Water Standards (see Table 9). The best available data for silver and cyanide in AFBC and Shawnee Steam Plant wastes contain no observations in excess of the "normal" minimum detectable amounts that are higher than the applicable standards.

The "worst case" concentrations of copper and iron result from high concentrations in the Ohio River and thus in the once-through condenser cooling water discharge. The high total residual chlorine concentrations at the mouth of the channel result from high concentrations in the condenser cooling water discharge.

Based on data taken by TVA for the Ohio River in the vicinity of Shawnee Steam Plant, the average concentrations of phenols, copper, mercury, nickel, and zinc in the river exceed the applicable standards; therefore, no mixing zone calculations were performed for these constituents. A mixing zone was also not calculated for total residual chlorine since the "worst case" concentration at the mouth of the channel was not caused by AFBC waste discharges. The total residual chlorine concentrations of spray pond blowdown will be limited to meet the EPA revised effluent limitations.[8] The calculated mixing zones for lead and unionized ammonia are respectively less than 30 meters (100 feet) in width by less than 1.6 kilometers (1 mile) in length and less than 30 meters (100 feet) in width by less than 1.0 kilometer (0.6 mile) in length.

Proposed Mitigative Measures

Solid Waste Disposal

AFBC wastes will be utilized to the extent practicable. Nevertheless, based on waste characteristics and projected regulatory requirements, a facility design for solid waste disposal could include a 0.31-meter (one-foot) thick clay liner with a minimum permeability of $1 \times 10^{-7}$ centimeters per second. This design will be adequate to attenuate any leachate produced at the site to the degree necessary to protect the ground water quality. Further, the base will be sloped to several sumps, and leachate collection wells will be installed. Leachate would be collected and treated as necessary.

Upon completion of the project or the useful life of the waste disposal area, the area would be closed and reclaimed in accordance with applicable State and Federal regulations. At a minimum this would include a 0.16-meter (6-inch) clay cap covered by 0.46 meter (18 inches) of topsoil. The site would then be graded to drain and seeded. Prior to closure the site would be tested to ensure that there is no unreacted calcium oxide sufficient to cause heat buildup or expansion of the spent sorbent disposal area.

Wastewater Treatment

The primary considerations of wastewater treatment alternatives are the characteristics of specific waste streams, effluent limitations and standards, demonstrated treatment technologies, and cost-effectiveness.[5]

179

The following are discussions of best available technologies proposed
for treating these individual and combined waste streams.

Construction Runoff and Concrete Mixing Plant Wastes.   The preferred
treatment of concrete mixing plant wastes will be to reuse the water in
processes where lower water quality is tolerated.   Nonrecycled concrete
mixing plant wastes and construction runoff would be treated by sedimentation to remove suspended solids.   Neutralization will be conducted for
pH adjustment if needed.   Holding ponds for construction runoff would be
designed to contain runoff from a 24-hour, 10-year rainfall event in
compliance with NPDES requirements.

Cooling Water Discharge.   Since the cycle of concentration of spray-
pond operation will be low and cooling water treatments such as
corrosion inhibitors and chlorine additions will be minimized, treatment
of spray-pond blowdown will most likely not be necessary to meet NPDES
permit limits.

If necessary, the residual chlorine discharge will be reduced or
eliminated by control options such as continuous mechanical condenser-
tube cleaning, the use of alternative biocides, or dechlorination by
adding sulfur dioxide or sodium sulfite to the blowdown.   Control of
corrosion inhibitors in blowdown would be made by lime precipitation;
however, the treatment need will be dependent upon the use of corrosion
inhibitors.

Metal Cleaning Wastes.   Metal cleaning wastes usually contain chelating
agents, and the selection of wastewater treatment processes is based on
the cleaning chemical used.   At present, methods have not been determined for preoperational and maintenance cleaning.   Chemical precipitation methods used by TVA power plants[9] would be employed to reduce
copper, iron, and phosphate concentrations to the effluent limitation
levels.

The treated metal cleaning wastes would be discharged to the existing
ash pond effluent ditch, and the final pH adjustment of the wastes would
be made while mixing with other waste streams.   The sludge would be
removed from the treatment ponds and transported to a State-approved
landfill if the sludge is determined to be nonhazardous.

Runoffs From Coal and Limestone Storage Piles.   Stormwater runoffs from
the coal and limestone piles and a small waste stream from the dust
suppression system would be collected and diverted to a lined storage
basin designed to treat a 24-hour, 10-year rainfall event, where the
coal and limestone fines settle out.   This enables neutralization of
acid coal pile drainage with alkaline limestone runoff and gains removal
of many heavy metals that are present in the runoffs.   Additional pH
adjustment will be provided if necessary to meet permit requirements and
to achieve maximum removal of heavy metals prior to discharge.

Runoffs From Fly Ash and Spent-Bed Material Storage/Disposal Areas.
Runoff waters from fly ash and spent-bed material storage/disposal areas
would be treated by collecting the storm runoffs in holding ponds
designed to treat a 24-hour, 10-year rainfall event, where sedimentation
to remove suspended solids will occur.   Adjustment of pH would be provided to meet permit requirements and reduce heavy metals in the
effluents.   The holding ponds would be provided with an impervious liner
or low permeability soil liner to prevent ground water contamination by
water percolation through the soil.

Miscellaneous Wastes.  Miscellaneous wastes, such as spent regenerants and those from laboratory and plant activities, would be diverted to sump collection ponds for suspended solids and oil and grease removal before discharge.  Sanitary wastes would be treated by biological means, and the final effluent would be chlorinated for disinfection prior to discharge.  Boiler blowdown normally does not exceed the effluent limitations; therefore, the blowdown can be discharged without treatment.

Conclusions

Process solid wastes (fly ash and spent sorbent) have been subjected to laboratory extraction tests, and the results indicate the AFBC wastes are not hazardous under the current RCRA regulations.  As the acceptable procedure for management of nonhazardous wastes, leachate control, runoff control, and monitoring will be incorporated in the design of the waste disposal site.  Also, TVA is studying the beneficial use of the waste material.

Many of the wastewater effluents from the AFBC plant are similar to those encountered in conventional steam-electric power plants.  Best available technologies will be employed for wastewater treatment, and the treated effluents are planned to be discharged via the existing power plant discharge channel.  Based on a "worst case" analysis, the total mixing zone in the Ohio River required to meet the Kentucky water quality standards for pollutants discharged from the plant will be limited to a very small area or volume.  Therefore, the wastewater discharged from the AFBC demonstration plant will not adversely affect the legitimate uses of the Ohio River water downstream from the plant.

References

1.  U.S. Environmental Protection Agency, "Hazardous Waste Management System--Identification and Listing of Hazardous Waste." Federal Register, 45, 33083 (1980).

2.  American Society for Testing and Materials, 1979 Annual Book of ASTM Standards--Part 31:  Water, 1258 (1979).

3.  U.S. Environmental Protection Agency, "Identification and Listing of Hazardous Waste--Advanced Notice of Proposed Rule Making." Federal Register, 43, 59022 (1978).

4.  Chu, T.-Y.J., et al., "Characteristics of Wastewater Discharges from Coal-Fired Power Plants." Proceedings of the 31st Annual Purdue Industrial Waste Conference, Ann Arbor Science, Ann Arbor, Michigan, 690 (1976).

5.  Chu, T.-Y.J., et al., "Wastewater Control Technology in Steam-Electric Power Plants." Progress in Water Technology. Pergamon Press, 10, 5/6, 801 (1978).

6.  State of Kentucky, "Kentucky Water Quality Standards." Department for Natural Resources and Environmental Protection (1979).

7.  U.S. Environmental Protection Agency, "National Interim Primary Drinking Water Regulations." Federal Register, 41, 28402 (1976).

8.  U.S. Environmental Protection Agency, "Effluent Limitations Guidelines, Pretreatment Standards and New Source Performance Standards Under Clean Water Act; Steam Electric Power Generating Point Source Category." Federal Register, 45, 68327 (1980).

9.  Steiner, G. R., et al., "Treatment of Metal Cleaning Wastes at TVA Power Plants." Presented at the 84th National Meeting of American Institute of Chemical Engineers, Atlanta, Georgia, February 26 to March 1, 1978. 44 pp.

Acknowledgements

Thanks are extended to Tracey L. Bell, Joe M. Castleberry, Kenneth D. Eisele, Lyman H. Howe III, and Patricia B. West for their assistance in the project.

Table 1.  Chemical Characteristics of AFBC Fly Ash and Spent Sorbent

| Constituent | Concentration (dry weight basis) | | |
|---|---|---|---|
| | Baghouse Fly Ash | Cyclone Fly Ash | Spent Sorbent |
| Aluminum, % | 2.07 | 1.26 | 1.05 |
| Calcium, % | 12.0 | 21.3 | 36.9 |
| Iron, % | 6.08 | 3.97 | 1.78 |
| Silicon, % | 8.09 | 5.22 | 4.70 |
| Sulfur, % | 5.65 | 3.47 | 7.91 |
| Antimony, µg/g | 5.0 | 1.77 | 0.80 |
| Arsenic, µg/g | 17 | 7 | 11 |
| Barium, µg/g | 243 | 352 | 125 |
| Beryllium, µg/g | 4.0 | 1.1 | 0.5 |
| Boron, µg/g | 313 | 151 | 115 |
| Cadmium, µg/g | 0.31 | 0.91 | 2.3 |
| Chromium, µg/g | 127 | 74 | 38 |
| Copper, µg/g | 66 | 30 | 27 |
| Lead, µg/g | 37 | 37 | 29 |
| Magnesium, µg/g | 7,160 | 3,350 | 5,825 |
| Manganese, µg/g | 246 | 142 | 163 |
| Mercury, µg/g | 0.70 | 0.06 | 0.28 |
| Nickel, µg/g | 72 | 35 | 20 |
| Selenium, µg/g | 2.8 | 2.6 | 0.8 |
| Silver, µg/g | 3.3 | 4.8 | 4.4 |
| Thallium, µg/g | <2.5 | <2.5 | <2.5 |
| Tin, µg/g | <100 | <100 | <100 |
| Titanium, µg/g | 2,140 | 1,088 | 1,050 |
| Zinc, µg/g | 222 | 141 | 185 |

Table 2. Chemical Characteristics of Leachates of Combined Fly Ash

| Constituent | Extraction EPA[a] | Extraction ASTM[b] | Extraction ASTM[c] | Column Percolation D.I. Water[d] | Column Percolation Buffer[e] | Maximum Extract Level[f] |
|---|---|---|---|---|---|---|
| Calcium, mg/l | 2,795 | 1,900 | 1,600 | 2,000 | 3,150 | - |
| Chloride, mg/l | 90 | 880 | 895 | 850 | 1,100 | - |
| Fluoride, mg/l | 2.4 | 3.8 | 43 | 1.6 | 2.0 | - |
| Magnesium, mg/l | <0.01 | 0.02 | 0.75 | 0.09 | 0.27 | - |
| Sulfate, mg/l | 1,200 | 1,250 | 12 | 1,467 | 1,200 | - |
| Total dissolved solids, mg/l | 11,940 | 7,010 | 5,100 | 5,123 | 8,750 | - |
| Aluminum, µg/l | <100 | <100 | <100 | <50 | 350 | - |
| Antimony, µg/l | <2 | <2 | <2 | <2 | <2 | - |
| Arsenic, µg/l | <2 | 4 | <2 | <2 | <2 | 5,000 |
| Barium, µg/l | 176 | 168 | 106 | 233 | 486 | 100,000 |
| Beryllium, µg/l | <5 | <5 | <5 | <5 | <5 | - |
| Boron, µg/l | <30 | 1,500 | 1,500 | 1,337 | 1,100 | - |
| Cadmium, µg/l | 1 | <1 | 8 | 8 | 8 | 1,000 |
| Chromium, µg/l | 19 | <5 | <5 | 7 | <5 | 5,000 |
| Copper, µg/l | 50 | 60 | 50 | 67 | 65 | - |
| Iron, µg/l | 8 | 8 | 819 | 16 | 68 | 5,000 |
| Lead, µg/l | 8 | <1 | <1 | 2 | 4 | - |
| Manganese, µg/l | <5 | <5 | 110 | <5 | <5 | 5,000 |
| Mercury, µg/l | <0.2 | <0.2 | 0.3 | 0.4 | 0.4 | - |
| Nickel, µg/l | 90 | <50 | <50 | <50 | <50 | 200 |
| Selenium, µg/l | 5 | 32 | 43 | 37 | 24 | 1,000 |
| Silver, µg/l | <10 | <10 | <10 | 48 | 93 | 5,000 |
| Thallium, µg/l | <50 | <50 | <50 | <50 | <50 | - |
| Tin, µg/l | <1 | <1 | <1 | <1 | <1 | - |
| Titanium, µg/l | <5 | <5 | <5 | <5 | <5 | - |
| Zinc, µg/l | 8 | 20 | 19 | 8 | 8 | - |
| pH (final), SU | 11.8 | 12.1 | 12.1 | 12.4 | 12.1 | - |

a. Average of duplicate samples (RCRA procedure).
b. Analysis of one sample (ASTM procedure using deionized water).
c. Average of duplicate samples (ASTM procedure using buffer solution).
d. Average of six samples.
e. Average of four samples (using pH 4.5 sodium acetate-acetic acid buffer solution as extractant).
f. Extract concentration designated by EPA above which original sample would be considered a hazardous waste.
g. The concentration in the blank exceeded the concentration in all samples.

Table 3. Chemical Characteristics of Leachates of Spent Sorbent

| Constituent | Extraction EPA[a] | Extraction ASTM[b] | Extraction ASTM[c] | Column Percolation D.I. Water[d] | Column Percolation Buffer[e] | Maximum Extract Level[f] |
|---|---|---|---|---|---|---|
| Calcium, mg/l | 2,800 | 950 | 1,650 | 1,535 | 2,800 | - |
| Chloride, mg/l | 8 | 22 | 29 | 31 | 91 | - |
| Fluoride, mg/l | 0.57 | <0.1 | 17 | <0.1 | 0.1 | - |
| Magnesium, mg/l | 0.37 | <0.01 | 0.09 | 8 | 0.07 | - |
| Sulfate, mg/l | 7,200 | 519 | 730 | 1,633 | 1,600 | - |
| Total dissolved solids, mg/l | 11,450 | 2,460 | 5,700 | 4,590 | 9,000 | - |
| Aluminum, µg/l | <100 | <100 | <100 | <100 | <100 | - |
| Antimony, µg/l | <2 | <2 | <2 | <2 | <2 | - |
| Arsenic, µg/l | 1.5 | <2 | 2 | 2 | <2 | 5,000 |
| Barium, µg/l | 201 | 143 | 181 | 168 | 318 | 100,000 |
| Beryllium, µg/l | <5 | <5 | <5 | <5 | <5 | - |
| Boron, µg/l | 210 | 110 | 133 | 612 | 2,225 | - |
| Cadmium, µg/l | 1 | <1 | <1 | <1 | 8 | 1,000 |
| Chromium, µg/l | 35 | 20 | 26 | 51 | 58 | 5,000 |
| Copper, µg/l | 60 | 30 | 45 | 51 | 108 | - |
| Iron, µg/l | 400 | <10 | 84 | <10 | 12 | 5,000 |
| Lead, µg/l | 1 | 1 | 9 | 6 | 4 | - |
| Manganese, µg/l | <5 | <5 | 9 | <5 | 1 | 5,000 |
| Mercury, µg/l | <0.2 | <0.2 | 0.9 | 0.2 | 0.5 | 200 |
| Nickel, µg/l | 135 | <50 | <50 | <50 | <50 | - |
| Selenium, µg/l | 1.3 | 3 | 3 | 2 | 6 | 1,000 |
| Silver, µg/l | 23 | <10 | <10 | <10 | 90 | 5,000 |
| Thallium, µg/l | <50 | <50 | <50 | <50 | <50 | - |
| Tin, µg/l | <1 | <1 | <1 | <1 | <1 | - |
| Titanium, µg/l | <5 | <5 | <5 | <5 | <5 | - |
| Zinc, µg/l | 8 | 4 | 8 | 8 | 8 | - |
| pH (final), SU | 11.8 | 12.2 | 12.1 | 12.3 | 12.4 | - |

a. Average of duplicate samples (RCRA procedure).
b. Analysis of one sample (ASTM procedure using deionized water).
c. Average of duplicate samples (ASTM procedure using buffer solution).
d. Average of six samples.
e. Average of four samples (using pH 4.5 sodium acetate-acetic acid buffer solution as extractant).
f. Extract concentration designated by EPA above which original sample would be considered a hazardous waste.
g. The concentration in the blank exceeded the concentration in all samples.

Table 4.  Dissolved Heavy Metals in Simulated Fly Ash and Spent
          Sorbent Runoffs

| | Concentrations, µg/l | | | | |
| | Fly Ash Runoff | | Spent Sorbent Runoff | | |
| Element | 0.5% Susp. Solids | 2% Susp. Solids | 0.5% Susp. Solids | 2% Susp. Solids | Process Blank |
|---|---|---|---|---|---|
| Antimony | <2 | <2 | <2 | <2 | <2 |
| Arsenic | 2 | 5 | <2 | <2 | <2 |
| Barium | 26 | 38 | 26 | 58 | <5 |
| Beryllium | <5 | <5 | <5 | <5 | <5 |
| Boron | 170 | 470 | <30 | <30 | <30 |
| Cadmium | <1 | <1 | <1 | <1 | <1 |
| Copper | 30 | 30 | 30 | 30 | <10 |
| Chromium | <5 | <5 | <5 | <5 | <5 |
| Iron | 53 | 26 | 13 | 11 | <10 |
| Lead | 2 | 2 | 7 | 2 | <1 |
| Manganese | <5 | <5 | <5 | <5 | <5 |
| Mercury | 0.3 | 0.4 | 0.3 | <0.2 | <0.2 |
| Nickel | <50 | <50 | <50 | <50 | <50 |
| Selenium | 5 | 14 | <1 | <1 | <1 |
| Silver | <10 | 20 | <10 | <10 | <10 |
| Thallium | <50 | <50 | <50 | <50 | <50 |
| Tin | <1 | <1 | <1 | <1 | <1 |
| Titanium | <5 | <5 | <5 | <5 | <5 |
| Zinc | 18 | 12 | 30 | 25 | <10 |


Table 5.  Organic Priority Pollutants Detected in Simulated Fly Ash
          Runoff

| | Concentration, µg/l | | | |
| Compound | 0.5% Susp. Solid | 2% Susp. Solid | 5% Susp. Solid | Process Blank |
|---|---|---|---|---|
| Trichlorofluoromethane | 1.6 | 3.4 | 1.4 | 0.97 |
| 1,1,1-Trichloroethane | 1.46 | 1.8 | 0.54 | 0.06 |
| Benzene | 0.29 | 0.82 | 0.13 | 0.01 |
| Methylbenzene | 0.19 | 1.2 | 0.05 | <0.01 |
| Phenol | 3.1 | 4.7 | 7.5 | 0.08 |
| 4-Chloro-3-methylphenol | 0.04 | <0.02 | 0.10 | <0.02 |
| Naphthalene | 0.20 | 0.11 | 0.14 | 0.05 |


Table 6.  Organic Priority Pollutants Detected in Simulated AFBC Spent
          Sorbent Runoff

| | Concentration, µg/l | | | |
| Compound | 0.5% Susp. Solid | 5% Susp. Solid | 10% Susp. Solid | Process Blank |
|---|---|---|---|---|
| Trichlorofluoromethane | 1.3 | 1.1 | 1.9 | 0.97 |
| 1,1,1-Trichloroethane | 0.11 | 0.13 | 0.46 | 0.06 |
| Benzene | 0.05 | 0.04 | 0.35 | 0.01 |
| Methylbenzene | 0.19 | 0.64 | 0.91 | <0.01 |
| Phenol | 0.62 | 0.37 | 0.13 | 0.08 |
| 2-Nitrophenol | 0.11 | 0.38 | 0.53 | <0.1 |
| Naphthalene | 0.40 | 0.69 | 1.8 | 0.05 |
| Acenaphthene | 0.05 | 0.09 | 0.11 | 0.02 |

Table 7.  Water Quality Criteria for Protection of Aquatic Life

| Constituent | State of Kentucky Numerical Criteria[a] for Aquatic Life | Maximum Concentration, µg/l (except where indicated) | | |
|---|---|---|---|---|
| | | 96-hr $LC_{50}$[b] | 0.44 times 96-hr | 0.01 times 96-hr $LC_{50}$ |
| Arsenic | 50 | 15,022 | 6,610 | - |
| Beryllium | 11 | 150 | 66 | - |
| Cadmium | 4 | 630 | 277 | - |
| Chromium (total) | 100 | 5,070 | 2,231 | - |
| Copper | 75 | - | 33 | c |
| Iron | 1,000 | - | - | - |
| Lead | - | 2,400 | 1,056 | 24 |
| Mercury (inorganic) | 0.05 | 35 | 15 | - |
| Nickel | - | 2,916 | 1,283 | 29 |
| Selenium | - | 2,060 | 906 | 21 |
| Silver | - | 64 | 28 | 0.64 |
| Zinc | - | 780 | 343 | 7.8 |
| Total residual chlorine | 10 | 190 | 80 | - |
| Cyanide | 5 | 74 | 33 | - |
| Phenols | 5 | 11,500 | 5,060 | - |
| 4-chloro-3-methylphenol | - | 30 | 13 | 0.3 |
| Naphthalene | - | 150,000 | 66,000 | 1,500 |
| Ammonia nitrogen (unionized) | 0.05 | - | - | - |
| Dissolved oxygen (mg/l) | 4 (min) | - | - | - |
| pH (SU) | 6 to 9 | - | - | - |

a.  Kentucky Water Quality Standards.
b.  Obtained from EPA Ambient Water Quality Criteria Development Documents, published in 1978 and the EPA Quality Criteria for Water, published in 1976.
c.  The criteria for nonaccumulative toxic substances such as copper is 0.1 of the 96-hr $LC_{50}$; therefore, the applicable standard for copper is 8 µg/l.

Table 8. Projected "Worst Case" Concentrations of Selected Constituents in AFBC and Shawnee Steam Plant Wastewaters

| Constituent | Maximum Concentration, µg/l (except where indicated) | | | |
| --- | --- | --- | --- | --- |
| | AFBC Waste Stream | Shawnee Steam Plant Ash Pond Effluent | Combined AFBC and Shawnee Ash Pond Discharge | Mouth of Condenser Cooling Water Channel (Ohio River Mile 947) |
| Arsenic | 70 | 77 | 77 | 34 |
| Beryllium | <10 | <10 | <10 | <10 |
| Cadmium | 14 | <1 | 3.9 | 4 |
| Chromium | 1,000 | 19 | 237 | 71 |
| Copper | 176 | 30 | 62 | 120 |
| Iron | 1,000 | 1,100 | 1,093 | 5,000 |
| Lead | 674 | <10 | 158 | 53 |
| Mercury | 0.25 | 0.8 | 0.76 | 0.30 |
| Nickel | 105 | 220 | 212 | 87 |
| Selenium | 2.9 | 14 | 13 | 9.7 |
| Silver | <10 | <10 | <10 | <10 |
| Zinc | 1,200 | 60 | 313 | 179 |
| Total residual chlorine | 98 | - | 30 | 830 |
| Cyanide | <10 | <10 | <10 | <10 |
| Phenols | 5 | - | 2 | 24 |
| 4-chloro-3-methylphenol | <0.02 | <0.02 | <0.02 | <0.02 |
| Naphthalene | <0.02 | <0.02 | <0.02 | <0.02 |
| Ammonia nitrogen (unionized) | 0.15 | 0.25 | 0.24 | 0.08 |
| Dissolved oxygen (mg/l) | 6.5 (min) | 6.0 (min) | 6.0 (min) | 5.8 (min) |
| pH (SU) | 8.9 | 10.5 | 10.0 | 8.6 |

Table 9. Projected "Worst Case" Concentrations of Constituents Exceed the Water Quality Criteria

| Constituent | Maximum Concentration, µg/l | | | | |
| --- | --- | --- | --- | --- | --- |
| | Mouth of Condenser Cooling Water Channel (Ohio River Mile 947) | State of Kentucky Numerical Criteria for Aquatic Life | 0.44 times 96-hr $LC_{50}$ | 0.1 times 96-hr $LC_{50}$ | 0.01 times 96-hr $LC_{50}$ |
| Ammonia nitrogen (unionized) | 0.08 | 0.05 | - | - | - |
| Copper | 120 | - | 33 | 8 | - |
| Iron | 5,000 | 1,000 | - | - | - |
| Lead | 53 | - | 1,056 | - | 24 |
| Mercury (inorganic) | 0.3 | 0.05 | 15 | - | - |
| Nickel | 87 | - | 1,283 | - | 29 |
| Phenols | 24 | 5 | 5,060 | - | - |
| Total residual chlorine | 830 | 10 | 80 | - | - |
| Zinc | 179 | - | 343 | - | 7.8 |

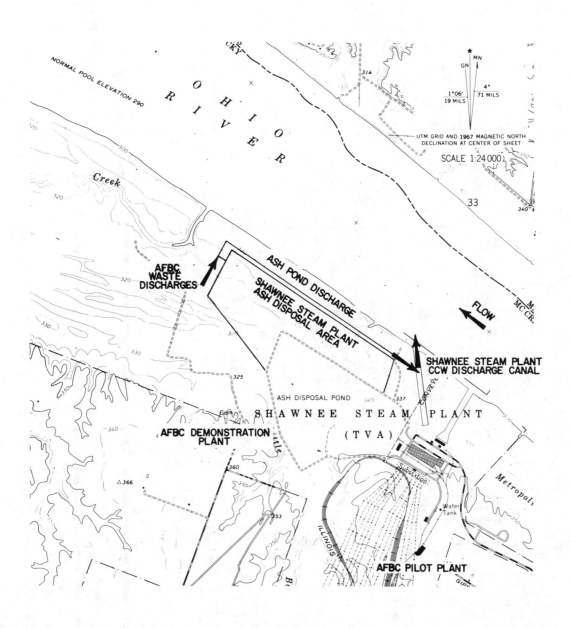

Figure 1.   Proposed AFBC Demonstration Plant Site

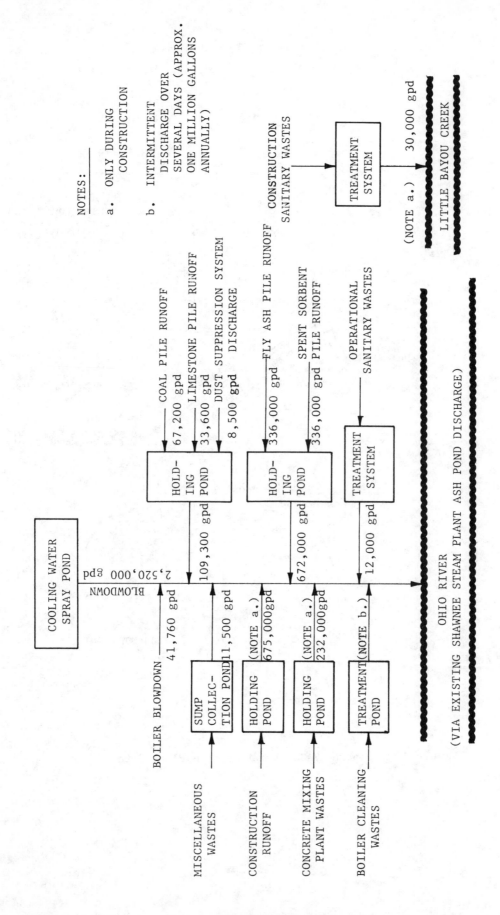

Figure 2. Proposed Schematic of Wastewater Discharge for the 200-MW AFBC Demonstration Plant.

188

# ENVIRONMENTAL CONSIDERATIONS
## OF COGENERATION-BASED DISTRICT HEATING

**Elliot P. Levine** and **Danilo J. Santini**
Argonne National Laboratory

Introduction

The U.S. Department of Energy and the Department of Housing and Urban
Development are encouraging the commercialization of district heating for
meeting heating demands in numerous urban areas.  Among the many district
heating technologies, this paper emphasizes cogeneration-based district
heating, which is a demonstrated mechanism for reducing oil and gas use
in the heating market.  For this discussion, we consider cogeneration-
district heating to be the use of hot water from electric power plants
which simultaneously provide electric and thermal energy for general
area-wide urban use in high density residential, commercial and certain
industrial applications.  The potential benefits of this technology
include:

  · Fuel savings, particularly that of oil and gas
  · Significant near-term employment opportunities within
    center cities for high- and low-skilled labor categories
    in the community, and
  · Reduced environmental pollution through the use of
    central plants with pollution control capabilities.

It is the latter two categories which this paper emphasizes.

Since the co-themes of this conference focus on both the economic and
environmental considerations in energy utilization, the discussion begins
with an introduction of the salient economic characteristics which distinguish district heating from several alternatives.  We then discuss
environmental considerations.  The most notable of these are the ambient
air quality and thermal discharge improvements and the enhanced employment effects.

Economic Considerations

Compared to the U.S., Europe currently makes more widespread use of district heating as shown in Fig. 1. Further, in the recently published World Coal Study,[1] all western European countries except Italy made explicit statements about expanding district heat (particularly coal based) during the next two decades (see Fig. 2). The U.S. component of this study, however, makes no statement about expanded use of district heating.[1] In order to understand the differences in the historical and future preferences of Europe and the U.S. toward district heating, a review of the conditions determining the economic attractiveness of district heating was conducted. Seven conditions have been identified as important determinants of the degree of use of district heating. They are: (1) many densely developed urban areas, (2) cold winters, (3) the technical ability to reduce oil and gas use by means of combined heat and power district heating systems, (4) lack of adequate domestic oil and gas supplies, (5) high consumer oil and gas prices, (6) unacceptable dependence on imported oil and, (7) winter peaking utilities. The first three conditions have long applied to much of the U.S. as well as Europe. They tend to determine spatial economics - i.e. the relative attractiveness of a location for cogeneration district heating. The first two conditions affect the heating demand density, a factor heavily influencing the substantial distribution costs involved in district heating. The third factor is related to the ability to use electric power plants for cogeneration. Here economics dictates that the power plant be within several miles of the demand center in order to hold down transmission costs.[2] The locations most commonly served by district heating are the downtown areas of major cities in northern climates. This pattern of district heating market location is found in both the U.S. and in Europe. The next three conditions apply to the ability of oil and gas to compete with district heating for the heating market. In Europe these factors have been a long-term historical fact-of-life, but have only become noteworthy in the U.S. in the last decade. The last condition applies to economic competition from electric heat. It has been absent from the U.S. for several decades but the analysis of recent electric heating trends indicates that many northern U.S. utilities will become winter peaking in the next two decades as indicated in Fig. 3.

The relative lack of district heating in the U.S. is explainable by the historical absence of the last four of the seven conditions. However, more recently, it is only the absence of the last condition which economically justifies the lack of U.S. plans for expansion of district heat. Our economic comparison of existing and new electric heat, district heating, and some synfuel alternatives in Fig. 4 indicates that U.S. electric utilities will be better able to serve the northern urban heating market in the future by using cogeneration based district heating rather than new electric capacity for heating only.[3] The economic analysis indicates that they will thus be able to provide heating services at a cost competitive with oil or gas derived from synfuel technologies. Further, the technical details of the system can allow a 35 to 55 percent reduction in fuel energy use and up to a total reduction of space heating oil and gas use if coal is used by the district heating power stations.

Environmental Effects of District Heating

The environmental benefits associated with cogeneration are largely due to the increase in system efficiency which results in the burning of less fuel and the consolidation of emissions to a few, large power gen-

erating sources.  Other environmental changes could occur because power
plants are generally more flexible in their ability to accomodate other
fuels.

The actual environmental impacts are site specific in nature and depend
upon additional factors beyond the facility's emissions e.g. ambient
background conditions, plant characteristics, meteorology, etc.  Much of
the documented information regarding the environmental effects of deploy-
ing a district heating system pertains to air and thermal pollution and
socioeconomic effects.  The remaining areas which are qualitatively
addressed - land use, water quality and solid waste are examined to
provide a perspective of the environmental considerations and an indication of the factors influencing these environmental changes.  The discussion is limited to the first order effects generally expected to
occur within the area served by the district heating system.

## Air Pollution

Air pollution is often the principal environmental impact resulting from
fossil fuel combustion, and for this reason it is important to examine
the air pollutant emissions and ambient air quality changes which could
occur by using cogenerating central-station power plants for meeting
electrical space heating and hot water demands in urban areas.  A district heating system may be deployed utilizing new generating units or by
retrofitting existing generating units to operate in a cogenerating mode.
The allowable level of emissions is sensitive to this difference.  In
general, the principal air pollution emission changes from district
heating reflect three factors.  These are:  (1) the additional fuel consumed at the power plants in place of the fuel discontinued at the space
heating boilers, (2) the emission characteristics of the substituted
fuels, and (3) the level of emission control at the power plants.

An analysis of the Boston city area examined the emission changes associated with the conceptualized district heating system.[5]  In the case
where existing power plant fuels (principally residual oil) are used to
generate the added electric and thermal requirements, emission changes
at the power plants increased by about 21% in proportion to the additional fuel used at the power plant.  Emission changes from heating
method shifts in the city area showed no significant change in emissions
of TSP (+2.2%), $SO_2$ (+5.9%), and $NO_x$ (-2.9%).  These results are illustrated in Fig. 5 for the Boston area.  In the Twin Cities of Minneapolis
and St. Paul, MN,[6] similar but reduced emission trends were estimated as
shown in Fig. 5.  These estimates reflect additional non-heating sources
within the city area.

The basis for the slight increase in emissions is attributed to the fact
that the fuels used at the power plants generate more emissions for each
Btu of fuel burned than would an equivalent amount of fuel burned in the
space heating boilers.  In this case study, this factor more than offsets
the emission benefits due to greater system efficiency which permits less
total fuel to be burned.  In the Boston area, power plant fuel is primarily residual oil with a 1% sulfur content; whereas the heating fuel
displaced in the residential, commercial and industrial sectors is a mixture of gas, distillate oil averaging 0.3% sulfur content, and residual
oil with a 0.5% sulfur content.  The emission coefficients (lbs of pollutant for each Btu of fuel burned) associated with these fuels are presented in Table 1, to demonstrate the wide variation in emissions among
different fossil fuels, different quality fuel oil, and different levels
of pollution control for coal.  It is apparent from this table that for
equivalent amounts of fuel, switching fuels will cause dramatic changes

in power plant emissions of TSP, $SO_2$, and $NO_x$, and only minor changes in CO and HC. The latter two pollutants are more affected by combustion factors than by any elements bound within the fuel, as are the other pollutants mentioned. The effect of power plant coal use is also of note. Controlled emissions of $SO_2$ and TSP from coal-fired power plants will vary dramatically with the level of pollution control, and potentially could match that of high quality distillate fuel oil. This last factor raises an additional point in favor of district heating. The relative economic advantage of controlling pollution on large sources affords large power generating sources the flexibility of accomodating various fuels including coal. All of these factors are, therefore, important when gas or distillate oil boilers are replaced by district heating.

The above analysis of emission discharge patterns reflects the geographical redistribution of heating related emissions from numerous area wide sources to a few large point sources. Of greater importance is how these emission changes translate into ambient air quality changes; which are a more accurate determinant of health and other air pollution related effects. In the Boston and Twin Cities studies and other studies,[7] the annual air quality levels were shown to be greatly improved by conversion to district heating, even though net emissions increased slightly.

The improvement in ambient $SO_2$ levels for the two study areas is illustrated in Figs. 6 and 7. In both studies the predicted average annual air quality showed large improvements in $SO_2$ levels. In the Boston case study, these annual average $SO_2$ improvements ranged from 20% to more than 80% of the original annual $SO_2$ levels attributed to fuel combustion for meeting the electric and thermal demand. In the Twin Cities area, $SO_2$ levels showed similar improvements, including the reduction of annual peak $SO_2$ levels from 59 $\mu g/m^3$ in Minneapolis and 52 $\mu g/m^3$ in St. Paul to 47 $\mu g/m^3$ and 41 $\mu g/m^3$, respectively. Also evident from these figures is that the level of improvement accrues in proportion to the amount of local space heating being displaced. The largest improvements occur in and around the downtown central business districts and the most densely populated areas of the city. Conversely, the range of district heating induced improvements trails off, as the district heating network is applied further from the downtown area into areas of lower heating demand. This dramatic improvement in ground-level air quality is the result of shifting fuel combustion for heating purposes from numerous, spatially spread demand sites to a small number of power plants which discharge pollutants at higher temperatures, with higher exhaust velocities and taller stacks. Emissions discharged from these sources, make a far smaller contribution to local air quality levels.

Despite this relatively large air quality improvement, it is important to recognize that actual air quality levels are influenced by both plant and non-plant related factors such as other sources and meteorology. As such, a switch to cogeneration systems should be preceeded by a careful review of the air quality impact.

## Water Quality

Another benefit of district heating is the potential for reduction of thermal discharges to lakes, rivers, and the atmosphere. The rejection of waste heat through once-thru-cooling systems (from retrofitted power plants) or from cooling towers will on the average decrease, and significantly decrease during the winter months when heating requirements are greatest.

At the power plant, wastewater loadings will be generated at a rate proportional to the increase in load. However, the use of water as a medium for providing district heat will require the addition of suitable feedwater to the boiler system. To maintain the integrity of the system, the feedwater must either be treated or conditioned to prevent the formation of scales and corrosion within the piping network. Softened or desalted water and chemical control of pH levels may be necessary depending upon the temperatures, pressures and size of the system.[8] The size of the water treatment system should consider the eventual size of the district heating system.

## Land Use

A district heating system requires an extensive underground piping network for supplying hot water to urban area customers. The installation of a distribution/transmission network requires a high degree of planning and coordination to minimize the impacts of the numerous activities involved. Construction of the entire system could be phased and take several years to fully complete. Much of the construction would occur in the downtown, high density commercial and residential zones. The major land use impacts will be the interference with existing land use activities, disruption of other utility services and disruption of traffic.

The potential for major problems exist due to interference with other utilities already in place underground. Most existing utilities are likely to be located between 2 and 10 feet underground, the same depth as appears to be most desirable for district heating pipes. Underground utility systems may include: electric cable, gas, water, sewer, subways, and other dedicated lines, e.g. western union cables. To successfully implant a hot water line is likely to require extensive coordination and cooperation, hand labor when excavating around other utility systems, and routing of piping around or under existing utility networks and major traffic arteries. Relocation of segments of existing utilities will also be periodically necessary due to physical or locational incompatibilities. Since disruption of other utility services such as a phone line to a commercial building is unlikely to be acceptable, a replacement segment must be in place before the old system may be removed. The actual impacts of a particular system are expected to be site-specific and may be mitigated by proper planning for construction and long-range operation.

Disruption of traffic is another major land use impact. Permit conditions imposed upon the contractor performing the installation would typically require that he be licensed for performing that type of work and limit the working hours so as to minimize the disruption to traffic or to commercial and residential establishments. Often, street work in an urban commercial area would be prohibited during rush hours, daylight, weekdays, or during holidays, such as the period between Thanksgiving and New Year. Only emergency repairs might be permitted.

The construction related noise impacts are another category of land use impact because of the potential for interference with existing land use activities. These are likely to occur in most areas of the city for varying periods of time. The time of day at which the construction occurs will determine the nature of the noise impact. For example, construction noise in a commercial area will cause different types of disruptions at different times of the day from construction noise in a high-density residential area. Noise impacts may be incompatible with certain activities for the duration of construction at a given location, particularly in areas near downtown hotels and residences. Limited mitigation potential may be available from the use of appropriate equipment and the erection of enclosures around stationary equipment, such as generators.

## Solid Waste

The generation of solid wastes will occur from both the combustion of fuel and the construction of the transmission/distribution network. At the power plant, ash and flue gas desulfurization residuals will increase in proportion to the increase in load. The actual increase will depend upon the specific fuels and fuel characteristics. In general, combustion of coal or residual oil generates considerably greater quantities of ash and sulfur waste than other fossil fuels commonly used for heating. Since the dominant portion of the displaced space heating boilers operate on natural gas or distillate oil[9] which are low-waste fuels, the overall production of waste from this activity is likely to increase.

If the district heating system is integrated with a municipal incineration system with heat recovery, then a reduction of land would be needed for landfill. Scarcity of land available for landfill operations has been a primary impetus behind the application of heat recoverable incineration in several European cities.

The greatest volume of solid wastes will be generated during the construction of the transmission and distribution network. This will necessitate the removal of the street surface followed by excavation of several feet of asphalt, concrete and earth at most locations. Clean backfill or sand may be required in some municipalities for backfilling the open trench. Clean fill is the preferred approach, since it reduces the potential of pipe damage by large rocks or rubble.[10] The use of clean backfill is not as necessary in rural areas where much of the piping will be laid in open countryside. Excavation wastes are of inert material and extraordinary disposal requirements are unnecessary.

## Employment

Our economic research indicates that when cogeneration district heating is compared to other alternatives for providing heating services, cogeneration results in greater social benefits by way of additional energy system construction employment opportunities. Several "intensity" factors influence the employment comparison and include: (1) labor intensity (jobs/dollar), (2) manual labor intensity (manual jobs/dollar), (3) construction intensity (construction dollars/Btu), and (4) urban intensity (share of construction dollars spent within the urban area).

For each of the four factors we have found that the district heating system has a greater intensity than for new systems providing oil and/or gas for heating. New electric heat, though having a higher construction intensity, is not economically competitive. In Figs. 5 and 6, we show the results of our earlier research comparing coal based synfuels to district heating.[2] As a result of factors 1 and 3, we find that district heating provides more total jobs per Btu of service. Adding factor 2 leads to the finding that more manual labor jobs per Btu are created. When factor 4 is also included into our estimates, we conclude that metropolitan and center-city construction employment will increase by a far greater amount if district heating is utilized. Fig. 5 shows that district heating potentially creates twice the number of construction jobs and creates a far greater number of center-city construction job opportunities. Fig. 6 shows essentially the same relationships, but specifically for lower skilled jobs. Here the positive job effects for the lower skilled and center-city residents are even more pronounced. Detailed comparisons with other heating technologies have not yet been performed. However, a quick look at oil shale and several other oil and gas production methods indicates that similar findings will occur when other comparisons are made.

References

1. Greene, R.P. and T.M. Gallagher, "Future Coal Prospects: Country and Regional Assessments - World Coal Study," Ballinger Publishing Company, Cambridge, MA (1980).

2. Santini, D.J. and E.P. Levine, "Indirect Benefits of Cogeneration District Heating," Preprint 80-672 of the Oct. 27-31, 1980 Convention and Exposition of the American Society of Civil Engineers, held at Hollywood, FL.

3. Davis, A.A. et al., "Analysis of Costs and Scarce Fuel Savings Associated with Nine Eastern and North Central Center City Conversions to a District Energy System," Informal Report ANL/CNSV-TM-12, Argonne National Laboratory, Argonne, IL (June 1979).

4. U.S. Bureau of the Census, "Statistical Abstract of the United States: 1979," (100th Edition) Washington, D.C. (1979).

5. Bernow, S.S. and E.P. Levine, "Annual Emissions and Air Quality Impacts of an Urban Area District Heating System -- Boston Case Study," ANL/CNSV-TM-36, Prepared by Energy Systems Research Group, Inc. and Argonne National Laboratory (Feb 1980).

6. Karnitz, M.A., F.C. Kornegay, H.A. McLain, B.D. Murphy, R.J. Raridon, and E.C. Schlatter, "Impact of a District Heating; Cogeneration System on Annual Average $SO_2$ Air Quality in the Twin Cities," ORNL/TM-6830/P11, Prepared by Oak Ridge National Laboratory, TN (August 1980).

7. Margen, P., et al., "District Heating/Cogeneration Application Studies for the Minneapolis-St. Paul Area," Informal Report ORNL/TM-6830/P3, Oak Ridge National Laboratory, Oak Ridge, TN (Oct 1979).

8. Mikkelsen, W., "Planning of Distribution Network to New District Heating Plants," Prepared by Bruun & Sorensen A/S, Aarhus, Denmark, Danish Board of District Heating (Undated).

9. Devitt, T. et al., "Population and Characteristics of Industrial/ Commercial Boilers in the U.S.," EPA-600/7-79-178a (August 1979).

10. Oliker, I. and J. Philipp, "Technical and Economic Aspects of District Heating Systems Supplied from Cogeneration Power Plants," Proceedings of the American Power Conference V40 (1978).

Acknowledgments

This work was sponsored by the Office of Building and Community Systems, Secretary for Conservation and Solar Energy, of the U.S. Department of Energy. The authors would like to thank Kathy Fischer for typing and other assistance with the manuscript.

Table 1 Emissions Coefficients for Space Heating Boilers[a]

| | Particulates | Sulfur Oxides | Nitrogen Oxides | Carbon Monoxide | Hydro-Carbons |
|---|---|---|---|---|---|
| **Space Heating Boilers[b]** | | | | | |
| **Residential[b]** | | | | | |
| Gas | 4.92 | 0.3 | 39.4 | 9.84 | 3.94 |
| Oil | 9.16 | 158.2 | 65.9 | 18.32 | 3.66 |
| **Commercial/ Institutional** | | | | | |
| Gas | 4.92 | 0.3 | 59.0 | 9.84 | 3.94 |
| Oil | 18.75 | 222.9 | 151.7 | 17.64 | 3.53 |
| **Industrial** | | | | | |
| Gas | 4.92 | 0.3 | 86.1 | 8.36 | 1.47 |
| Oil | 18.28 | 220.5 | 148.8 | 17.67 | 3.53 |
| **District Heating Utility Boilers** | | | | | |
| Gas | 5.0 | 0.3 | 350.0 | 8.50 | 0.50 |
| Oil 0.3%S | 7.5 | 158.0 | 80.5 | 18.50 | 3.50 |
| Oil 0.5%S | 27.5 | 272.5 | 205.5 | 17.00 | 3.50 |
| Oil 1.0%S | 44.5 | 544.5 | 205.5 | 17.00 | 3.50 |
| Oil 2.2%S | 85.5 | 1198.0 | 205.5 | 17.00 | 3.50 |
| Coal (NSPS) | 50.0 | 600.0 | 350.0 | 20.80 | 6.25 |
| Coal Max Control | 16.7 | 79.0 | 350.0 | 20.80 | 6.25 |

[a]Coefficients were derived from Tables 1.1-2, 1.3-1, 1.4-1 in AP-42 (U.S. EPA).

[b]Fuel characteristics for the Boston city area

Residential oil: 100% distillate;

Industrial oil: 45% distillate 55% residual:

Commercial oil: 43% distillate 57% residual.

Distillate oil has 0.3% sulfur.
Residual oil has 0.5% sulfur.

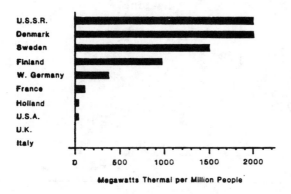

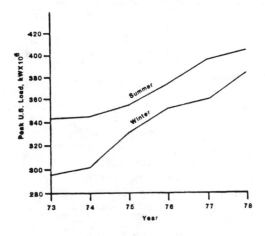

Fig. 1   District Heating Capacity
          Per Capita in Various
          Countries (Ref. 1)

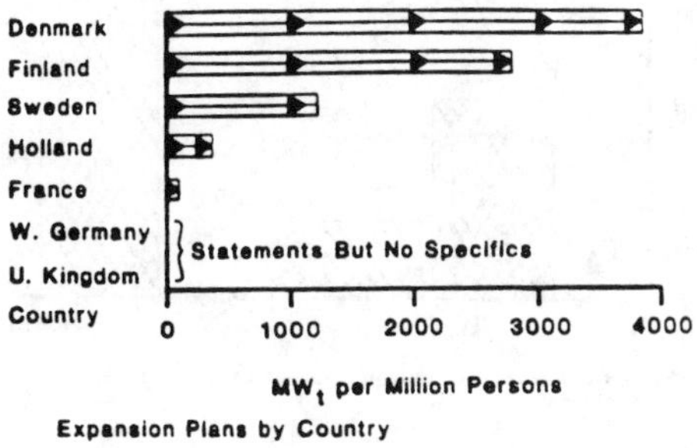

Fig. 2   European District Heating,
          Expansion Plan, by Country
          (Ref. 1)

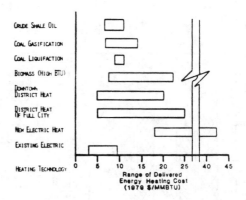

Fig. 3   Recent National Trends in
          Peak U.S. Electric Utility
          Load (Ref. 4)

Fig. 4   Rough Cost Comparison
          of Selected Heating
          Energy Sources

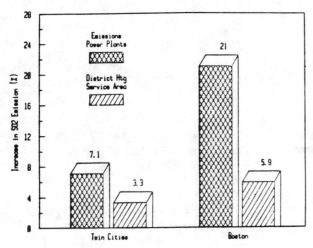

Fig. 5   Annual Average
         Increase in
         SO2 Emissions
         due to District
         Heating

ASSUMES THAT A NATURAL GAS CURTAILMENT PLAN HAS SUBSTITUTED OIL FOR MANY EXISTING
USES OF GAS HEAT

Fig. 6   Ambient SO$_2$
         Levels in
         Boston from
         the Heating
         Sector

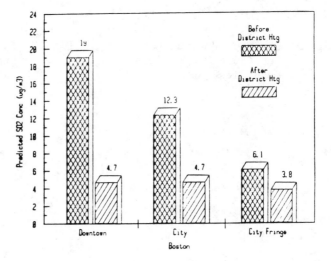

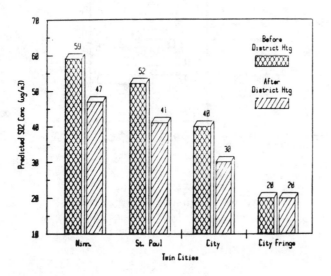

Fig. 7   Ambient SO2
         Levels in the
         Twin Cities
         from Heating
         and Other Large
         Point Sources

198

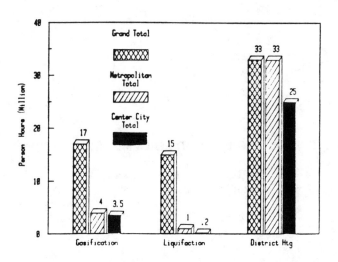

Fig. 8   A Comparison of Total Construction
         Job Opportunities with Equal Heating
         Service to Detroit

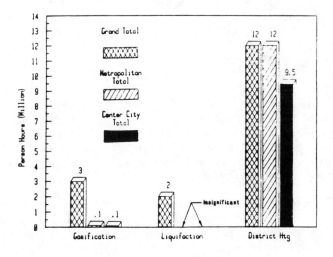

Fig. 9   A Comparison of Lower Skill Job
         Construction Opportunities with
         Equal Service to Detroit

# A STANDARDIZED COSTING METHOD FOR EMERGING TECHNOLOGIES

**G. R. Miller** and **M. S. Massoudi**
Teknekron Research, Inc.

Introduction

Accurate assessments of economic feasibility are vital to the commercialization of emerging energy technologies.  Cost estimations for these technologies must therefore be treated with consistent engineering and economic judgment.  The problems of estimation, however, are numerous: these
technologies are at various stages of development; each has its own uncertainties; no full-scale plant exists from which costs can be derived;
and, importantly, the cost estimates for each technology often come from
a proponent and lobbyist for that technology.

We have developed a systematic approach for developing capital cost estimates on the basis of a refined and detailed division of each technology
into process-related cost subaccounts.  Because of common elements and
consistent assumptions, this approach affords a more accurate ranking of
the emerging technologies and a better comparison of these technologies
with each other and with conventional processes.  We have applied this
approach to several technologies.

In this paper we present the results of one application in which we examined the wind, fuel cell, solar thermal, solar photovoltaic, and wet-
steam geothermal technologies.  The paper also provides a description of
the methodology, which is similar in structure to the Concept-5[1] approach developed by Oak Ridge National Laboratory for conventional technologies (coal-fired and nuclear steam electric generating units).  Our
approach was derived initially as an extension of Concept-5 for application to nonconventional technologies.  In order to construct a logical
breakdown of each technology into appropriate subprocesses, we developed
a structure in which the status of each was assessed.  In the case of each
technology, the emphasis was placed on pinpointing the elements of cost
uncertainties.

## Methodology

Capital cost estimates are usually derived from the physical principles and engineering designs upon which each technology is based. These principles and designs are embodied in the various processes constituting the components of the technology. We have constructed a detailed breakdown of each emerging technology into several process-related cost accounts.

Our objective has been to ensure that all components that contribute to the total capital cost are accounted for and that the cost estimates for the various technologies are made on a consistent basis. In addition to consistent technical and economic assumptions, common engineering components can be identified and examined.

Standardized cost account structures have previously been developed for conventional energy technologies, such as coal- and nuclear-based systems. The efforts have included the Federal Power Commission's Unified Code of Accounts and the similar accounts developed by the Atomic Energy Commission and the Concept-5 cost accounting system.[1,2] Because the Concept-5 system provides a method for manipulating capital costs as a function of various factors and also provides a structure for categorization, our methodology was based on a modification of Concept-5. We have named our cost acquisition and computational system TRICOST. Our modifications to Concept-5 include the addition of new accounts and the elimination of accounts unique to the conventional technologies.

TRICOST is based on the premise that, for any given technology, any central-station power plant involves approximately the same major cost components regardless of the plant's location or date of initial operation. We have specified detailed cost models for each plant type at a reference condition through use of size-, time-, and location-dependent cost algorithms. The reference cost model is modified to produce a specific capital cost estimate for conditions different from those of the reference case.

The TRICOST computer program also requires input of the following data: (1) construction schedule; (2) expenditure for each account during construction; (3) detailed labor-category and material-usage estimates for each account. In addition, historical data based on Engineering News Record information for these labor and material categories are contained as files in the computer package. These parameters allow the computer program to calculate capital costs at times and locations other than the reference ones and also to calculate accurate AFDC estimates.

Detailed estimates of construction schedules, time-dependent account expenditures, and account-specific labor and material allocations are not usually supplied with the conceptual designs that are made for an emerging technology. However, we have been able to develop these data by analyzing the commonalities that exist between processes, particularly with respect to construction. The treatment of emerging technologies also requires an understanding of the time frame in which price maturity will be achieved. Again, such understanding is not available in the standard engineering report, nor can it be developed from the principles of cost engineering. We have developed estimates by classifying technology components by their respective developmental status and examining the problems hindering the movement of any given component from one stage of development to the next.

## Model Application

In this section we describe the successive steps taken in applying our standardized costing method to each technology. These steps include: selection of technologies for cost model applications; assessment of developmental status; selection of reference systems for each technology; development of cost structure; cost data acquisition and screening; and model application.

### Selection of Technologies

The original scope of work included thirteen nonconventional technologies. In this paper we restrict the discussion to five: wind, solar thermal, photovoltaics, fuel cells and geothermal. These five were chosen so that a wide spectrum of developmental status would be represented. There is no other particular significance to the selection of any of these technologies, since the costing approach is general in nature and hence applicable to any central-station generating technology.

### Assessment of Developmental Status

First, a five-phase framework was developed for assessing technological status. The five phases of status development relating to commercialization, as shown in Figure 1, include: concept exploration; screening; product/system development; market or application testing; and commercial introduction.

The first two phases--concept exploration and screening--together constitute a broader category, "technical development." The culmination of technical development--technical readiness--requires meeting such achievement milestones as stable reproducible performance, consistency, mass production feasibility, and the accomplishment of a successful subscale demonstration unit. The screening phase involves engineering prototypes and tests, as well as design improvements. The next two phases--product/ system development and market testing--together constitute the process of the technology's "industrialization" and comprise the activities needed to bring the technology into a state of readiness prior to commercial introduction. The final phase--full-scale commercial introduction--involves determining the economics of scale, ensuring the existence of a delivery and distribution system, and maintaining a stable price.

We have summarized the current status of each technology according to this five-phase structure. The status of wind technology, given in Table 1, serves as an example.

### Selection of Reference Systems

Within any specific emerging technology there are variations in component design, component arrangements, and plant configuration. Only a limited number of these components will reach maturity (full-scale commercial stage). An important step in the economic analysis is to identify the reference system or systems that will represent the ultimate configuration for any given technology. These reference systems serve as the basis of our cost estimations. One or two reference systems were selected for each technology by considering the following elements: state of development and availability of each component, efficiency, economics, grid compatibility, and utility confidence.

Table 2 lists the reference systems selected in this study.

## Development of Cost Structure

The reference systems were examined in terms of their engineering processes and components. Then, through the use of flow charts, they were divided into subsystems on the basis of process boundaries. The subsystems constituted the first level of breakdown of the direct cost accounts. This method differs from the FPC system of accounts in that the FPC accounts are basically equipment related.

Each subsystem was then further divided into its processes and the components comprising them, a procedure that created subaccounts at the second and third level of detail.

Also considered in the creation of cost accounts were process elements common to the emerging technologies and conventional ones. Whenever common accounts--say, for land, sites, structures, or turbine--could be created, this was done, so that the accounts could be compared across technologies. The indirect cost items, common to all technologies, consisted of indirect field costs, home office engineering costs, field engineering costs, and owner's cost. As an example, the capital cost breakdown into first-level subaccounts for the wind reference system (69 MOD-2 Wind Turbines) is shown in Figure 2. A further breakdown of account 22 is shown in Figure 3.

## Cost Data Acquisition and Screening

Cost data for the components of each technology were acquired through a search of the published literature, an identification of the experts on each technology, and private communications with these experts to obtain their estimates and unpublished data. This search revealed the many difficulties that had led us to undertake the study. That is, costs were reported on differing bases, included differing components, and revealed differing biases. A major part of the effort in data acquisition was devoted to resolving these discrepancies.

The benefit of our cost breakdown approach and methodology revealed itself in that missing or inconsistent figures became apparent. The cost information we obtained for each technology was placed on a consistent basis with that for the other technologies. For instance, care was taken to ensure that all required roads, fencing, and yardwork were included. The pricing of conventional equipment and materials was verified through common cost estimation procedures and references. Corrections were made to those elements that had been incorrectly costed or for which costs were inconsistent across technologies.

Further elements guided our cost-gathering effort:

- All costs gathered were collected in the greatest possible detail. Where further disaggregation was possible, so that the costs for specific components could be calculated, this was done. Thus, wherever possible, our cost estimates for each system are based upon estimates for each component of the system.

- Only estimates for mature plants were retained, although estimates for first-of-a-kind plants were often used to derive disaggregations for aggregated mature-plant costs.

- A mature plant was defined as one in which market-penetration efforts and the overcoming of all significant learning curves had achieved cost reduction.

- The year of the dollars in which estimates were reported was retained with the magnitude of the estimates.

In order to estimate expected costs, a contingency account was applied to each two-digit cost account. The addition of contingency is a standard practice in cost estimation. A contingency of 10 percent is a fair average value for the construction of complex, conventional electric generating plants such as coal and nuclear steam electric units. For innovative components of uncommercialized technologies, contingencies ranging from 10 percent to 35 percent were assigned. The higher contingencies reflect the generality with which the nonconventional technologies have been designed.

The sums, then, of these contingencies and the design costs formed the accounts from which the expected costs were estimated. However, the capital cost of a generating unit is greater than the sum of the labor, material, and equipment expenses associated with that unit. In keeping with utility practice, real escalation and an allowance for funds used during construction (AFDC) were estimated as additions to obtain the total capital cost. The TRICOST escalation calculations are based upon cash flow curves showing the relative distribution of equipment, labor, and materials associated with each two-digit account. The TRICOST package contains a file of the Engineering New Record cost indices for each of these cost factors over the past 15 years for 23 locations in North America. By subjecting these cost files to a linear regression and projecting that regression into the future, TRICOST calculates the future prices of these elements. Only real escalation was projected in this study, all costs being expressed in 1980 dollars. By specifying the date of the plant's completion, the program escalates the costs of each component only to the date at which expenditure takes place. The computer program can optionally be programmed with specified escalation assumptions.

The calculation of AFDC in TRICOST was based upon the cash flow curves in the reference cost files and the input AFDC rate of 9 percent. The capital costs developed in the preceding steps also assumed full maturity of cost. For most technologies, however, the date at which the technology is projected to be commercially available precedes the date of mature cost levels. That is, the early commercial plants will cost more than plants constructed after realization of all the maturing effects of mass production, market completion, and surmounted learning curves. To correct for costs encountered before the date of full maturity, we note that previous observations of new technologies have shown an exponential decline between these costs and the mature costs. Thus, lacking any more sophisticated model, we have constructed a maturity factor based on these observations.

Results and Discussion

The results of applying the TRICOST model to the eight reference systems for the five technologies considered are shown in Table 3. All costs are in 1980 dollars and include the assumptions of escalation and interest rates as explained. These costs projections represent the total expected capital investment, including indirect costs, interest, and escalation charges. By combining these capital costs with estimates of operating and maintenance and fuel costs, and by applying an appropriate economic equation, the cost of electricity can be estimated.

Besides its usefulness in developing cost estimates for comparison across technologies, our model provides a framework for further analysis of the effects of differing assumptions regarding economic variables and technological advances. These assumptions can be evaluated for their implications regarding the cost and prospective utilization of the emerging technologies.

References

1.  Hudson, C.R. II, CONCEPT-5 User's Manual. ORNL-5470 Oak Ridge National Laboratory, Oak Ridge, Tenn. January 1979.

2.  Federal Power Commission, Uniform System of Accounts. Washington, D.C. 1963.

3.  Boeing Engineering and Construction. MOD-2 Wind Turbine System Concept and Preliminary Design Report. DOE/NASA 0002-80/2. Seattle, Wash. July 1979.

4.  Bechtel National, Inc. Terrestrial Central Station Array Life-Cycle Analysis Support Study. DOE/JPL 954848-78/1. Jet Propulsion Laboratory, Pasadena, Calif. August 1978.

5.  Rittlemann, P.R. "Residential photovoltaic module and array requirements study." Presented at the U.S. Department of Energy Semiannual Program Review of Photovoltaic Technology Development. Burt Hill Kosar Rittleman Associates, Butler, Penn. November 1979.

6.  Bechtel Corporation. Advanced Design and Economic Considerations for Commercial Geothermal Power Plants at Heber and Niland, California, Final Report. NTIS SAN-1124-2. U.S. Department of Energy, Washington, D.C. October 1977.

7.  United Technologies Corporation. Energy Conservation Alternative Study - ECAS. NASA-CR-124955. National Aeronautics and Space Administration, Washington, D.C. 1976.

8.  Arthur D. Little, Inc. Assessment of Fuels for Power Generation by Electric Utility Fuel Cells. Vol. 2. EM-695. Electric Power Research Institute, Palo Alto, Calif. March 1978.

9.  Westinghouse Electric. Solar Thermal Repowering Utility Value Analysis. Final Report. SERI Contract SH-0-8016-1. East Pittsburgh, Penn. December 1979.

10. Martin-Marietta Corporation. Conceptual Design of Advanced Central Receiver Power System. Final Report. DOE Contract E6-77-C-03-1724. Denver, Colo. September 1978.

## Table 2

### Reference Power Plant Systems

| Plant Description | Plant Size (MW) | Capacity Factor (%) | Plant Lifetime |
|---|---|---|---|
| Wind Farm, MOD-2 (3) | 165 | 38 | 30 |
| Photovoltaic, Flat-Plate Central (4) | 200 | 25 | 30 |
| Photovoltaic, Flat-Plate Rooftop (5) | 0.01 | 24 | 30 |
| Geothermal Double Flash (6) | 50 | 85 | 30 |
| Fuel Cell, Molten Carbonate/ Coal Gasifier (7) | 860 | 90 | 20 |
| Fuel Cell, Phosphoric Acid (8) | 27 | 70 | 20 |
| Solar Thermal, Tower Repowering (9) | 100 | 40 | 30 |
| Solar Thermal, Tower Stand Alone (10) | 100 | 65 | 30 |

## Table 3

### Capital Cost Estimates for Reference Systems
### (All Costs in 1980 Dollars per kW)

| System | 1980 | 1985 | 1990 | 1992 | 2000 |
|---|---|---|---|---|---|
| Wind Farm | N/A | 1,920 | 1,502 | 1,534 | 1,670 |
| Photovoltaic Central Station | N/A | 5,041 | 1,732 | 1,762 | 1,949 |
| Photovoltaic, Distributed | N/A | 5,448 | 2,589 | 2,619 | 2,768 |
| Geothermal, Liquid | 1,189 | 1,283 | 1,363 | 1,370 | 1,532 |
| Fuel Cell, Molten Carbonate | N/A | N/A | N/A | 3,684 | 2,044 |
| Fuel Cell, Phosphoric Acid | N/A | N/A | 1,405 | 1,237 | 1,420 |
| Solar Thermal, Repowering | N/A | 3,330 | 2,810 | 2,575 | 2,397 |
| Solar Thermal, Stand Alone | N/A | N/A | 4,329 | 3,264 | 3,124 |

N/A = Not commercially available.

**Table I**

**Assessment of Current Status of Alternative
Electrical Generation Technologies**

**Wind**

| Phase | Technology | Current Status of Technology | Milestones Achieved | Milestones Aimed For |
|---|---|---|---|---|
| 1. Concept exploration | | | | |
| 2. Screening | | | | |
| 3. Product/system development | Wind | | | |
| | A. HAWT: Advanced MW(e)-scale machines (MOD-5) | A. DOE/NASA funding to General Electric and Boeing for development work | A. Contracts awarded | A. COE of $0.03/kWh (1977$) – marketable 1986 |
| | B. VAWT: 50-kW(e) machines utility applications | B. Demonstration project by DWR at San Louis Dam, Pachecho Pass, Calif. | B. Purchase made of first unit | B. On grid April 1981 |
| | C. HAWT: Advanced 0.2-0.5-MW(e) scale machines (MOD-6) | C. Proposed for funding for development work | | C. Multi-purpose machine |
| | D. HAWT: (50-100 kWe) mass produced machine | D. Plan to test for DWR at Pacheco Pass, Calif. | | D. COE $0.05/kWh (1980$) |
| | E. VAWT: 300-500-kW(e) machines – utility application | E. 1) Planned installation of 6 machines by Windfarms Ltd. in Hawaii | E. 1) Plans announced | E. Profitable energy scales for utility |
| | | 2) Installation of 1 machine at San Gorgonio Pass, Calif. | 2) Purchases made | |
| 4. Market or mass applications testing | Wind | | | |
| | A. HAWT: Multi-MW(e) scale machines | A. Various demonstration projects under construction: 1) Hills, Wash., Boeing, MOD-2 2) Medicine Bow, Wyo., Hamilton-Standard, WTS-4 3) San Gorgonio Pass, Calif., Bendix, 3 MW(e) | | A. Proof of concept, energy costs, reliability, grid intertie |
| | B. HAWT: 1-MW(e) machine | B. Demonstration project, Boone, N.C., MOD-1 | | |
| | C. HAWT: 200-kW(e) machines | C. Various demonstration projects in existence: 1) Clayton, N.M., MOD OA 2) Block Island, R.I., MOD OA 3) Culebra Island, P.R., MOD OA 4) Oahu, Hawaii, MOD OA | | |
| | D. SWECS: | D. Various concepts are being tested. DOE has coordinated test center at Rocky Flats. | | |
| 5. Commercial introduction | | | | |

Figure 1

Current Status Assessment
Five-Phase Framework

| | Phase | Status | Milestones |
|---|---|---|---|
| 1. | Concept exploration | Lab models | Technical feasibility<br>• Reproducible performance |
| 2. | Screening | Experimental prototypes | System readiness<br>• Subscale demonstration |
| 3. | Product/system development | Technical demonstrations | Product readiness<br>• Production prototypes |
| 4. | Market or application testing | Commercial demonstration | Market readiness<br>• Major markets defined |
| 5. | Commercial introduction | Full-scale implementation | Commercial availability<br>• Economies of scale<br>• Distribution<br>• Price stability |

Figure 2

**Wind Turbine Cost Accounts**

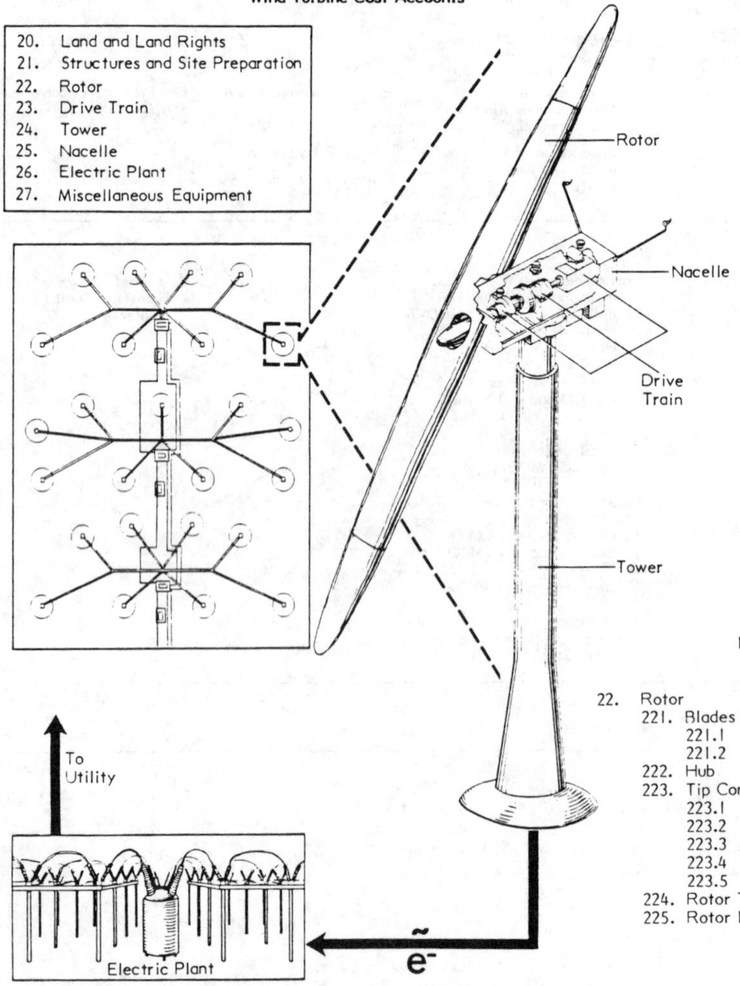

20. Land and Land Rights
21. Structures and Site Preparation
22. Rotor
23. Drive Train
24. Tower
25. Nacelle
26. Electric Plant
27. Miscellaneous Equipment

— Rotor

— Nacelle

Drive
Train

— Tower

To
Utility

Electric Plant

$e^-$

Figure 3

**Rotor Account Detail**

22. Rotor
   221. Blades
      221.1 Midsection Assembly
      221.2 Tip Assembly
   222. Hub
   223. Tip Control
      223.1 Blade Tip Actuator
      223.2 Blade Hydraulics
      223.3 Pitch Control — Low-Speed Shaft
      223.4 Hydraulic Reservoir
      223.5 Electric Cables
   224. Rotor Transport and Packing
   225. Rotor Installation

# FEDERAL ENVIRONMENTAL LAWS AND EMERGING COAL TECHNOLOGIES

**James E. Mann** and **Thomas C. Ruppel**
U.S. Department of Energy

Introduction

This paper presents a general discussion of Federal environmental laws
and more specifically the Clean Air Act of 1970, Clean Water Act of
1972, Resource Conservation and Recovery Act of 1976, Safe Drinking
Water Act of 1974, and the Toxic Substances Control Act of 1976 which
have and will have a significant effect on emerging coal combustion and
conversion processes.  The regulatory constraints to commercialization
of these technologies are identified with respect to the associated unit
operations.

The present body of Federal environmental laws consists of a
semi-consistent, complicated, overlapping, and often confusing set of
constraints to previously unrestricted energy production, chemical
manufacture, waste disposal, and other industrial activities.  While the
same general assessment can be made for other forms of law, the
situation for environmental law is exacerbated by the relatively short
time span of enactment of most of the laws.

Stated from a restrictive viewpoint, environmental laws define what
cannot be done.  Stated more generally, environmental laws signify an
attempt to internalize more of the true costs of industrial activity and
thereby synergistically integrate industrial activity into social and
natural processes on earth.  Along with capital, labor, raw material
costs, etc., the costs of pollution control, waste disposal, and
resource reclamation should be included in the life cycle of a product.
Thus the overall cost will increase.  But this is the true cost to
society.

The U. S. Environmental Protection Agency was created by Executive Order
on December 2, 1970 within the Executive Office of the President, from
the then-existing Federal Water Quality Administration, National Air
Pollution Control Administration, Environmental Control Administration,

and the Pesticides Research and Standards Setting Program of the Food and Drug Administration. The Pesticides Registration Authority of the Department of Agriculture was also absorbed into the EPA.[1,2]

The EPA's mission is to control and abate pollution in the areas of air, water, solid waste, noise, radiation, and toxic substances. The Agency's mandate is to mount a coordinated attack on environmental pollution in cooperation with State and local governments. In carrying out its role, the EPA has become the chief economic growth regulator in the Country.

The following discussion attempts to describe the observed and anticipated impacts of the Clean Air Act of 1970, Federal Water Pollution Control Act of 1972, Resource Conservation and Recovery Act of 1976, Safe Drinking Water Act of 1974, and the Toxic Substances Control Act of 1976 on emerging coal combustion, gasification, and liquefaction technologies. Of the many Federal environmental laws, the above five Acts have the greatest influence on the emerging coal technologies.

Specifically, EPA's Congressional mandates concerning these Acts are the following:[3]

o  Clean Air Act:  to protect the public health and welfare from the harmful effects of air pollution, and ensure that existing clean air is protected from significant deterioration by controlling and preventing harmful substances from entering the ambient air.

o  Federal Water Pollution Control Act:  to restore and maintain the chemical, physical, and biological integrity of the Nation's waters, with 1977 amendments to the law giving added emphasis to toxic water pollutants.

o  Resource Conservation and Recovery Act:  to assure that both hazardous and nonhazardous solid wastes are disposed of in environmentally sound ways and to conserve natural resources both directly and through resource recovery from wastes.

o  Safe Drinking Water Act:  to protect and improve the quality of the Nation's drinking water supply by establishing drinking water standards that specify maximum permissible contaminant levels.

o  Toxic Substances Control Act:  to develop adequate data and knowledge on the effects of chemical substances and mixtures on health and the environment, and assure that those chemicals which present an unreasonable risk of injury are regulated to reduce that risk.

State and local environmental laws, while equally binding to the company which wishes to construct or modify a facility, are not discussed due to their diverse nature. Surveys of state environmental laws are available.[4-6]

Federal Environmental Laws

The first of what may be considered to be Federal laws to protect the environment was passed by Congress in 1886. The River and Harbor Act of that year laid the foundation for and was recodified as The Rivers and Harbors Act of 1899. Also called the Refuse Act, it regulated obstruction to navigation and refuse dumping in navigable waterways. From this beginning, it has been possible to identify more than 45 environmental

Acts of Congress (Figure 1). Proceeding through the important Food,
Drug, and Cosmetics Act of 1906, the environmental activity of Congress
remained sporadic until about 1960, when a flurry of environmental laws
signalled that the environmental movement had begun. This activity was
a manifestation of certain realizations. The Donora, Pennsylvania air
pollution episode of 1948 and the Los Angeles, California smog problems
of the 1950's focused the concern of people on air pollution. Several
river foaming problems caused by nonbiodegradable detergents in the
early 1960's and fish-kill episodes emphasized the threat of surface
water pollution.

But the environmental Act of environmental Acts appeared in 1969. The
National Environmental Policy Act of that year set the tone for the next
decade and beyond. "NEPA compliance" is standard language for environmental engineers. NEPA, a short Act of a few pages, created the Cabinet
level influential Council on Environmental Quality and the Environmental
Impact Statement. A deluge of environmental laws followed NEPA, outdo-
ing the significant Federal activity of the sixties. Although there has
been some Congressional amendment activity since 1977, the next anticipated significant environmental milestone is the 1981 debate covering
the Clean Air Act amendments. The exponential increase in the number of
laws will probably yield to a sigmoidal or growth function, as indicated
by the short extrapolated dashed line in Figure 1. Natural or counter-
forces tend to balance an activity at some point, whether the activity
be biological, demographic, or cultural.

Only the environmental laws which significantly and directly affect the
emerging coal technologies will be discussed. It may be debated which
laws the above qualifications really imply, since so many laws affect
the emerging coal technologies. The last several Acts in Figure 1 are
the morphological results of previous Acts. Thus the Clean Air Act of
1977, the Clean Water Act of 1977, the Resource Conservation and
Recovery Act of 1976, the Safe Drinking Water Act of 1974, the Surface
Mining Control and Reclamation Act of 1976, and the Toxic Substances
Control Act of 1976 constitute more comprehensive and far-reaching
pieces of legislation than their amended more-piecemeal predecessors.

The Surface Mining Control and Reclamation Act of 1976 will not be
discussed since it is desired to concentrate on the emerging coal
technologies themselves.

Nevertheless, other laws could become significant in special cases. For
example, NEPA requires that an Environmental Assessment be written by
the sponsoring Government agency before a major Federal action significantly affecting the quality of the human environment is actually
initiated. If it is found during the process of writing the statement
that, for example, a species of flora or fauna, which is listed among a
Federal or State's endangered species, will possibly be displaced, the
Endangered Species Act could be invoked to delay or terminate a project.
The same is true of archaeological finds on a proposed construction
site. Federal, State, or local environmental laws are equally binding,
and any one of them could delay a project. The interface between the
emerging coal technologies and the myriad environmental laws constitutes
the interesting focal point for the following discussion.

Environmental Laws with a Major Impact on Coal Combustion and Conversion
Processes

There is a uniformity of impact of the environmental laws among coal
combustion and conversion processes. The environmental concerns of any

211

major industrial activity are air quality, water quality, solid wastes, health and safety, ecology and socioeconomic factors, and product use. These issues are controlled in the main by the Clean Air Act, Clean Water Act, Safe Drinking Water Act, Resource Conservation and Recovery Act, National Environmental Policy Act, and the Toxic Substances Control Act, respectively. Thus Congress has succeeded in the nonnuclear area in passing environmental legislation which covers the range of environmental concerns. Further legislation will certainly be passed to refine and moderate the present laws, but the comprehensiveness is obvious.

The effects of the environmental laws on coal combustion and conversion unit operations will be discussed concurrently in order to emphasize the uniformity of process impacts. Also there are many references which describe the laws individually and concurrently in detail at the process level.[5],[8-17]

Before proceeding to individual processes, it is instructive to note the holistic view of the life cycle of a chemical (manufactured substance) as perceived by EPA.[18] Figure 2 depicts the Federal environmental laws which are brought to bear on a chemical from the chemist's creation in glass to the ultimate disposal in a properly secured form (e.g., land-fill, incineration). The comprehensive coverage is obvious. The underlying nature of TSCA is apparent in that at every point where the raw or processed material enters or reenters commerce, TSCA applies to the transaction.

The anticipated impacts of the environmental laws on the several emerging coal technologies are presented in Figures 3 through 6 and should be viewed collectively in the following discussion. The similarities of the impacts are far more apparent than their differences. For example, in all processes, coal pile storage rain run-off could be regulated by either the CWA (Secs. 304, 307, 311), SDWA (Sec. 1421), or RCRA (Secs. 3001, 3002, 3004, 3005). Leachates from landfill, including bottom ash or slag, sludge, and spoil piles, also fall under these regulations. The actual Act and sections which would be invoked would be decided by the agency with jurisdiction over the facility. This could be the U.S. EPA, the appropriate Regional EPA Office (there are ten Regions nationally), or a State agency which has been approved by the U.S. EPA to administer the particular sections of the chosen Act.

Similarly, stack emissions from new or modified combustion processes, which could also be combustion unit operations associated with steam raising in coal conversion processes, should be regulatable under Section 111 of the Clean Air Act (New Source Performance Standards).

Oil spills have been handled by agencies with Sec. 311 of the Clean Water Act. This would apply to Coal-Oil-Mixture mixing, storage, and transport operations; coal gasification waste water treatment; and of course to several unit operations of coal liquefaction.

Fugitive emissions from leaking valves, lagoon evaporations, or dust-causing operations are regulated separately by the Clean Air Act and the Occupational Safety and Health Act. Sections 109 (NAAQS) and 163, 166 (PSD) of CAA assure the maintenace of a predetermined level of area air quality; and OSHA sections, depending upon the particular in-plant pollutant, attempt to assure worker safety.

New chemical substances (those manufactured for entry into commerce for the first time after January 1975 or which have not been included in a

massive Initial Inventory and supplement) cannot be sold until a Pre-
manufacture Notice (PMN) is submitted to U.S. EPA for approval, inaction, or qualified or unqualified rejection.  Inaction is tantamount to
approval.  This PMN is the all-important step in obtaining EPA approval
under Sec. 5 of TSCA.  The environment, health, and safety sections of
the PMN questionnaire constitute the present gauntlet for coal liquefaction technology.  Demonstration-size plants will probably fall within
the TSCA laws.  It appears that the low, medium, and high Btu gaseous
products from coal gasification will be approved for manufacture in
commerce by U.S. EPA.  Whether they are new or old chemical substances
under the TSCA definition is academic.

The sections of the Acts cited above, and shown in Figures 3 through 6,
are briefly discussed in Table 1.

Table 1. - Environmental Laws with a Major Impact on Coal
Combustion and Conversion Processes[a]

| Law | Section | Description |
|---|---|---|
| Clean Air Act | 109 | Requires the establishment of "national primary and secondary ambient air quality standards" (NAAQS).  See Reference 7, pp. 3-22 and 3-35 for standards, and 36 FR 15486 (1971). |
| | 111 | Requires the establishment of "new source performance standards" (NSPS) to limit the emission of primary[b] pollutants from a select list of 28 new or modified major stationary sources.[c, 19-22] |
| | 163,166 | If the AQCR meets the primary and secondary NAAQS for a Criteria Pollut-ant (CO, hydrocarbons, $NO_2$, Pb, photo-chemical oxidants, $SO_2$, and TSP), the region is designated an attainment or PSD area by EPA.  BACT may be used to control that pollutant.  If the NAAQS are not met for a Criteria Pollutant, LAER must be employed without regard for cost. |
| Clean Water Act | 304(e) | Requires the control, through best man-agement practices, of toxic pollutants resulting from ancillary industrial activities and regulates plant site run-off, spillage or leaks, and sludge or waste disposal. |
| | 307(a) | Regulates on a chemical-by-chemical basis substances (esp. pesticides) which can be   proved to have toxic effects on identified organisms in affected waters. |
| | 311 | Establishes an extensive regulatory scheme for dealing with accidental or |

intentional discharges of oil and hazardous substances.

402  "...the Administrator (EPA) may, after opportunity for public hearing, issue a permit for the discharge of any pollutants...upon condition that such discharge will meet either all applicable requirements under Secs. 301, 302, 306, 307, 308, and 403 of this Act, or...such conditions as the Administrator determines are necessary to carry out the provisions of this Act."

Resource Conservation and Recovery Act

3001  Waste generators must determine if waste is hazardous (ignitible, corrosive, reactive, or toxic).

3002  If hazardous, generator must keep records concerning quantity, composition, and disposition; label appropriately any (appropriate) containers used; inform persons transporting, treating, storing, or disposing waste of its composition; use a manifest system to assure proper treatment, storage, or disposal (other than facilities on the premises where waste is generated); report to EPA or state agencies when requested concerning quantities and disposition of waste. 43 FR 58969 (Dec. 18, 1978), 40 CFR Part 250, Subpart B.

3004  Owners and operators of hazardous waste treatment, storage, and disposal facilites must, in addition to record-keeping and reporting as required in Sec. 3002, locate, design, and construct facilities and treat, store, or dispose of waste in a manner satisfactory to EPA; have contingency plans to minimize unanticipated damage; and plan for continuation of operation regarding ownership, personnel training, and financial responsibility. Ground water and leachate monitoring is included in the contingency plans and will integrate with environmental goals set forth in the Clean Water Act (this table) and the Safe Drinking Water Act (this table). 43 FR 58994 (Dec. 18, 1978), 40 CFR Part 250, Subpart D.

3005  Requirements for obtaining permits to treat, store, or dispose of hazardous waste.

| Safe Drinking Water Act (Also 40 FR 248 (1975) and 45 FR 33290 (1980)) | 1421 | Establishes maximum groundwater contaminant levels for inorganic chemicals, including arsenic, barium, cadmium, chromium, lead, mercury, selenium, and fluoride. Maximum contaminant levels for some organic compounds, including pesticides are also established. |
|---|---|---|
| Toxic Substances Control Act | 5 | "... no person may manufacture a new chemical substance ... or process any chemical substance for a use which the Administrator (EPA) has determined ... is a significant new use, unless such person submits to the Administrator ... a notice (PMN) ... of such person's intention to manufacture or process such substance ..." |

---

a. Other environmental Acts impact in varying degrees on coal combustion and conversion processes, as can be seen by an inspection of Figure 1. However, the CAA, CWA, RCRA, SDWA, and TSCA have an overriding impact.

b. A primary pollutant is one which is emitted directly into the atmosphere from a source, e.g., $SO_2$, $NO$, $NO_2$, and particulates from a stack or coal grinding operation, while secondary pollutants (e.g., photochemical oxidants and sulfates) are formed by chemical processes in the atmosphere.

c. Emerging coal technologies ( COM, MHD, coal gasification and liquefaction, are not yet specifically listed among the 28 sources, but this is expected to change.

Regulatory Approach and Perceived Trend of EPA with Respect to the Emerging Coal Technologies

There is no question that EPA is feverishly active, albeit in a measured and experimental manner, in implementing its Congressional mandate as protector of the National environment. It is continually assessing the myriad environmental impacts of industrial and Governmental activity through an enormous amount of contractor support. It promulgates a continuing stream of preliminary, interim, proposed, and final regulations in the Federal Register. The Agency is regularly embroiled in endless litigation concerning its implementation and enforcement of environmental laws. In short, the Environmental Protection Agency is one of the most visible and controversial agencies within the U. S. Government.

There is also no question that regulations seriously affect technology choice, development, and siting. Yet-to-be-developed regulations based on existing Federal mandates may have a major impact that could, individually or in concert, terminate particular technologies. An annotated list of a number of industrially planned projects which were terminated because of environmental impediments has been compiled.[13] There are an ample number of reports attempting to intuit the mind of EPA concerning emerging coal technologies. References 5, 10, 12, 13, 15, and 17 give one a good flavor of the possibilities. However. the periodicals

"Inside E.P.A."[23] and "Synfuels"[24] provide a running commentary on what may be expected in specific regulations and technologies.

Thus, by the process of preliminary to final regulations, EPA carries out its Congressional mandate to implement the Federal environmental laws. Virtually every office in EPA is involved in coordinating the development of environmental guidelines for new synfuels plants with the Presidential and Congressional wills expressed in the Energy Security Act of 1980. The adversarial nature of the energy-environment interface tends to obscure this point.

References

1. J. E. Heer and D. J. Hagerty, Environmental Assessments and Statements, Van Nostrand Reinhold Co., New York, NY, 1977, p. 59.
2. J. G. Arbuckle, G. W. Frick, M. L. Miller, T. F. P. Sullivan, and T. A. Vanderver. Environmental Law Handbook, Government Institutes, Inc., Wash., DC, 1979, p. 221.
3. J. B. Ritch, Jr., "Protecting public health from toxic chemicals", Env. Sci. & Tech. 13, No. 8, 922 (1979).
4. Major Legislative & Regulatory Impediments to Conventional and Synthetic Fuel Energy Development, p. C-1, Am. Petroleum Inst., 2101 L St., N.W., Wash., DC 20037, Mar. 1, 1980.
5. Impediments to Synthetic Liquids Development, TRW Energy Systems Planning Div., 8301 Greensboro Dr., McLean, VA. 22102, Sept. 10, 1979, p. 4-15.
6. J. Golden, R. P. Ouellette, S. Saari, and P. N. Cheremisinoff, Environmental Impact Data Book, Ann Arbor Science Pub., Inc., 230 Collingwood, Ann Arbor, MI 48106, 1979.
7. J. G. Rau and D. C. Wooten, eds., Environmental Impact Analysis Handbook, McGraw-Hill Book Co., New York, NY, 1980.
8. Environmental Development Plan, Magnetohydrodynamics, U. S. Dept. of Energy, Wash., DC 20545, May 1979, DOE/EDP-0045.
9. P. Matray and G. Huddleston, "MHD emissions and their controls", Env. Sci. & Tech. 13, No. 10, 1208 (1979).
10. Production of Synthetic Liquids from Coal: 1980-2000. A Preliminary Study of Potential Impediments, Final Report to U. S. Dept. of Energy, Wash., DC 20545 by Bechtel National Inc., San Francisco, CA, Dec. 1979, No. FE-3137-T1.
11. Environmental Development Plan, Coal Gasification Program, U. S. Dept. of Energy, Wash., DC 20545, 1979, No. DOE/EDP-0043.
12. R. V. Steele et al., Leading Trends in Environmental Regulation that Affect Energy Development, Final Report to U. S. Dept. of Energy, Wash., D. C. 20545 by Flow Resources Corp. and Internat. Res. & Technol. Corp., Aug. 1979, pub. Jan. 1980, No. DOE/EV-01682.
13. Major Legislative & Regulatory Impediments to Conventional and Synthetic Fuel Energy Development, Am. Petroleum Inst., 2101 L St., N.W., Wash., DC 20037, Mar. 1, 1980.
14. Achieving a Production Goal of 1 Million B/D of Coal Liquids by 1990, prepared for U. S. Dept. of Energy, Wash., DC 20545 by TRW Energy Systems Planning Div., McLean, VA; Bechtel National, Inc., San Francisco, CA; Mechanical Technology, Inc., Latham, NY; Monsanto Res. Corp., Miamisburg, OH, Mar. 1980, No. DOE/FE/10490-01.
15. Synthetic Fuels and the Environment: An Environmental and Regulatory Impacts Analysis, U. S. Dept. of Energy, Wash., DC 20545, June 1980, No. DOE/EV-0087.
16. Environmental Development Plan, Coal Liquefaction, U. S. Dept. of Energy, Wash., DC 20545, Aug. 1980, No. DOE/EDP-0012.

17. Guide to Key Federal Environmental, Health, and Safety Requirements Applicable to Synthetic Fuels Projects, prepared for U. S. Dept. of Energy, Wash., DC 20545 by Sobotka & Co., Inc., 2501 M St. NW, Suite 550, Wash., DC 20037, Oct. 27, 1980.
18. T. Temple, "Controlling toxics", EPA J. 5, No. 7, 12 (1979).
19. ERT Handbook on Industrial Expansion and the 1977 Clean Air Act Amendments, Environmental Research and Technology, Inc., 696 Virginia Rd., Concord, MA 01742, June 1979.
20. G. F. Hoffnagle and R. Dunlap, "Industrial expansion and the 1977 clean air act amendments", Poll. Eng. 10, 36 (Dec. 1978).
21. J. R. Kruse and G. F. Hoffnagle, "Air quality monitoring for PSD Permits", Poll. Eng. 12, 43 (Apr. 1980).
22. B. J. Goldsmith and J. R. Mahoney, "Implications of the 1977 clean air act amendments for stationary sources", Env. Sci. and Tech. 12, No. 2, 144 (1978).
23. Inside E.P.A., Inside Wash. Pubs., P. O. Box 7167, Ben Franklin Station, Wash., DC 20044, weekly periodical.
24. Synfuels, McGraw-Hill, Inc., 1221 Avenue of the Americas, New York, NY 10020, weekly periodical.

List of Abbreviations

| | |
|---|---|
| AQCR | Air Quality Control Region |
| BACT | Best Available Control Technology |
| CAA | Clean Air Act |
| CEQ | Council on Environmental Quality |
| COM | Coal-Oil-Mixture |
| CWA | Clean Water Act |
| DOE | Department of Energy |
| EPA | Environmental Protection Agency |
| LAER | Lowest Achievable Emission Rate |
| MHD | Magnetohydrodynamics |
| NAAQS | National Ambient Air Quality Standards |
| NEPA | National Environmental Policy Act |
| NPDES | National Pollutant Discharge Elimination System |
| NSPS | New Source Performance Standards |
| OSHA | Occupational Safety and Health Act |
| PMN | Premanufacture Notice |
| PSD | Prevention of Significant Deterioration |
| RCRA | Resource Conservation and Recovery Act |
| SDWA | Safe Drinking Water Act |
| SMCRA | Surface Mining Control and Reclamation Act |
| SNG | Substitute Natural Gas |
| TSCA | Toxic Substances Control Act |
| TSP | Total Suspended Particulates |

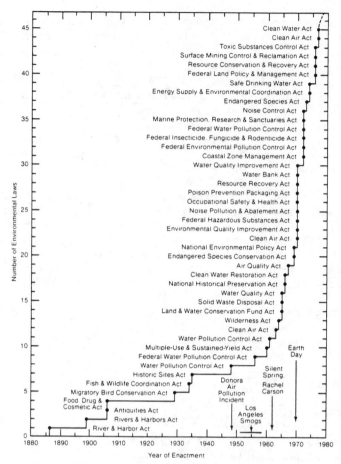

FIGURE 1.—A CENTURY OF ENVIRONMENTAL LAWS

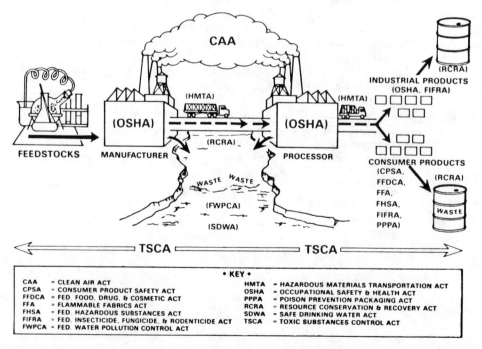

| • KEY • | | |
|---|---|---|
| CAA = CLEAN AIR ACT | HMTA | = HAZARDOUS MATERIALS TRANSPORTATION ACT |
| CPSA = CONSUMER PRODUCT SAFETY ACT | OSHA | = OCCUPATIONAL SAFETY & HEALTH ACT |
| FFDCA = FED. FOOD, DRUG, & COSMETIC ACT | PPPA | = POISON PREVENTION PACKAGING ACT |
| FFA = FLAMMABLE FABRICS ACT | RCRA | = RESOURCE CONSERVATION & RECOVERY ACT |
| FHSA = FED. HAZARDOUS SUBSTANCES ACT | SDWA | = SAFE DRINKING WATER ACT |
| FIFRA = FED. INSECTICIDE, FUNGICIDE, & RODENTICIDE ACT | TSCA | = TOXIC SUBSTANCES CONTROL ACT |
| FWPCA = FED. WATER POLLUTION CONTROL ACT | | |

EPA JOURNAL

FIGURE 2.— LEGISLATIVE AUTHORITIES AFFECTING THE LIFE CYCLE OF A CHEMICAL

218

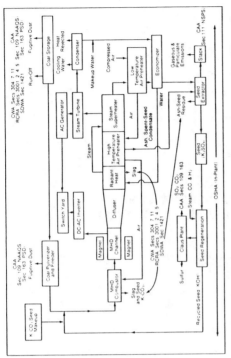

FIGURE 4 – OPEN-CYCLE MAGNETOHYDRODYNAMIC (MHD) POWER PLANT & THE ENVIRONMENTAL LAWS

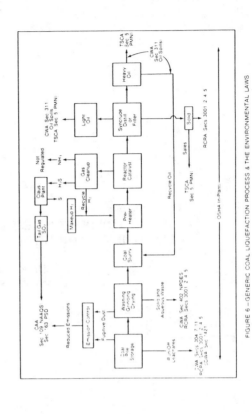

FIGURE 6 – GENERIC COAL LIQUEFACTION PROCESS & THE ENVIRONMENTAL LAWS

FIGURE 3 – COAL-OIL-MIXTURE TECHNOLOGY & THE ENVIRONMENTAL LAWS

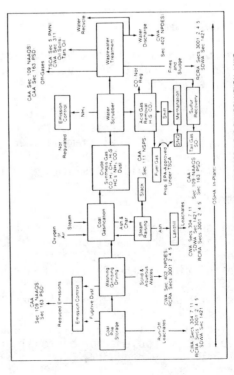

FIGURE 5 – GENERIC COAL GASIFICATION PROCESS & THE ENVIRONMENTAL LAWS

219

# THE IMPACT OF THE RESOURCE CONSERVATION AND
# RECOVERY ACT (RCRA) ON COAL-FIRED FACILITIES

**Dave Burstein**
Engineering—Science

## Acknowledgement

The author wishes to acknowledge the assistance and input from the following individuals who have contributed to the project which is reported in this paper:

- o  Mr. Val Weaver, U.S. Department of Energy
- o  Mr. James Johnson, U.S. Department of Energy
- o  Mr. William Webster, Webster & Associates
- o  All of the individuals and utility companies which provided assistance in the case study activities

## Introduction

Engineering-Science was retained by the U.S. Department of Energy (DOE) to evaluate the impact on coal-fired facilities resulting from regulations issued by the U.S. Environmental Protection Agency (EPA) for implementation of the Resource Conservation and Recovery Act of 1976 (RCRA).

The National Energy Plan (NEP) has focused recent attention on conversion of utility and industrial boilers from oil and natural gas to coal and coal-based fuels. The increased coal use recommended by the NEP will pose larger problems for disposal of coal by-products than for oil or natural gas by-products.

The overall purpose of this study is to assess RCRA's impacts on coal-fired electric generating facilities as a means to understanding the implications of proposed RCRA regulations on the NEP and utilities, and understanding RCRA's implications on fossil energy development programs. Specifically, the project has four major objectives:

1. Analyze RCRA Impacts on Coal-Fired Utilities.

2. Assess Implications on Emerging Coal Technologies.

3. Assess Implications on Implementation of the Fuel Use Act of 1978.

4. Develop a range of potential national impacts which might result from implementation of RCRA on major coal-fired facilities.

The project is being conducted as a three phase effort. Phase I deals with the utility sector; Phase II deals with the emerging coal technologies (advanced combustion, coal conversion and advanced environmental control); Phase III consists of an analysis of financial, economic, institutional and programmatic impacts on a nationwide basis. This paper presents an overview of the methodology used for the entire project as well as a summary of results obtained to date.

Work completed to date consists primarily of an analysis of potential costs to the utility sector based on the RCRA regulations proposed by EPA on December 18, 1978. Work is presently in progress to update these cost estimates in accordance with the final RCRA regulations promulgated in 1980 and to increase the accuracy of these estimates by expanding the data base.

General Methodology

The scope of activities undertaken to accomplish the project objectives is shown in Figure 1, along with the sequence and interelationships among the various major tasks.

## Methodology for Phase I

As previously described, Phase I of the project consists of an assessment of potential RCRA impacts on the utility sector. The overall approach is based on Engineering-Science assuming the role of a consultant retained by various utilities for the purpose of designing waste disposal facilities and estimating their costs (capital and O&M). While it is somewhat unconventional for government studies of this type, this "hands-on" approach was selected because it was felt that the results would best reflect the true costs of implementing the regulations. Furthermore, this approach is one that is most commonly used by industry and would thus be most likely to be accepted by this sector.

Cost Estimating Procedure. Figure 2 presents the various steps utilized in the cost estimating process. This procedure is being conducted for the following scenarios under which coal combustion wastes might be disposed:

1. Current Disposal Methods

    - Wet Disposal

- Dry Disposal

2. Non-hazardous Disposal as specified in RCRA Sections 1008, 4004.

3. Hazardous Disposal as specified in RCRA Section 3004.

Idealized Designs. The concept of "idealized designs" was developed in order to establish (1) the lowest probable cost of compliance and (2) the effects of economies of scale on waste disposal costs. These idealized designs consist of a number of design assumptions which would correspond to a site which is virtually ideal for waste disposal. This includes such factors as clay availability, climate, population density, etc. Four idealized designs were developed for each waste disposal scenario based on waste disposal volumes of 800,000 to 60,000,000 cubic feet per year. The economies of scale are thus determined by varying only the disposal volume, while maintaining the same assumptions for all other variables. The cost data generated from the idealized designs is then used to formulate mathematical modes of capital and O&M costs is a function of waste disposal volume.

Case Studies. One of the key elements in this approach is the selection of 24 case study sites that represent the range of factors which impact utility waste disposal costs (e.g., plant size, coal characteristics, use of $SO_2$ scrubbers, site characteristics and potential siting restrictions). Each case study site was (or will be) visited and the following information collected:

1. Sources of Coal

   o Supplier
   o Quantity
   o District Number
   o Seam
   o General Characteristics

2. Boiler And Plant Operating Characteristics

   o Capacity
   o Age
   o Capacity Factors
   o Current Coal Burn
   o Emission Control Technology

3. Waste Disposal

   o Current Waste Production (Fly Ash, Bottom Ash, Scrubber Sludge, Water Treatment Sludge And Other Misc. Wastes)
   o Projected 1985 Waste Production
   o Current Waste Disposal Cost
   o Projected 1985 Disposal Cost

4. Schematic Flow Diagram Of Facility

5. Site Characteristics

   o Topography
   o Soils, Geologic, And Climatic Data
   o Potential Disposal Site Locations And Configurations

Preliminary designs and cost estimates are prepared for each case study site based on the above data and the conceptual designs for each waste disposal scenario. These site-specific costs are then used to adjust the mathematical cost models developed from the idealized designs.

Regional and National Cost Estimates. Having developed the cost estimating models, the next step is to use these models to estimate regional and national cost impacts. This is done by recording the coal burned by each coal-fired power plant with greater than 25 MW capacity. An estimate of waste volume is then made based on the coal characteristics, the type of $SO_2$ scrubber employed (if any) and the method of disposing of the waste material. The waste volumes are entered into the cost models to obtain capital and O&M cost estimates for each of the approximately 400 plants in the U.S. with greater than 25 MW capacity. Regional costs are estimated by summing the disposal costs for all plants within a given Federal Energy Regulatory Commission (FERC) Region and making appropriate adjustments based on projected modifications to current practice. National costs are developed by summing the regional cost estimates. The results of these calculations are estimates of waste disposal costs for:

(1) current practice (both wet and dry disposal),

(2) disposal as a non-hazardous waste (RCRA Sections 1008 and 4004) and

(3) disposal as a hazardous waste (RCRA Section 3004).

Potential Modifications to Current Practice. As a result of recent increases in the costs of waste disposal and raw materials, a number of processes are being developed as a means of reducing these costs. These include:

o   combustion modification
o   process modification
o   waste treatment
o   waste reuse
o   resource recovery

These processes will be evaluated to assess their potential for ameliorating waste disposal costs by (1) reducing the volume of waste to be disposed and/or (2) altering the characteristics of the waste in order to reduce the unit cost of disposal.

Methodology for Phase II

The objective of Phase II is to assess the implications of RCRA on a number of emerging coal technologies involving advanced combustion, coal conversion and advanced environmental control processes. As shown in Figure 1, the approach is similar to that employed for Phase I with the following exceptions:

Case Studies. Unlike the Phase I case studies, many of the 12 emerging technology sites to be visited are pilot rather than full scale operations. This means that definitive, steady-state operating data is generally not available because of the scale differences and the ongoing variability of operating conditions that are typical of pilot facilities. Therefore, the objective of the Phase II case study visits is to obtain enough information to project the probable waste quantities and characteristics of similar full scale units.

223

Waste Disposal Cost Estimates. The number, size and location of commercial scale facilities utilizing the emerging technologies studied cannot be accurately determined at this time. Cost estimates made on a plant-by-plant basis (such as those in Phase I) are therefore not possible. Instead, scenarios will be developed for groups of hypothetical full scale facilities based on process data obtained from the emerging technology case studies and disposal site data obtained from the utility case studies. Waste disposal costs are then estimated using the cost models developed in Phase I.

Evaluation of Commercialization Impacts. Because the different emerging technologies affect waste generation in significantly different ways, some of these technologies are likely to be much more heavily impacted by RCRA than others. This project therefore includes a comparison of the relative waste disposal costs for various competing technologies in order to assess the probable impact of RCRA on their commercialization. Additionally, the projected commercialization rates of various emerging technologies will be factored into estimated costs of waste disposal for conventional (Phase I) power plants.

## Methodology for Phase III

The primary objective of Phase III is to evaluate the results of Phases I and II on a nationwide, macroeconomic basis so that the DOE can formulate and implement policies which are consistent with the National Energy Pla while recognizing legitimate environmental requirements. The major elements of Phase III are shown in Figure 1 and described below.

Economic Background Data. This activity will involve the compilation of various types of economic background information for use in the remainder of Phase III.

Utility Sector Evaluation. The potential impacts of RCRA on the utility sector will be evaluated in terms of:

- o impacts on projected coal use
- o allocation of RCRA costs among electricity users and stockholders
- o capital crowding impacts
- o impacts of tax policies
- o consumer income impacts

Non-quantifiable RCRA Impacts. A number of the potential impacts of RCRA cannot be quantified in terms of direct costs. Items to be evaluated will include:

- o jurisdictional impediments to RCRA implementation
- o impacts resulting from delays created by RCRA
- o difficulties in permitting waste disposal sites
- o strategic factors resulting from changes in coal demand
- o local, regional and national geo-political impacts
- o factors affecting insurance, financial and governmental institutions

Potential RCRA Benefits. Potential benefits resulting from RCRA will be assessed. This assessment will include:

- o historical documentation of adverse environmental, health and economic effects

224

    o  potential damage costs of groundwater contamination
    o  impacts on waste utilization and resource recovery
    o  changes in land use and land values
    o  cost/benefit and/or cost effectiveness comparisons

<u>Sensitivity Analysis</u>.  The critical assumptions made during Phases I, II and III will be examined to determine their effects on the conclusions which were derived.

<u>Programmatic Impacts</u>.  The results of Phases I, II and III will be summarized in terms of the overall impacts of RCRA on DOE programmatic objectives.

Preliminary Results Obtained to Date

Although work is presently in progress on all three phases, preliminary results are available only for Phase I (utility power plants) and, to a lesser degree, Phase II (emerging coal technologies).  These are presented in an Interim Report prepared by Engineering-Science and submitted to DOE in November 1979.

## Preliminary Results of Phase I

Idealized designs and cost estimates were prepared for each of the four waste disposal scenarios considered (i.e., current wet disposal, current dry disposal, non-hazardous and hazardous disposal were defined according to the proposed RCRA regulations published in 1978 (see Figures 3 and 4). Preliminary designs and cost estimates were also prepared for six case study sites which were visited in 1979 (see Figure 5) using the same four waste disposal scenarios.  The basic assumptions used in these designs and cost estimaes are summarized in Figure 6.  These cost estimates resulted in a series of mathematical models of capital and O&M cost associated with each waste disposal scenario (Figure 7).  These cost models were applied to 1977 coal burn data from Federal Power Commission Form 67 to obtain regional and national cost estimates.  (An example of this calculation is  shown in Figure 8).  A summary of regional and national cost impacts based on this interim analysis is presented in Figure 9.

## Preliminary Results of Phase II

The November 1979 Interim Report presented a cursory assessment of the comparative costs of various emerging coal combustion and conversion technologies.  Because case study site visits of various emerging technologies were not possible within the time frame of the Interim Report, wast production volumes and characteristics were estimated based on available literature and discussions with individuals knowledgeable of the various processes.  This characterization data, coupled with the interim Phase I cost models, were used to estimate relative disposal costs, resulting in the values shown in Figure 10.  It is important to note that the values presented in Figure 10 are not intended to represent actual costs, but only the relative disposal costs of one process versus another.

Conclusions

Although the project is still in progress and most of the work has yet to be completed, certain conclusions are possible based on the results obtained to date.  Some of the most significant findings are summarized below:

1. Presently available data regarding utility waste disposal is often incomplete, inaccurate and/or contradictory.

2. In most cases, current practices for disposal of coal combustion wastes do not conform to RCRA, whether or not the wastes are declared to be hazardous.

3. Captial costs for disposal of coal combustion wastes can be accurately modeled using a log-log ($y = ax^b$) relationship between costs and waste volume generated.  O&M costs can be accurately modeled using a slope-intercept ($y = ax + b$) relationship.

4. Based on requirements presented in the 1978 draft RCRA regulations, utility waste disposal costs can be expected to increase the cost of electricity by approximately 0.8 to 3.0 mils per kwh, depending upon whether these wastes are declared hazardous or non-hazardous.

5. Siting of waste disposal facilities in environmentally sensitive areas (e.g., wetlands, floodplains,  active fault zones, habitats of endangered species, etc.) can drastically increase disposal costs as well as permitting difficulties.

6. Relative costs for disposal of residues from various emerging coal technologies can vary considerably from one technology to another.  These cost differences appear to be of sufficient magnitude to affect the rates of commercialization of various competing technologies.

7. At the time the project was begun (January 1979), there was relatively little comprehension of RCRA by states and utility owners as it impacts coal-fired facilities.  Although this has been changing over the past few months, there is still a general underestimation of the potential magnitude of this impact.

FIGURE 1
## SCOPE OF RCRA IMPACT ASSESSMENT

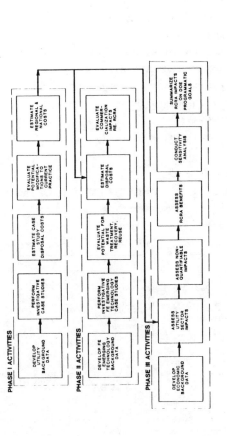

FIGURE 2
## COST ESTIMATING PROCEDURE

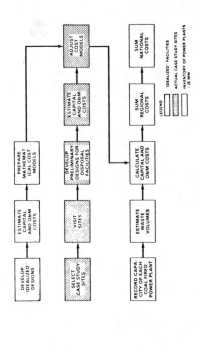

FIGURE 3
## TYPICAL* HALF SECTION THROUGH NON-HAZARDOUS DISPOSAL CELL

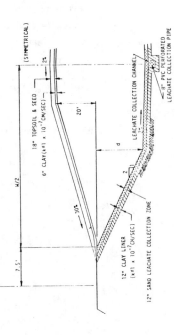

*BASED ON PROPOSED RCRA REGULATIONS.

FIGURE 4
## TYPICAL* HALF SECTION OF HAZARDOUS WASTE DISPOSAL BASIN

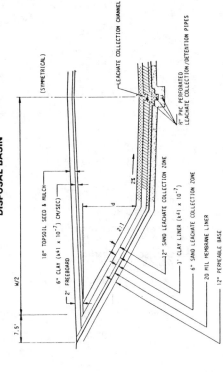

*BASED ON PROPOSED RCRA REGULATIONS.

227

FIGURE 5
## CASE STUDY SITE LOCATIONS
### (Interim Report)

COLSTRIP

EDDYSTONE

CONESVILLE

BOWEN

TOMBIGBEE

MARTIN LAKE

FERC REGIONS
1. NEW ENGLAND
2. MIDDLE ATLANTIC
3. EAST NORTH CENTRAL
4. WEST NORTH CENTRAL
5. SOUTH ATLANTIC
6. EAST SOUTH CENTRAL
7. WEST SOUTH CENTRAL
8. MOUNTAIN
9. PACIFIC — PLUS ALASKA & HAWAII

FIGURE 6

## BASIC ASSUMPTIONS EMPLOYED FOR INTERIM ASSESSMENT

- ANALYSES BASED ON PROPOSED RCRA REGULATIONS

- ANALYSES INCLUDED DIRECT COSTS ONLY (1979 DOLLARS)

- DISPOSAL SITES LOCATED TO MINIMIZE COSTS

- REGIONAL COST ESTIMATES BASED ON LIMITED DATA

- NATIONAL COSTS ESTIMATES BASED ON WEIGHTED AVERAGE

- COSTS FOR ENVIRONMENTALLY SENSITIVE AREAS (ESA'S) NOT INCLUDED

- NO CONSIDERATION GIVEN TO WASTE SEGREGATION

- ANALYSES INCLUDED ALL PLANTS > 25 MW

- STATE/LOCAL, JURISDICTIONAL AND OTHER IMPEDIMENTS NOT CONSIDERED

FIGURE 7
## EQUATIONS FOR DETERMINING COSTS OF VARIOUS DISPOSAL OPTIONS
### (Based On 1978 Proposed RCRA Regulations)

| Cost Functions | Current Disposal Practices Wet Disposal | Dry Disposal | Non-Hazardous Disposal | Hazardous Disposal |
|---|---|---|---|---|
| **Capital Costs** | | | | |
| Idealized Design | $Y_1 = 190 \ X^{.509}$ | $Y_1 = 68.3 \ X^{.538}$ | $Y_1 = 39.1 \ X^{.693}$ | $Y_1 = 248 \ X^{.647}$ |
| Region 2 | ----- | $Y_1 = 200 \ X^{.538}$ | $Y_1 = 84.8 \ X^{.693}$ | $Y_1 = 570 \ X^{.647}$ |
| Region 3 | $Y_1 = 205 \ X^{.509}$ | $Y_1 = 69.8 \ X^{.538}$ | $Y_1 = 76.3 \ X^{.693}$ | $Y_1 = 447 \ X^{.647}$ |
| Region 5 | $Y_1 = 194 \ X^{.509}$ | ----- | $Y_1 = 61.7 \ X^{.693}$ | $Y_1 = 468 \ X^{.647}$ |
| Region 6 | $Y_1 = 216 \ X^{.509}$ | ----- | $Y_1 = 63.0 \ X^{.693}$ | $Y_1 = 412 \ X^{.647}$ |
| Region 7 | ----- | $Y_1 = 115 \ X^{.538}$ | $Y_1 = 48.6 \ X^{.693}$ | $Y_1 = 304 \ X^{.647}$ |
| Region 8 | $Y_1 = 453 \ X^{.509}$ | ----- | $Y_1 = 57.8 \ X^{.693}$ | $Y_1 = 396 \ X^{.647}$ |
| Nationwide | $Y_1 = 235 \ X^{.509}$ | $Y_1 = 106 \ X^{.538}$ | $Y_1 = 69 \ X^{.693}$ | $Y_1 = 451 \ X^{.647}$ |
| **O&M Costs** | | | | |
| Idealized Design | $Y_2 = .0016 \ X + 45000$ | $Y_2 = .043 \ X + 69000$ | $Y_2 = .042 \ X + 134000$ | $Y_2 = .070 \ X + 151000$ |
| Region 2 | ----- | $Y_2 = .043 \ X + 224000$ | $Y_2 = .042 \ X + 502000$ | $Y_2 = .070 \ X + 578000$ |
| Region 3 | $Y_2 = .0016 \ X + 67000$ | $Y_2 = .043 \ X + 107000$ | $Y_2 = .042 \ X + 322000$ | $Y_2 = .070 \ X + 519000$ |
| Region 5 | $Y_2 = .0016 \ X + 74000$ | ----- | $Y_2 = .042 \ X + 178000$ | $Y_2 = .070 \ X + 234000$ |
| Region 6 | $Y_2 = .0016 \ X + 43000$ | ----- | $Y_2 = .042 \ X + 212000$ | $Y_2 = .070 \ X + 337000$ |
| Region 7 | ----- | $Y_2 = .043 \ X + 293000$ | $Y_2 = .042 \ X + 606000$ | $Y_2 = .070 \ X + 610000$ |
| Region 8 | $Y_2 = .0016 \ X + 58000$ | ----- | $Y_2 = .042 \ X + 143000$ | $Y_2 = .070 \ X + 417000$ |
| Nationwide | $Y_2 = .0016 \ X + 62000$ | $Y_2 = .043 \ X + 150000$ | $Y_2 = .042 \ X + 283000$ | $Y_2 = .070 \ X + 422000$ |

$Y_1$ = Capital Cost, 1979 dollars/year
$Y_2$ = O&M Cost, 1979 dollars/year
$X$ = Waste Generation $Ft^3$/year
Total Annual Cost = $Y_1 + Y_2$

228

# FIGURE 8

## TYPICAL COMPUTATION OF REGIONAL DISPOSAL COSTS

*HAZARDOUS DISPOSAL – RCRA SECTION 3004    ****FERC REGION 7 (WEST SOUTH CENTRAL)

NOTE:

ASH QUANTITIES USED ARE BASED ON 1977 REPORTED VALUES
ALL COSTS ARE EXPRESSED IN 1979 DOLLARS
ALL QUANTITIES ARE EXPRESSED ON AN ANNUAL BASIS
ASH DENSITY USED = 75 LBS/CF

| STATE | PLANT OWNER | PLANT NAME | COAL BURN 1000 TON | ASH PCT | ASH 1000 TON | ASH 1000 CF | O&M COST x $1000 | CAP COST x $1000 | TOTAL ANNUALIZED COST x $1000 |
|---|---|---|---|---|---|---|---|---|---|
| TX | SAN ANTIONI PUB SERVICE BD | DEELY | 516 | 10.500 | 54 | 1444 | 711 | 2923 | 3634 |
| | SOUTHWESTERN ELEC PWP | WELSH | 1073 | 10.500 | 112 | 3004 | 820 | 4693 | 5513 |
| | SOUTHWESTERN PUB SERVICE | HARRINGTON | 1053 | 10.500 | 110 | 2948 | 816 | 4636 | 5462 |
| | TEXAS POWER & LIGHT | BIG BROWN | 5026 | 10.500 | 527 | 14072 | 1592 | 12737 | 14329 |
| | | MARTIN LAKE | 2854 | 10.500 | 299 | 7991 | 1167 | 8834 | 10002 |
| | | MONTICELLO | 5988 | 10.500 | 628 | 16766 | 1780 | 14264 | 16044 |
| OK | OKLAHOMA GAS & ELECTRIC | MUSKOGEE | 438 | 5.400 | 23 | 630 | 654 | 1710 | 2365 |
| | ***** REGIONAL TOTALS | | 16948 | 10.368 | 1757 | 46858 | 7542 | 49799 | 57341 |

COST EQUATION:

$$y = [304.203 \times (\text{CUBIC FT. OF ASH})^{0.46667}] + [0.0697612 \times (\text{CUBIC FT. OF ASH}) + 610508]$$

WHERE y = TOTAL ANNUALIZED COST

TOTAL COST $ = [CAPITAL COST] + [O&M COST]

---

# FIGURE 9

## SUMMARY OF INTERIM REGIONAL AND NATIONAL COST IMPACTS

| FERC REGION | TOTAL NO. OF FACILITIES | POWER PRODUCTION (KWH x 10^9) | CURRENT PRACTICE (MILS/KWH) | NON-HAZARDOUS* DISPOSAL (MILS/KWH) | HAZARDOUS* DISPOSAL (MILS/KWH) |
|---|---|---|---|---|---|
| I | 1 | 2 | 0.31 | 1.10 | 3.2 |
| II | 41 | 95 | 0.37 | 1.63 | 4.8 |
| III | 127 | 304 | 0.20 | 1.19 | 3.1 |
| IV | 75 | 118 | 0.29 | 1.17 | 3.4 |
| V | 61 | 186 | 0.22 | 0.97 | 3.3 |
| VI | 38 | 143 | 0.21 | 0.97 | 2.8 |
| VII | 7 | 35 | 0.18 | 0.65 | 1.6 |
| VIII | 27 | 91 | 0.28 | 0.76 | 2.4 |
| IX | 1 | 11 | 0.16 | 0.86 | 2.4 |
| NATIONAL TOTAL | 378 | 985 | 0.24 | 1.09 | 3.2 |

*ASSUMES DISPOSAL SITES ARE NOT LOCATED IN AN ENVIRONMENTALLY SENSITIVE AREA

---

# FIGURE 10

## PROJECTED WASTE DISPOSAL COSTS FOR EMERGING COAL TECHNOLOGIES*

| POWER GENERATION TECHNOLOGY | WASTE DISPOSAL COST (1979 mils/kwh) | | |
|---|---|---|---|
| | EXISTING PRACTICE | NON-HAZARDOUS DISPOSAL | HAZARDOUS DISPOSAL |
| CONVENTIONAL W/FGD | | | |
| 250 MW | 0.57 | 2.30 | 6.83 |
| 500 MW | 0.36 | 1.80 | 5.25 |
| 1000 MW | 0.27 | 1.43 | 4.07 |
| CONVENTIONAL W/O FGD | | | |
| 250 MW | 0.35 | 1.46 | 4.37 |
| 500 MW | 0.25 | 1.12 | 3.33 |
| 1000 MW | 0.18 | 0.87 | 2.13 |
| SRC II LIQUEFACTION | | | |
| 250 MW | 0.14 | 0.77 | 2.12 |
| 500 MW | 0.13 | 0.76 | 2.08 |
| 1000 MW | 0.13 | 0.74 | 2.03 |
| LOW-BTU GASIFICATION | | | |
| 250 MW | — | — | — |
| 500 MW | — | — | — |
| 1000 MW | 0.16 | 0.79 | 2.29 |
| ATMOSPHERIC FLUIDIZED-BED | | | |
| 250 MW | 0.58 | 2.77 | 8.19 |
| 500 MW | 0.42 | 2.18 | 6.31 |
| 1000 MW | 0.32 | 1.74 | 4.90 |
| PRESSURIZED FLUIDIZED-BED | | | |
| 250 MW | 0.49 | 2.24 | 6.68 |
| 500 MW | 0.35 | 1.75 | 5.13 |
| 1000 MW | 0.26 | 1.39 | 3.97 |

*INTERIM REPORT, PHASE I - UTILITY SECTOR (NOVEMBER, 1979)

# PSD AND ENERGY DEVELOPMENT: THE FIRST FIVE YEARS

**D. B. Garvey** and **S. B. Moser**
Argonne National Laboratory

Introduction

One goal of air quality management in the U.S. is the protection of
areas where the air is currently cleaner than the National Ambient Air
Quality Standards from undue additions to the pollutant levels.
Regulations to prevent the significant deterioration of air quality
have the potential for conflicting with national energy goals of
decreased dependence on oil and increased use of coal. This paper
examines the interactions between these energy and environmental goals
through an examination of recent decisions for the siting of new
fossil-fueled utility boilers under PSD regulations, based on data from
EPA permit files.

PSD Permits for Utilities

In a review of PSD and energy development, it was learned that 81 PSD
permits had been approved (as of July 1980) for a total addition to
generating capacity of 63,240 MW. The total includes 2,155 MW of oil
and/or gas-fired facilities. Thirty-one permits are pending (as of
July 1980), for a further addition to generating capacity of 25,215 MW
(including 365 MW of oil-fired units). Within that pending permit
total are 6,000 MW (or 25%) covered by permits that have been pending
since 1977. Only two permits, covering a total of 675 MW, were determined to have been formally withdrawn. No permit for a fossil-fueled
power plant that was reviewed had been denied. However, three permits
(two in Kansas and one in Missouri) for facilities greater than 3,000
MW to be constructed over a period of 10-15 years, had been denied for
the total capacity but approved for one of the smaller units.

## Regional Trends

The approved and pending PSD permits reviewed have been grouped by Federal Region, as shown in Table 1. (See Fig. 1 for the boundaries of these Federal Regions.) Regions I and X have had no permit action that was discovered in the review. Regions IV and VI have experienced the largest growth in approved generating capacity, as displayed in Fig. 2. Region IV has approved 18 PSD permits for 19,480 MW of fossil-fueled generating capacity, or 31% of the total approved additions for all regions. Moreover, Region IV has 7,430 MW of capacity covered by pending permits, or 30% of potential additions to national generating capacity. Region VI has approved 22 permits for 16,920 MW of new fossil-fueled capacity, or 27% of the national total of approved additions. The region contributes 4,275 MW or 17% to the total of new capacity covered by pending PSD permits. Regions V and VIII have roughly comparable capacity additions covered by PSD permits, although permits in Region VIII are typically for larger facilities (9 permits for 9,605 MW) than in Region V (14 permits for 8,835 MW). Region VIII has only one pending PSD permit for a 600 MW unit.

The state-level contributions to these regional totals are presented in Figs. 3 and 4, with counties shaded according to approved and pending additions to capacity. Texas has approved 15 PSD permits for a total of 12,550 MW of new fossil-fueled generating capacity and is reviewing permits for an additional 1,400 MW. Kentucky ranks second in new utility construction with 7,000 MW of approved capacity additions and 3,500 MW under review. Florida, Georgia, Indiana, South Carolina and Utah each have approved permits covering approximately 3,500 MW of new generating capacity.

## BACT Determinations

According to the legislation and regulations, a new major source siting in a PSD area is to be required to use the best available control technology, or BACT, as determined on a case-by-case basis. New Source Performance Standards are to be viewed as guidelines for BACT determinations, setting minimum acceptable emission limitations, but not necessarily reflecting new or improved control technology, as BACT should. A BACT determination is to take into consideration energy, economic and environmental factors.

The BACT determinations for particulate matter (PM) range from the most stringent level on a PSD permit reviewed of 0.02 lb PM/$10^6$Btu to the 0.1 lb PM/$10^6$Btu of the pre-1979 NSPS. The highest percentage removal required was 99.9%, to be achieved by a combination of an electrostatic precipitator and a baghouse.

BACT determinations for $SO_2$ exhibit considerably more variation and complexity than those for PM. The timing of the permit approval relative to the revised NSPS for coal-fired utility boilers (June 1979) seems to be significant. Table 2 summarizes the emission limitations on the PSD permits in comparison to the 1.2 lb $SO_2$/$10^6$Btu limit required by NSPS prior to the 1979 regulations. More than half of the approved new capacity will be required to meet the 1.2 lb. limit, while 33% will need to achieve more stringent emission limitations. Nine of the 81 approved permits were completed in 1980; eight reflect the revised NSPS and require a percent reduction in $SO_2$ emissions. Of the pending permits, 40% of capacity is expected to achieve 1.2 lb $SO_2/10^6$Btu and 34% will be required to meet more stringent emission limits. We do not have information on the remaining 27% of capacity with pending permits.

An alternative approach to summarizing BACT determinations is presented in Tables 3 and 4 where the method of $SO_2$ control is reviewed across the sizes of facilities. The column headed "low-sulfur coal" indicates that the use of the fuel alone, without post-combustion clean-up, was determined to be BACT. Thirty-two percent of approved capacity will be allowed simply to burn low sulfur coal as a permit condition. Approximately 33% of the facilities approved that are larger than 1,000 MW were reviewed before the 1979 NSPS revisions and received this low-sulfur coal BACT determination. As seen in Table 4 only 3% of total capacity covered by a pending permit, and no facility larger than 1,000 MW, will be permitted to use only low-sulfur coal as a method of compliance with $SO_2$ emission limitations.

BACT determinations for $SO_2$ will require the use of FGD systems with less than 85% removal on 18,000 MW of approved capacity, or 29% of the total. Similar control requirements will be placed on 27% of the capacity covered by pending permits, or 6,800 MW. The installation of FGD systems with greater than 85% reduction will be needed by 22% of approved capacity and 11% of capacity still subject to review. These totals do not reflect the instance where low-sulfur coal and an FGD for more than 85% removal of $SO_2$ are both required. The latter occurs, for example, in Montana, where a Class I increment is to be protected.


Increment Consumption

PSD regulations established increments for $SO_2$ and TSP placing an absolute ceiling on new emissions in an area. The consumption of those increments can clearly have an impact on the siting of new major sources. The baseline and the increments have changed since EPA's initial promulgation of regulations in December 1974. The permit applications we reviewed were submitted throughout the time period 1975-1980 and were covered by different regulations. We have used the increment consumption analysis as provided in the PSD permits with only one minor exception -- if EPA's 1974 increments (e.g., 100 $\mu g/m^3$ for $SO_2$, 24-hr) had been used in the permit, we examined the consumption relative to the increments provided by the 1977 CAAA and subsequent regulations (or 91 $\mu g/m^3$ for the $SO_2$, 24-hr increment).

In the permits reviewed, increments were usually allocated on a first-come, first-served basis. Typically, a power plant was the first major source in an area, BACT was set at NSPS levels, and the new source was allocated as much increment as needed, as long as a violation did not occur. The only exception to this procedure was in the case of permit applications for large energy parks of more than 3,000 MW. These construction projects, planned to be built in phases over 10-15 years, were not allowed to save the increment for future use. In one case, 250 MW were approved out of a proposed 2,340 MW, and in another example, 630 MW were approved out of 3,000 MW. Applications for the additional capacity at these sites have not yet been submitted. The future availability of increments may limit such large energy facilities.

Regional SO₂ Increment Consumption

Increment consumption for $SO_2$ has been examined for the 112 PSD permits (pending and approved) that were located in the study. The 24-hour increment was chosen as the determining standard, based on the conclusion that the short-term increment was the most restrictive. Permits were flagged if the increment analysis indicated that more than 50% of the available increment would be consumed. This consumption could be either as a result of the emissions of a single new major source, or as the result of the impact of other sources in addition to the new utility.

There are 41 permits where more than 50% of the 24-hour $SO_2$ increment has been allocated. Overall, 34,610 MW of new capacity (or 39% of the total) will be sited in counties resulting in the use of more than 50% of the increment. In Regions V and VI, 60% of the capacity reviewed under PSD will consume more than 50% of the $SO_2$ increment. In contrast, Regions IV and VII will use up half the increment in the siting of approximately 25% of new capacity.

Figure 5 displays those counties where pending and approved PSD permits will consume more than 50% of the $SO_2$ 24-hour increment. This map should <u>not</u> be viewed as displaying all counties in the U.S. where such increment consumption has occurred, but only those where PSD permits for utilities have been reviewed and the increment consumption identified. Other areas may be similarly affected by other major sources.

Types of Increment Consumption

<u>Class II, Single Source.</u> An individual new power plant can easily consume more than 50% of the $SO_2$, 24-hour increment, depending on the BACT determination. For example, in Illinois, a new 1,300 MW utility, using 3.5% sulfur coal, with a FGD system designed for 82% $SO_2$ removal, to comply with a 1971-NSPS emission limit of 1.2 lb $SO_2/10^6$ Btu, will use up 47 $\mu g/m^3$ of the statutory 91 $\mu g/m^3$ increment. A 1,500 MW utility (two 750 MW units) in Texas is predicted to result in a maximum 24-hr concentration of 76.5 $\mu g/m^3$, when emissions are limited to 1.2 lb $SO_2/10^6$Btu.

Site specific characteristics have a significant impact on the modeled increment consumption. For example, the 3,000 MW Intermountain Power Project in Utah is projected to use up 50 $\mu g/m^3$ of the 91 $\mu g/m^3$, although the plant will remove 90% of the $SO_2$ from coal with 0.79% sulfur content. The site is located in hilly terrain, where the plume tends to be less dispersed. In contrast, Georgia Power Company's new 3,275 MW plant, composed of four new units is predicted to use up only 32.4 $\mu g/m^3$ of the 24-hour increment, with emissions limited to 1.2 lb $SO_2/10^6$Btu by the use of low-sulfur coal.

<u>Class I Increments.</u> Three PSD permits that were reviewed involved Class I $SO_2$ increment usage. In Arizona a new 350 MW coal-fired plant was predicted to result in a maximum concentration of 2.6 $\mu g/m^3$, or more than 50% of the 5 $\mu g/m^3$ of the Class I, 24-hour $SO_2$ increment. In South Carolina two new 280 MW units, limiting emissions to 1.2 lb $SO_2/10^6$Btu, were initially projected to violate the Class I increment of the Cape Romain Wilderness area. The utility revised

the amount of flue gas to be scrubbed and the removal efficiency for a maximum allowable emission rate of 0.6 lb $SO_2$/ $10^6$Btu, to achieve a predicted ambient impact of 3.9 $\mu g/m^3$. The Class I increment was not violated, but there would be little room left for another major $SO_2$ source to site within the vicinity of Cape Romain.

In Montana the 1,500 MW additions to the Colstrip power plant will use up 4.8 $\mu g/m^3$ of the Class I increment of 5, despite the use of 95% scrubbing on 1.0% sulfur coal to achieve an emission rate of 0.18 lb $SO_2/10^6$Btu. The Class I increments are clearly readily consumed and the area surrounding these pristine areas will support minimal new emissions.

Increment Shared With Other Sources.    In a number of the counties identified as having more than 50% of the increment used, the consumption was the result of the projected emissions of the new utility in addition to other, previously approved, PSD sources. For example, in Oklahoma a new 1,100 MW addition to a coal-fired power plant had a predicted maximum 24-hour concentration of 25 $\mu g/m^3$. The impact of other PSD sources resulted in a concentration of 65 $\mu g/m^3$, or more than 50% of the increment. Similarly, in Louisiana, a 1,600 MW plant only contributed 5 $\mu g/m^3$ to a total increment consumption of 78. The permit review lists eleven other $SO_2$ sources within the impact area.

In the case of Indianapolis Power and Light's proposed 1,950 MW plant, approval was based on permit conditions of a 91% scrubber, 90% available, and a demonstration that the coal supply would never exceed 3.47% sulfur, to achieve a 0.55 lb $SO_2/10^6$Btu maximum emission rate. The utility had to share the increment with two other power plants in Kentucky, a distillery, and a glass manufacturing company. The power plant is located in an area that is expected to see major industrial and utility growth -- the Ohio River Basin. Under the stringent $SO_2$ controls of the permit, the plant has been approved for construction, leaving only 4.8 $\mu g/m^3$ of the increment available for additional major sources.

Energy Parks.    Three permits were identified where $SO_2$ increment consumption issues resulted from the effort to site multiple units of a coal-fired utility. In Pottawatomie County, Kansas, for example, a PSD permit was issued for two 720 MW additions to two existing units of 720 MW each. The Jeffrey Energy Center had been planned as a major (3,000-5,000 MW) power park. Unit #1 was not covered by PSD review and its emissions were included in the baseline air quality. Unit #2 was exempt from review, but because construction took place after January, 1975, the emissions counted against the increment. Units 3 and 4 have undergone PSD review with a BACT determination of 60% scrubbing on 0.48% sulfur coal, to achieve a maximum emission rate of 0.5 lb $SO_2/10^6$Btu. Units 2, 3, and 4 together will consume 84.9 $\mu g/m^3$ of the increment of 91. Thus, the original plans for additional units (if demand warranted further expansion) will be virtually impossible within the increment still available.

Conclusion

The idea of preventing significant deterioration of air quality in areas that are cleaner than the standards was a product of the environmental enthusiasm of the early 1970's. As a result of legal action by

234

the Sierra Club and an affirmative decision by the Supreme Court, regulations were deemed necessary to ensure that the air quality in clean air areas of the U.S. would not be allowed to deteriorate to the level of the NAAQS.

PSD regulations have been issued, disputed, and revised, but the concept has not changed since the promulgation of the first set of rules in December 1974. PSD has been castigated by industrial spokesmen for being a "no growth" policy. On the other hand, PSD has been heralded as necessary to preserve the aesthetic integrity of certain natural wonders of the nation, such as the Grand Canyon, and to restrict the adverse air quality effects of unlimited growth. Based on our review of permits, we have concluded that, during this period of declining energy demand and general economic recession, major siting constraints on new generating capacity from increment availability have not occurred. As moderate industrial and energy growth takes place, however, incidents of increment constraint may occur with increasing frequency.

Acknowledgments

This work was sponsored by the Regulatory Analysis Division, Office of Environmental Assessments, Assistant Secretary for Environment, U.S. Department of Energy. DOE project officer is Doug Carter. The opinions expressed are those of the authors, and should not be construed as representing ANL or DOE policy.

Table 1  PSD Permits for Fossil-Fueled Utility Boilers
as of July 30, 1980: Regional Totals

| Federal Region | Approved (#) | Capacity (MW) | % of Total Capacity | Pending (#) | Capacity (MW) | % of Total Capacity |
|---|---|---|---|---|---|---|
| I | 0 | – | – | 0 | – | – |
| II | 1 | 900 | 1 | 2 | 2,350 | 9 |
| III | 2 | 430 | <1 | 3 | 1,850 | 7 |
| IV | 18 | 19,480 | 31 | 6 | 7,430 | 30 |
| V | 14 | 8,835 | 14 | 3 | 4,410 | 18 |
| VI | 22 | 16,920 | 27 | 4 | 4,275 | 17 |
| VII | 8 | 4,650 | 7 | 8 | 1,590 | 6 |
| VIII | 9 | 9,605 | 15 | 1 | 600 | 2 |
| IX | 7 | 2,560 | 4 | 4 | 2,710 | 10 |
| X | 0 | – | – | 0 | – | – |
| TOTAL | 81 | 63,380 | | 31 | 25,215 | |

Table 2  BACT Determinations Compared to NSPS[a] for $SO_2$

| Status of Permit | Capacity (MW) Subject to Emission Limitations | | | | | | |
|---|---|---|---|---|---|---|---|
| | NSPS[a] | | More Stringent than NSPS | | N/A[b] | | Total |
| Approved | 34,465 | (56%) | 20,275 | (33%) | 6,365 | (10%) | 61,105 |
| Pending | 9,580 | (39%) | 8,470 | (34%) | 6,800 | (27%) | 24,850 |
| Total | 44,045 | (51%) | 28,745 | (33%) | 13,165 | (15%) | 85,955 |

[a]NSPS = 1.2 lb $SO_2/10^6$Btu.

[b]Not available.

Table 3  BACT Determinations for $SO_2$:  Approved PSD Permits for Coal-Fired Utility Boilers, as of July 30, 1980

| Facility Size (MW) | Capacity (MW) to be Controlled | | | | |
|---|---|---|---|---|---|
| | Low Sulfur Coal | FGD <85% | FGD > 85.1% | N/A[a] | Total |
| 0-499 | 1,170 | 1,850 | 905 | 1,205 | 5,130 |
| 500-1,000 | 5,030 | 5,045 | 2,750 | 4,355 | 17,180 |
| >1,000 | 12,255 | 11,090 | 9,850 | 5,600 | 38,795 |
| TOTAL | 18,455 (32%) | 17,985 (29%) | 13,505 (22%) | 11,160 (18%) | 61,105[b] |

[a]Not available.

[b]Does not include 2,155 MW of non-coal-fired capacity additions.

Table 4  BACT Determinations for $SO_2$:  Pending PSD Permits for Coal-Fired Utility Boilers, as of July 1980

| Facility Size (MW) | Capacity (MW) to be Controlled | | | | |
|---|---|---|---|---|---|
| | Low Sulfur Coal | FGD <85% | FGD > 85.1% | N/A[a] | Total |
| 0-499 | 140 | 350 | 560 | 495 | 1,545 |
| 500-1,000 | 650 | 1,000 | 880 | 2,500 | 5,030 |
| >1,000 | - | 5,450 | 1,300 | 11,520 | 18,275 |
| TOTAL | 790 (3%) | 6,800 (27%) | 2,740 (11%) | 14,520 (58%) | 24,850[b] |

[a]Not available.

[b]Does not include 365 MW of non-coal-fired capacity additions.

236

Fig. 1  Boundaries of Federal Regions

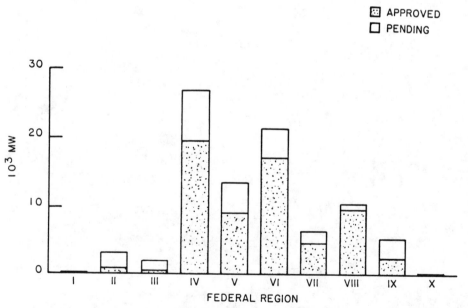

Fig. 2  Additional Fossil-Fueled Generating Capacity
Based on PSD Permits up to July 1980

237

Fig. 3  Approved Additions to Generating
Capacity, as of July 1980

Fig. 4  Pending Additions to Generating
Capacity, as of July 1980

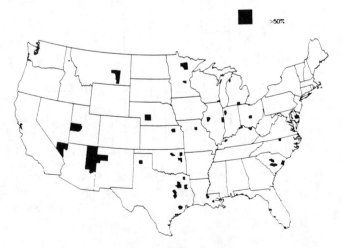

Fig. 5  PSD SO$_2$ 24-Hour Increment
Usage, as of July 1980

238

# IMPACT OF RECENT ENERGY AND ENVIRONMENTAL LEGISLATION ON THE PETROLEUM REFINING INDUSTRY

**Doan L. Phung**
Oak Ridge Associated Universities

## Introduction

As the energy crisis of the 1970's is essentially a mismatch between petroleum supply and demand, we expect to see the petroleum industry subjected to strong political and economic forces.  Indeed, in the United States, there have been no fewer than two dozen major laws designed to assist, control, regulate, and reorganize the industry.  These laws can be broadly classed into four categories: economic, land use and siting, environmental, and energy (Figure 1).

Today the industry is vigorous and continues to play the central role in the energy drama.  The upstream sector, which consists of exploration and production activities, is engaged in more drilling and rig construction than at any other time in history.  The downstream sector, which consists of refining and retailing, is also to undergo a large upgrading program, the outcome of which is yet to be seen.  At a demand level of 16-18 million barrels per day (MMB/D), crude price at upwards of $30 per barrel ($/B) and refined products at upwards of $40/B, the petroleum industry will have dealt with revenues of over $250 billion in 1980, more than any other sectors of the economy (Figure 2).  Oil companies presently account for approximately 40% of the total profits of the manufacturing sector, in sharp contrast to 18% in 1977 and 15% in 1972.[1]

In this paper we shall review and determine the impact of some pieces of legislation and their ensuing regulations on the refining sector of the oil industry.  For lack of space, only the main features will be discussed.

Structure of the Petroleum Refining Industry

There are basically three groups of petroleum refiners: the multi-nationals, the domestic majors, and the domestic independents. The first two groups are characterized by large, integrated operations, multiple sources of crude oil supply, and multiple criteria for profit maximization of which petroleum refining profit is only one factor. The domestic independents are characterized by small operations, service to the local markets, and frequent dependence on the majors and the government for crude oil supply.

The distribution of oil refineries according to geography, size, and complexity is shown in Figure 3. As of January 1979, these refineries had a gross book value of $24 billion ($1350/B/D), a replacement cost of $65 billion ($3730/B/D). They processed approximately 14.6 MMB/D (the balance of demand being supplied by imports) at an operations and maintenance cost of $2.29/B, of which half was for fuels.[2,3]

Winds of Change

There are several active trends which could bring about dramatic changes in the U.S. refining industry in the 1980's. These changes include the following:

Decontrol: By October 1981 all price controls on domestic crude oil and petroleum refined products will be removed.[4] As price and allocation controls of one form or another have been in effect on the oil industry for over 60 years, this decontrol will force a major rerestructuring of the industry.

Leveling of petroleum demand: Supply shortage, price increases, and legislative as well as consumers' reaction to those factors have leveled petroleum demand. The refining industry will not need crude distillation addition as much as upgrading facilities that will enable it to increase the yield of transportation fuels.

Lower-quality crude oil supply: The crude oil supplied to U.S. refineries is getting higher in sulfur and residuum. This requires the addition of facilities (vacuum distillation, cracking, desulfurization).

More stringent refined products: Environmental regulations require phasing down lead as a gasoline additive and reducing the level of sulfur in residual fuel oil. This also requires adding facilities (cracking, reforming, desulfurization).

Oil backout: Legislation is being introduced to back residual oil out of electricity generation. To upgrade this heavy and high-sulfur oil into lighter distillates requires facility additions (cracking, coking, desulfurization).

Environmental and safety regulations: Compliance with environmental laws such as clean air, clean water, and resource conservation and recovery continue to add costs. Occupational regulations on benzene and noise may also present additional burdens on refinery operations.

Impacts of Economic Legislation

Far outstripping other categories of legislation, economic laws and regulations have affected the industry most profoundly and will continue to.

240

At the heart of these laws are the questions of economic stability and equity--between the petroleum industry and other economic sectors, between domestic products and imports, between producers and refiners, and between groups of refiners.

Figure 4 illustrates some effects of major economic laws in the past several decades. The petroleum industry organized its own cartel in the 1920's and 1930's to maintain price of crude (at about $1/B) when production was larger than demand. The effect of this was a squeeze on the profit margin of refiners; thus, small independent refiners would not be able to compete with large refiners of petroleum companies who could afford to use the upstream profits to subsidize downstream operations.

As the share of cheaper foreign crude oil increased, the Mandatory Oil Import Program (MOIP) was instituted to protect domestic production. This program distributed quota tickets of imports to all refiners, with a percentage bias towards small and inland refineries. Since then, the number of small refineries has increased somewhat.

The price control of domestic crude and refined products, since 1971, has favored refiners with access to domestic crude. The Entitlement Program was supposed to correct this, but in the process, caused further construction of small, inefficient refineries and Caribbean refineries. Between 1974 and 1979, the capacities of 63 of the 64 grass-roots refineries built in the United States fell below 40,000 B/D, frequently below 10,000 B/D.[5] Price controls also depressed prices of California heavy crude, discouraged refinery innovation, and encouraged consumption of residual fuel oil for electricity production.

President Carter launched a decontrol program to begin by October 1981, as provided by EPCA.[4] Since the industry has been profoundly influenced by controls and other economic programs over the years, decontrol will not be without trauma. Small, independent refiners will no longer have the subsidies of the Entitlement Program. In addition, they will have to upgrade their facilities to manufacture products with more stringent specifications. It is possible that many will have to join forces, to sell out to larger refiners, or to exit from the industry.

Even large domestic refiners will also face difficult competition from imports. It is known that Caribbean and overseas refiners can process crude oil more cheaply than U.S. refineries. In addition, they do not have to observe the Jones Act on shipping. Legislation is now being considered either to establish a quota on the import of refined products, or to provide domestic refineries certain subsidies.[6]

Energy Legislation

Energy laws affecting the petroleum refining industry fall into three broad classes: energy conservation, oil backout, and synthetic fuel production.

The Energy Supply and Environmental Coordination Act of 1974, the Energy Policy and Conservation Act of 1975, and the National Energy Conservation Act of 1977 provide a legal mechanism for the government to establish energy efficiency standards and goals for energy-consuming facilities and equipment. Two of the most noteworthy programs have been the Corporate Average Fuel Economy (CAFE) Program and the Voluntary Industrial Energy Conservation Program.

241

CAFE requires the average fleet fuel efficiency of any single automobile manufacturing company to be 18.0 miles per gallon (mpg) by 1978, 20.0 mpg by 1980, and 27.5 mpg by 1985. Although the automobile industry had resisted such requirements for a number of years, shift in consumers' preference to small cars (due to sharp gasoline price increases) has launched the industry into a massive investment program for producing fuel-efficient cars. To date, all CAFE targets have been met, and it appears that most major car manufacturers will exceed the 1985 CAFE target of 27.5 mpg by a substantial margin.

Although the use of petroleum in industry (including the utility industry) increased from 4.42 MMB/D in 1974 to 5.20 MMB/D in 1979 (18%), much of that increase (2.2 quads) was for replacing natural gas (13.21 quads in 1974, 11.32 quads in 1979, a decrease of 1.9 quads).[7] Industry also responded diligently to oil price increases by instituting voluntary energy conservation programs. According to data monitored by the Department of Energy, the manufacturing industry achieved savings of between 10% and 16% per unit of production from 1972 to 1978.[8]

Data for the residential and commercial sectors reveal much the same story. Between 1974 and 1979, oil and gas consumption in these sectors essentially leveled in spite of the fact that the number of building stocks and floor space increased by an average of 10% during that time. Most projections on oil demands currently establish a leveling off of U.S. petroleum demand at 16-18 MMB/D in the 1980's, even accounting for economic growth. The impact on refineries is that new refineries will not be much needed except where logistics dictates such addition at the expense of the capacity elsewhere.

The industry, however, will still need and has embarked upon a major upgrading program to process lower quality crude oil, to upgrade residual fuel oil into higher quality products, and to turn out refined products with more stringent specifications. Later in the decade, adjusting some refineries to process synthetic liquid oils may also be a possibility.

Processing lower-quality crude: The crude oil slate of U.S. refineries is getting heavier and more sour (high in sulfur content). In 1978, the crude slate was estimated to be 46% sour, only half of which was light (less than 15% residuum after being heated to 1050°F).[2] With increasing shares of tertiary oil, California oil, Alaskan oil, and heavy Middle East oil, the crude slate by 1985 is expected to be 54% sour, more than half of which would be heavy. Additional facilities are needed in vaccum distillation, cracking, coking, desulfurization, hydrogen production, and sulfur reduction. Figure 5 indicates that an investment program of about $11 billion (1980) would be needed to double the capability to process high-sulfur heavy crude (from the current 2.2 MMB/D to 5.0 MMB/D).[2,9,10]

Upgrading residual fuel oil: 10% of the crude throughput, or about 1.7 MMB/D, becomes residual fuel oil (very heavy and high sulfur). This oil is only good for burning in ship bunkers or in electric generating power plants. The technology exists to further upgrade this oil to transportation fuels, such as gasoline, No. 2 oil, and diesel oil. Figure 6 shows three scenarios studied by Purvin and Gertz for upgrading residual fuel oil.[10] Legislation recently proposed would subsidize existing oil-fired power plants to back out of oil.[11]

More stringent specifications for refined products: This topic is discussed below.

Environmental Legislation

The environmental movement of the 1970's has made significant improvement in the air quality around the sites of oil refineries. Heavy smog and foul-smelling air are no longer permanent features of these areas. Since 1970, the industry has invested approximately $8.3 billion in air pollution abatement and $4.4 billion in water pollution abatement (1980 dollars).[12] Sheppard [13] estimates that by 1990, this amount would increase to a total of $30 billion (1980 dollars) for compliance with anticipated regulations of the Clean Air, Clean Water, Safe Drinking Water, and Resource Conservation and Recovery Acts.

Figure 7 shows the emissions to air from a model 120 MB/D refinery under three abatement scenarios: with reasonably available control technology (RACT), with best available control technology (BACT), and with BACT and the lowest achievable emission rate (LAER) for new emissions. Upgrading programs of refineries to process lower-quality crude and residual fuel oil may give rise to more pollution at the refinery site. In 1979, 76% of all refineries were located in non-attainment areas for hydrocarbon, 48% for particulates, 27% for carbon monoxide, 24% for sulfur dioxides, and only 9% for nitrogen oxides. Strict applications of the New Source Performance Standards (NSPS) for refineries within these non-attainment areas would incur additional costs to the upgrading programs. However, the Bubble and Banking concepts recently promoted by the Environmental Protection Agency (EPA) appear to ease these costs.

The Clean Water and Safe Drinking Water Acts impact refineries mostly in the control of condensed water effluents, storm water collection and discharge, ballast water discharge, wastewater discharge, and oil spills. The measurements and control of trace elements, such as benzene, cadmium, and cyanide, continue to exact more costs.

Recently EPA issued proposed regulations for implementing the Resource Conservation and Recovery Act (RCRA) of 1977. A number of refinery wastes are listed as hazardous, requiring a specified series of tests, manifests, and disposal. Figure 8 gives an estimate of solid waste sources and quantities for U.S. refineries.[14,15] SCS Engneers,[16] a consulting firm for the American Petroleum Institute, estimated that the investment costs for complying with RCRA would be in the order of $2.7 billion (1979 dollars). Sheppard et al [13] estimated the investment costs to be about $3.6 billion under the strict interpretation of the law.

The indirect impact of the Clean Air Act Amendments of 1977 on U.S. refineries is also dramatic. These amendments require that refined petroleum products must have low sulfur contents (0.7% for residual fuel oil, 0.5% for distillates, and 0.04% for gasoline) and that lead must be phased out as a gasoline octane booster. Many states such as California have even stricter sulfur and lead content standards.

Removing sulfur further from refined petroleum products requires desulfurization and hdyrogen facilities. Both types are quite expensive. For example, equipment for hydrogen production by partial oxidation has a complexity factor of 2; equipment for sulfur reduction has a complexity factor of 85. The investment cost is about $0.9-1.9 per cubic foot per day for hydrogen and $215,000 per short ton of sulfur per day. [9,10]

Phasing down lead in gasoline would require other avenues and/or additives to provide adequate octane for motor gasoline. Facilities needed are reforming, cracking, treating, alkylation, and isomerization. In order to comply with the current standard of 0.5 gram of lead per gallon of the gasoline pool, the industry needs an investment of $3.6 billion.[13] However, much of that investment is underway or has been completed.

Figure 9 summarizes our estimates of the costs of environmental laws to the petroleum refining industry and the reflection of these costs on refined products.

Conclusion

Being the principal actor in the energy scene, the petroleum industry has perhaps been subject to more legislation and regulations than any other industry. The refining sector of the industry was squeezed in the past by several economic laws that were designed to profit the upstream sector (exploration and production) or to protect consumers' interests. The industry structure was also deformed in the direction of small, inefficient refineries by the small refiners' bias in several pieces of economic legislation. Energy and environmental legislation also encouraged or required refineries to embark on a major investment program in spite of the fact that the total crude oil throughput will not grow significantly in the 1980's.

The industry is, however, alive and strong. Decontrol will return the industry to a free enterprise instead of a utility. We expect a major restructuring of the industry. Several small independent refineries will have to join forces, to sell out to the majors, or to close down. Even larger integrated refineries will feel the pinch from competition of refined products and imports.

The importance of these changes is demonstrated vividly by several pieces of legislation currently pending in the U.S. Congress.[6,11,17]

References

1. Business Week, "How 1200 companies performed in 1979" (March 17, 1980) and "Implications of oil company profit" (August 18, 1980).

2. National Petroleum Council, "Refinery Flexibility," an interim report, Vols. I an II, Washington, D.C. (1979).

3. The Oil and Gas Journal, "Annual Refining Report" (March 24, 1980).

4. U.S. Department of Energy, "National Energy Plan II," a report to the Congress required by Title VIII of the Department of Energy Organization Cut (May 1979).

5. E.L. Peer, F.V. Marsik and J.F. Hutchins, "Trends in refining capacity and utilization," DOE Assistant Secretary for Resource Application, Report DOE/RA-0010, Washington, D.C. 20461 (1979).

6. The U.S. Senate, Committee on Energy and National Resources, Hearing before the Subcommittee on Energy Regulation on the Domestic Refinery Development and Improvement Act of 1979 (September 11-12, 1979).

7. U.S. Department of Energy, Energy Information Administration, "Annual Report to Congress, Vol. 2," DOE/EIA-0173(79), Washington, D.C. (1980).

8. U.S. Department of Energy, "Industrial Energy Efficiency Improvement

Program," Assistant Secretary for Conservation and Solar Applications, Washington, D.C. 20545 (1978)

9.  P.R. Watters, "Bottom-of-the-barrel processing requirements," paper presented at the Energy Bureau's Conference on Petroleum Refining, Houston, Texas (October 6-7, 1980).  (Also available at the Pace Company, 5251 Westheimer, P.O. Box 53473, Houston, Texas 77052.)

10. Purvin and Gertz, Inc., "An analysis of potential for upgrading domestic refining capacity," report prepared for the American Gas Association (March 1980).

11. The U.S. Congress, The Oil Backout Legislation; S. 2470 (Ford); H.R. 6930 (Staggers); H.R. 7809 (Dingel) (1980).

12. American Petroleum Institute, "Environmental expenditures of the U.S. petroleum industry 1969-1978," Pub. No. 4314, Washington, D.C. 20037 (1980).

13. W.J. Sheppard et al., "The cost of environmental regulations to the petroleum industry," work prepared for the American Petroleum Institute by the Battelle Columbus Laboratory (July 1980).  Also available in a paper by the same author and title presented at the American Chemical Society Meeting, Las Vegas, Nevada (August 25-29, 1980).

14. American Petroleum Institute and National Petroleum Refiners Association, 1976 API-NPRA Conference Proceedings, Dallas, Texas, pp. 41 et seq. (January 31-February 1, 1980).

15. Jacobs Engineering Co., "Assessment of hazardous waste practices in the petroleum refining industry," EPA Publ. PB-259 097 (June 1976).

16. SCS Engineers, "Assessment of petroleum industry cost of compliance with proposed hazardous waste regulations," work done for the American Petroleum Institute (April 1979).

17. Pending bills that would affect the petroleum refining industry include S. 1371, Domestic Energy Policy Act; S. 1470, Mandatory Oil Import Control Act; H.R. 2608, Emergency Oil Import Quota Bill; H.R. 4985, Priority Energy Act; H.R. 6130, Oil Independence Act; S. 2202, Imported Oil Reduction Act; and S. 2412, Used Oil Recycling Bill.

Acknowledgement

This paper was compiled partially under the research carried out under Contract No. DE-AC05-760R00033 between the Technical Assessment Division, Office of Environmental Impact, Department of Energy, and Oak Ridge Associated Universities, and by letter agreement between Oak Ridge Associated Universities and Oak Ridge National Laboratory.

245

## Figure 1

Environmental

- Clean Air
- Clean Water
- Resource Conservation and Recovery
- National Environmental Policy
- Marine Protection
- Endangered Species
- Safe Drinking Water

Energy

- National Energy (FUA, PURPA ...)
- Energy Policy and Conservation
- Energy Supply and Environmental Coordination
- National Energy Conservation Policy
- Energy Security (Synfuels)
- Oil Backout (Pending)

Economic

- Jones
- Emergency Petroleum Allocation
- Economic Stabilization
- Petroleum Marketing Practices
- Windfall Profits
- Divorcement (Pending)

Land Use and Siting

- Federal Land Policy and Management
- Wilderness
- Antiquity
- Coastal Zone Management
- Outer Continental Shelf
- Deepwater Ports
- Crude Oil Transportation
- Alaska Lands

MAJOR PIECES OF LEGISLATION AND PROGRAMS
AFFECTING THE PETROLEUM INDUSTRY

Figure 1

## Figure 2

| | 1979 Sales ($ billion) | 1979 Return on Common Equity (%) |
|---|---|---|
| Aerospace | 44 | 21.7 |
| Airlines | 27 | 6.8 |
| Appliances | 11 | 9.3 |
| Automotive | 159 | 11.3 |
| Banking | 81 | 15.2 |
| Beverage | 16 | 14.6 |
| Building Materials | 20 | 15.8 |
| Chemicals | 84 | 17.1 |
| Conglomerates | 73 | 18.0 |
| Containers | 20 | 13.4 |
| Drugs | 38 | 20.8 |
| Electrical & Electronics | 72 | 19.7 |
| Food Processing | 115 | 15.4 |
| Food & Lodging | 13 | 17.1 |
| General Machinery | 31 | 17.5 |
| Instruments | 12 | 15.9 |
| Leisure time Industries | 23 | 18.0 |
| Metals & Mining | 28 | 19.4 |
| Miscellaneous Manufacturing | 44 | 19.3 |
| Natural Resources[1] | 407 | 21.5 |
| Non-bank Financial | 81 | 17.1 |
| Offices & Computers | 56 | 19.8 |
| Oil Services & Supply | 19 | 20.6 |
| Paper & Forest Products | 46 | 17.7 |
| Personal Care Products | 25 | 18.2 |
| Publishing | 12 | 20.6 |
| Radio & TV | 7 | 22.0 |
| Railroads | 21 | 12.9 |
| Real Estate & Housing | 4 | 21.0 |
| Retailing (Food) | 69 | 15.5 |
| Retailing (Non-food) | 110 | 14.5 |
| Savings & Loan | 4 | 15.4 |
| Service Industries | 96 | 19.3 |
| Special Machinery | 22 | 16.5 |
| Steel | 48 | 5.4 |
| Textiles & Apparels | 27 | 13.5 |
| Tire & Rubber | 21 | 7.8 |
| Tobacco | 28 | 20.5 |
| Trucking | 8 | 16.9 |
| Utilites[2] | 160 | 12.8 |
| All-Industry Composite | 2180 | 16.6 |

[1] Include oil, gas, and coal.
[2] Include electricity, gas, and water.

MAJOR U.S. INDUSTRIES
1979 SALES & RATES OF RETURN ON EQUITY

Figure 2

CAPACITY AS OF JANUARY 1, 1979*

| By Size | No. of Refineries | MMB/D |
|---|---|---|
| < 10 MB/D | 57 | .4 |
| < 30 MB/D | 117 | 1.9 |
| < 50 MB/D | 151 | 3.3 |
| <100 MB/D | 192 | 6.3 |
| <175 MB/D | 220 | 9.9 |
| <600 MB/D | 287 | 17.3 |

| By Complexity** | No. of Refineries | MMB/D |
|---|---|---|
| < 3 MB/D | 90 | 1.3 |
| < 5 MB/D | 127 | 2.8 |
| < 7 MB/D | 178 | 8.2 |
| < 9 MB/D | 220 | 14.2 |
| < 11 MB/D | 233 | 15.7 |
| < 11+ MB/D | 244 | 16.9 |

| By Location** | No. of Refineries | MMB/D |
|---|---|---|
| PAD I | 28 | 1.9 |
| PAD II | 62 | 4.1 |
| PAD III | 83 | 7.3 |
| PAD IV | 24 | .6 |
| PAD V | 47 | 3.0 |
| | 244 | 16.9 |

CAPABILITY AS OF JANUARY 1980†

| | MMB/D |
|---|---|
| • Crude distillation | 18.7 |
| • Vacuum distillation | 6.7 |
| Thermal cracking | 1.6 |
| Catalytic cracking | |
| Fresh feed | 5.3 |
| Recycle | 0.8 |
| Catalytic hydrocracking | 0.9 |
| Catalytic hydrorefining | 2.0 |
| Catalytic hydrotreating | 6.3 |
| Catalytic reforming | 3.9 |
| • Alkylation | 1.0 |
| Aromatic/Isomerization | 0.5 |
| Lube oil | 0.2 |
| Asphalt | 0.8 |
| Coke | 48,000 t/d |
| H2 | 1.6 x 10^9 SCF/d |

\* Source: Reference 2
\*\* Results of survey by NPC did not include non-respondents.
† Source: Reference 3

STRUCTURE AND CAPABILITY OF U.S. PETROLEUM REFINING INDUSTRY

Figure 3

| Economic Legislations and/or Program | Major Provisions | Adverse Effects on the Petroleum Refining Industry |
|---|---|---|
| 1915 Jones Act | Forbids shipping goods between two American ports on foreign ships. | Adds $0.60 to $1.00 per barrel to deliver refined products from the Gulf Coast to the East Coast (1980 $). Caribbean refiners have the advantage. |
| 1932 Connally Hot Oil Act Market-demand pro-rationing | Controls and allocates oil production to maintain price. | High acquisition costs. |
| 1959 Mandatory Oil Import Program | Limits cheap crude oil imports and allocates quota tickets to refiners. | Northern-tier double dip Brownsville shuffle Proliferation of small refineries |
| 1971-1973 Price Controls Emergency Petroleum Allocation Act | Sets limits on prices of domestic crude oil and refined products. Allocates supplies. | Squeeze on refiners' profit margin. |
| 1975 Energy Policy and Conservation Act (EPCA) Entitlement Program | Institutionalizes price controls. Favors small independent refiners. | Proliferates small refineries and refineries in the Caribbean. Discourages innovation. |
| 1979 Decontrol Program | President Carter phases out control by October 1981 according to provision of EPCA. | Small independent refiners would meet tough competition. More inroads of refined products import. |

SOME ADVERSE EFFECTS ON ECONOMIC LEGISLATION ON THE PETROLEUM REFINING INDUSTRY

Figure 4

EXISTING FACILITIES (1979)

| | |
|---|---|
| MMB/D | 17.4 |
| Book Value | $1350/B/D ($24 billion total) |
| Replacement Value | $3730/B/D ($65 billion total) |

ESSENTIAL FACILITIES FOR UPGRADING*

| | $/B/D | MB/D Needed |
|---|---|---|
| Vacuum | 750 | 570 |
| Cracking | 9,000 | 400 |
| Coking | $80,000/ST/D | 11,000 ST/D |
| S Reduction | $215,000/ST/D | 4,200 ST/D |
| H$_2$ Production | $1.9/CF/D | 800 MMCF/D |
| Others | -- | 50% of above |
| | Total | $11 billion |

* Based on NPC assumption for 2.8 MMB/D additional high-sulfur heavy crude capability. Unit costs based on Pace, Inc. (Reference 9), and Purvin & Gertz (Reference 10).

UPGRADING TO COPE WITH LOW-QUALITY CRUDE

Figure 5

| | Tons/Year | | BACT/ |
|---|---|---|---|
| Pollutant | RACT | BACT | LAER |
| Total Suspended | | | |
|   Particulates (TSP) | 4,250 | 4,050 | 6 |
| Sulfur Dioxide (SO$_2$) | 5,300 | 1,100 | negl. |
| Carbon Monoxide (CO) | negl. | negl. | negl. |
| Nitrogen Oxides (NO$_x$) | 8,200 | 8,200 | 4,100 |
| Hydrocarbon (HC) | 1,250 | 830 | 550 |

RACT = Reasonably available control technology
BACT = Best available control technology
LAER = Lowest achievable emission rate

Source: Reference 13

EMISSIONS TO AIR FROM MODEL REFINERY -- 120 MB/D

Figure 7

| | Scenario 1 | Scenario 2 | Scenario 3 |
|---|---|---|---|
| Status | | | |
| 1979 Crude Throughput | 1.35 | 1.35 | 14.6 |
| 1979 Residual Oil | 0.57 | 0.57 | 1.8 |
| Objective | | | |
| Reduce Resids | 0.2-0.3 | 0.4-0.5 | 1.6-1.7 |
| Gasoline Increase | 0.1-0.2 | 0.2 | 0.5 |
| Diesel Increase | 0.1 | 0.2 | 0.5-0.6 |
| Low-Btu Gas | -- | 0.2x10$^{12}$ Btu/D | 1.3x10$^{12}$ Btu/D |
| Capacity Additions | | | |
| Vacuum | 0.6 | 0.6 | 0.8 |
| Cracking | 0.4 | 0.4 | 0.8 |
| Coking/Gasification | -- | 0.2 | 1.2 |
| Reforming | -- | -- | 0.2 |
| H$_2$ Production (MMCF/D) | 40 | 150 | 850 |
| Total Investments | | | |
| $ billion (1979) | 2.3 | 4.7 | 18.0 |

Source: Reference 10

RESIDUAL OIL BACKOUT (MMB/D)

Figure 6

## Figure 9 — Costs of Environmental Laws on Petroleum Refining

| | Investments ($ billion, 1979) | | Unit Cost* $ (1979)/B | |
| --- | --- | --- | --- | --- |
| | Up to 1980 | Proj. 1990 | 1980 | 1990 |
| Clean Air | 8.3 | 19.0 | 0.25 | 0.60 |
| Clean Water | 4.4 | 7.5 | 0.14 | 0.23 |
| Solid Wastes | -- | 3.6 | -- | 0.11 |
| Subtotal | 12.7 | 30.1 | 0.39 | 0.94 |

| | O & M ($ billion, 1979) | | Unit Cost $ (1979)/B | |
| --- | --- | --- | --- | --- |
| | 1980 | Proj. 1990 | 1980 | Proj. 1990 |
| Clean Air | 1.0 | 3.0 | 0.15 | 0.45 |
| Clean Water | 0.1 | 0.9 | 0.02 | 0.14 |
| Solid Wastes | -- | 0.7** | -- | 0.11 |
| Subtotal | 1.1 | 4.6 | 0.17 | 0.70 |
| Total | | | 0.56 | 1.64 |

\* Fixed charge at 20%/yr.
\*\* Assumed same as capital charge.

COSTS OF ENVIRONMENTAL LAWS ON PETROLEUM REFINING

Figure 9

## Figure 8 — Estimates of Refinery Solid Waste Streams and Trace Metals

| Stream | Estimated Annual Metal Weight (MT/Yr) | |
| --- | --- | --- |
| | By API | By Jacobs |
| 1. Waste FCC Catalyst | 36.5 | 21.1 |
| 2. API Separator Bottoms | 28.2 | 43.4 |
| 3. Storm Water Silt | 15.9 | 24.3 |
| 4. Air Flotation Units | 12.2 | 15.8 |
| 5. Once-Through Cooling Sludge | 9.1 | 17.8 |
| 6. Waste Biological Sludge | 8.1 | 33.6 |
| 7. Cooling Tower Sludge | 8.1 | 0.6 |
| 8. Treating Clays | 6.8 | 12.7 |
| 9. HF Alkylation Sludge | ? | ? |
| 10. Leaded Tank Bottoms | ? | ? |
| Total | 125.0 | 169.0 |

| Trace Metal | Distribution in Refinery Solid Waste | |
| --- | --- | --- |
| 1. Zinc | 28.00% | 33.00% |
| 2. Chromium | 26.00% | 39.00% |
| 3. Vanadium | 15.00% | 9.00% |
| 4. Nickel | 12.00% | 8.00% |
| 5. Lead | 10.00% | 4.00% |
| 6. Copper | 8.00% | 6.00% |
| 7. Arsenic | 0.30% | 0.70% |
| 8. Cadmium | 0.30% | 0.10% |
| 9. Selenium | 0.20% | 0.20% |
| 10. Mercury | 0.05% | 0.20% |

Sources: References 14 and 15

ESTIMATES OF REFINERY SOLID WASTE STREAMS AND TRACE METALS

Figure 8

# THE URANIUM MYSTIQUE: A SEARCH FOR IDENTITY

**Stephen Schermerhorn**
Impact, Ltd.

To pursue any discussion on modeling uranium mill tailings, one should first consider what mill tailings really are.  Several studies have addressed this issue although not to a degree necessary to define truly the behavior of tailings in the atmosphere.  Probably the most definitive study to date on mill tailings and their relationship or contribution to dose was performed by Battelle.  Unfortunately, the study was performed on a carbonate leach process which has tailings considerably different from those found in most acid leach mill tailings ponds.  Further, the tailings pond analyzed uses mechanical cyclonic separation in the tailings stream to concentrate larger particulates on the dike and smaller particulates and slimes in the center of the pond.

While the study did confirm that radioactivity seems to be differentially associated with smaller particulates and increases more as a function of the surface area of a particle rather than its diameter, and while that study did add substantial data on mill tailings to the library, the bottom line was simply that the study was inconclusive.  The Battelle study states "Ideally the source terms should be calculated for each particle size of interest".

Let us look for a moment at the way particulates are handled in the most recent NRC tailings routine.  The theory essentially begins with a calculation of a wind velocity profile near the surface according to Bagnold which results in a threshold shear velocity.  However, the initial calculation of the threshold shear velocity is completed using an average grain diameter of 300 microns.  To this is added a modification of the threshold velocity based on moisture content according to Belly.  Moisture in the tailings is assumed to be 0.1 percent.  Moving then through a rate of horizontal particle movement by saltation a modified relationship is proposed by Lettan and reported by

Gillette. After this modification Gillette applies a coefficient of proportionality to convert from threshold velocities for the original 300 micron particle to vertical fluxes for 20 micron particulates. Travis further modifies this equation assuming that vertical fluxes are always associated with horizontal fluxes to finally arrive at an equation which allows one to specify vertical flux for 0-100 micron particulates of a specific radionuclide in terms of vertical fluxes, specific activities and activity fractions for 20 micron particulates. This is then applied to a tailings mix that is assumed to contain 30% of the radioactivity in tailings particles of 5 microns with the remaining 70% of the activity associated with 35 micron particulates. One only hopes by this treatise that you have now reached mental chaos - so goes the state of the Union.

If one compares an actual particle analysis taken from the surface of a tailings pond, one finds that the size fractions are quite different from those assumed by NRC in the GEIS. When one compounds the size fraction problems with questions of density, radiological activity, deposition velocity, or vertical flux, and when large tailings areas are involved causing the tailings portion of the radiological dose to be the greatest contributor, significant compliance considerations hang on the characterization of the particle releases from a tailings pond.

In an effort to determine which of these characteristics was most significant a sensitivity analysis was performed on each of the constants affecting the results of tailings dose calculation. The results are summarized in the table which follows:

| Parameter Change from BASECASE | Lung Dose to Adult Due to Inhalation | Percent of BASECASE |
|---|---|---|
| BASECASE - Default values used for all parameters | 1.61E-02 | 100% |
| TPSIZ = 1.25 default = 2.5 | 8.05E-03 | 50 |
| TPSIZ = 3.75 default = 2.5 | 2.42E-02 | 150 |
| PTSZ20 = 0.25 default = 0.5 | 3.22E-02 | 200 |
| PTSZ20 = 0.75 default = 0.5 | 1.07E-02 | 66 |
| DM = 0.015 default = 0.03 | 9.31E-02 | 578 |
| DM = 0.045 default = 0.03 | 5.29E-03 | 33 |

Table 1

Sensitivity Analysis of Parameters Used in
the Calculation of the Mass Flux of Radioactive
Particles from Tailings Sources
MILDOS Subroutine TAILPS

We find, for example, that one of the significant factors called PTSZ20 in the MILDOS Code or simply $F_{20}$ in the GEIS is simply a correction factor to convert 20 micron theory to the 0-100 micron range. If the total amount of radioactivity is defined by the source term, and the portion of that activity which appears in the small tailings is established relative to the portion which appears in the large tailings, increasing the amount of activity attributable to the less than 20 micron particulate has the effect of decreasing the overall source term size. This can be shown from the sensitivity curve graph attached. Remember that some of the particulate theory originally derived from an average tailings diameter of 300 microns. Drastic changes can be observed when the mean diameter (DM) is adjusted up or down by 50%. A linear relationship can be observed for the size of the tailings area. A linear relationship also exists for some of the other factors such as the ratio of tailings dust activity in the less than 80 micron fraction to that of the tailings, the isotopic activities, and selected model conversion constants.

With regard to the physical description of tailings, one finds the model extremely simplistic yet one must still ask at what cost compared to the cost of controls you would be able to actually model a spectrum of tailings sizes, densities, depositional velocities and activities.

Let us move for a moment to a discussion of the radiological activity of tailings. For the NRC model mill and mills recently licensed isotopic activities were derived based on assumptions of ore secular equilibrium and extraction process efficiencies. This results in concentrations of Radium-226 and Lead-210 more than fourteen times greater than Uranium-238. Tailings istopic analyses do not support the relationship. A composite of data from existing operations is compared to the model values below. Radium-226 and Lead-210 values exceed Uranium-238 valus by factors of ten and six, respectively. It seems that one must account for those differences.

|  | MILDOS | | TAILINGS COMPOSITE | |
|---|---|---|---|---|
|  | Activity | U-238 Ratio | Activity | U-238 Ratio |
| U-238 | 75.6 | -- | 39 | -- |
| Th-230 | 1026 | 13.6 | 266 | 6.8 |
| Ra-226 | 1078 | 14.3 | 395 | 10.1 |
| Pb-210 | 1078 | 14.3 | 223 | 5.7 |

Table 2

Radiological Activity in Tailings
(pCi/gm)

Although the discussion of the treatment of the movement of tailings in air will be limited in this paper, it should suffice to say that the treatment of tailings movement in air is a continuation of the power plant argument that has gone on for years with regard to movement of sulfur dioxide and other emissions. Gaussian diffusion was not a representative modeling technique in 1970 and it is not a reasonable modeling technique today. This is particularly true if uranium mills are to be held to the stringent standards that have been proposed by

EPA. The assumptions in the Gaussian model, of course, are quite simplistic. Dust cloud behavior is not. To account for increased deposition of large particulates by tilting a perfectly conical plume is a gross misapplication of the theory and is far from the state of the art which includes both finite difference and stochastic models.

I will speak briefly to the costs that result from this kind of analysis for the establishment of absolute levels of control. Not only have control level requirements more than doubled to keep up with advances in modeling but the control costs also have more than doubled due to inflation and other factors. When the regulations we are currently analyzing were originally developed, control cost for compliance for an example facility was estimated to be approximately 250,000 dollars. Today's cost due to inflation and other factors alone for that same installation would be 400,000 dollars according to government estimates or as much as twice that based on vendor sampling. However, since improvements in the theory have made the model more "precise" the models now tell us that more control efficiency is required. Due to modeling differences alone the cost for compliance more than doubles again to nearly 2,000,000 dollars.

All of this is simply to say that modeling is a good technique for estimating, for design, and for determining the most cost-effective methods of mitigation. But modeling as it exists today without more field verification and with no more definitive information than we have seen is simply not an appropriate tool for enforcing extremely rigid standards on the uranium milling industry.

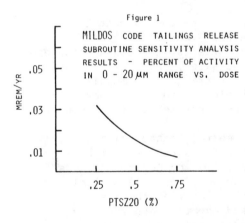

Figure 1

MILDOS CODE TAILINGS RELEASE SUBROUTINE SENSITIVITY ANALYSIS RESULTS - PERCENT OF ACTIVITY IN 0 - 20 μM RANGE VS. DOSE

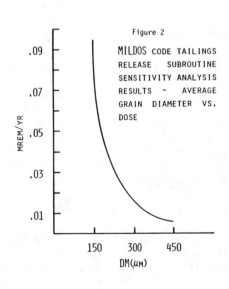

Figure 2

MILDOS CODE TAILINGS RELEASE SUBROUTINE SENSITIVITY ANALYSIS RESULTS - AVERAGE GRAIN DIAMETER VS. DOSE

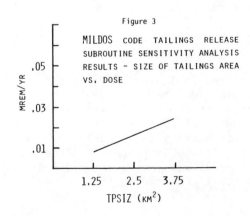

Figure 3

MILDOS CODE TAILINGS RELEASE SUBROUTINE SENSITIVITY ANALYSIS RESULTS - SIZE OF TAILINGS AREA VS. DOSE

## DESIGN OF ENVIRONMENTAL CONTROL SYSTEMS
## FOR COAL SYNTHETIC FUELS PLANTS

**Edward C. Mangold**
MITRE Corporation

Introduction

The MITRE Corporation has been evaluating the environmental effects
of proposed synthetics fuels plants for several years. We are
interested in the siting requirements of the plants, the consumption of
coal, water and other natural resources, the safety and health hazards
of these plants to both the workers and the surrounding communities.
The adequacy of the pollution control technologies for atmospheric
emissions, water and the disposal of ash and other solids is of concern.

Presented here is an overview of the control technologies that have
been proposed for seven coal based synthetic fuels plants that have good
chances of being constructed by 1985. The pollution control systems for
water and air pollutants, as proposed by the developers in preliminary
design documents, will be described. Although many of these systems
have been successfully used in petroleum and chemical applications,
their successful operation in synthetic fuels plants of the size and
complexity now being designed will be first demonstrated in the
following plants.

Environmental Control Process Description

The processes that are reviewed are listed in Table 1. The first
two processes produce substitute pipeline gas, together with smaller
quantities of byproducts and use two versions of the Lurgi gasification
process. The U-Gas process produces an industrial fuel gas with a
fluidized bed gasifier. A Texaco gasifier is already in operation
providing hydrogen for a TVA ammonia plant. The COGAS process produces
both liquid and gaseous products from coal pyrolysis and gasification of
the char. The SRC-I and II processes produce liquid and solid
hydrocarbons by direct liquefaction.

The American Natural Gas Company plant using the dry ash Lurgi process is schematically illustrated in Figure 1. These figures have been drawn to illustrate only the environmental control process flows and omit many of the energy process steps. Even the environmental processes are simplified, showing only the major stream flows. Many minor flows in the sulfur and ammonia recovery systems and separate control systems for coal pile runoff and other minor processes are not shown. Condensate water from the gaseous streams is collected at numerous locations. The principal product is pipeline gas, with sulfur, ammonia and phenols as byproducts.[1] Table 2 lists the significant features of the environmental control system.

A schematic of the second plant, the CONOCO slagging Lurgi, will not be presented since it is similar to the first plant.[2] Although the mass flows in the process are altered the prinicpal qualitative differences are shown in Table 3. The slagging Lurgi produces a quenched slag while the dry ash produced by the conventional Lurgi process contains considerably larger quantities of carbonaceous material. Tar is recycled to the gasifier. Because the steam requirements are less, the oil, naphtha and phenols are available as by products and are not consumed as boiler fuel. A regenerable Wellman Lord process producing elemental sulfur is used to treat the sulfur plant tail gas instead of the Stretford process. Both plants are designed to consume all process water internally and discharge only a stream of saline water to a deep well.

The sulfur removal system of the U-Gas plant illustrated in Figure 2 does not feature the extremely high sulfur removal efficiency of the previous SNG plants. Nevertheless the Selexol process removes 99.5% of the sulfur while leaving 95% of the $CO_2$ remaining in the industrial fuel gas. The gasification temperature of the fluidized bed reactor is high enough to destroy all heavy hydrocarbons in the gas which minimizes the water treatment required. Powdered carbon is added to the effluent to absorb trace organics as noted in Table 4.[3]

Environmental measurements from the Texaco gasifier illustrated in Figure 3 have been obtained since operations began last summer. As noted in Table 5 the Texaco process is unique in that the coal is slurried with water for injection into the gasifier. Because the plant produces the feed gas for an ammonia plant the ammonia stripped from the scrubbing water is recycled back to the coal slurry to neutralize acidity. Water from downstream units in the gas processing train is circulated to the start of the process where organic and mineral material is forced to pass through the high temperature slagging gasifiers before being treated for release.[4]

Figure 4 illustrates the most complex of the environmental control systems, planned for the COGAS plant. Because of the pyrolysis process, both liquid products and pipeline gas are produced, this plant features gasification of the distillation bottoms to produce hydrogen and plant fuel gas. As described in Table 6 both the gasification char and ash from the distillation bottoms are subjected to high temperature gasification to decompose hydrocarbons. Phenols are evaporated before incineration. The plant has the only Sulfinol ADIP sulfur removal system and treats all plant flue gas in a single sulfur dioxide scrubber using the dual alkali process.[5]

Direct liquefaction processes slurry pulverized coal with recycle oil to break down the coal into a wide spectrum of gaseous, liquid and solid products. Repeated flashing and distillation separates a range of products from light hydrocarbon gases and unreacted hydrogen, through naphtha and fuel oil distillates, to vacuum tower bottoms containing ash and solid SRC. In the SRC-I process of Figure 5 separation of the ash through antisolvent deashing produces the SRC stream and the ash and heavy hydrocarbons which are gasified to produce plant hydrogen.[6] The gasification destroys the high molecular weight hydrocarbons and reduces the toxicity of the slag as shown in Table 7. The SRC is solidified for use as a fuel producing a contaminated cooling water stream, hydrocracked into distillates, and processed into anode coke, requiring an additional scrubber for the tail gas.

The SRC II plant of Figure 6 uses a more severe hydrotreatment process to produce gaseous and liquid products and no solid fuel product. The vacuum tower bottoms are gasified to produce hydrogen and plant fuel gas. An extensive waste water treatment plant noted in Table 8 is required.

In summary Table 9 shows that the Rectisol and Selexol processes will be used extensively for acid gas removal with the Stretford and Claus processes used for sulfur recovery. Table 10 shows that when ammonia is not incinerated in the Claus plant it is recovered by the Phosam W process. Phenols are recovered by the Phenosolvan process or incinerated. A brief description of the operating principle of each commercial process is given in Table 11. Extensive waste water treatment facilities are required to produce an acceptable discharge if the water cannot be consumed or evaporated in the cooling towers. Construction of these systems will enable measurements of the emissions to be performed that will determine emissions levels, toxicity of the material and the satisfactory performance of the control systems.

References

1.  Department of Energy, Final Environmental Impact Statement, Great Plains Gasification Project, Mercer County, North Dakota, August 1980, pp. 27-42, Appendix B.

2.  Continental Oil Company, "Phase I: The Pipeline Gas Demonstration Plant, Design and Evaluation of Commercial Plant," 1978.

3.  Memphis Light, Gas and Water Division, "Industrial Fuel Gas Demonstration Plant Program, Conceptual Design and Evaluation of Commercial Plant, Volume II - Commercial Plant Design," December 1978.

4.  Bell, L.R., C.E. Bohoc and W.C. Lee, Tennessee Valley Authority, "Environmental Development Plan: Ammonia From Coal Project, Volumes I and II, October," 1979, pp. 12-19.

5.  Eby, R.J., Illinois Coal Gasification Group, "Pipeline Gas Demonstration Plant, Phase I, Record Environmental Analysis Report," September, 1979.

6.  Southern Company Services, Inc. "SRC-I Phase Zero Deliverables, Conceptual Commercial Plant Design," July 30, 1979.

## TABLE 1

### COAL SYNFUELS PROCESSES REVIEWED

| SPONSOR LOCATION | PROCESS | MAIN PRODUCTS BY PRODUCTS |
|---|---|---|
| American Natural Gas Mercer County, N. Dakota | Dry Ash Lurgi | Substitute Pipeline Gas Sulfur, Ammonia, Phenols |
| CONOCO Montgomery County, Illinois | Slagging Lurgi | Substitute Pipeline Gas Sulfur, Ammonia, Phenols, Naphtha, Oil |
| Memphis Light, Gas & Water Memphis, Tennessee | U-Gas | Industrial Fuel Gas Sulfur, Ammonia |
| Tennessee Valley Authority Muscle Shoals, Alabama | Texaco | Hydrogen for Ammonia Plant Sulfur, Carbon Dioxide |
| Illinois Coal Gasification Group Perry County, Illinois | COGAS | Substitute Pipeline Gas Sulfur, Ammonia, Naphtha, Fuel Oil |
| Southern Company Services Newman, Kentucky | SRC-I | SRC Fuel Oil, Naphtha, LPG Sulfur, Ammonia, Phenols |
| Pittsburg & Midway Coal Mining Co. Fort Martin, W. Virginia | SRC-II | Substitute Pipeline Gas, LPG, Fuel Oil Sulfur, Ammonia, Phenols |

## TABLE 2

### AMERICAN NATURAL GAS COMPANY
### SIGNIFICANT FEATURES OF ENVIRONMENTAL CONTROL SYSTEM

- Low temperature Lurgi process produces 7% of energy output in byproducts.
- Dry ash from Lurgi contains hazardous hydrocarbons that may require a more secure landfill.
- Because of remote North Dakota location, naphtha, tar and oil are burned as low sulfur boiler fuel.
- Tail gases from raw gas water scrub and Rectisol sulfur removal are incinerated.
- High percentage (99.9$^+$%) sulfur removal Rectisol system required to avoid poisoning of methanation catalyst.
- Salts in Stretford purge stream are concentrated for deep well disposal, or dried for landfill.
- Process has zero water discharge, all excess water evaporated in cooling tower.

## TABLE 3

### CONOCO SIGNIFICANT FEATURES
### OF ENVIRONMENTAL CONTROL SYSTEM

- Low temperature Lurgi process produces large quantity of hydrocarbon byproducts.
- Slagging Lurgi gasification process consumes 12% as much steam as dry ash.
- Tar recycled to gasifier.
- Quenched slag from gasifier contains almost no harmful hydrocarbons.
- Oil and naphtha available for sale as byproducts.
- Wellman-Lord process for flue gas desulfurization of sulfur plant and coal fired boilers.

## TABLE 4

### U-GAS
### SIGNIFICANT FEATURES OF ENVIRONMENTAL CONTROL SYSTEM

- Raw gas cooler condensate and water scrubber produce separate streams for water treatment.
- Powdered carbon dispersed in effluent for final organic cleanup of water discharge.
- Catalytic hydrolysis of COS prior to Selexol acid removal.
- Selexol system removes 99.5$^+$% of hydrogen sulfide and carbonyl sulfide.
- Selexol system removes only 5.4% $CO_2$ which is satisfactory for industrial fuel gas.

## TABLE 5

### TVA Texaco
### SIGNIFICANT FEATURES OF ENVIRONMENTAL CONTROL SYSTEM

- Selexol system removes 99.9% sulfur to avoid poisoning ammonia plant catalyst.
- $CO_2$ removed in Selexol system and used as byproduct in urea plant.
- Unique Texaco coal water slurry permits process water to be cycled through the high temperature gasifier which minimizes hydrocarbon content.
- Water treatment system includes both aerobic and anaerobic digestion of wastes.
- Ammonia stripped from waste water stream is recycled back to process because output gas feeds ammonia plant.

## TABLE 6

### ILLINOIS COAL GASIFICATION GROUP
### SIGNIFICANT FEATURES OF ENVIRONMENTAL CONTROL SYSTEM

- Process hydrogen generated from H-Oil hydrotreator bottoms.
- $H_2S$ and $CO_2$ removed in seperate steps using Sulfinol and Benfield processes.
- Residual hydrocarbons from gasifier destroyed in thermal oxidizer.
- Claus plant tail gas incinerated to $SO_2$ before desulfurization.
- Single dual alkali flue gas desulfurization system serves gasifier, Claus plant tail gas, auxiliary boiler, and phenolic incineration products.
- Phenols concentrated by evaporation, then decomposed in thermal oxidizer.
- Inorganic wastes in plant water evaporated to dryness.

## TABLE 7

### SOLVENT REFINED COAL-I
### SIGNIFICANT FEATURES OF ENVIRONMENTAL CONTROL SYSTEM

- Gas produced in liquefaction reactor treated in Selexol system for recovery of recycle hydrogen and light hydrocarbons.
- Ash seperated from heavy liquids by solvent deashing process. All ash passes through high temperature gasifier with underflow to provide hydrogen and plant fuel gas.
- Solvent Refined Coal is further treated to produce anode coke and light hydrocarbons. Acid gas is sent to DEA sulfur removal system.
- Cooling water for solid SRC product may become contaminated with hydrocarbons.
- Anode coke off gas requires incineration and scrubbing with soda ash.

## TABLE 8

### SOLVENT REFINED COAL-II
### SIGNIFICANT FEATURES OF ENVIRONMENTAL CONTROL SYSTEM

- Acid gas from liquefaction reactor treated by Selexol system.
- Claus plant tail gas treated by Super SCOT system and incinerator.
- Product separation bottoms gasified to reduce hydrocarbon content of ash.
- Single water treatment system for plant recovers ammonia and phenols.
- Waste water is evaporated to produce sludge which is incinerated prior to landfill.

TABLE 9

# Sulfur Removal Systems

| Process | Acid Gas Removal | Sulfur Recovery |
|---|---|---|
| Dry Ash Lurgi | Rectisol | Stretford |
| Slagging Lurgi | Rectisol | Stretford |
| U-Gas | Selexol | Claus-Beavon |
| Texaco | Selexol | Stretford |
| COGAS | Sulfinol - ADIP | Claus-Dual Alkali |
| SRC-I | Selexol | Claus-Beavon |
| SRC-II | Selexol Dimethanol Amine | Claus-Super SCOT |

TABLE 10

# Ammonia Removal Systems

| Process | Ammonia Recovery |
|---|---|
| Dry Ash Lurgi | Phosam W |
| Slagging Lurgi | Phosam W |
| U-Gas | Incineration in Claus plant |
| Texaco | Recycled to process |
| COGAS | Phosam W |
| SRC-I | Incineration in Claus plant |
| SRC-II | — |

TABLE 11

PROCESS DESCRIPTION

Sulfur Recovery

| | |
|---|---|
| STRETFORD | Chemical oxidation/reduction in solution of vanadium salts |
| CLAUS | Combustion of $H_2S$ and catalytic reduction of $H_2S$ with $SO_2$ |
| BEAVON | Incineration of $H_2S$ to $SO_2$ followed by solvent absorption |
| Super SCOT | Catalytic conversion of Claus Tail gas into $H_2S$ and removal in selective amine sorbent |

Acid Gas Removal

| | |
|---|---|
| SELEXOL | Absorption in dimethyl ether of polyethylene glycol |
| RECTISOL | Refrigerated methanol solvent for $H_2S$ and $CO_2$ |
| ADIP | Selective absorption of $H_2S$ in methyl di-ethinol amine |
| SULFINOL | Physical/chemical absorption in di isopropanol amine |

Ammonia Recovery

| | |
|---|---|
| Phosam W | Abosrption of ammonia in ammonium phosphate solution |

Phenol Recovery

| | |
|---|---|
| Phenosolvan | Solvent extraction in di isopropyl ether |

Flue Gas Treatment

| | |
|---|---|
| Dual Alkali | Reaction system using sodium and calcium sulfites |
| Wellman-Lord | Regenerable sbsorption of $SO_2$ in sodium sulfite solution followed by reduction to elemental sulfur |

## FIGURE 1

# American Natural Gas Environmental Control Process Flow

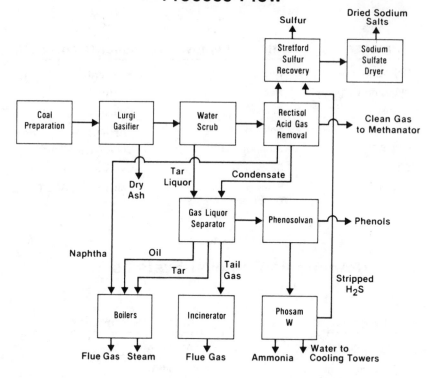

## FIGURE 2

# U-Gas Environmental Control Process Flow

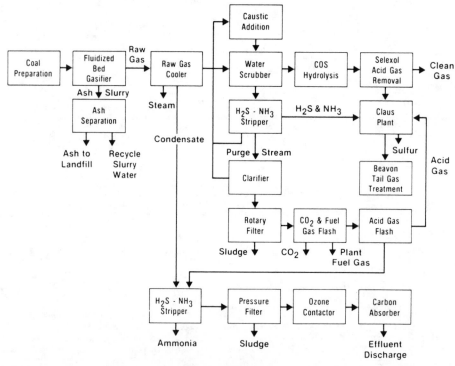

FIGURE 3

# TVA Texaco Ammonia Plant Environmental Control Process Flow

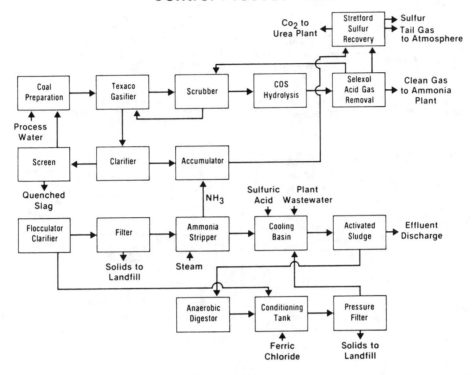

FIGURE 4

# Illinois Coal Gasification Group Environmental Control Process Flow

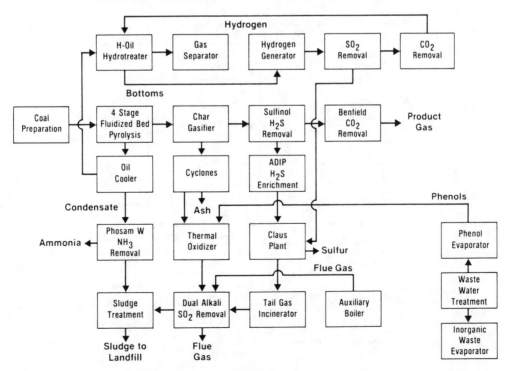

## FIGURE 5

# SRC-I Environmental Control Process Flow

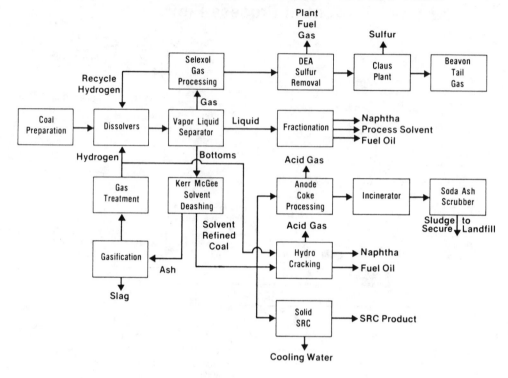

## FIGURE 6

# SRC-II Environmental Control Process Flow

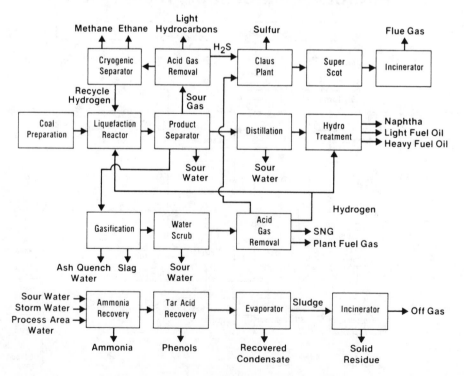

# METHODOLOGY FOR COMPARISON OF ALTERNATIVE PROCESSES USING LIFE-CYCLE CONSUMPTION

**J. C. Uhrmacher**
Carltech Associates

Summary

The current fuel scarcity is forcing reevaluation of national priorities, including waste treatment methods to be used. New and old treatment processes, therefore, require examination from the viewpoint of energy effectiveness. Several methods are available for this analysis; however, a need exists for a consistant method that can be used to compare alternative treatment technologies based upon energy use over the lifetime of the project.

A consistant methodology has been developed which combines the classical energy balance with life cycle costing methods. This form of life cycle energy analysis includes energy demands for acquisition (construction, site preparation, etc.); operation (fuel, power, chemicals, etc.); and support (maintainance, waste disposal, teardown, etc.).

The applicability of the life cycle energy method was illustrated by applying the method to three alternative measures for sewage sludge disposal. Life cycle energy demands were calculated for each of the three alternative processes; results were ranked in the order of increasing life cycle energy requirement. Calculated life cycle energy results showed the life cycle method to be dependent on estimation assumptions. Current studies were performed using top-down data. Future work will emphasize gathering of data on embodied energy of chemicals and construction materials and performance of additional technology analyses using a bottom-up approach.

Introduction

The increasing price of petroleum-based fuel is causing great concern in all sectors of the United States. This concern and the expected scarcity of fuel has been forcing a reevaluation of all plans and priorities. One area that is being examined and reevaluated is the choice of technologies for waste treatment.

Existing local state, regional, and Federal laws require reduction in emissions and discharges from all facilities, both public and private. Such laws usually require achievement of some (numerically) specified amount of emissions. The choice of technology to achieve this level of emission or discharge may be left to the owner of the treatment facility. The owner, then, is required to choose the technology to be employed. There are, usually, several technologies or technology variations which can achieve a specified level of pollutant reduction. Facility owners have historically selected methods for construction based on consideration of capital and operating costs. Recently, however, energy consumption has been suggested as a criterion for choice of technology. What is required, therefore, is a consistent method to compare energy utilization of alternative treatment technologies, that produces results to be used in decision-making processes. This, the first, of a series of papers concerns itself with the development of a methodology to perform an energy balance around a waste treatment plant that employs a particular technology. This energy balance includes energies used in acquisition and construction of plant equipment, operation of the treatment plant, and plant teardown. It is performed on a life cycle basis.

The paper, itself, is organized in the following fashion:

The Methodology section presents and explains the derivation and principals underlying the energy balance. The Results section presents results of applying the life cycle energy method to an example consisting of three technologies for sewage sludge disposal. The last section presents conclusions, recommendation, and future work to be performed in this area and related areas.


Methodology

The classical energy balance given in equation (1) below:

(1)  $\Delta E$ accumulated $= \sum E$ input $- \sum E$ output $+ Q - W$

where $\Delta E$ is a function of the physical and chemical properties of the reactants and products. However, this equation entirely neglects the various physical, mechanical, and chemical energies used in providing the raw materials at the locations and in the forms necessary to be the input to the energy balance. For example, in the case of a sewage treatment plant, this energy balance does not reflect any energy associated with mining, transportation and handling of iron chloride, petroleum (for polymers), limestone etc. We, therefore, propose to use the following energy definition:

(2)  $B = \sum \Delta E$ accumulated

where:  B is the embodied energy representing accumulation of all forms of energy expended (by deliberated action) and summed over all operations.

264

Applying basic principles of life cycle cost analysis, we further sub-divide energy requirements over the life of a system into the energies of:

- Operation - actions used in producing the product as well as raw material requirements.
- Acquisition - actions used in obtaining the physical elements of the system performing the operation.
- Support - actions necessary to assure the performance of the acquired system, including maintainance later system modification and teardown.

Each of these terms is illustrated in Exhibit 1 and discussed briefly below.

The energy of operation is described by the classical energy balance (Equation (1)) with addition of the embodied energy of the inputs, and is given by equation (3) below:

$$(3) \quad Eo = Qo - Wo + \sum Bo$$

where:  $Eo$ = Energy expended in operation
$Qo$ = Heat expended in operation
$Wo$ = Work performed during operation, and
$Bo$ = Embodied energy of raw material.

The energy of acquisition is defined by equation (4) below:

$$(4) \quad E_A = Q_A - W_A + \sum B_A$$

where: $E_A$ =  Energy expended in acquisition of equipment for waste treatment.

$Q_A$ = Heat expended in acquisition of treatment equipment.

$W_A$ = Work expended in acquisition of treatment equitment.

$B_A$ = Embodied energy of all acquisition materials included equipment, cement, structural steel, piping, construction machinery, etc.

The energy of support is defined by equation (5) below:

$$(5) \quad E_s = Q_s - W_s + B_s$$

where: $Q_s$, $W_s$ and $B_s$ refer to heat, work and embodied energy associated with support operations.

Equation (1) can now be rewritten as:

$$(6) \quad \Delta E \text{ accumulated} = E_0 + E_A + E_S - E_D$$

where $E_D$ is the energy of the treatment process discharge.

The discharge of a treatment process is any matter that leaves the boundaries of a treatment plant.  Treated wastewater and waste sludge are examples of discharges from a wastewater treatment plant.

Equation (2) defines B, the embodied energy of a given operation. Applying Equation (2) to discharges and combining the result with Equation (6) gives:

$$(7) \quad B_D = E_O + E_A + E_S$$

where $B_D$ is the embodied energy of treatment process residues. Summing $B_D$ for all discharges from the treatment process gives the following equation:

$$(8) \quad \sum B_D = \sum E_O + \sum E_A + \sum E_X.$$

The right side of equation (8) represents the total energy consumed in acquisition (erection), operation, and support of the treatment process. Processes having large values of $\sum B_D$ represent a larger expenditure of energy during the process life. Thus values of $\sum B_D$, when compared among several processes can potentially reflect the overall energy uses of the processes and serve as a meaningful parameter for comparison.

Some consideration should be given to availability of data to calculate $\sum B_{\overline{D}}$. This discussion follows below and is organized by topic: acquisition, operation, and support. Acquisition energies are expended during the time of plant construction. Some data are available for embodied energies of concrete steel, etc.[1] used for construction. Factors have been developed relating dollar cost of construction and equipment[2] to the embodied energy of materials and the energy of construction. Data on operational energies are usually available as rates, namely on a monthly or annual basis. These types of data include fuels, chemicals, and purchased electric power. The rates of usage can be expected to prevail over the entire project life. Energy usage rates cannot be added to a bulk (one-time) acquisition energy expenditure unless one of the terms is converted. The life cycle energy method converts the rates to a single quantity of multiplying by the project life cycle as illustrated below in equation (9).

$$(9) \quad E_O = \sum (q_{oi} + w_{oi} + \sum b_{oi}) t_1$$

where: $E_O$ is the energy of operation

$q_{oi}$ is the rate of heating (BTU per year)

$w_{oi}$ is the rate of work (Horsepower)

$b_{oi}$ is the rate of use of embodied energy in chemicals or other added materials, and

$t_1$ is the life cycle time of the treatment process.

Support energy data are usually available as rates, and are treated in a manner similar to operating energy. Support energy also includes the energy for teardown. Teardown energy, like acquisition energy is a one-time expenditure. The author was not able to find sources of data on teardown energy and assumed that it was proportional to acquisition energy.

Assumption of the life cycle time is important in performing life cycle energy analysis. Results of the analysis are dependent on the length of time assumed for project life. The times chosen should be realistic and correspond to the planned life of the facility. The life cycle should also be the same as that used in economic analysis of alternative treatment processes, in order to assure meaningful comparison of results.

266

Setting the basis for embodied energy values has an important effect on life cycle energy results.[1,2] Embodied energy values obtained from two or more different sources can have differences in basis which can severely affect comparison of results. The most important part of embodied energy bases is the so-called "datum level" or point of zero embodied energy. Embodied energy values are all estimated with respect to this level. If data sources use different datum levels, then reported values must be adjusted before use in comparison. Therefore, all sources of embodied energy data should be checked for basis and adjusted prior to use, if necessary.

The life cycle energy methodology has been developed to compare alternative treatment processes with respect to energy demands. All aspects of project life have been included in the method, including acquisition (construction), operation, support, and teardown. Nonetheless, an example is presented in which the method is applied to a case in order to examine the results for utility. The following section discusses results of applying the life cycle energy method to alternative methods of sewage sludge disposal.

Results and Discussion

Sewage sludge disposal processes were chosen for this example as a topic of current concern. Three alternative methods were examined, namely: multiple-hearth sludge incineration, area landfilling, and windrow composting.

These treatment alternatives were analyzed to illustrate the life cycle energy method. Several subalternatives exist within each alternative type; there are wide variations in energy usage among subalternatives. Alternatives chosen for this example were extremes in high or low energy usage and were found in the same reference.[1]

Assumptions

Several assumptions were made during the application of the life cycle energy method to the example described.

Overall assumptions (Exhibit 3) include combined primary and secondary sludge feeds from a ten million gallon per day (MGD) sewage treatment plant. All three alternative methods were assumed to be usable at the treatment plant site.

Other properties such as sludge characteristics were obtained from Reference 1 and were held constant throughout the analysis in order to assure comparability of energies calculated. Assumptions applying to each alternative treatment are discussed below.

Incineration. A multiple-hearth incinerator was selected for use in this example. Other types of incinerator technologies are available, which may use less energy than this example. However, a fairly standard[1] multiple-hearth unit was chosen since it provides a simple example of high energy use. Operating times of 7 days per week and 20 hours per day were obtained from Reference 1 and used to estimate incinerator operating energy usage. Fuel costs used to convert operating costs were obtained directly from Reference 1 (as were the operating costs).

Incinerator waste disposal energies were estimated at seven percent of area landfill operating energy. This figure is based upon reduction of 2/3 of dry solids via incineration.

Landfilling. An area fill was selected for use in the example. As with other technologies area filling was assumed to be employed at the wastewater plant site. Landfill parameter assumptions used to estimate energies (Exhibit 3) were obtained directly from Reference 1. Teardown energies of 25 percent of acquisition energies were assumed, to maintain uniformity with the other two cases considered.

Composting. Windrow composting was chosen for use in this example. Assumed energy values were obtained from Reference 1, with two exceptions. An overall value of 26,000 BTU per     cost $ (1978) used to estimate acquisition energy was obtained from Reference 2, and is believed representative of the embodied energy of the purchased equipment. Operating energy data were not directly available for composting. Therefore, all non-labor dollar expenditures were assumed to have been spent for diesel fuel. Both of these assumptions are believed to be conservative tending to overstate actual composting energy use for windrow operation. Energy costs for removal of composted sludge were not considered, nor were energy credits taken for recovered use of composted sludge. This assumption was made because of extreme variability in energy uses and credits applied to composted sludge. It was, therefore, assumed that energy used to transport compost from the site would be equal to energy saved via land reclamation (e.g. construction stabilization) or from displacement of energy-intensive synthetic fertilizer (e.g. ammonium sulfate).

Results

Results of applying life cycle energy analysis to example technologies show the following ranking of processes based on increasing life cycle energy use (Exhibit 4):

- Landfilling
- Composting
- Incineration

The life cycle energy methodology applied in this example shows a large difference between energy use for incineration vs. landfilling or composting. This type of result is physically valid, since incineration uses large quantities of fuel oil to evaporate the 80 percent water content of the sludge. Composting and landfilling do not attempt to remove water from the sludge. The life cycle methodology shows that the largest contribution to incineration energy comes from equipment acquisition. The method also shows that incineration energies for acquisition operation and teardown are more than two orders of magnitude larger than similar energies calculated for landfilling and composting.

Area landfilling and windrow composting energies require nearly identical life cycle energies. Landfilling has directionally lower energies than does composting. Acquisition energies for landfilling are larger because the larger land requirement[1] for a sludge landfill results in increased diesel fuel and gasoline use for land clearing. Operational energy uses are larger for composting, however, because windrow turning and screening require larger use of fuel than does sludge dumping and covering in a landfill. Overall, the larger operational energies for windrow composting result in a somewhat larger (twenty-year) life cycle energy use.

268

## Discussion

These results show that for the simple case of on-site sludge disposal, the life cycle energy method produces results showing differences among types of disposal/recovery technologies. This example also shows that results are sensitive to assumptions used in estimation. For example, if sludge were transported by truck either for composting or for land-filling, life cycle energies would be increased by approximately one billion BTU's per mile transported.[1] Previous studies [1,2,3] have shown that economic analyses are equally as sensitive to assumptions. Thus, life cycle energy analysis should be conducted with the same care as would be used to conduct economic analysis. Results also show that methodological separation of acquisition from operation/support effects, make the life cycle energy method capable of analysis of technology alternatives involving multiple renewals (replacement of worn-out equipment) and expansions.

## Conclusions Recommendations and Future Work

Results presented to illustrate use of life cycle energy methodology show that the method can produce meaningful results for comparison of energy usage of alternative treatment technologies. Methodology results are highly dependent on estimation assumptions, as are economic analyses, thus life cycle energy analysis must be used with care. Use of life cycle energies is currently limited because of limited availability of data on embodied energies of chemicals and equipment, requiring use of top-down (cost-to-energy) conversion factors. Future work, therefore, will emphasize accumulation (or if necessary generation) of data on embodied energy having a common datum level, and performance of additional life cycle energy analyses using a bottom-up (without cost to energy conversions) approach. Future analyses will also include resource and energy recovery methods of waste treatment as these are receiving considerable current interest.

## References

1. US EPA, "Innovative and Alternative Technology Assessment Manual" Office of Water Program Operations, Document No. EPA 43019-78-009, Washington, D.C. (February, 1980)

2. US EPA Unpublished Data (1980)

3. J.C. Uhrmacher and B.L. Pollack, "Environmental Decision Making Based on Energy Conservatism Criteria - Part 1 Water Pollution Control," Presented at the XVI UPADI Convention, Mexico City, (October, 1980)

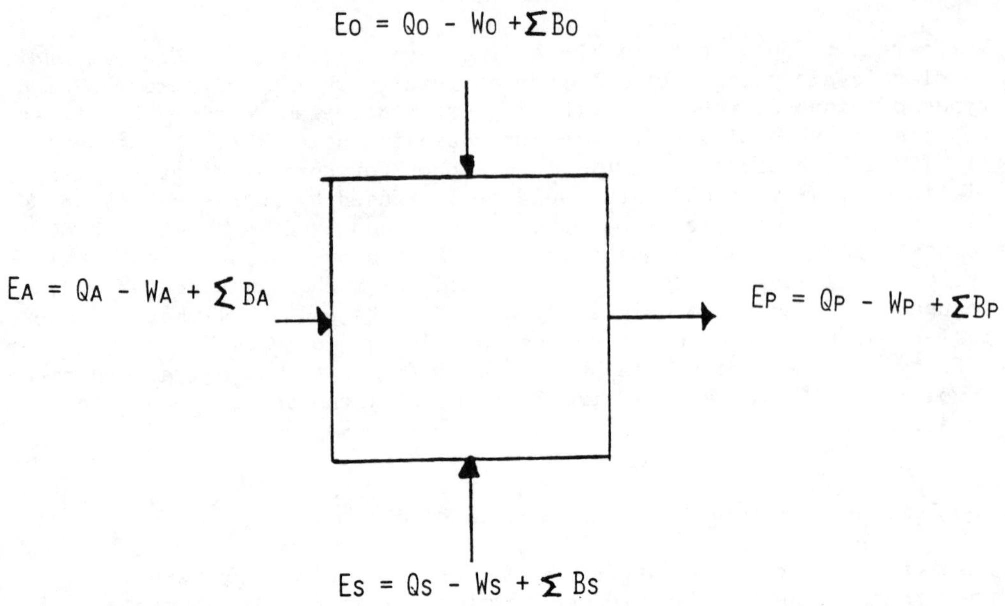

$$E_O = Q_O - W_O + \Sigma B_O$$

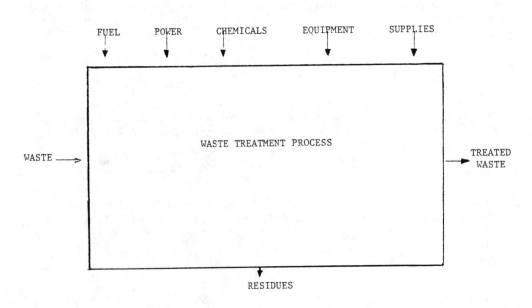

$$E_A = Q_A - W_A + \Sigma B_A$$

$$E_P = Q_P - W_P + \Sigma B_P$$

$$E_S = Q_S - W_S + \Sigma B_S$$

FIGURE 1.   CLASSICAL ENERGY BALANCE

FUEL     POWER     CHEMICALS     EQUIPMENT     SUPPLIES

WASTE →

WASTE TREATMENT PROCESS

→ TREATED WASTE

RESIDUES

FIGURE 2.   SYSTEM FOR OVERALL ENERGY BALANCE WASTE TREATMENT

270

| ENERGY USE, $10^6$ BTU | SLUDGE INCINERATOR* | ONSITE SLUDGE LANDFILL** | ONSITE SLUDGE COMPOSTING‡ |
|---|---|---|---|
| Aquisition (Construction) | 64,512,000 | 144,540 | 8,736 |
| Operation | | | |
| • Fuel | 4,135,320 | 22,995 | 218,560 |
| • Power | 22,180 | | |
| • Chemicals | | | |
| • Waste Disposal | 1,610 | | |
| • Other | 80,000  Negligible | | |
| • Subtotal | 4,239,110 | 22,995 | 218,560 |
| • Teardown | 16,128,000 | 36,135 | 2,184 |
| Total | 84,879,110 | 203,670 | 229,480 |

\* Multiple-Hearth
\*\* Area Fill
‡ Windrow

FIGURE 4.  LIFE-CYCLE ENERGY USE FOR SEWAGE SLUDGE DISPOSAL

Overall
- Source: Sewage Treatment Plant Capacity. MGD — Reference 1 — 10
- Sewage Sludge Amount, Pounds per MGD — 900
- Teardown Energy, Percent of Acquisition Energy — 25
- Project Life, years — 20
- Sludge, Percent Solids — 20
- Landfilling Operating Energy, BTU/Dry Ton Sludge — 700,000

Incineration
- Multiple Hearth Furnace
- Conversion Factor, 1978 Construction Cost $ to BTU, BTU per $ — $38,400^2$
- Operating times:[1]
  - Days per week — 7
  - Hours per day — 20
- Operating Cost Conversions:

| | Annual $[1] | Energy Cost $/$10^6$ BTU[x] | Life Cycle Years | Life Cycle Energy $10^6$ BTU |
|---|---|---|---|---|
| Fuel | 550,000 | 2.66[1] | 20 | 4,135,320 |
| Power | 65,000 | 5.86[1] | 20 | 22,184 |
| Other | 25,000 | 6.25[2] | 20 | 80,000 |

- Waste Disposal Energy, Percent of Landfill Operating Energy

On-Site Landfilling
- Area Fill
- Fill Equal 50% of Total Site
- Acquisition $ to Energy Conversion, BTU per 1978 Cost $ — $80,000^2$

On-Site Composting
- Windrow Method
- Uses Woodchips as Bulking Agent
- Acquisition $ to Energy Conversion Factor, BTU per 1978 Cost $ — 26,000
- All Operating Costs Except Labor Are Spent For Diesel Fuel at $0.41[1] Per Gallon
- BTU per Gallon of Diesel Fuel — 140,000

FIGURE 3.  ASSUMPTIONS TO LIFE-CYCLE ENERGY ANALYSIS

271

# SIMULTANEOUS HEAT RECOVERY AND POLLUTION CONTROL
## WITH FLUIDIZED–BED HEAT EXCHANGERS

**G. John Vogel** and **Paul Grogan**
Argonne National Laboratory

## Introduction

The objective of this paper is to characterize environmental impacts associated with fluidized-bed, waste-heat recovery technology. These impacts, which largely affect air quality, will likely be a net benefit to the environment because of the ability of fluidized-bed heat exchangers to limit sulfur oxides, ($SO_x$) and nitrogen oxides ($NO_x$) emissions while simultaneously recovering waste heat. The report focuses on a particular configuration of fluidized-bed, heat exchanger for a hypothetical industrial application. The application is a lead smelter where a fluidized-bed, waste-heat boiler (FBWHB) is used to control environmental pollutants and to produce steam for process use.[1]

A fluidized bed consists of a bed of rapidly moving solid particles that are held in suspension by a continuous flow of gas (or liquid) up through the bed. Heat energy can be added or removed from the bed by circulating a heat-transfer medium through tubes in the bed. Because the heat=transfer rate is rapid, the heat exchanger is smaller than it would be for other types of heat-transfer systems, thus allowing the use of a physically smaller boiler. The high heat-transfer rate is maintained because the scouring action of the rapidly moving particles prevents fouling of the tubes. The temperature of the bed is uniform, which is important when a fluidized bed is used to perform chemical reactions and which accounts for the many applications of fluidized beds in the chemical and petroleum process industries. A disadvantage of the system is that the energy required to continually force a gas or a liquid through the bed can be greater than the energy requirements in other types of heat-transfer equipment.

In the most probable applications of FBWHB systems, heat is transferred from a hot, contaminated gas (such as that exhausted from an internal combustion engine or a smelter) to water in order to produce hot water, saturated steam, or superheated steam. Industrial FBWHBs are available for this application. The largest known FBWHB system, in use since 1976, delivers 8,000 lb/hr of 200-psig steam by extracting heat energy from the exhaust gas of a 12,000-hp diesel engine mounted in a ship.[2]

In other FBWHB applications, heat energy in a liquid solid could be transferred to another medium - gas, liquid, or solid.[3] Fluidized-bed equipment for pilot plants has been developed for removing heat energy from a hot, chemically contaminated geothermal fluid and transferring that energy to clean water. Systems have been developed that transfer heat energy from hot solids to liquids and from hot solids to gases. Only gas-fluidized beds are discussed in this paper, however, because this type of FBWHB has the greatest potential for practical applications.

The main components of a fluidized-bed, waste-heat boiler are (1) a gas distributor plate through which the gas is distributed over the cross-sectional area of the bed, (2) a bed of particulate solids, (3) a bundle of tubes that carry the heat-transfer fluid, and (4) a disengaging space that allows particles carried out of the bed to disengage from the fluidizing gas and fall back into the bed (see Fig. 1). Also, a system for removing particulate solids from the gas leaving the FBWHB is necessary when the gas has to be cleaned to meet the state or federal regulations.

Materials of construction will be selected on the basis of the temperature and pressure of the system in each application, as well as the corrosive contaminants in the waste gas.[4] Corrosion data indicate that austenitic alloys are suitable for construction of the heat-transfer tubes at metal temperatures of 700°C or less and that low-chromium steels can be used in tubes where the metal temperature is less than 400°C. Relative costs of transferring a unit of heat (using 51-mm tubing carrying fluid at 180 atm) indicate that mild steel is favored at fluid temperatures to 240°C, 2.25% chrome steel to 430°C, and Niomic PE16 to 550°C. Qualitatively, no operating or technical difficulties are envisioned for employing fluidized-bed units in waste-heat-boiler applications. Fluidized-bed units generally require a minimum of operating attention and maintenance.

A potential application for a FBWHB was identified in a process in which secondary lead is smelted in a reverberatory furnace.[1] The smelter currently meets local air pollution standards by using a combination of dilution air (to meet $SO_2/NO_x$ standards) and fabric filters (to meet particulate standards). No attempt is made to recover the heat content in the waste gas. Conventional methods would recover the heat energy in a waste-heat boiler (or another type of heat recovery unit) and control sulfur ($SO_2$) emission by liquid scrubbing the waste gas after it had exited the boiler.

Environmental Implications

SO$_2$ Control

Processes for removal of $SO_2$ may be classified as wet or dry. In wet processes, the gas is contacted with a solution or slurry of a compound that reacts with the $SO_2$. In dry processes, the reaction is with a dry solid. Wet or dry processes can also be classified as regenerable or

273

throwaway. In the regenerable process, the reactive compound is regenerated and reused; in the throwaway process it is discarded and replaced with fresh compound. The sulfur products of a regenerable process can be sulfur, $SO_2$, or sulfuric acid. Although wet processes can be applied in FBWHB applications, they are not considered in this study because of the perceived advantages of $SO_2$ control by reaction with particles comprising the solid media in a FBWHB.

Dry processes for removing $SO_2$ from a gas can be divided into the following four categories:

1. Absorption processes. $SO_2$ is reacted with simple or complex oxide compounds ($CaO$, $CuO$, $Mn_xO_y$, titanates, and aluminates) to obtain a nonvolatile sulfate compound. The oxide compounds are regenerated by thermal or by reductive decomposition (the exception is $MnSO_4$ which can also be regenerated in an aqueous $NH_4OH/O_2$ solution). Characteristics of the major absorption $SO_2$ removal processes are summarized in Table 1.[5-17]

2. Adsorption processes. $SO_2$ is first adsorbed and oxidized to $SO_3$ in the pores of a solid particle, and, in the presence of moisture sorbed in the particles. The $SO_3$ is then converted to sulfuric acid ($H_2SO_4$). The $H_2SO_4$ can be recovered by washing the particles or the $H_2SO_4$ can be converted to concentrated $SO_2$ by reaction with carbon or to sulfur by reaction with $H_2S$. Characteristics of the major processes are summarized in Table 2.[16,17]

3. Catalytic oxidation processes. $SO_2$ is catalytically converted to $SO_3$ which reacts with water vapor in the contactor to form sulfuric acid mist which is the product recovered in a separate vessel.[16-18]

4. Gas-phase reaction processes. $SO_2$ is reacted with a gas such as $CO$ or $NH_3$ to obtain a solid compound which is recovered by filtration from the gas stream.[16,17,20-24]

NO_x Control

$NO_x$ removal processes, like $SO_2$ removal processes, can be classified as wet or dry.[25,26] The dry methods reviewed here are decomposition, chemical reduction, adsorption, and irradiation. Thermal decomposition of $NO_x$ to $N_2$ with or without a catalyst is impractical because of low achievable decomposition rates and does not hold much promise for application to FBWHB technology at the present time.[27-29]

Chemical reduction methods may be classified by the selectivity of the reduction reaction.[25] In selective reduction only $NO_x$ reacts with the reductant whereas in nonselective reduction other compounds in the exhaust gas (e.g., $O_2$, $SO_2$) are removed in addition to $NO_x$. Chemical reduction methods can also be classified according to whether or not a catalyst is used. Most developed processes comprise the selective, catalytic type.[25] In these processes over 90% of the $NO_x$ is removed by using $NH_3$ as the selective reagent along with a catalyst. Nonselective catalytic processes utilize reducing gases such as $H_2$, $CO$, or hydro-

274

carbons to react with $NO_x$.[25,30-31] However, these reagents will also react with $SO_2$ and $O_2$ which are usually present in flue gases and therefore the chemical consumption is high, and therefore, costly. Over 90% $NO_x$ removal can be achieved however. In the selective non-catalytic processses, the reaction is completed at a higher temperature and at a higher molar ratio of $NH_3/NO_x$ than when the catalyst is present. $NO_x$ removals have been generally less than 90%.

Adsorption processes utilize activated char to trap $NO_x$ as well as other pollutants in the flue gas.[25] Sorbed $NO_x$ is converted later to $N_2$ and $O_2$ by heating the char to 650°C (1202°F). While high, 90%, removal of $NO_x$ can be obtained by this process, it is usually applied to removing $SO_2$ at a temperature where only approximately half of the $NO_x$ is removed from the gas stream.

Electron beam irradiation of a gas, a relatively new technique, is being studied by a foreign company.[25] Over 90% removal of $NO_x$ as well as $SO_2$ has been achieved. Capital cost of the process, however, is high.

A comparison of the features, status, and costs of the various $NO_x$ removal processes are presented in Table 3.

## Particulate Control

Particulate emission standards can be met readily by using a bag filter or an electrostatic precipitator to clean the gas exiting a FBWHB. Under some conditions particulate in the waste gas stream entering the FBWHB can affect its operation. Particles smaller than about 5 microns can be trapped and removed from the waste gas by agglomerations with the solids forming the fluidized bed. Removal efficiencies can be higher than 90% but the gas velocity in the bed must be low, less than 1.5 ft/sec, a velocity which may not be practicable in most FBWHB applications since the area of the bed, and thus the cost, is inversely proportional to the gas velocity. Particulate removal can be enhanced in an electrofluidized bed in which the particles are electrically charged by electrodes. Particulate removal may be advantageous in some applications, but there may be disadvantages if solids surfaces become coated and the heat transfer characteristic of the bed media is adversely affected.

## Results

A potential application for a FBWHB is shown in Fig. 2. The application proposed is a secondary lead smelter reverberatory furnace. The smelter currently meets local air pollution control standards using a combination of dilution (to meet $SO_2$ and $NO_x$ standards) and fabric filters (to meet particulate standards).

A FBWHB designed to simultaneously recover waste heat and control $SO_2$ emissions from the smelter waste gas is shown in Fig. 3. Approximately 3,060 scfm of gas (combustion products and air leakage into the smelter) leave the smelter at a temperature of 1204°C (2200°F) and are diluted with 2079 $m^3$/min (7,340 $ft^3$/min) of air which reduces the gas temperature to 511°C (951°F). The hot gas enters the lower bed of the FBWHB which is maintained at a temperature of 232°C (450°F). Heat is removed from the bed by producing 150 psig steam from the water entering the heat exchange tubes immersed in the bed. The gas exits the lower bed at a temperature of about 232°C (450°F) and enters the upper bed which is maintained at a temperature of 164°C (327°F). The gas leaves the upper bed at a temperature of 164°C (327°F) sufficiently cooled to pass through a fabric filter. Water, at a temperature of 16°C (60°F), enters the upper bed heat exchanger and is heated to 154°C (310°F) by passage through it. It then goes to the lower bed heat exchanger where it is heated to 185.2°C (365.4°F) and converted to saturated steam.

$SO_2$ is controlled by reaction with MnO particles that make up the solid media of the upper bed.

Conclusions

Various dry methods for controlling $NO_x$, $SO_2$, and particulate solids emissions in waste gas streams have been examined with respect to their possible application to FBWHB systems. The principal conclusion is that the control technologies are sufficiently advanced and can probably be applied to FBWHB systems. However, optimization of $SO_2/NO_x$ control strategies will require further experimentation.

Although control technologies are sufficiently advanced and can be applied to the design of FBWHB systems, additional experimental information would allow systems to be optimized. Most information has been obtained at temperatures which are suitable for removing $SO_2/NO_x$ from the flue gas of utility power plants. Temperatures of 350°C (662°F) to 400°C (752°F) (the temperature of the flue gas leaving the economizer) or about 150°C (302°F) (temperature of the flue gas leaving the air preheater and entering the stack) are commonly used in experimental studies. However in a FBWHB system, although the bed temperature is fixed, the selected design bed temperature can range from 150°C (302°F) to 400°C (752°F) (or higher) depending on such factors as the quantity of heat energy removed from the waste gas and whether hot water, steam or hot air is produced. Therefore, it is essential to know the behavior of adsorbents/absorbents over a temperature range in order to select an optimum bed temperature and an optimum bed particle composition. To insure this, the following program needs can be identified.

$SO_2$ Control

Currently $SO_2$ control methods rely on either absorption or adsorption techniques. In FBWHR units $SO_2$ can be absorbed in limestone ($CaCO_3$), CuO, or MnO depending on the temperature of the fluid bed. Considerable design information is available for using $CaCO_3$ at 850°C in a fluidized bed, but little published information is available for CuO at 400°C and even less for MnO at 150°C. Similarly, little has been published on regenerating CuO and MnO in fluidized-bed units.

For adsorption of $SO_2$, activated carbon is the usual choice and considerable information is available on its sorption/desorption properties. However, its sorption efficiency drops at temperatures above 175°C. In many FBWHR units the temperature of the fluidized bed will be higher than 175°C. Consequently, compounds that sorb $SO_2$ at higher temperatures should be investigated.

Based on the above the following are recommended:

1. Continue laboratory-scale studies to determine absorption, adsorption and regeneration characteristics of selected compounds. Thermo-balances or small fixed beds can be utilized with addition of analytical equipment such as $SO_2/NO_x$ analyzers, porosimeters, surface area analyzers, x-ray and electron micro-probe analyzers as required. The following areas should be emphasized:

Absorption Systems

    a. Determine absorption characteristics of selected compounds as a function of temperature.

    b. Determine thermal and also reductive decomposition characteristics. Particular attention should be reserved for thermal decomposition studies for two reasons. First, a hot waste gas is available for regenerating a compound by thermal decomposition. Secondly, the costs and safety problems associated with handling hydrogen or other reducing agent in the reductive decomposition method are eliminated. If thermal decomposition appears unworkable (as it has so far in the $CuSO_4/CuO$ system) the reasons why should be elucidated.

    c. Characteristics of particles fabricated by impregnating a compound on an alumina substrate should be determined so that possible high-cost materials may be used effectively.

Adsorption Systems

    a. Determine the adsorption characteristics of selected compounds other than carbon as a function of temperature. The behavior of carbon is known.

    b. Determine the regeneration characteristics of adsorbates.

2. Test promising absorption and adsorption materials in bench-scale units using a gas which simulates the composition expected to be processed in FBWHR units.

3. Model the process using experimental data.


$NO_x$ Control

Considerable information has been published for the most widely used process, selective catalytic reduction, for controlling $NO_x$ but lesser information is available on the other processes. Experimentation on $NO_x$ control -- laboratory scale, bench-scale, and modeling -- should be done in the following areas.

    a. Catalytic behavior of absorbent compounds for the $NH_3/NO_x$ reaction as a function of temperature should be investigated.

277

b. Determine the removal/regeneration behavior of $NO_x$ adsorbents that also are useful for adsorbing $SO_2$.

References

1. The Rules and Regulations of the County of Los Angeles Air Pollution Control District, Air Pollution Engineering Manual, Environmental Protection Agency, J.A. Danielson, ed., 2nd Ed. (May 1973).

2. Virr, M.J., Industrial Coal Fired Fluidized Bed Boilers and Waste Heat Boilers, Proc. 5th Int. Conf. on Fluidized Bed Combustion, Washington, D.C., Dec. 1977, MITRE Corp., M78-68 (Dec. 1978).

3. Vogel, G.J., P.J. Grogan, and A.R. Evans, An Analysis of the Application of Fluidized-Bed Technology to the Recovery of Waste Heat, ANL/CNSV-TM-34 (August 1979).

4. Cook, J.J. and E.A. Rogers, High Temperature Corrosion of Metals and Alloys in Fluidized-Bed Combustion Systems, Proc. 5th Int. Conf. on Fluidized Bed Combustion, Washington, D.C., Dec. 1977, MITRE Corp., M78-68 (Dec. 1978).

5. Lowell, P.S., and T.B. Parsons, Identification of Regenerable Metal Oxide $SO_2$ Sorbents for Fluidized-Bed Coal Combustion, Radian Corp., EPA-650/2-75-065 (PB 244 402) (July 1975).

6. Cusumano, J.A., and R.B. Levy, Evaluation of Reactive Solids for $SO_2$ Removal During Fluidized-Bed Coal Combustion, Catalytica Associates, Inc., EPRI TPS75-603 (Oct. 1, 1975).

7. O'Neill, E.P., and D.L. Keairns, Selection of Calcium-Based Sorbents for High-Temperature Fossil Fuel Desulfurization, presented at the 80th American Institute of Chemical Engineers National Meeting, Boston (Sept. 10, 1975).

8. Lowell, P.S., K. Schwitzgebel, T.B. Parsons, and K.J. Sladek, Selection of Metal Oxides for Removing $SO_2$ from Flue Gas, Industrial Engineering Chemistry, Process Design and Development, 10(3):384-390 (1971).

9. Welty, A.B., Flue Gas Desulfurization Technology, Hydrocarbon Processing 50:104-107 (Oct. 1971).

10. Bienstock, D., and J.H. Field, Bench-Scale Investigation on Removing Sulfur Dioxide from Flue Gases, Journal Air Pollution Control Association, 10(2):121-125 (1960).

11. Bienstock, D., J.H. Field, and J.G. Myers, Process Development in Removing Sulfur Dioxide from Hot Flue Gases, Part 1, Bench-Scale Experimentation, Bureau of Mines, U.S. Department of Interior, Report 5735 (1961).

12. Vogel, R.F., B.R. Mitchell, and F.E. Massoth, Reaction of $SO_2$ with Supported Metal Oxide-Alumina Sorbents, Environmental Science and Technology, 8(5):432-436 (May 1974).

13. Uno, T., et al., Scale-Up of an SO$_2$ Control Process, Sulfur and SO$_2$ Developments, Chemical Engineering Progress Technical Manual, American Institute of Chemical Engineers, pp. 73-77 (1976).

14. McCrea, D.H., A.J. Forney, and J.G. Myers, Recovery of Sulfur from Flue Gases Using a Copper Oxide Absorbent, J. Air Pollution Control Assn., 20(12):819-824 (Dec. 1970).

15. The Status of Flue Gas Desulfurization Applications in the United States, Bureau of Power, Federal Power Commission (July 1977).

16. Chemical Reactions as a Means of Separation: Sulfur Removal, B.L. Crynes, ed., Marcel Dekker Inc., New York and Basel (1977).

17. Slack, A.V., Air Pollution: The Control of SO$_2$ from Power Stacks, Part 3 -- Process for Recovering SO$_2$, Chemical Engineering, 74:188-196 (Dec. 1967).

18. Zawadski, E.A., Removal of Sulfur Dioxide from Flue Gases: the BCR Catalytic Gas Phase Oxidation Process, Transactions of the American Institute of Mining, Metallurgical and Petroleum Engineers, 232:241-246 (Sept. 1965).

19. Hunter, W.D. Jr., J.C. Fedoruk, A.W. Michener, and J.E. Harris, The Allied Chemical Sulfur Dioxide Reduction Process for Metallurgical Emissions, Sulfur Removal and Recovery from Industrial Processes, J.B. Pfeiffer, Ed., Advances in Chemistry Series 139, American Chemical Society (1975).

20. Henderson, J.M., and J.B. Pfeiffer, Reduction of Sulfur Dioxide to Sulfur, The Elemental Sulfur Pilot Plant of ASARCO and Phelps Dodge Corp., Sulfur Removal and Recovery from Industrial Processes, J.B. Pfeiffer, Ed., Advances in Chemistry Series 139, American Chemical Society (1975).

21. Querido, R., and W.L. Short, Removal of Sulfur Dioxide from Stack Gases by Catalytic Reduction to Elemental Sulfur with Carbon Monoxide, Industrial Engineering Chemical Process Design and Development, 12(1): 10-17 (1973).

22. Okay, V.C., and W.L. Short, Effect of Water on Sulfur Dioxide Reduction by Carbon Monoxide, Industrial Engineering Chemical Process Design and Development, 12(3):291-194 (1973).

23. Khalafalla, S.E., and L.A. Haas, Dual Catalyst Beds to Reduce Sulfur Dioxide to Elemental Sulfur in the Presence of Water Vapor, Sulfur Removal and Recovery Processes, J.B. Pfeiffer, Ed., Advances in Chemistry Series 139, American Chemical Society (1975).

24. Cortelyou, C.G., Commercial Processes for SO$_2$ Removal, Chemical Engineering Progress, 65(9):69-77 (Sept. 1969).

25. Faucett, H.L., J.D. Maxwell, and T.A. Burnett, Technical Assessment of NO$_x$ Removal Processes for Utility Application, Tennessee Valley Authority PB 276 637, WPA-600/7-77-127 (November 1977).

26. Environmental Control Implications of Generating Electric Power from Coal, 1977 Technology Status Report, Tennessee Valley Authority from Argonne National Laboratory, ANL-ECT-3, Appendix B (December 1977).

27. Shellef, M., K. Otto, and H. Gander, The Hetrogeneous Decomposition of Nitric Oxide on Supported Catalysts, Atmospheric Environment, 3:107-122 (1969).

28. Shellef, M., K. Otto, and H. Gandhi, The Behavior of Nitric Oxide in Heterogeneous Catalytic Reactions, Chemical Engineering Progress Symposium Series, 67(115):74092 (1971).

29. Edwards, H.W., and R.M. Harrison, Catalysis of NO Decomposition by MN₃O₄, Environmental Science and Technology, 13(6):673-676 (June 1979).

30. Hammons, G.A., and A. Skopp, NOx Formation and Control in Fluidized-Bed Coal Combustion Processes. Paper presented at ASE Winter Annual Meeting, Washington, D.C. (1971).

31. Armitage, J.W., and C.F. Cullis, Studies of the Reaction Between Nitrogen Dioxide and Sulfur Dioxide, Combustion and Flame, 16:125-130 (1971).

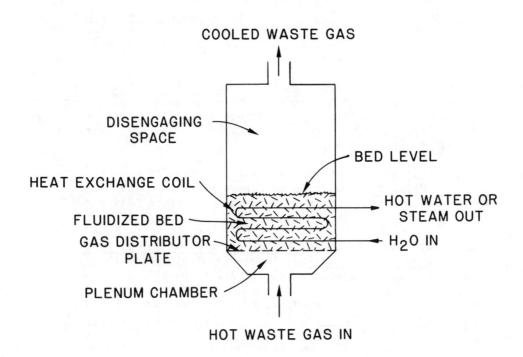

Fig. 1. Fluidized-Bed, Waste-Heat Boiler

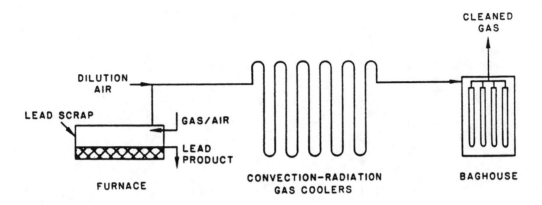

Fig. 2. Schematic of a Secondary-Lead Smelting Facility

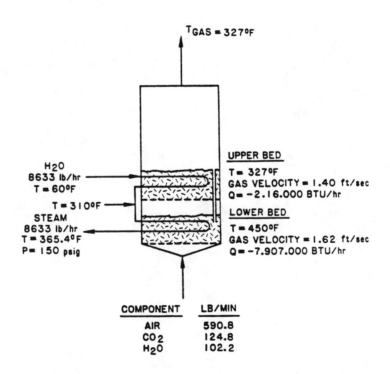

Fig. 3. Details of FBWHB for Recovering Waste
from Secondary Smelter Offgas

Table 1. Summary of Sulfation-Regeneration Information
for Major $SO_2$ Absorption Systems

| Sorbent | Process Step | Reaction | Typical Reaction Temperature (°C) | $SO_2$ Removal Efficiency, Particle Attrition, and Effect Of Variables | Status of Process |
|---|---|---|---|---|---|
| MnO | Sulfation | $MnO_x \cdot iH_2O + SO_2 + 1/2(2-x)O_2 \longrightarrow MnSO_4 + iH_2O$ | 150 | Sulfation rate decreases gradually until about a third of particle is sulfated, after which the sulfation rate decreases rapidly. Greater than 90% removal efficiency was demonstrated in the entrained-bed reactor. | Processed 25,000 $ft^3$/min of power plant flue gas in an entrained-bed reactor. |
| | Regeneration | $MnSO_4 + NH_4 \longrightarrow Mn(OH)_2 + (NH_4)_2SO_4$ $Mn(OH)_2 + 1/2(x-1)O_2 + (i-1)H_2O \longrightarrow MnO_x \cdot iH_2O$ | Liquid System | | |
| CuO | Sulfation | $2CuO + 2SO_2 + O_2 \longrightarrow 2CuSO_4$ | 400 | In the parallel passage reactor, the removal efficiency is 100% initially and decreases with time. Conversion to sulfate increases with temperature until at 550°C the reaction is significantly reversed. Copper content of the copper-impregnated alumina carrier affects the conversion to sulfate. About 4% copper content gave the best results. Poorer results were obtained with higher and lower copper concentrations. | 20,000 $ft^3$/min of power plant flue gas was processed in a fixed-bed, parallel passage reactor. |
| | Regeneration (with $H_2$) | $CuSO_4 + 2H_2 \longrightarrow Cu + SO_2 + 2H_2O$ | 400 | Increasing the regeneration temperature increases the rate of the regeneration reaction. Too high a temperature reduces the capacity of the particle to reabsorb $SO_2$ in the subsequent $SO_2$ removal step. | |
| | Regeneration | $CuSO_4 \longrightarrow CuO + SO_2 + 1/2 O_2$ | 750 | Higher temperature increases thermal decomposition rate. | |

Table 1. (Cont'd)

| Sorbent | Process Step | Reaction | Typical Reaction Temperature (°C) | SO₂ Removal Efficiency, Particle Attrition, and Effect of Variables | Status of Process |
|---------|--------------|----------|-----------------------------------|-------------------------------------------------------------------|-------------------|
| CaO | Sulfation | $CaO + SO_2 + 1/2\ O_2 \longrightarrow CaSO_4$ | 850 | Increasing temperature increases the SO₂ removal-efficiency until a maximum is reached. After that, increasing the temperature causes a decrease in the removal efficiency. | Sulfation characteristics were determined in a fluidized-bed combustion, 300,000-lb/hr steam plant. |
| | | | | Increasing the gas velocity decreases the SO₂ removal efficiency. | |
| | | | | Increasing the Ca/S ratio increases the SO₂-removal efficiency until, at a ratio of 3.0, only small incremental improvement results. | |
| | | | | Increasing the system pressure or oxygen partial pressure increases SO₂-removal efficiency. | |
| | | | | Physical properties of naturally occurring limestones and dolomites affect SO₂-removal efficiency. Increased efficiency is obtained with coarser-grained, smaller particles of high surface area and with particles containing iron. Water vapor adversely affects efficiency. A slow decarbonation (release of CO₂ from the limestone) rate aids SO₂-removal efficiency. Ninety percent efficiency demonstrated. | |
| | | | | High attrition (greater than 10%) results if limestone particles are decarbonated rapidly. Sulfation of the limestone particle hardens it. | |
| | Regeneration | $CaSO_4 + H_2 \longrightarrow CaO + SO_2 + H_2O$ | 1100 | Increasing the particle residence time or increasing the reaction temperature increases the extent of regeneration. About 2% attrition if CaO is regenerated in a fluidized-bed reactor. | Process has been tested on a 1-ton/day plant. |

283

Table 1. (Cont'd)

| Sorbent | Process Step | Reaction | Typical Reaction Temperature (°C) | SO$_2$ Removal Efficiency, Particle Attrition, and Effect of Variables | Status of Process |
|---|---|---|---|---|---|
| Titanates (Ca or Ba) | Sulfation | $CaTiO_3 + SO_2 + 1/2\ O_2 \longrightarrow (CaSO_4 + TiO_2)$ | 850 | Over 90% removal efficiency demonstrated. Attrition rate was low but greater than 0.3%/h. | Sulfation and regeneration steps were demonstrated in a 1-MW fluidized-bed coal combustor with attached regeneration bath operated at a pressure of 1000 kPa. |
| | Regeneration | $(CaSO_4 + TiO_2) + CO \longrightarrow CaTiO_3 + SO_2 + CO_2$ | 1000 | | |
| Aluminates (Ca or Na) | Sulfation | $CO \cdot nAl_2O_3 + SO_2 + 1/2\ O_2 \longrightarrow (CaSO_4 + nAl_2O_3)$ | 850 | Increasing the reaction temperature or the SO$_2$ or O$_2$ concentrations in the gas increases the SO$_2$ sulfation rate. Pellets of calcium aluminate broke up at a rate of less than 0.3%/h. | Pellets were tested in a 1-MW fluidized-bed combustor plant (approximately 3000 ft$^3$/min gas). |
| | Regeneration | $(CaSO_4 + nAl_2O_3) + H_2 \longrightarrow CaO \cdot nAl_2O_3 + SO_2 + H_2O$ | 850 | Conversion increases with increasing temperature at 900°C and with increasing hydrogen concentration in the reducing gas mixture. | |

Table 2. Summary of Sulfation-Regeneration Information for Major SO$_2$ Absorption Systems

| Process Step | Process Developer | Reaction | Typical Reaction Temperature (°C) | Effect of Variables | Status of Process |
|---|---|---|---|---|---|
| Adsorption | All | $SO_2 + 1/2\ O_2 \rightarrow SO_3$ (sorbed)<br>$SO_3$ (sorbed) + $H_2O$ (sorbed) $\rightarrow H_2SO_4$ | 140 | Increasing temperature decreases the rate of SO$_2$ removal. Increasing SO$_2$, O$_2$, or H$_2$O concentrations increases rate of SO$_2$ removal. If NO$_x$ concentration is lower than 50 ppm, the rate is increased. | * |
| Regeneration | Lurgi | The H$_2$SO$_4$ product is continuously washed from the pores of the carbon particles. | | | Treated 1000 ft$^3$/min of dust-free flue gas from an oil-fired boiler (approximately 0.3 MW). |
| | Hitachi | Same as Lurgi except washing is done in batches. | | | Treated gases from a 2-MW oil-fired plant. |
| | Reinluft | $C + 2H_2SO_4 \xrightarrow{\text{heat}} 2SO_2 + CO_2 + 2H_2O$ | 371 | | Used in treating flue gas from a 10-MW power plant. |
| | Westvaco | $3H_2S + H_2SO_4 \rightarrow 4S + 4H_2O$ | 149 | Reaction rate is faster at higher temperatures, higher H$_2$S concentrations, and sulfur loading, i.e., weight sulfur/weight carbon. | |
| | BF/FW | $2H_2SO_4 + C \rightarrow CO_2 + 2H_2O + 2SO_2$<br>$SO_2 + C \rightarrow S + CO_2$ | 650 | | Treated 20,000 ft$^3$/min of flue gas from combustion of 3% sulfur oil. |

Table 3. Process Conditions and Economics for the Dry SCR $NO_x$ Removal Processes

| Company | Development Status (MW equiv) | Removal Efficiency (%) | | Catalyst | Reaction Temp. (°C) | Mol $NH_3$ per mol $NO_x$ | Reported Capital Investment ($/kW)[a] | Reported Revenue Requirement (mills/kWh)[a] |
|---|---|---|---|---|---|---|---|---|
| | | $NO_x$ | $SO_2$ | | | | | |
| Eneron | Pilot (1.5) | 65 (oil) 85 (gas) | – | 0.03% Pt by wt | 280-305 255 | 2.0-2.5 | 11[b] | 0.2[b] |
| Exxon[c] | Bench (0.003) | 70-95 | 90-95 | – | 315-370 | 0.7-1.0 | – | – |
| Hitachi Ltd. | Comm. (170) | >90 | – | – | 300-400 | 0.9-1.1 | 45 | – |
| Hitachi Zosen | Comm.[d] (275) | >90 | – | – | 300-400 | 0.8-1.2 | 16 | 1.5 |
| JGC Corp. | Proto. (23) | >95 | – | – | 380-420 | 1.1-1.3 | 27 | – |
| Kobe | Bench (0.3) | 90 | – | Base metal | 350-400 | 1.0-1.1 | 12-21 | 1.3-1.7 |
| Kurabo | Proto. (10) | >90 | – | $CuO$[e] | 350-400 | 0.9-1.0 | 35 | 1.4 |
| Kureha | Pilot (1.6) | 90 | – | – | 150 | 1 | – | – |
| Mitsubishi Heavy Ind. | Pilot (1.3) | 90 | – | – | 350-400 | 1 | – | – |
| Mitsubishi KK | Pilot (4.7) | >90 | – | $Fe_2O_3 \cdot H_2O$ | 400-450 | 1.0-1.3 | 36 | 1.8 |
| Mitsubishi Petro.[f] | – | 95 | – | Titanate | 300-600 | 1 | – | – |
| Mitsui E&S | Comm. (67) | 90 | – | Base metal | 350 | 1 | 10 | – |
| Mitsui Toatsu | Proto. (30) | >90 | – | – | 350-400 | 1.0-1.2 | 49 | 1.6 |
| Sumitomo Chemical | Comm. (100) | 90 | – | Base metal | 300-350 | 1 | 80[g] | 1.2[g] |
| Sumitomo Heavy Ind. | Bench (0.5) | >90 | – | Metal oxide | 270-370 | 1 | 30 | 2.3 |
| Sumitomo Heavy Ind. | Pilot (3.3) | 85-90 | 95 | Act. carbon | 200-230 | 2.0-2.5 | 46 | 5.9 |
| Takeda | Pilot (3.3) | 90 | 80 | Act. carbon | 210-250 | 0.7-1.2[h] | 90 | 0.65 |
| Ube | Pilot (3.3) | 90 | – | – | 350-400 | 1.1-1.3 | – | – |
| Unitika | Bench (0.07) | >90 | – | – | 320-410 | 1.0-1.1 | 24 | 2.1 |
| Unitika | Pilot (1.5) | >90 | >90 | Act. carbon | 200-250 | –[i] | 60 | 6.8 |
| UOP | Proto.[j] (40) | >80 | 90 | $CuO/CuSO_4$ | 400 | 1.0-1.2 | 131[k] | 5.0[k] |
| UOP | Proto.[l] (40) | >90 | – | $CuSO_4$ | 400 | 1.0-1.2 | 31[k] | 1.4[k] |

[a]Unless otherwise noted, cost based on 1976 dollars and Japanese location; see each detailed process description for additional bases.

[b]Based on 1975 dollars and U.S. location.

[c]Exxon has studied both simultaneous $NO_x$-$SO_2$ removal and $NO_x$-only removal.

[d]Being tested on 0.07-MW equiv coal-fired unit.

[e]Kurabo has newly developed iron-based catalyst which is more suitable for coal-fired flue gas.

[f]Offering catalyst only.

[g]Capital investment is based on treatment of dirty gas using an ESP. The revenue requirement is in 1973 dollars and based on treatment of clean gas at the Higashi Nihon Methanol plant.

[h]This is mol $NH_3$/mol ($NO_x$ + $SO_2$).

[i]Mols $NH_3$ – (0.8) mols NO + (0.3-0.5) mols $SO_x$.

[j]UOP has performed tests with coal-fired flue gas on 0.6-MW equiv pilot plant in FGD mode.

[k]These costs are based on 1977 dollars and U.S. locations.

[l]$NO_x$-only removal has been tested at the same 40-MW equiv treatment facility which normally operates with simultaneous $SO_x$ and $NO_x$ removal.

Source: Ref. 25.

# USE OF NAHCOLITE FOR COAL-FIRED POWER PLANTS

**Dennis E. Lapp** and **Eric Samuel**
Envirotech Corporation

**Theodore G. Brna**
U.S. Environmental Protection Agency

**Navin D. Shah** and **Ben Weichman**
Multi-Mineral Corporation

**Ronald L. Ostop**
City of Colorado Springs

Introduction

Nahcolite is a naturally occurring sodium bicarbonate ($NaHCO_3$) mineral. Nahcolite occurs either in the form of a brown crystalline substance along with oil shale or in the white, bedded form by itself. Vast deposits of nahcolite are found in the Piceance Creek Basin of Northwest Colorado. Total reserves of nahcolite in the Piceance Creek Basin are estimated to be approximately 27 billion metric tons (30 billion short tons).[1] Tables 1 and 2 show the physical and chemical properties of a typical nahcolite sample.

In the dry state, $NaHCO_3$ in nahcolite is fairly reactive with sulfur dioxide ($SO_2$) in the flue gas generated from burning sulfur containing fuels (e.g., coal). At a temperature above 121°C (250°F) nahcolite reacts with $SO_2$ according to the reaction equations shown in Table 3. $NaHCO_3$ in nahcolite first decomposes to sodium carbonate ($Na_2CO_3$) with the evolution of carbon dioxide ($CO_2$) and water vapor (see Table 3).[2,3,4] This results in a highly porous and reactive $Na_2CO_3$ which then reacts with $SO_2$ to form sodium sulfate ($Na_2SO_4$). Theoretically, to remove 0.454 kg (1 lb) of sulfur (0.907 kg or 2 lb of $SO_2$), it takes 3.4 kg (7.5 lb) of nahcolite. The process of using nahcolite in the dry state for removing $SO_2$ from the flue gas is termed "Dry Flue Gas Desulfurization" (FGD). Before getting into the details of the dry FGD process, the basis for utility interest in nahcolite and dry FGD will be briefly discussed.

Before 1978, Environmental Protection Agency (EPA) regulations for $SO_2$ emissions from coal-fired power plants were such that utilities had to control their $SO_2$ emissions to 520 mg/J (1.2 lb $SO_2/10^6$ Btu) of heat input. Figure 1 shows the EPA regulations (before 1978) in the form of $SO_2$ removal requirements as a function of percent sulfur in coal for 23,240 kJ/kg (10,000 Btu/lb) coal. It is clear that utilities burning coal with less than 0.6 percent sulfur (10,000 Btu/lb coal) did not have to worry about $SO_2$ control. Since most of the western coal falls into this category, most utilities in the West did not have to remove any $SO_2$. In the East, utilities burning high sulfur coal started using wet $SO_2$ scrubbers to meet the 1971 EPA regulations. Part of the reason for selecting the wet scrubber was that wet $SO_2$ scrubbing was the best available technology to utilities during the early and mid-1970s.

In 1978, EPA proposed changes to the regulations for $SO_2$ emissions. The new regulations promulgated in June 1979 for $SO_2$ emission are summarized in Figure 1. In addition, the EPA particulate emission regulation which used to be 43 mg/J (0.1 lb/10^6 Btu) was lowered in 1978 to 14 mg/J (0.03 lb/10^6 Btu). Utilities were able to meet the 0.1 lb/10^6 Btu standard, using the electrostatic precipitator (ESP) and wet scrubbing, but with the 0.03 lb/10^6 Btu standard, the utilities (particularly those burning low sulfur coal) started thinking about using a baghouse instead of an ESP. Thus, changes in EPA regulations have impacted utility planning in two ways:

- No matter what the sulfur content of the coal is, at least 70 percent $SO_2$ removal is required for new plants.

- Tightening of particulate standards requires serious consideration of a baghouse as an alternative to an ESP.

When a utility is considering the use of a baghouse for particulate control, dry injection of nahcolite into the flue upstream of the baghouse is a potential dry $SO_2$ control approach.

The dry FGD system using nahcolite and a baghouse is a simpler process compared with the complicated wet scrubber and ESP. In Table 4, a comparison of advantages and disadvantages of FGD by dry injection is summarized.

Problems associated with the disposal of water soluble wastes are still the primary hindrance to using sodium compounds in dry injection FGD. EPA is supporting (through Buell Emission Control Division of Envirotech Corporation) a study by Battelle-Columbus Laboratories of the "Sinterna[R]5 process for insolubilization of sodium salt/fly ash dry FGD wastes. The results of this study, which will also provide estimates of costs of insolubilizing the wastes and identify potential uses for the pelletized wastes, are expected to be available in late December 1980. Preliminary estimates indicate that the insolubilization process would cost about $20/ton (compared to about $15/ton for disposing of lime spray dryer wastes via landfill) less any offset when the utilized waste is sold as a concrete filler material or for other purposes.

A number of bench and pilot scale studies[5-8] have been conducted in the past to examine the feasibility and define the optimum operating parameters of the nahcolite dry FGD system. The most noteworthy tests have been those conducted by EPRI[9] and EPA[10]. EPRI has conducted a systematic bench-scale study of the dry FGD system using nahcolite and has concluded that the dry FGD system appears to be a commercially viable alternative to wet scrubbers, particularly for the western part of the country. As a result, full-scale demonstration tests of the nahcolite dry FGD process are currently being conducted by EPRI/Public Service Company of Colorado/

Multi-Mineral Corporation at the 22 MW Cameo Station of the Public Service Company of Colorado.

This paper summarizes the results of the tests conducted by EPA in cooperation with Buell Emission Control Division of Envirotech Corporation at the City of Colorado Springs' Martin Drake Station. These tests sought to evaluate, in addition to nahcolite, trona (a naturally occurring mineral containing predominately $Na_2CO_3$), and refined trona (bicarbonate-enriched trona) as dry adsorbents for FGD.

Experimental

Buell has constructed at the Martin Drake Station, operated by the City of Colorado Springs, a pilot facility consisting of two slipstreams attached to the Unit 6 boiler. The primary slipstream consists of a 0.660-m (26-in.) diameter duct feeding a 9.44-$m^3$/s (20,000-cfm) spray dryer, a cyclone dust collector, and a booster fan. The secondary slipstream consists of a smaller 0.394-m (15.5-in.) diameter duct feeding a 1.42-$m^3$/s (3,000-cfm) pilot baghouse and a booster fan. Both systems extract flue gas through two isokinetic scoops installed in both the hot and the cold sides of the Unit 6 air heater and interconnected through butterfly dampers.

Variations in flow rate and gas temperature are achieved by adjustments of the butterfly dampers. Both slipstreams exhaust into the inlet of a 2,376-bag, 12-compartment, 189-$m^3$/S (400,000-cfm) Buell main baghouse serving Unit 6. The secondary slipstream was used to perform the test program to evaluate FGD by the dry injection of sodium compounds. A schematic flow diagram of the test setup is shown in Figure 2. Pilot scale testing of dry FGD using the spray dryer connected to the primary slipstream is described elsewhere.[12]

Unit 6 is served by a pulverized coal-fired boiler. Natural gas serves as the start-up and secondary fuel. Steam from the boiler operates an 85 MW turbine-generator. Design drum pressure is $1.03 \times 10^7$ pascal (1500 psig) while rated steam flow is 311,000 kg/hr (685,000 lb/hr). Normal flue gas volume is 189 $m^3$/s (400,000 cfm) at approximately 177°C (350°F).

The Martin Drake boiler fires Colorado subbituminous coals whose sulfur content is in the range of 0.4 to 0.6 percent and gives a $SO_2$ concentration of 300 to 600 ppm by volume (or 0.76 to 1.14 lb/$10^6$ Btu).

The pilot baghouse used in these tests is a single-compartment unit containing 16 full-size bags, 0.30 m in diameter by 9.14 m long (1 ft in diameter by 30 ft long), identical to those used in the main baghouse. The bags are of Menardi-Southern Teflon[R]-impregnated, woven fiberglass having a weight of 0.36 kg/$m^2$ (10.5 oz/$yd^2$). The pilot filter is equipped with both reverse air and mechanical shaker capability. Flue gas enters the pilot baghouse through the hopper. The flow rate through this baghouse is maintained constant with the aid of a flow controller deriving flow measurement from a venturi (fabricated to ASME specifications) connected to a pressure transducer equipped with a square-root extractor.

The sodium compounds were ground to 95 percent minus 200 mesh and then fed by a vibration screw feeder into a star wheel rotary airlock for pneumatic conveying into the flue gas slipstream with the aid of a blower. Injection was countercurrent into the slipstream to increase turbulent mixing.

Both slipstreams at the facility were well instrumented for continuous measurement and recording of the important operational parameters. All equipment was housed in a 3-m by 7-m (10-ft by 24-ft) trailer. Some of the key features of the system are:

- Automatic data acquisition with 60-channel data logger.

- Offsite, self-contained computer processing using data logger magnetic tape.

- On-line monitors with hard copy records (e.g., temperatures, $\Delta P$, opacity).

- Automatic gas species sampling ($SO_2$, $O_2$, $NO_x$).

- Manual flue gas sampling with resident source testing crew.

(Figure 3 describes the instrumentation specific to the secondary slipstream and the dry injection program.)

For the continuous measurement of $SO_2$, $NO_x$, and $O_2$, York Research, Inc. was contracted to design, install, and maintain an analyzing system. York's configuration, depicted in Figure 4, can be described as a "dry" system utilizing conditioning apparatus at the sampling points and uninsulated, unheated sample lines.

The gas was sampled at two locations in the secondary slipstream: at the baghouse inlet, 6 m (20 ft) downstream of the cold-side scoop, and at the baghouse outlet, about 1 m (several feet) before exiting into the main baghouse inlet duct. Table 5 lists the instrumentation used in gas species measurement.

Results and Discussion

### $SO_2$ Removal Efficiency

The effectiveness of nahcolite, trona, and refined trona as $SO_2$ suppressants, as determined through a series of 44 parametric tests covering a range of stoichiometric ratios, flue gas temperatures at the adsorbent injection point, fabric filter compartment (baghouse) temperatures, and air-to-cloth ratios (see Table 6) are compared in Figures 5, 6, and 7. The results of these tests indicate that all three compounds were effective in removing $SO_2$ from the slipstream flue gas.

Nahcolite was the best performer, yielding approximately 67 percent $SO_2$ removal at a stoichiometric ratio (SR) of 1.0. The baghouse temperature was 163°C (325°F), and injection was into the duct approximately 30.5 m (100 ft) upstream of the baghouse inlet. Increasing the baghouse temperature to 260°C (500°F) decreased the amount of removal to 38 percent at an SR of 1.0. For continuous injection over a 30-minute test, an SR of 1.05 was required to achieve 70 percent removal; increasing to an SR of 2.0 yielded 91 percent $SO_2$ removal.

Trona was also capable of reducing $SO_2$, although removal rates were considerably lower than with nahcolite. At an SR of 1.0, trona yielded an $SO_2$ reduction of 23 percent from the same upstream injection location and at the 163°C (325°F) baghouse temperature. Increasing the temperature to 260°C (500°F) increased the $SO_2$ removal to 32 percent at an SR of 1.0. Raising the SR to 1.6 improved removal efficiency to approximately 50

290

percent. No further increase in SO$_2$ removal occurred as the stoichiometric ratio was raised from 1.6 to 2.75.

Refined trona, a bicarbonate-enriched form of the same ore, performed much better than trona. At an SR of 1.0, it demonstrated 45 percent removal at the 163°C (325°F) baghouse temperature. Efficiency was slightly less (same trend as nahcolite) when the temperature was elevated to 260°C (500°F), showing 41 percent removal at an SR of 1.0. An SR of 1.6 was required to achieve 70 percent SO$_2$ removal. An SR of 1.75 yielded a maximum efficiency of 79 percent.

Varying the air-to-cloth ratio from 0.44 to 0.91 m/min (1.44 to 2.99 ft/min) had no noticeable effect on SO$_2$ reduction.

Nahcolite and refined trona are thought to be more effective in SO$_2$ adsoprtion (when compared to trona) because of their higher sodium bicarbonate composition (Table 7). In addition, the mineral structure of nahcolite may have enhanced SO$_2$ removal with nahcolite, relative to refined trona, via greater thermal decomposition of nahcolite at flue gas temperatures to produce porous Na$_2$CO$_3$ with high surface area capable of reacting quickly and effectively with SO$_2$.

### Effect of Dry Injection on Baghouse Pressure Drop

Table 8 shows the increase in baghouse inlet grain loading that may be expected during injection of nahcolite at an SR of 1.0 and trona at an SR of 1.5, assuming an inlet SO$_2$ concentration of 400 ppm, a baghouse temperature of 177°C (350°F), and 70 percent SO$_2$ removal.

Figure 8(a) compares the measured pressure drop over time for the Martin Drake Unit 6 baghouse with a computer model prediction. The computer model uses for input the pressure/time characteristics of the pilot baghouse. Table 9 shows that the predictions of the computer model are in reasonable agreement with measurement. Table 9 also shows that the effect of a variable grain loading from 0.0035 to 0.0049 kg/m$^3$ (1.53 to 2.12 gr/ft$^3$) is much less than the effect of other changes in inlet conditions such as particle size and air-to-cloth ratio. A time between successive cleanings of 60 minutes appears typical.

The increase in grain loading due to dry injection is large enough that it must be factored into the fabric filter system design. The computer model referred to above predicts an average cleaning time in the range of 35 to 42 minutes during dry injection of nahcolite whose bicarbonate content varies in the range of 46.8 to 80 percent, respectively, assuming the same size distribution for the nahcolite as for the inlet fly ash. Pressure drop measurements during the pilot-scale experiments using the lower grade nahcolite (46.8 percent bicarbonate) generally showed lower rates of increase with time than those predicted by the computer model.

### Economic Comparison of Dry vs Wet FGD

An economic comparison between dry nahcolite FGD systems and wet FGD systems and spray dryers was performed using methods adopted by EPRI.[13] The results of the economic comparison for a typical 500 MW power plant are summarized in Figures 9 and 10. Figure 9 compares the capital costs, while Figure 10 indicates the operating costs for different FGD approaches. Using nahcolite, an SR of 1.0 was assumed to effect 70 percent SO$_2$ removal as seen in Figure 8. It must be emphasized that the curves relating to the nahcolite injection process in Figures 9 and 10 are based on hypothetical nahcolite costs and do not include the costs, both capital and operational, of disposing of the water-soluble off-product.

The accuracy of cost estimates relating to the various FGD systems is usually in the range of -40 to +50 percent.[14,15] In order to illustrate the above uncertainty in the cost estimates, Figures 9 and 10 also show the results of two other EPA-sponsored studies by Burnett et al.[14] of the TVA relating to the lime spray dryer and wet limestone processes and by Lutz et al.[15] of TRW, Inc. relating to the nahcolite dry FGD process. Both results include the cost of waste disposal. The results taken from Lutz et al. pertain to a 500 MW plant firing 24,400 kJ/kg (10,500 Btu/lb) coal. The capital costs from the Lutz study, originally quoted in 1977 dollars, have been adjusted to 1982 using a levelized inflation rate of 14.7 percent.[14] The operating costs for the nahcolite process from the Lutz study assumed a nahcolite cost of $29.48/metric ton ($32.50/ton) and a waste disposal cost of $5.44 metric tone ($6/ton). Here these costs have been adjusted to reflect nahcolite costing $81.65/metric ton ($90/ton) and waste disposal $18.14/metric ton ($20/ton).[11] Figures 9 and 10 also show the results of an economic comparison by Genco et al.[5] updated using the procedure described above to update the results of Lutz et al. While the absolute accuracy in the economic analyses is in the range of -40 to +50 percent, the relative accuracy in comparing the different FGD methods is smaller; in the range of ±10 percent[14], if the same costing methods are used throughout.

In spite of the spread in the cost estimates and their accuracy (-20 to +50 percent), the dry FGD process using nahcolite appears favorable in capital costs for low sulfur coal (~1 percent S). The apparent attractiveness of the dry injection process in terms of capital cost, as evident from Figure 9, must be viewed with some caution in the light of the preliminary basis for the available cost of waste disposal (i.e., $20/ton). The economic and operational viability of the proposed methods[5,15] of waste disposal has not been fully demonstrated. The operating costs, on the other hand, for all FGD processes appear to be in the same range for about 1 percent sulfur coal. In addition, transportation costs may limit the nahcolite process to only Western and some Midwestern sites in close proximity to nahcolite mines. Lower coal sulfur contents, more land availability for waste disposal, and lower rainfall are other considerations consistent with the above limitation.

Future Demand for Nahcolite

The nahcolite requirement, in tons per day* for a given power plant, may be calculated as follows:

$$D = 0.11 \frac{C\ S\ R}{E\ B\ F}$$

where

$D$ = nahcolite required, tons/day

$C$ = plant capacity, W

$S$ = fractional sulfur content in coal by weight

$R$ = stoichiometric ratio for required $SO_2$ removal level

*To convert to metric tons per day, multiply the value for D in the following equation by 0.907.

E = overall efficiency (fraction) of converting heating
value of coal to electrical energy

B = heating value of coal as fired, Btu/lb (2.234 kJ/kg)

F = weight fraction of $NaHCO_3$ in nahcolite

Figure 11 summarizes nahcolite requirements using the above equation based on $C = 500 \times 10^6$, $E = 0.3$, $B = 10,000$, and $F = 0.5$. The stoichiometric ratios given in Figure 11 for specified $SO_2$ removal efficiencies were obtained from Figure 5 and from results of previous bench-scale studies.[9]

The future demand for nahcolite may be assessed from the results of a recently published study of recent trends in utility FGD by Smith et al.[16] (Tables 10 and 11 are from Reference 16.) Table 10 shows the total capacity of present and planned FGD systems, while Table 11 summarizes the distribution of present and planned FGD systems by process. A total capacity of 2,047 to 19,000 MW is obtained for planned FGD systems in which nahcolite could be used. The lower estimate is based on the assumption that the nahcolite dry FGD process has the same total capacity as that already committed to $Na_2CO_3$-based FGD. The higher estimate is based on the assumption that all planned FGD systems in the Western states could be available to the nahcolite dry FGD process. Assuming that the percentage of sulfur in the coal is about 1 percent, and that 70 percent $SO_2$ removal can be effected with an SR of 1.0, the above range of total capacity potentially available to the nahcolite dry FGD process translates to an average nahcolite demand in the range of 0.4 to 3.0 million tons of nahcolite per year (with $E = 0.33$ and $F = 0.6$).

References

1.  Beard, T.N., D.B. Tait, and J.W. Smith, "1974 Nahcolite and Dawsonite Resources in the Green River Formation, Piceance Creek Basin, Colorado," Rocky Mountain Association of Geologists Guidebook, pp 101-109 (1974).

2.  Oil Shale Department, "The Use of Nahcolite for Removal of Sulfur Dioxide and Nitrogen Oxides from Flue Gas," The Superior Oil Co., 97 pp (1977).

3.  Stern, F.R., "Bench Scale Study of Sulfur and Nitrogen Oxides Adsorption by Nahcolite and Trona," M.S. Thesis, The University of North Dakota, Grand Forks, North Dakota, 81 pp (1978).

4.  Ness, H.M. and S.J. Selle, "Control for Western Power Plant Sulfur Dioxide Emissions: Development of the Ash-Alkali FGD Process and Dry Adsorption Techniques at the Grand Forks Energy Technology Center," Proceedings of DOE Symposium on Environmental Control Activities, Department of Energy, Washington, D.C., 20 pp (1978).

5.  Genco, J.M., H.S. Rosenberg, M.Y. Anastas, E.C. Rosar, and J.M. Dulin, "The Use of Nahcolite Ore and Bag Filters for Sulfur Dioxide Emission Control," JAPCA, Vol. 25, No. 12, p 1244 (1975).

6.  Estcourt, V.F., R.O.M. Gruttler, D.C. Gehri, and H.J. Peters, "Tests of a Two-Stage Combined Dry Scrubber/$SO_2$ Absorber Using

7.  Doyle, D.J., "Fabric and Additive Remove $SO_2$," <u>Electrical World</u> (February 15, 1977).

8.  Konianoff, C., "Pollution Control Improvements in Coal-Fired Electric Generating Plants: What They Accomplish, What They Cost," <u>JAPCA</u>, Vol. 30, No. 9, p 1031 (1980).

9.  Muzio, L.J., J.K. Arand, and N.D. Shah, "Bench-Scale Study of Dry $SO_2$ Removal with Nahcolite and Trona," presented at Second Conference on Air Quality Management in the Electric Power Industry, <u>University of Texas</u>, Austin, Texas (1980).

10. Furlong, D.A., T.G. Brna, and R.L. Ostop, "$SO_2$ Removal Using Dry Sodium Compounds," presented at 89th Annual Meeting, <u>American Institute of Chemical Engineers</u>, Portland, Oregon, 20 pp (1980).

11. Samuel, E.A. and D.E. Lapp, "Testing and Assessment of a Baghouse for Dry $SO_2$ Removal - Task 4: $SO_2$ Removal Using Dry Sodium Compounds," forthcoming EPA report, <u>U.S. Environmental Protection Agency</u>, Industrial Environmental Research Laboratory, Research Triangle Park, N.C., prepared under Contract No. 68-02-3119.

12. Parsons, Jr., E.F., L.F. Hemenway, O.T. Kragh, T.G. Brna, and R.L. Ostop, "$SO_2$ Removal by Dry FGD," presented at the Sixth Symposium on the Flue Gas Desulfurization, <u>U.S. Environmental Protection Agency</u>, Houston, Texas (1980).

13. EPRI Economic Premises, <u>Electric Power Research Institute</u>.

14. Burnett, T.A., K.D. Anderson, and R.L. Torstrick, "Spray Dryer FGD: Technical Review and Economic Assessment," presented at Sixth Symposium on Flue Gas Desulfurization, <u>U.S. Environmental Protection Agency</u>, Houston, Texas (1980).

15. Lutz, S.J., R.C. Christman, B.C. McCoy, S.W. Mulligan, and K.M. Slimak, "Evaluation of Dry Sorbents and Fabric Filtration for FGD," EPA-600/7-79-005 (NTIS No. PB 289921), <u>U.S. Environmental Protection Agency</u>, Industrial Environmental Research Laboratory, Research Triangle Park, N.C., 145 pp (January 1979).

16. Smith, M.P., M.T. Melia, R.A. Laseke, Jr., and N. Kaplan, "Recent Trends in Utility Flue Gas Desulfurization," presented at the Sixth Symposium on Flue Gas Desulfurization, <u>U.S. Environmental Protection Agency</u>, Houston, Texas (1980).

TABLE 1.  PHYSICAL PROPERTIES OF NAHCOLITE

| | |
|---|---|
| TRUE DENSITY | $= 2082 \text{ kg/m}^3 \ (130 \text{ lb/ft}^3)$ |
| BULK DENSITY <br> (Granules smaller than <br> 0.953 cm (0.375 in.)) | $= 1281 \text{ kg/m}^3 \ (80 \text{ lb/ft}^3)$ |
| SURFACE AREA <br> (-400 mesh) | $= 5 \text{ m}^2/\text{g} \ (2.4 \times 10^4 \text{ ft}^2/\text{lb})$ |
| WEIGHT PERCENT LOSS ON HEATING <br> (At 149°C (300°F)) | = 30 |
| WATER SOLUBILITY | = 2.6 g/100 g water |
| pH OF NAHCOLITE SOLUTION | = 8.4 |
| WEIGHT PERCENT OIL | = 1 |
| OIL/NAHCOLITE | = 6.9 l./tonne <br> 2 gal/ton |

TABLE 2.  CHEMICAL PROPERTIES OF NAHCOLITE

| | |
|---|---|
| SODIUM (Na) | = 20.00% |
| BICARBONATE ($HCO_3$) | = 53.30% |
| CARBONATE ($CO_3$) | = 3.30% |
| CALCIUM (Ca) | = 2.20% |
| MAGNESIUM (Mg) | = 0.70% |
| IRON (Fe) | = 0.30% |
| ALUMINUM (Al) | = 0.50% |
| CHLORIDE (Cl) | = 0.04% |
| NITRATE ($NO_3$) | = 0.00% |
| PHOSPHATE ($PO_4$) | = 0.00% |
| SULFATE ($SO_4$) | = 0.00% |
| TOTAL WATER SOLUBLES | = 80.34% |
| TOTAL WATER INSOLUBLES | = 19.66% |

TABLE 3.  REACTIONS OF NAHCOLITE
WITH SULFUR COMPOUNDS

$$\underset{32}{S} + \underset{32}{O_2} \rightarrow \underset{64}{SO_2}$$

$$\underset{240}{NAHCOLITE^*} \rightarrow \underset{168}{2NaHCO_3}$$

$$\underset{168}{2NaHCO_3} \rightarrow \underset{106}{Na_2CO_3} + \underset{18}{H_2O} + \underset{44}{CO_2}$$

$$\underset{106}{Na_2CO_3} + \underset{64}{SO_2} + \underset{16}{\tfrac{1}{2}O_2} \rightarrow \underset{142}{Na_2SO_4} + \underset{44}{CO_2}$$

0.454 kg (1 lb) of S = 3.4 kg (7.5 lb) of NAHCOLITE*
0.907 kg (2 lb) of $SO_2$ = 3.4 kg (7.5 lb) of NAHCOLITE*

*70% $NaHCO_3$

TABLE 4.  ATTRIBUTES OF DRY FGD
BY NAHCOLITE INJECTION

ADVANTAGES

SIMPLE PROCESS

$SO_2$ + PARTICULATE CONTROL IN ONE ADDITIONAL EQUIPMENT ITEM

UTILITY FAMILIARITY WITH DRY SYSTEM (E.G., COAL HANDLING)

LOWER CAPITAL AND OPERATING COSTS

GREATER RELIABILITY AND LOWER MAINTENANCE

ENERGY SAVINGS

NO WATER CONSUMPTION

MEETS OPACITY REGULATIONS

POTENTIAL FOR REMOVING $SO_3$ AND $NO_x$

DISADVANTAGES

WASTE DISPOSAL

NOT YET AVAILABLE IN COMMERCIAL QUANTITY

LACK OF OPERATING EXPERIENCE

TABLE 5.  TEST INSTRUMENTATION
(REF. 10 AND 11)

| PARAMETER | INSTRUMENT |
|---|---|
| SO$_2$ | TECO MODEL 40 PULSED FLUORESCENT |
| NO/NO$_x$ | TECO MODEL 10 CHEMILUMINESCENT |
| O$_2$ | TELEDYNE MODEL 9500X ELECTRO-CHEMICAL |
| SODIUM COMPOUND FLOW | GRAB SAMPLE, TIMED WEIGHT |
| FLUE GAS FLOW | ASME VENTURI |
| TEMPERATURE | THERMOCOUPLES (TYPE J) |

TABLE 6.  TEST VARIATIONS
AND THEIR RANGES
(REF. 10 AND 11)

| VARIABLE | RANGE |
|---|---|
| SODIUM COMPOUND | NAHCOLITE, TRONA, REFINED TRONA |
| INJECTION TEMPERATURE | 204 TO 327°C (400 TO 620°F) |
| BAGHOUSE TEMPERATURE | 163 TO 260°C (325 TO 500°F) |
| STOICHIOMETRIC RATIO | 0.7 TO 2.1 |
| AIR-TO-CLOTH RATIO | 0.46 TO 0.91 M/S (1.5 TO 3.0 FT/MIN) |

TABLE 7.  CHEMICAL COMPOSITION
OF SODIUM COMPOUNDS
(REF. 10 AND 11)

| | NAHCOLITE | | REFINED TRONA | TRONA |
|---|---|---|---|---|
| TEST RUN | N-5A | N-5B | B-1 | T-1 |
| CONSTITUENT (%) | | | | |
| NaHCO$_3$ | 52.76 | 54.01 | 59.32 | 26.43 |
| Na$_2$CO$_3$ | 3.64 | 3.13 | 14.22 | 41.39 |
| NaCl | 0.49 | 0.46 | 1.27 | 4.14 |
| Na$_2$SO$_4$ | 0.40 | 0.31 | 1.19 | 5.19 |
| Na$_2$SO$_3$ | 0.001 | 0.002 | <0.001 | <0.001 |
| NaNO$_3$ | 0.02 | 0.03 | <0.01 | <0.01 |
| INSOLUBLES | 40.86 | 40.01 | 15.34 | 9.72 |
| WATER | 1.83 | 2.04 | 8.65 | 13.12 |

TABLE 8. EFFECT OF DRY INJECTION ON THE
BAGHOUSE INLET GRAIN LOADING

| ADSORBENT | STOICHIO-METRIC RATIO | BICARB, % B | CARB, % C | BAGHOUSE INLET LOADING | | | | % INCREASE IN LOADING |
|---|---|---|---|---|---|---|---|---|
| | | | | $gr/ft^3$ | | $kg/m^3$ $(x10^3)$ | | |
| | | | | ADSORBENT | + FLY ASH | ADSORBENT | + FLY ASH | |
| NAHCOLITE | 1.0 | 46.8 | 3.9 | 1.21 | 2.91 | 2.77 | 6.66 | 71 |
| | 1.0 | 80.0 | 4.0 | 0.74 | 2.44 | 1.69 | 5.58 | 44 |
| TRONA | 1.5 | 28.2 | 39.8 | 1.05 | 2.75 | 2.40 | 6.29 | 62 |

TABLE 9. PRESSURE DROP CHARACTERISTICS
OF MARTIN DRAKE UNIT 6 BAGHOUSE

| TEST NO. | AIR-TO-CLOTH RATIO | | BAGHOUSE INLET LOADING | | MEASURED | | | PREDICTED | | |
|---|---|---|---|---|---|---|---|---|---|---|
| | | | | | $\Delta P_{MIN}$ | | $\Delta t$, min | $\Delta P_{MIN}$ | | $\Delta t$, min |
| | ft/min | m/s $(x10^3)$ | $gr/ft^3$ | $kg/m^3$ $(x10^3)$ | in., Water | Pascal | | in., Water | Pascal | |
| 1 | 1.616 | 8.209 | 1.802 | 4.124 | 3.1 | 772 | 118 | 3.1 | 772 | 120.5 |
| | | | | | 3.5 | 871 | 68 | 3.5 | 871 | 84.5 |
| 2 | 2.683 | 13.63 | 1.531 | 3.504 | 3.1 | 772 | 69 | 3.1 | 772 | 140.5 |
| | | | | | 3.6 | 896 | 60 | 3.6 | 896 | 88.5 |
| 3 | 1.691 | 8.590 | 2.115 | 4.840 | 3.5 | 871 | 66 | 3.5 | 871 | 84.5 |
| | | | | | 3.5 | 871 | 64 | | | |
| | | | | | 3.5 | 871 | 93 | | | |
| 4 | 1.611 | 8.184 | 1.602 | 3.666 | 4.1 | 1021 | 14 | 4.1 | 1021 | 40.5 |
| | | | | | 4.1 | 1021 | 19 | | | |
| | | | | | 4.2 | 1046 | 12 | | | |
| | | | | | 4.0 | 996 | 30 | | | |
| 5 | 1.637 | 8.316 | 1.760 | 4.028 | 4.0 | 996 | 28 | 3.8 | 946 | 104.5 |
| | | | | | 4.0 | 996 | 32 | 4.0 | 996 | 74.5 |
| | | | | | 3.8 | 946 | 43 | | | |
| 6 | 1.651 | 8.387 | 1.991 | 4.556 | 4.2 | 1046 | 12 | 4.15 | 1033 | 20.5 |
| | | | | | 4.1 | 1021 | 14 | | | |

$\Delta P_{MIN}$ = Baghouse pressure drop just after completion of cleaning cycle
(cleaning cycle initiated when pressure drop reaches 4.5 in., water)

$\Delta t$ = Time between successive cleanings

TABLE 10. NUMBER AND TOTAL CAPACITY OF FGD SYSTEMS - AUGUST 1980 (REF. 16)

| | NO. OF UNITS | TOTAL CONTROLLED CAPACITY, MW[a] | EQUIVALENT SCRUBBED CAPACITY, MW[b] |
|---|---|---|---|
| OPERATIONAL | 73 | 27,155 | 24,765 |
| UNDER CONSTRUCTION | 39 | 17,855 | 16,854 |
| PLANNED: | | | |
| CONTRACT AWARDED | 29 | 13,769 | 12,919 |
| LETTER OF INTENT | 7 | 5,590 | 5,590 |
| REQUESTING/EVALUATING BIDS | 15 | 8,424 | 8,424 |
| CONSIDERING ONLY FGD | 40 | 24,200 | 23,980 |
| TOTAL | 203 | 96,993 | 92,532 |

[a]Total controlled capacity represents the gross capacities of coal-fired units brought into compliance by FGD systems, regardless of the percent of the flue gas treated.

[b]Equivalent scrubbed capacity represents the effective capacities of the FGD systems (in equivalent MW), based on the percent of the flue gas treated.

TABLE 11. DISTRIBUTION OF FGD SYSTEMS BY PROCESS (REF. 16)

| PROCESS | FGD EQUIVALENT SCRUBBED CAPACITY, MW | | | | |
|---|---|---|---|---|---|
| | OPERATIONAL | UNDER CONSTRUCTION | PLANNED | TOTAL | % OF TOTAL |
| LIMESTONE[a] | 11,172 | 8,816 | 16,164 | 36,152 | 53.1 |
| LIME[b] | 9,869 | 4,940 | 6,035 | 20,844 | 30.6 |
| LIME/SPRAY DRYING | 0 | 1,120 | 1,907 | 3,027 | 4.5 |
| LIME/LIMESTONE | 20 | 0 | 475 | 495 | 0.7 |
| SODIUM CARBONATE | 925 | 330 | 250 | 1,505 | 2.2 |
| MAGNESIUM OXIDE | 0 | 574 | 750 | 1,324 | 1.9 |
| WELMAN LORD | 1,540 | 534 | 0 | 2,074 | 3.0 |
| DUAL ALKALI | 1,181 | 0 | 842 | 2,023 | 3.0 |
| AQUEOUS CARBONATE/ SPRAY DRYING[c] | 0 | 540 | 0 | 540 | 0.8 |
| CITRATE[d] | 60 | 0 | 0 | 60 | 0.1 |
| TOTAL | 24,767 | 16,854 | 26,423[e] | 68,044 | 100.0 |

[a]Includes alkaline fly ash/limestone and limestone slurry process design configurations.

[b]Includes alkaline fly ash/lime and lime slurry process design configurations.

[c]Includes nonregenerable dry collection and regenerable process design configurations.

[d]This system is operating at the St. Joseph Zinc Co., F.G. Wheaton Plant, and is listed as a utility FGD system because the plant is connected by a 25-MW interchange to the Duquesne Light Company.

[e]Because the processes of all planned systems are not known, the totals in this status category are less than those in Table 10.

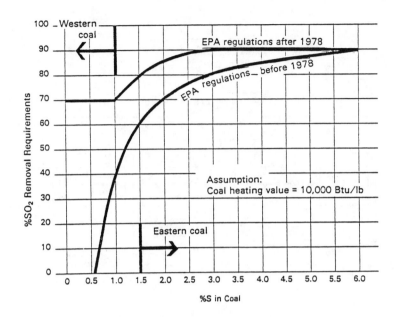

FIG. 1 - EPA REGULATIONS FOR SO₂ CONTROL FROM
NEW COAL-FIRED UTILITY ELECTRICAL POWER PLANTS

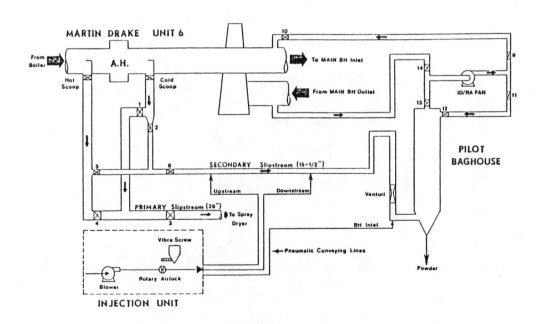

FIG. 2 - SCHEMATIC FLOW DIAGRAM OF THE BUELL PILOT FGD TEST FACILITY
AT THE MARTIN DRAKE STATION OPERATED BY THE CITY OF COLORADO
SPRINGS (REF.10 AND 11). NOTE: NUMBERS INDICATE VALVE POSI-
TIONS. VALVES 7 AND 8 BELONG TO SPRAY DRYER PILOT AND ARE NOT
SHOWN. THE ABBREVIATIONS A.H., B.H., I.D. AND R.A. STAND FOR
AIR HEATER, BAGHOUSE, INDUCED DRAFT AND REVERSE AIR RESPECTIVELY.

299

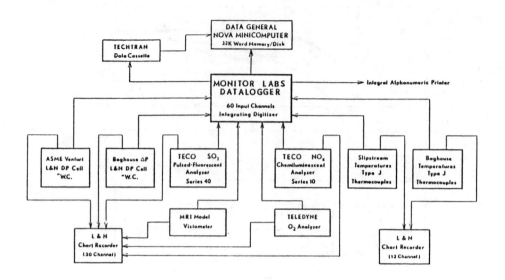

FIG. 3 - DRY FGD DATA COLLECTION HARDWARE AT THE
BUELL PILOT FGD TEST FACILITY AT COLORADO
SPRINGS (REF.10 AND 11)

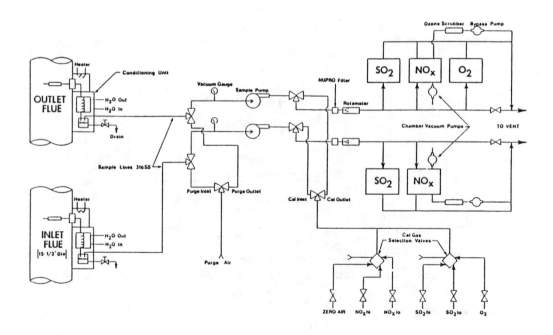

FIG. 4 - DRY FGD GAS SAMPLING SYSTEM AT THE
BUELL PILOT FGD TEST FACILITY AT
COLORADO SPRINGS (REF.10 AND 11)

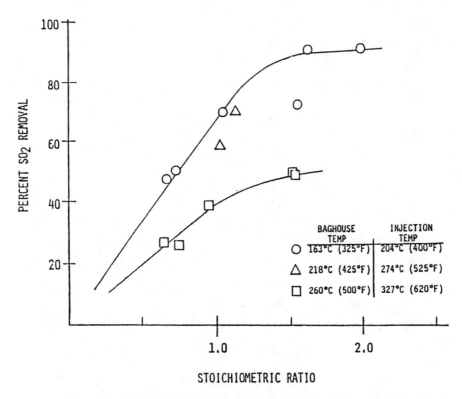

FIG. 5 - PILOT SCALE EXPERIMENTS: VARIATION OF AVERAGE SO₂ REMOVAL EFFICIENCY WITH TOTAL STOICHIOMETRIC RATIO AND TEMPERATURE DURING CONTINUOUS INJECTION OF NAHCOLITE (REF. 10 AND 11)

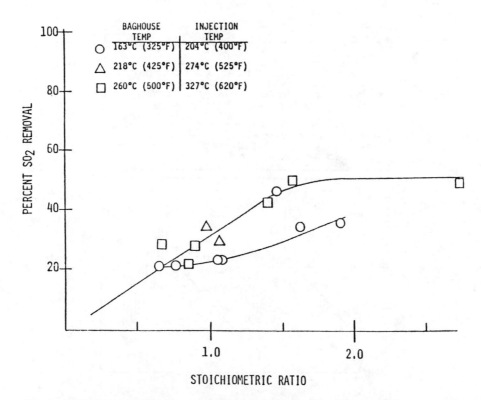

FIG. 6 - PILOT SCALE EXPERIMENTS: VARIATION OF AVERAGE SO₂ REMOVAL EFFICIENCY WITH TOTAL STOICHIOMETRIC RATIO AND TEMPERATURE DURING CONTINUOUS INJECTION OF TRONA (REF. 10 AND 11)

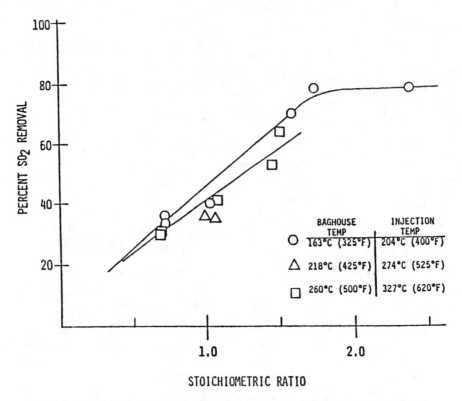

FIG. 7 - PILOT SCALE EXPERIMENTS: VARIATION OF AVERAGE $SO_2$ REMOVAL EFFICIENCY WITH TOTAL STOICHIOMETRIC RATIO AND TEMPERATURE DURING CONTINUOUS INJECTION OF REFINED TRONA (REF. 10 AND 11)

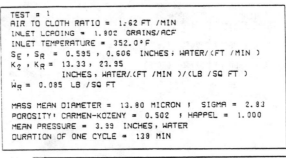

```
TEST = 1
AIR TO CLOTH RATIO = 1.62 FT /MIN
INLET LOADING = 1.902 GRAINS/ACF
INLET TEMPERATURE = 352.0°F
SE , SR = 0.595 , 0.606 INCHES, WATER/(FT /MIN )
K2 , KR = 13.33 , 23.95
          INCHES, WATER/(FT /MIN )/(LB /SQ FT )
WR = 0.085 LB /SQ FT

MASS MEAN DIAMETER = 13.80 MICRON , SIGMA = 2.83
POROSITY: CARMEN-KOZENY = 0.502 , HAPPEL = 1.000
MEAN PRESSURE = 3.99 INCHES, WATER
DURATION OF ONE CYCLE = 138 MIN
```

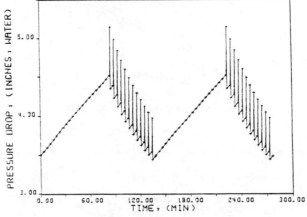

FIG. 8(a) - PRESSURE DROP WITH TIME ACROSS MAIN BAGHOUSE
PREDICTED BY COMPUTER MODEL FOR INLET CONDITIONS
MEASURED DURING TEST #1 (REF. 10 AND 11)

302

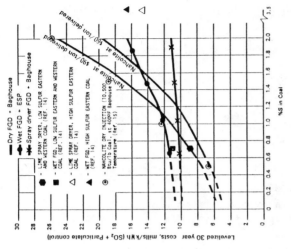

FIG. 10 - COMPARISON OF OPERATING COSTS: NAHCOLITE DRY FGD, WET FGD, AND SPRAY DRYER FGD

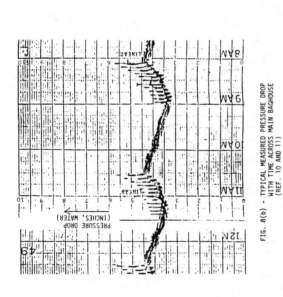

FIG. 8(b) - TYPICAL MEASURED PRESSURE DROP WITH TIME ACROSS MAIN BAGHOUSE (REF. 10 AND 11)

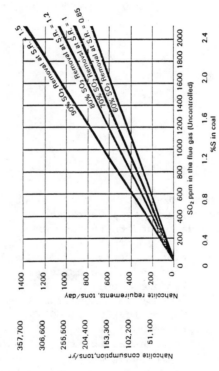

FIG. 11 - NAHCOLITE CONSUMPTION FOR A 500-MW COAL-FIRED POWER PLANT

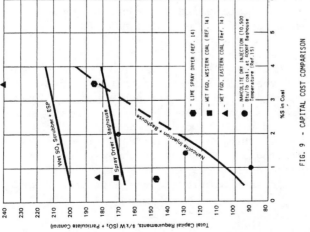

FIG. 9 - CAPITAL COST COMPARISON

303

# HIGH TEMPERATURE PARTICULATE REMOVAL
# BY GRANULAR BED FILTRATION

**Larry L. Moresco, Jerry Cooper** and **John Guillory**
Combustion Power Company, Inc.

## Introduction

Recent developments in the available energy sources of the world have directed research to investigate utilizing fuels which are less desirable but abundant, and recovering previously wasted energy.  Particulate control plays an important role in these energy developments, both from an air pollution control standpoint and, as in this case, an essential component in the power cycle itself.  In the combined cycle power plant, two power cycles are coupled.  The open cycle or hot combustion gas cycle uses the combustion gases from a pressurized fluid bed.  Combustion gases at typically 870 C (1600 F) are passed through several stages of particulate removal as required by the expansion gas turbine.  The high temperature granular bed filter would be the final cleaning stage before the gases enter the hot gas turbine of the cycle.  Particulate removal is necessary to prevent erosion and fouling of the turbine components.  Erosion in combination with corrosion would result in catastrophic failure of the turbine.  Exhaust gases of the turbine are then used to generate steam for the coupled steam powered closed cycle.  Combined, the gas turbine and steam turbine power cycles provide a more efficient means of converting raw coal to electric power.

Granular filtration, used for many years in its static or packed-bed form[1], occurs when a particle-laden gas is forced through a bed of closely packed granules.  With a typical voidage of 35-40% the bed forms passages between the granules similar to the jet ports of an impactor classification device.  As the gas passes through the bed, particles collide with the granules of the bed, removing a percentage of the particulate from the gas.  Collected particulate also serve as collection sites, improving the overall collection of the bed similar to the dust cake formed on other filtration devices.

Continuous cleaning and recirculation of the granules in the CPC moving granular bed filter allows steady operation, unlike other granular filters requiring intermittent shut downs for "puff-back" cleaning or complete offline servicing.

The absence of any completely static surfaces in the filtration area prevents plugging of the filter caused by high temperature "sticky" (adhesive, cohesive) particulate. During pneumatic transport at 870 C (1600 F) sufficient cleaning of the granular collectors (media) occurs, allowing them to be returned to the bed continuously.

Typically, the recirculation rate amounts to a granular plug flow velocity in the bed of 24 cm/hr (9.5 in/hr). After a once-through pass in the granular bed filter, the high temperature gas is sufficiently cleaned to inlet turbine particulate size and loading requirements such as they are tentatively stated today.

Experimental Methods

A schematic of the high temperature moving granular bed filter is provided in Figure 1. The GBF is axisymmetric starting from the top of the media flow annuli down to the bottom of the hopper where the dirty granular media leaves the filter by gravity. There are no static screened surfaces retaining the granular collectors and no mechanical moving parts except for the flow of the collectors themselves.

Tracing the two flow paths of the GBF starting with the granular collector flow; the "cleaned" granules enter the GBF via the media return legs, fill the media flow annuli, flow down the bin and hopper sections of the GBF, collect particulate and finally leave the filter where the granules are cleaned during transport, and returned - completing the loop. At the bottom of the inner and outer annular sleeves the granular collectors form the free surface through which the hot gases pass as they leave the GBF. Hot combustion gases enter the GBF as shown in Figure 1, travel down the gas inlet duct, reverse direction of flow after entering the granular bed, flow counter currently to the moving granular collectors, depositing hot particulate in the moving granular bed, and finally, the cleaned hot combustion gases leave the GBF at the top of the granular bed.

During the 1000 hour test series, the GBF configured as shown in Figure 1 was coupled to a conical bottom 0.46 m$^2$ (5 ft$^2$) atmospheric fluid bed burning a mixture of Illinois #6 coal and dolomite. The complete pilot unit is referred to as Model 4, as shown in Figure 2. Propane inline air heaters were used to maintain temperatures in the small subpilot system. Due to its physical size, heat losses are significant when attempting to maintain 870 C.

Metered streams of coal and dolomite enter the spouting fluid bed with the fluidizing air. Design of the fluid bed was intended to produce particulate rather than combustion efficiency. Particulate-laden combustion gases pass first through the recycle cyclone where large particles are returned to the fluid bed. The gases then pass through a primary ("separator") cyclone, removing large particulate and providing a more representative size distribution to the GBF. Gases are then directed through the particle sampling duct and into the GBF. Cleaned combustion gases leave the GBF at 840 C (1550 F), pass through the exhaust particle sampling duct and out to the atmosphere.

Nearly spherical granular (2 mm high alumina refractory) collectors flow down counter-currently to the hot gas flow in the GBF. Granules with collected particulate flow by gravity to the pneumatic lift pipe. During transport to the disengagement vessel, sufficient cleaning of the granules takes place. Previously, the upper fluid bed was used to clean the granular collectors, but during the 400 hour test of the 1000 hour test series, experiments were run without the upper fluid bed working and no degradation to the GBF performance was evident. Pneumatic transport gases laden with captured particulate are cooled and transferred to a small baghouse for final disposal.

Cascade impactors were used to determine the inlet and outlet particulate size distributions. A modified EPA Method 5 procedure was used for total loading measurements and was compared to the total catch of the impactors on the inlet and outlet streams of the GBF. Figure 3 shows the sample port and sample train arrangements for these experiments. Cooling air was introduced in both sampling ports to lower the temperature of the isolation valves. At the inlet to the GBF, the impactor or the EPA 5 filter were located behind a heated sampling probe inside a temperature controlled oven. An S-pitot tube was used to measure the local velocity in the duct. A ball valve able to go to higher temperatures was located on the GBF outlet sampling port, allowing a closer coupling of the impactor to the sampled stream. Both inlet and outlet sample streams were passed through four stainless steel impingers in an ice bath and then through a flow controller, ensuring isokinetic sampling rates. Impactors were operated between 430-500 C (800-930 F) while sampling the inlet and outlet gas streams.

Typical operating conditions of the Model 4 subpilot unit are detailed in Table I. Four test segments amounting to continuous operation of the GBF for 100, 200, 300, and 400 hour periods for a total of 1000 hours were run burning Illinois #6 coal and dolmite in the fluid bed.

Results

Performance of the high temperature granular bed filter did not show any signs of deterioration during any of the long duration tests with results similar to the previous hot filtration parametric tests.[2] Table II contains all the basic performance data from the test series. At a pressure drop between 6.0-7.0 kPa (24-28 IW) the overall particle capture efficiency of the GBF remained near 99% for all tests with an inlet gas particulate loading of 6.0 $g/m_o^3$ (0.63 gr/acf) at a gas flow rate of 0.18 $m_o^3$/sec (1600 acfm) at 840 C (1540 F).

Particulate classification data typical of the test series for inlet and outlet gas flows of the granular bed filter are presented in Figure 4. Andersen impactors were used to obtain the aerodynamic particle size distributions. Impactor size classification data converted to fractional efficiency of the granular bed filter are presented in Figure 5. Upper and lower limits of the efficiency band represent measured impactor fractional efficiencies from the test series. Repetitive agreement of the experimental data in Figure 5 depicts the internal consistency of the measurement methods. One should, however, be cautioned as to the absolute value of the aerodynamic particle diameter. Current work is underway to verify impactor classification data at these elevated temperatures near 540 C (1000 F). The minimum in fractional collection efficiency near 0.7μm of Figure 5 is predictable from the theory of particle capture[3,4], where inertial impaction capture is decreasing with particle size and diffusional capture is beginning to become a dominant mechanism of capture. The location of this minimum provides further credence to the measured aerodynamic diameters.

Momentary upsets of the Model 4 subpilot unit resulting from pertubations of the fluid bed, recycle cyclone, or primary cyclone did not produce significant degradation of the filtration capabilities of the high temperature granular bed filter. Continuous opacity measurements of the cooled GBF exhaust stream allowed instantaneous evaluation of such effects (e.g. cyclone ash line plug) on the GBF. Comparison of EPA 5 total loading measurements and opacity values taken during these measurements after the first few hundred hours of testing assured test personnel of a reasonable online measure of the GBF performance using the instantaneous opacity value.

Redesign of the Model 4 fluid bed components will hopefully eliminate these subsystem fluctuations and allow a more in-depth scrutiny of the moving granular bed filter performance in future tests.

It should be noted that solids build-up in the duct before the granular filter approximately 1 cm (0.4 in.) thick after 1000 hours could conceivably present a problem for all combustion systems dealing with coal and dolomite in the future.

Conclusions

After the completion of the 1000 hour, long duration performance evaluation of the high temperature moving granular bed filter, the following was determined:

- High temperature moving granular bed filters (as currently designed without retaining screens) can operate successfully, filtering coal combustion particulate and reacted dolomite.

- No sign of irreversible plugging of the GBF at 840 C (1540 F) occurred during these tests.

- With the tentative information available from gas turbine manufacturers[5], a theoretical prediction of granular filtration at high pressure (10 atm) and the experimental data of these atmospheric tests, a gas turbine can be protected from erosion and fouling with the use of a high pressure - high temperature moving granular bed filter of the design utilized in the described test series.

- Evaluation of the heat input versus particulate output of the atmospheric system provides a further benefit of high temperature granular filtration besides hot gas equipment protection. During the test series the Model 4 output ranged from 0.024 - 0.038 kg/GJ (0.05 - 0.08 lb/10^6 Btu), nearly meeting the EPA new source performance standard of 0.014 kg/GF (0.03 lb/10^6Btu). On a real system a secondary cyclone would be located before the GBF, using the granular filter as a final cleanup device for particulate removal, making the overall system even closer to the NSPS.

Acknowledgement

This work is currently funded by the United States Department of Energy under contract number DE-AC21-77ET-10373.

307

References

1. Marchello, J.M. and Kelly, J.J., eds., Gas Cleaning for Air Quality Control, Vol. 2, Dekker, New York, 1975.

2. J.L. Guillory, "High Temperature Particulate Removal by Moving Bed Granular Filtration", 25th Annual International Gas Turbine Conference, New Orleans, Louisiana, March 1980.

3. L.L. Moresco, P. Ngai, J. Cooper, "Granular Bed Filter Development Program Technical Progress Report", FE10373-01, (DOE), October, 1979.

4. Rao, A.K. et al, "Particulate Removal from Gas streams at High Temperature/High Pressure", PB245-858 (EPA), August 1975.

5. D.F. Ciliberti, M.M. Ahmed, N.H. Ulerich, M.A. Alvin, and D.L. Keairns: Experimental/Engineering Support for EPA's FBC Program: Final Report - Volume II, EPA-600/7-80-015b, January 1980.

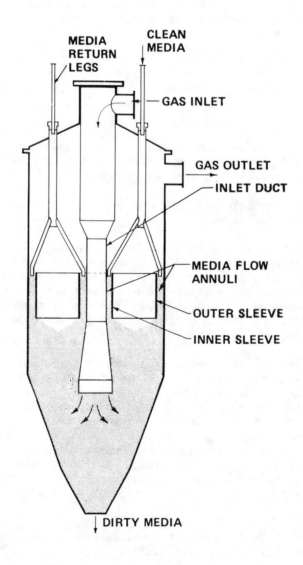

**Figure 1 High Temperature Moving Granular Bed Filter**

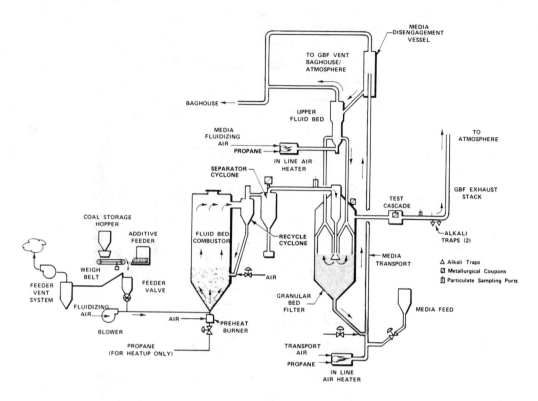

Figure 2   Model 4 High Temperature Atmospheric Moving Granular Bed Filter Pilot Unit Schematic

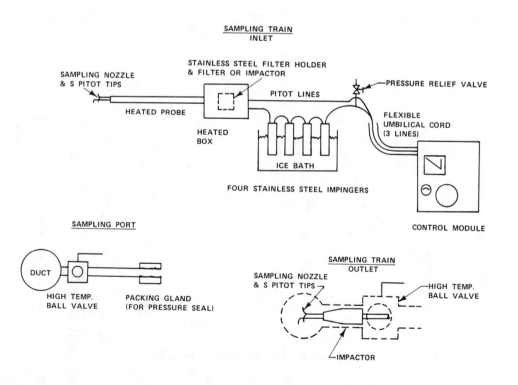

Figure 3   CPC Particulate Sampling Port and Train

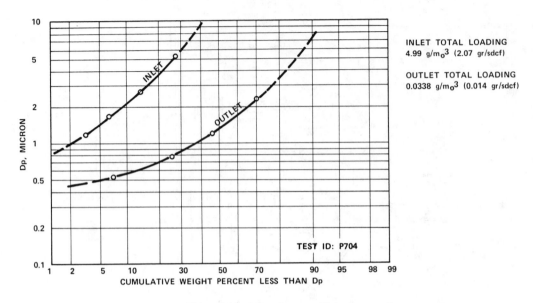

Figure 4 Andersen Impactor Data

NOTE: Dp = CALCULATED 50% CUTOFF AERODYNAMIC DIAMETER (DENSITY = 1 g/cc)

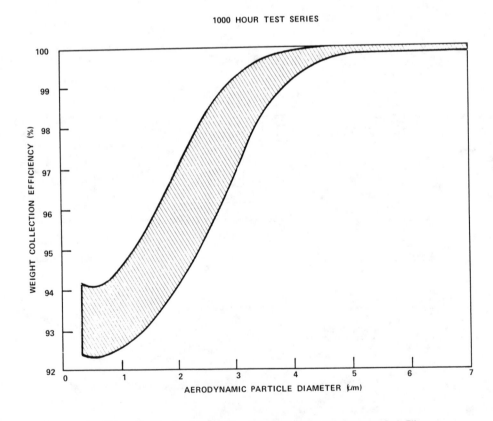

Figure 5 Fractional Efficiency, Counterflow Moving Granular Bed Filter

## TABLE I

### 1000—HOUR GBF TEST CONDITIONS

| | | |
|---|---|---|
| SYSTEM TEMPERATURE | 1120 K | (1550 F) |
| INLET GBF PRESSURE | 122 kPa | (17.6 PSIA) |
| GRANULAR BED DEPTH | 0.61 m | (2 FT) |
| INLET PARTICULATE LOADING | 6.3 $g/m_o^3$ | (0.66 GR/ACF) |
| GAS FLOW RATE | 0.18—0.20 $m_o^3$/sec | (400—450 SCFM) |
| GRANULAR COLLECTOR DIAMETER | 2 mm | (0.075 IN.) |
| GRANULAR COLLECTOR FLOW RATE | 0.19 kg/sec | (25 LB/MIN) |
| COAL FEED RATE (ILLINOIS NO. 6) | 16 g/sec | (2 LB/MIN) |
| DOLOMITE FEED RATE* | 6.5 g/sec | (0.86 LB/MIN) |

| TEST SEGMENT TITLE | P701 | P702 | P703 | P704 |
|---|---|---|---|---|
| TEST SEGMENT DURATIONS | 100 HR | 200 HR | 300 HR | 400 HR |

* DOLOMITE USED FOR SULFUR SUPPRESSION AND TO INCREASE PARTICULATE CONCENTRATION

## TABLE II
### 1000—HOUR TEST SERIES

| | | | P701 | | P702 | | P703 | | P704 | |
|---|---|---|---|---|---|---|---|---|---|---|
| GAS FLOW | — $m_o^3$/sec | (ACFM) | .18 | (1594) | .18 | (1590) | .17 | (1500) | .18/.19 | (1600/1800) |
| MEDIA RATE | — kg/sec | (LB/MIN) | .19 | (25) | .23 | (31) | .19 | (25) | .19 | (25) |
| PRESSURE DROP | — kPa | (IW) | 6.3 | (25.3) | 6.8 | (27.1) | 6.9 | (27.5) | 6.0 | (24) |
| INLET TEMPERATURE | — K | (F) | 1110 | (1538) | 1110 | (1538) | 1120 | (1550) | 1100 | (1520) |
| INLET LOADING | — $g/m_o^3$ | (GR/ACF) | 3.6 | (0.38) | 6.9 | (0.72) | 6.1 | (0.64) | 5.7 | (0.6) |
| OUTLET LOADING | — $g/m_o^3$ | (GR/ACF) | .077 | (0.008) | .048 | (0.005) | .057 | (0.006) | .067 | (0.007) |
| OVERALL EFFICIENCY — % | | | 97.9 | | 99.2 | | 99.1 | | 98.8 | |
| TOTAL HOURS ON COAL | | | 183 | | 179 | | 328 | | 423 | |
| TOTAL HOURS ON STEADY STATE | | | 95 | | 107 | | 295 | | 388 | |
| CASCADE EXPOSURE HOURS | | | 135 | | 177 | | 312 | | 407 | |

HIGH TEMPERATURE GRANULAR BED FILTER TEST DATA SUMMARY

311

## HYDROCARBON EMISSION CONTROL WITH ENERGY SAVINGS
## AND A POSITIVE PAYBACK

**James H. Mueller**
REECO—Regenerative Environmental Equipment Co., Inc.

It strikes me as highly appropriate that I should have the opportunity to address this auspicious gathering just twenty-three (23) days before Christmas. Like those gaily wrapped packages which will be awaiting us under the tree, I hope you consider what I have to say as a gift or contribution.

The chemical industry has always been the frontier in the western world for offering new technology that helps mankind. However, we are far removed from the days when shooting from the hip was acceptable. We are now in a much more delicate, restricted era made more and more complicated by conflicting demands.

Consumers demand more & more quality. Government wants energy conservation and evironmental control. We, of industry, seem desperate. But there is an answer!

Oftentimes, the common belief is that the required pollution control equipment is expensive to purchase, install, and operate. Also, that process changes will be required which will have a detrimental effect on the quality of the products being produced.

Many of us delayed making a move to "clean up the air" because we felt that the government just didn't have its act together. However, we must realize Pollution Controls Are Here To Stay! We are nearly four (4) years into the Clean Air Act of 1977 and the National Oxidant Ambient Air Quality Standards were promulgated in the Federal Register on February 8, 1979.

But fortunately, for all of us in industry, there is a practical solution which is well within the scope of present-day technology. Pollution can be controlled by using a highly energy-efficient form of regenerative incineration that has demonstrated a heat recovery efficiency of 85%, 90% and 95%. This regenerative incineration system can offer an excellent ROI (Return On Investment) of 50% 100% and even a

312

spectacular 192%! The 192% ROI, achieved at FMC Corporation's Middleport, New York agricultural plant, is actually saving more than 960,000 gallons of fuel oil per year.

This form of pollution control is simple and straightforward. Energy recovered from hydrocarbons is used to incinerate and destroy fumes, with very little or "NIL" fuel usage. Even without solvents, the fuel cost of a basic regenerative system is only one tenth of a common after-burner or one fifth of a unit having a 40% thermal energy recovery.

The regenerative unit is very simple and reliable. This means that people can concern themselves with production rather than spending their time keeping a complex chemical system working. One may ask, haven't incinerators traditionally had a bad name for fuel consumption? They have -- in historical times -- but a modern thermal oxidation system with efficient energy recovery and recycling is different! Regenerative incineration can be tremendously efficient, both in terms of cleaning the air and in terms of self-sustaining operation.

The objective of this paper is to show the plant manager, plant engineer, plant environmentalist and others concerned with pollution and energy, that practical, straightforward, effective incineration and heat recovery is available. Regenerative incineration is being used by approximately fifty (50) manufacturers to reduce plant energy consumption while meeting stringent environmental code requirements. In many cases complete environmental compliance is possible, with the cost of the installation being paid from plant energy savings alone. A regenerative incinerator is an energy recovery unit that operates on the mass/surface area principle of heat recovery.

Figure 1, RE-THERM® operation, is an illustration of a regenerative incinerator.

The equipment has 3, 5 or 7 chambers which allows continuous air flow while each chamber periodically reverses flow. The reversing cycle, together with the energy storage and energy recapture, can be seen by looking at only two chambers. It is important to also realize that each "chamber" has thousands of special stones which capture and then release the energy. The base efficiency of this energy capture/release is 85%, 90% and even 95% energy efficient.

Having seen the results of an energy efficient system, it may be helpful to briefly review how the regenerative incinerator operates. As hydrocarbon-laden exhaust air travels toward the central incineration chamber, it passes through a smaller chamber containing previously warmed stoneware elements. The fumes, thus preheated, can be inci-nerated in a fraction of a second, with little or no auxiliary fuel needed in addition to the fuel value of the solvent fumes themselves.

The superheated clean air is routed through another chamber also con-taining the special stoneware. Most heat energy is stored in these sto-neware elements for use in preheating the next cycle. A minimal amount of energy is exhausted. This exhaust energy normally can and should, if at all possible, be used in the plant or process.

Subsequently, the flow is reversed and fumes enter through the other chamber. Thus, a back and forth cycle is established to capture the energy of incineration and reuse the energy for the incineration process. This feature reduces the auxiliary fuel requirement to approximately 10 to 15% of that required by a common afterburner.  This energy-efficient system has a number of benefits.  It offers a total system fume and odor control and is fuel-efficient through the a wide load operating range.  It can self-sustain on a fume concentration as low as 5% of the L.E.L.

It can be installed with a minimal amount of rework to retrofit existing lines.  It eliminates complicated controll and instrumentation as well as high maintenance catalytic or tube-type preheat exchangers.  Autoignition problems are eliminated.  Another important consideration is Product Quality.  Product Quality is essential for staying in business.  Industry must continue to concentrate on maintaining and/or improving quality in a cost competitive manner.

Emission Control! Regulations! Quality! What about costs?  The cost of VOC control equipment is misunderstood.  A recent article for a given application stated that "costs" for control equipment run deep into six digits.  Not one word was said about savings . . . cash savings . . . energy savings.  Also, it was indicated that central incineration units do not perform as well as individual incineration units.  This is contrary to fact.  Central incineration units are not only giving outstanding performance but this is now being done with very attractive financial results.  Further, the performance of a central incineration unit is improved and simplified with multiple sources.

A recent financial analysis was published by WAPORA and presented at a Flexible Packaging Association meeting in Chicago on May 31 - June 1, 1979.  While this analysis uses incineration with 70% heat recovery as a portion of its analysis, it neglected to incorporate current "state of the art" technology.  If one were to use the cost figures developed within this report, he may conclude that control would cost $0.80 to $1.00 per pound of solvent to control that solvent.

The magnitude of error, using these numbers, is so great (i.e., 5,000 - 10,000% error) that is important to start from ground zero to establish realistic costs.

The first item to address is that most installations have not used sound engineering techniques to manage air flows. Generally speaking, equipment has been installed with little concern about either the amount of air used for drying or the amount of energy required to heat that air.  Once there has been an agreement to use a fundamentally sound air and energy engineering approach, then it is possible to develop a realistic financial analysis.

The following is a model line analysis.  The selected operational requirements are reasonably typical of a gravure operation having a good (not great, but good) design for air and energy management.  This analysis will summarize all costs associated with the purchase and operation of this current technology equipment.

PLANT EQUIPMENT

| | |
|---|---|
| Press or Dryer | 15,000 SCFM |
| Coater or Floor Sweep | 5,000 SCFM |
| Total Exhaust | 20,000 SCFM |
| Solvent in Exhaust | 60 GAL/HR |
| Solvent energy @ 100,000 BTU/GAL | 6,000,000 BTU/HR |
| Average Temperature of Exhaust | 240°F |
| Annual Operation | 3,600 HR/YR |
| Fuel Energy Cost | $3.00/MM BTU |

The equipment which requires control is as listed. The exhaust flows, temperatures, fuel costs and annual usage are based on averages. From the above model, the cost of common incineration would result in unnecessarily excessive costs and requires considerable auxiliary energy for operation. If one were to stop at this point, he would feel that such equipment is expensive, it wastes energy and perhaps, he feels that the only means to reduce the cost/energy is through complex energy techniques.

Fortunately, there are simple, proven methods for both reducing the exhaust flows and for increasing the L.E.L.'s. The most desirable method, for a given plant, can only be determined by an engineering study. It is very likely, however, that recirculation with a reduced exhaust flow rate will be the solution.

Also, the emission control equipment to be used, for our example, is a standard regenerative incineration system having 85% thermal energy recovery. Such a system is capable of self-sustaining at a "NIL" (essentially zero) fuel usage, using only the solvents in the exhaust as an energy source, while at the same time providing the necessary dryer heat energy. Such equipment has been installed and is operating in various locations throughout the U.S.A., such at Mead Corporation in the Los Angeles basin.

Figure 2 is a diagram of a typical press/coater installation.

FINANCIAL ANALYSIS

The objectives of the financial analysis is to determine the extent to which a requirement for control equipment impacts the "typical" firms profitability and working capital. Also, to estimate the cost per pound of the pollutant removed from the atmosphere . . . that is, to measure the cost effectiveness of the control equipment to society. Plant operators are interested in the former. Pollution control authorities are interested in the latter.

Total costs comprised of fuel, maintenance, property taxes and insurance, interest expense and depreciation can be calculated for each year of the ten (10) year depreciation life. Table 1 is a detailed "Financial Analysis Work Sheet".

The bottom line of the financial analysis is a combination of cash flow and pollution control cost.

For this example, this results in:

| | |
|---|---|
| Ten (10) Year Cummulative Cash Flow | + $142,020 |
| Solvent Controlled/YR | 1,512,000 #/YR |
| Cost of VOC Control<br>$ per pound of VOC Controlled | $0.012/# |
| $ per ton of VOC Controlled | $23.67 |

Figure 3 illustrates the cummulative cash flow for a gravure operation.

To evaluate any system, full engineering and financial evaluation is involved.  Such evaluations are freely available from my company and from other sources.

Another more simple comparison of an auto industry oven with and without regenerative incineration equipment may also be useful.  The following comparison is of an oven as presently operated versus the same oven with a regenerative incineration system return to process and oven feedback. The oven temperature, solvent concentrations and exhaust volume are as established by a major auto manfuacturer.

To simplify, the following assumptions are used:

1. Both systems will use existing solvent based materials.

2. There will be no basic change to the process.

3. Energy is the main cost factor and the key of the evaluation.

4. The energy cost for heating the product is the same for both systems - this may be omitted for this comparison.

5. The energy cost for oven wall losses, solvent vaporization, etc. is esstially the same for both - this may be omitted for both systems.

6. When the solvents, from the oven exhaust, are destroyed, 70% of the resulting clean warm air can be used for oven make-up (or a boiler or plant heat or dry off, etc.).

7. The example oven exhausts 53,200 SCFM @ 70°F.

Using the above, a simple cost evaluation for heating the make-up air for the oven and exhausting it to atmosphere gives these results:

| | |
|---|---|
| OVEN AIR HEATING COST | $373,740/YR |
| AIR HEATING COST WITH REGENERATIVE<br>INCINERATION @ 90% THERMAL ENERGY<br>RECOVERY AND FEEDBACK | $ 84,240/YR |
| SAVINGS IN AIR HEATING COST | $289,500 |

Figures 4, 5 and 6 illustrates this comparison.

The above comparison may seem simplistic. In many respects it is. The results, nonetheless, show that with existing technologies, existing equipment, existing coatings and most important existing quality, a substantial savings - energy savings - is possible with regenerative incineration.

As a case history, the SupraCote Division of Reliance Steel installed a regenerative incinerator in September 1979 on a coil coating line. Existing incinerators meet EPA requirements. The exhaust volume was 30,000 SCFM with solvent concentrations ranging from 0 to 25% of the L.E.L. The problem was energy cost.

Prior to the installation, the plant was using 225,000 therms of natural gas per month. The current use is 124,000 therms per month for the same production. Lawrence Dwyer, President of SupraCote, a Director and former Chairman of the National Coil Coaters Association Energy Committee, has stated that at 40¢/therm ($4.00/mm or oil at 60¢/gallon), the savings is $489,600/yr. With projected fuel rates, the anticipated return on investment is 29 months for the completely installed system.

Each system is different. Each system has its own economics. (There is no panecea). Each system must stand the scrutiny of hard business evaluation.

Figure 7 shows two regenerative incinerators providing air pollution control and energy conservation in the graphics industry.

In summary, regenerative incineration equipment has been and is being installed which meets the environmentalists needs. This equipment gives industry a financially attractive solution to the ENERGY ENVIRONMENT problem. While each application is different, operators have written that their ROI (Return On Investment) averages from 30% to over 100%.

I suggest that we ... as an industry ... can help ourselves. The government wants us (and OPEC is doing its bit to encourage us) to save energy ... Is this not the time to muster forces and find a better way to answer all needs? I believe it is!

In conclusion, I suggest consideration of these steps:

1. Establish existing operating conditions.

2. Exhaust flow management through recirculation and incineration.

3. Installation of quality energy efficient equipment.

4. Obtain government supported projects through low interest bonds.

Taking these steps will result in costs and cash flows that are practical. This will give a partial solution to the pressing environmental and energy problems.

And most of all, remember that thermal energy waste is money up the stack and that it is practical to have energy conservation with pollution control as the bonus.

317

TABLE 1

FINANCIAL ANALYSIS WORK SHEET

"TYPICAL GRAPHICS" COMPANY - GRAVURE

VOC EMISSION CONTROL

| | YEAR 1 | YEAR 2 | YEAR 3 | YEAR 4 | YEAR 5 | YEAR 6 | YEAR 7 | YEAR 8 | YEAR 9 | YEAR 10 |
|---|---|---|---|---|---|---|---|---|---|---|
| **COST** | | | | | | | | | | |
| Fuel (Savings) | (42,250) | (47,320) | (52,998) | (59,358) | (66,481) | (74,459) | (83,394) | (93,401) | (104,609) | (117,166) |
| Maintenance | 2,500 | 2,650 | 2,809 | 2,977 | 3,156 | 3,346 | 3,546 | 3,759 | 3,985 | 4,224 |
| Taxes and Insurance | 11,500 | 11,500 | 11,500 | 11,500 | 11,500 | 11,500 | 11,500 | 11,500 | 11,500 | 11,500 |
| Depreciation | 57,500 | 57,500 | 57,500 | 57,500 | 57,500 | 57,500 | 57,500 | 57,500 | 57,500 | 57,500 |
| Interest Expense | 36,800 | 33,120 | 29,440 | 25,760 | 22,080 | 18,400 | 14,720 | 11,040 | 7,360 | 3,680 |
| TOTAL COST | 66,050 | 57,450 | 48,251 | 38,379 | 27,755 | 16,287 | 3,872 | ( 9,602) (Savings) | ( 26,264) (Savings) | ( 40,262) (Savings) |
| **CASH FLOW** | | | | | | | | | | |
| Total Cost or Saving (+) or (-) | (66,050) | (54,450) | (48,251) | (38,379) | (27,755) | (16,287) | ( 3,872) | 9,602 | 29,264 | 40,262 |
| Loan Payment (-) | (46,000) | (46,000) | (46,000) | (46,000) | (46,000) | (46,000) | (46,000) | (46,000) | ( 46,000) | ( 46,000) |
| Tax Credit (+) | 115,000 | 0 | 0 | 0 | 0 | 0 | 0 | 0 | 0 | 0 |
| Depreciation (+) | 57,500 | 57,500 | 57,500 | 57,500 | 57,500 | 57,500 | 57,500 | 57,500 | 57,500 | 57,500 |
| Tax Saving or Cost (+) or (-) | 33,025 | 27,225 | 24,125 | 19,189 | 13,877 | 8,143 | 1,936 | ( 4,801) | ( 14,632) | ( 20,131) |
| CASH FLOW | 93,475 | (15,725) | (12,626) | ( 7,690) | ( 2,378) | 3,356 | 9,564 | 16,301 | 26,112 | 31,631 |
| CUMMULATIVE CASH FLOW | 93,475 | 77,750 | 65,124 | 57,434 | 55,056 | 58,412 | 67,976 | 84,277 | 110,389 | 142,020 |

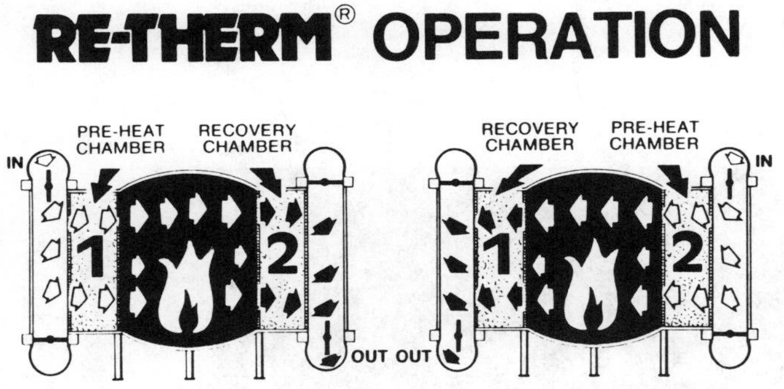

AIR HEATING COST
WITH RE-THERM & FEEDBACK

EXHAUST
13,200
@ 371°F

40,000 SCFM
@ 371°F

53,200 SCFM
@ 257°F

53,200 SCFM
371°F

SOLVENT

400°F

RE-THERM®     $14.04/HR

OVEN

MAKE-UP AIR
13,200 SCFM
@ 70°F

FUEL COST:
@ 1.6 MM BTU/HR   RE-THERM®
@ $4.00/MM BTU
@ 6000 HR/YR      $84,240

Figure 5

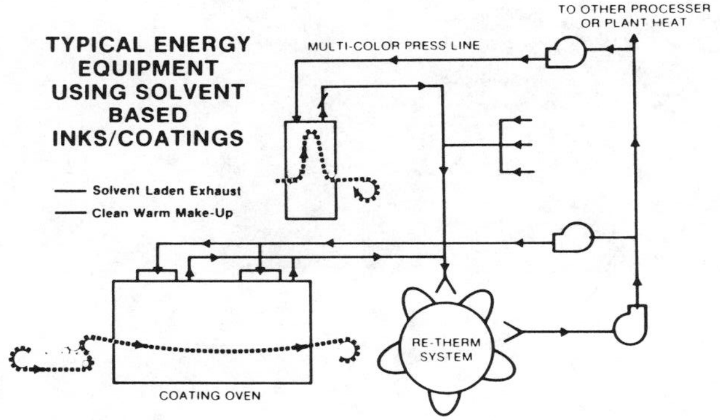

CUMULATIVE CASH FLOW
(Gravure)

$142,020

$55,056

$93,475

YEAR   0   2   4   6   8   10

140,000
120,000
100,000
80,000
60,000
40,000
20,000

Figure 3 – Cumulative Cash Flows for High Energy Efficiency

Gravure VOC Control Installations

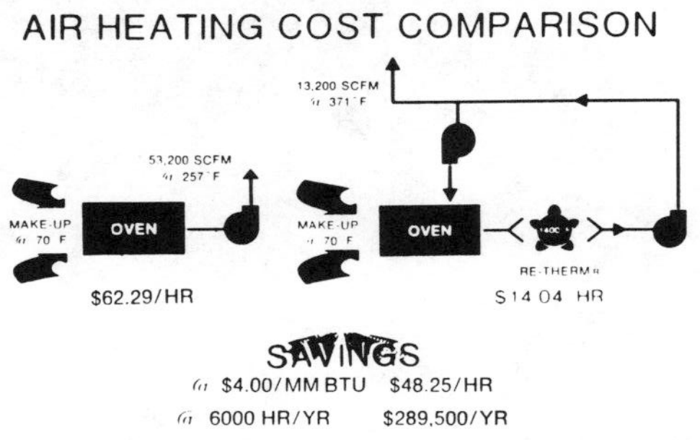

AIR HEATING COST COMPARISON

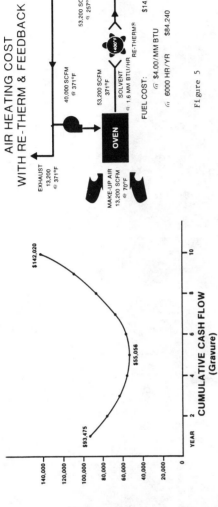

OVEN
13,200 SCFM
@ 371°F

RE-THERM
$14.04 HR

MAKE-UP
@ 70°F

53,200 SCFM
@ 257°F

OVEN

$62.29/HR

MAKE-UP
@ 70°F

SAVINGS

@ $4.00/MM BTU   $48.25/HR
@ 6000 HR/YR     $289,500/YR

Figure 6

AIR HEATING COST

EXHAUST
53,200 SCFM
@ 257°F

257°F

OVEN

$62.29/HOUR

MAKE-UP AIR
53,200 SCFM
@ 70°F

FUEL COST:     $373,740
@ $4.00/MM BTU
@ 6000 HR/YR

Figure 4

RE-THERM® OPERATION

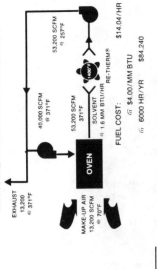

PRE-HEAT
CHAMBER

RECOVERY
CHAMBER

IN

RECOVERY
CHAMBER

PRE-HEAT
CHAMBER

IN

OUT OUT

Figure 1 – RE-THERM® Operation

(Regenerative Incinerator)

TYPICAL ENERGY
EQUIPMENT
USING SOLVENT
BASED
INKS/COATINGS

TO OTHER PROCESSER
OR PLANT HEAT

MULTI-COLOR PRESS LINE

RE-THERM
SYSTEM

COATING OVEN

—— Solvent Laden Exhaust
—— Clean Warm Make-Up

Figure 2 – Typical Press/Coater Installation

319

9,000 SCFM For
Gravure Press

15,000 SCFM From
Gravure Press, Floor
Sweep and Ink Room

Figure 7 – RE-THERM® Energy Recovery Fume Incinerators

Designed to Operate on a "NIL" Fuel Usage

# FLUE GAS DESULFURIZATION IN MOLTEN SALT
# ELECTROCHEMICAL CELL: PRELIMINARY EXPERIMENTS

**Omar E. Abdel-Salam**
Cairo University

## Introduction

The emission of sulfur oxides, mainly as $SO_2$, in the stack gases of power plants is one of the most serious air pollution problems worldwide. $SO_2$ causes several problems including bronchitis , plant damage, and corrosion of metal structures. With the increasing demand of energy, larger amounts of high-sulfur coals are expected to be used in the future. Flue gas desulfurization (FGD) processes will remain, at least for the next decade, the best approach for meeting the standards set by the environmental agencies.[1]

Many FGD processes have been proposed but only few of them have been used.[1] Generally speaking, these processes are either expensive or lead to the formation of massive wastes which need considerable efforts for their disposal. For that reason, none of these processes has been completely acceptable to meet the requirements of a universal control method. The capital investment of an FGD unit installed on a power plant is in the range of $33 to $197/kW of the plant capacity (1975 estimates) and the operating costs can run between $0.002 and $0.007/kWh.[1]

A novel process for $SO_2$ removal from stack gases has been proposed.[2] This process is based on the electrochemical reactions of $SO_2$ in molten salts. It works on the principle that when a DC potential is applied between the terminals of a cell containing molten electrolyte it is possible to remove $SO_2$ from the stack gases which pass at the cathode and recover it in a concentrated form at the anode. The same concept has been previously applied for the removal of $CO_2$ from gas streams for possible utilization in the environmental control of the atmosphere in manned space cabins.[3] By controlling the cell current, the desired reactions can be utilized in the most efficient way to produce a concentrated stream which can be subsequently treated to recover sulfur or sulfuric acid.

Description of the Process

The electrochemical desulfurization cell is shown schematically in Fig.1. It consists of two porous electrodes made of suitable materials and separated by a matrix made of a mixture of a molten electrolyte, such as a sulfate eutectic, and a filling inert material such as $LiAlO_2$. The flue gas is passed into the cathode compartment where cathode reactions convert the sulfur oxides into sulfite and sulfate ions. The following reactions might all contribute to the removal mechanism : -

$$SO_2 + \tfrac{1}{2} O_2 + 2e \quad \longrightarrow \quad SO_3^{2-} \qquad \qquad \ldots (1)$$

$$SO_3 + \tfrac{1}{2} O_2 + 2e \quad \longrightarrow \quad SO_4^{2-} \qquad \qquad \ldots (2)$$

$$CO_2 + \tfrac{1}{2} O_2 + 2e \quad \longrightarrow \quad CO_3^{2-} \qquad \qquad \ldots (3)$$

$$H_2O + \tfrac{1}{2} O_2 + 2e \quad \longrightarrow \quad 2\,OH^{-} \qquad \qquad \ldots (4)$$

$$SO_2 + CO_3^{2-} \quad \longrightarrow \quad SO_3^{2-} + CO_2 \qquad \qquad \ldots (5)$$

$$SO_2 + 2OH^{-} \quad \longrightarrow \quad SO_3^{2-} + H_2O \qquad \qquad \ldots (6)$$

$$SO_3^{2-} + \tfrac{1}{2} O_2 \quad \longrightarrow \quad SO_4^{2-} \qquad \qquad \ldots (7)$$

The equilibrium constants of reactions (5) and (6) are very high.[4] It is possible to obtain almost complete removal of the sulfur oxides without appreciable transfer of $CO_2$(or $H_2O$) by controlling the current density as is explained in Appendix 1. This will maintain a high current ($SO_x$ removal) efficiency.

On the passage of current, sulfite and sulfate ions migrate to the anode where they discharge as follows :-

$$SO_3^{2-} \quad \longrightarrow \quad SO_2 + \tfrac{1}{2} O_2 + 2e \qquad \qquad \ldots (8)$$

$$SO_4^{2-} \quad \longrightarrow \quad SO_3 + \tfrac{1}{2} O_2 + 2e \qquad \qquad \ldots (9)$$

A stream of preheated air flows at a small rate into the anode compartment to carry away concentrated sulfur oxides for further processing to obtain sulfuric acid or sulfur.

For the desulfurization of the stack gases of a power plant, the electrochemical cell needs installation at an intermediate point between the combustion chamber and the steam generating section where the temperature is sufficiently high to keep the cell electrolyte in the molten state. Most molten salts give a useful operating temperature range of 200 °C above their melting points when they are mixed with an inert ceramic material. Operating the cell at a high temperature has the advantage that we avoid the excessive cooling of the stack gases with the corresponding loss of their buoyancy and the formation of acidic mists in the case of aqueous scrubbers.

In the case of coal burning, the stack gases contain varying amounts of particulates which must be removed prior to desulfurization. This requires the use of "hot" electrostatic precipitator. These precipitators are efficient removing 99.5% of the particulates at a temperature around 500°C.

## Experimental Apparatus

The removal mechanism as described above was studied on a small bench-scale apparatus. The cell design is not typical of large-scale molten-salt electrochemical cells but it allows the verification of the removal mechanism and the identification of the most important factors which must be looked for in the design of practical systems. A gold alloy was selected as a suitable electrode material to minimize the possible interference of the corrosion reactions with the desired removal reactions.

The cell is shown schematically in Fig.2. It consists of a vessel made of stainless steel 316 with an inside diameter of 5 cm. It is separated into two compartments, the cathode's and the anode's by a stainless steel sheet which is perforated at the bottom and covered on each side by a 100 mesh stainless steel screen. This prevents the intermixing between the catholyte and anolyte. The cell is partially filled with $(K,Li)_2SO_4$ eutectic mixture (m.p.535°C). Gas streams of both compartments are introduced through side tubes so as to bubble into the electrolyte against the electrodes. The electrodes are made of gold 21K sheets with an area of 1.6 $cm^2$ and are welded to gold wires. The distance between the electrodes is close to 2 cm. The electrodes are welded to stainless steel current collectors which pass through the cell cover via suitable ceramic insulators. The outlet tubes are welded to the cell cover. The cell is contained in a pot furnace and the temperature is controlled to $\pm$ 5°C using a variable transformer.

The experimental setup is shown schematically in Fig.3. Air from a compressor is cleaned, dried and measured. In some experiments $CO_2$ is added to air at predetermined propotions. The mixture is then passed over a ceramic boat which contains 5 gm of cobalt sulfate and is placed at the center of a stainless steel tube which is placed in a horizontal cylindrical furnace. The temperature is controlled using a variable transformer. At high temperatures, $CoSO_4$ decomposes according to the following reaction : -

$$CoSO_4 \longrightarrow CoO + SO_2 + \tfrac{1}{2} O_2 \qquad\qquad ... (10)$$

By changing the temperature, the concentration of $SO_2$ in the gas stream can be varied. Mixtures of air and $SO_2$ with and without added $CO_2$ are introduced into the cathode chamber. Another air stream is introduced into the anode chamber.

A stabilized DC power supply of the Farnell L 20-5 type is used to supply constant current to the cell. The cell current and potential are measured using two Hioki Model 3205 digital multimeters.

Samples of the inlet and outlet gases are analyzed for $SO_2$ by absorption into standard $KMnO_4$ solutions and back titration with oxalic acid.

## Experiments and Results

A number of experiments were made to verify the removal mechanism. $SO_2$ concentration was varied in the range of 0.18 - 0.57% which is typical of the composition of stack gases of power plants. The current density was varied in the range of 3-25 mA/$cm^2$. This range was selected so that the cell potential would not reach high values which might accelerate the attack of the electrodes by the cell electrolyte. The flow rates of the cathode and the anode gases were fixed at 1 $cm^3$/sec each. The cell temperature was maintained at 630°C. In some experiments, $CO_2$ was added to the cathode gas to study its effect on the $SO_2$ removal and current efficiency.

Results to date are summarized in Fig.4 through Fig.8. The current efficiency was calculated on the basis that theoretically one mole of $SO_2$ should be transferred for each two Faradays of electricity. From these results we can reach the following conclusions :

1. The removal efficiency and current efficiency increase as $SO_2$ concentration (in the cathode gas) increases.

2. The removal efficiency increases and the current efficiency decreases as the current density increases within the experimental range.

3. Adding $CO_2$ to the cathode gas decreases both the $SO_2$ removal efficiency and current efficiency.

4. The cell potential is quite low (about 3 volts at a current density of 25 mA/$cm^2$). This takes place with an electrode separation of 2 cm. Further reduction is possible by decreasing the electrode separation and by reducing the cell overpotential.

5. The visual appearance of the electrodes shows little attack by the molten electrolyte.

Discussion

Preliminary results show that, in principle, the process of $SO_2$ removal is feasible. Current density of 25 mA/$cm^2$ was reached at a reasonable cell potential. These experiments were performed using planar electrodes. Porous electrodes allow much higher current densities because they provide higher surface area for the electrode reactions to take place. Although the concentration of $CO_2$ was one to two orders of magnitude greater than that of $SO_2$ in those experiments where $CO_2$ was added to the cathode gas, the $SO_2$ removal efficiency and current efficiency were not much less than those without $CO_2$ as long as the current density was low. The effect of $CO_2$ becomes more pronounced as the current density increases. This is in agreement with our expectation based on the values of the equilibrium constants as explained in Appendix 1.

A preliminary design was made on the basis of state-of-the-art high temperature fuel cell technology.[2] This was followed by a preliminary economic analysis which showed that the process can have a capital investment of about $20/kW of the plant capacity as compared with $33-197/kW for other FGD processes.[1] The total cost of stack gas desulfurization is about $0.0006/kWh as compared with $0.002-0.007/kWh for other FGD processes. The design and economic analysis are summarized in Appendix 2. Although the present results were obtained with a small bench-scale system, they predict that the process is feasible and can be tested on a larger scale.

Acknowledgment

This work was assisted by the U.S. National Science Foundation under Grant INT 78-01472. Thanks are due to Dr. Jack Winnick for useful suggestions and to Mr. Mohamed Ibrahim for this help in running the experiments.

Appendix 1. Effect of Current Density

As mentioned previously, the partial pressures of $SO_2$ and $SO_3$ are negligibly small in the presence of significant amounts of $CO_2$ above molten carbonate/sulfate mixtures.[4] Therefore, if the rate of gas phase mass transfer of $SO_2$ and $SO_3$ to the cathode reaction area is equivalent to the rate of transport of $SO_4^{2-}$ and $SO^{2-}$ ions from the cathode region to the anode region, there will be no formation of $CO_3^{2-}$ at the cathode. The $CO_3^{2-}$ ions which will form by the reaction of $CO_2$ and $O^{2-}$ at the cathode will subsequently react with $SO_2$ and $SO_3$. Since the cations do not participate in the electrode reactions, the current is only carried by the anions. This puts an upper limit on the current density used. As the current density increases from zero upwards, the concentration gradients of $SO_2$ and $SO_3$ in the gas phase will correspondingly increase. A certain current density will be reached at which the concentrations of $SO_2$ and $SO_3$ at the electrode reaction area approach zero. Above this limiting current density, which depends on the concentration and mass transfer coefficient of $SO_x$ in the gas phase, there will be a net gain of $CO_3^{2-}$ ions at the cathode and $CO_2$ will be simultaneously transferred from the cathode to the anode region. This situation should be avoided in order to maintain high current ($SO_x$ removal) efficiency.

Appendix 2. Sample Design and Economic Analysis

A detailed sample design has been given to illustrate how the electrochemical desulfurization process can be used in the cleaning of stack gases of power plants.[2] This sample design is summarized here. For a 1000 MW power plant burning coal containing 3.3% of sulfur the amount of $SO_2$ to be removed is 101 gmole/sec (800 lbmole/hr). A suitable design of large-scale high temperature electrochemical cells is that of IGT.[5] The electrolyte matrix is formed by hot pressing into thin sheets and the electrodes are made of porous metal sheets. Assuming a current density of 54 mA/cm$^2$ (50 A/ft$^2$) and a current ($SO_2$ removal) efficiency of 80%, the electrode area requirements is calculated as 46513 m$^2$(500,000 ft$^2$) which corresponds to 500 cells. Although the present experiments did not show this high performance, it is possible with porous electrodes to reach these levels if they are carefully designed. The cathode gas velocity and the mass transfer coefficient can be calculated. These values indicate that the chosen current desity does not exceed the limiting value. The calculated pressure drop is about 6.5 cm ($2\frac{1}{2}$ in.) of water which is reasonable.

Based on these values and the cost figures of the IGT cells[5] it is estimated that the capital investment of the FGD unit for a 1000 MW power plant is less than $20,000,000 (1975 prices). This includes a penalty of $6,000,000 for using a hot ESP instead of the commonly used ESP. Based on the energy requirements, labor, maintenance, materials, and depreciation: the total cleaning cost is estimated as $0.0006/kWh.[2]

References

1.    J. Jesefson, Stack Gas Cleaning : A 1976 Update , Environmental Science and Technology, 10 (5), 416 (1976)

2.    O.E. Abdel-Salam and J. Winnick, Flue Gas Desulfurization in Molten Salt Electrochemical Cell, Paper presented at the Conference on Electrochemistry as an Intermediate Technology, London (June 1979)

3.    O.E. Abdel-Salam and J. Winnick, Molten-Carbonate $CO_2$ Concentrator :

325

Preliminary Experiments, Eighth Intersociety Conference on Environmental Systems, Paper No. 78-ENAs-2, San Diego, Calif. (July 1978)

4.    R.A. McIlroy, G.A.Atwood, and C.J.Major, Absorption of Surlfur Dioxide by Molten Carbonate, Environmental Science and Technology, 7(11), 1022 (1973)

5.    D.Y.C. Ng, H.C.Maru, H.Feldkirchner, B.S.Baker, and N.P.Cochran, An Engineering Study of Fuel Cell Power Supply for Electrothermal Stage of the Hygas Process, Paper presented at Intersociety Conference on Energy Conversion, Las Vegas, Nevada (Aug.1970).

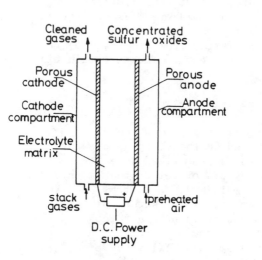

Fig.1   Schematic of process

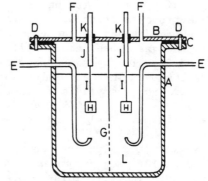

A Cell body        G Stainless steel diaphragm
B Cell cover       H Cell electrodes
C Gasket           I Gold wires
D Bolts            J Stainless steel terminals
E Inlet tubes      K Insulators
F Outlet tubes     L Electrolyte

Fig. 2    The experimental cell

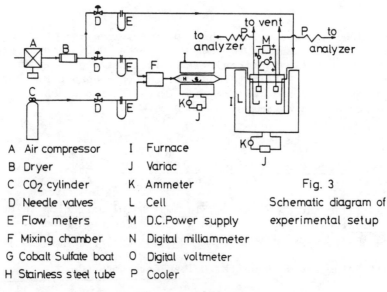

A  Air compressor      I  Furnace
B  Dryer               J  Variac
C  CO2 cylinder        K  Ammeter
D  Needle valves       L  Cell
E  Flow meters         M  D.C.Power supply
F  Mixing chamber      N  Digital milliammeter
G  Cobalt Sulfate boat O  Digital voltmeter
H  Stainless steel tube P  Cooler

Fig. 3

Schematic diagram of experimental setup

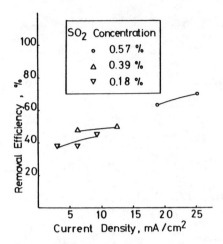

Fig.4 Effect of $SO_2$ concentration and current density on $SO_2$ removal efficiency.

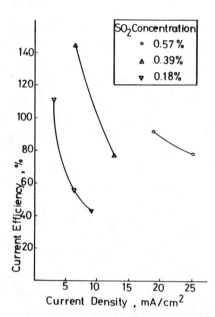

Fig.5 Effect of $SO_2$ concentration and current density on current efficiency.

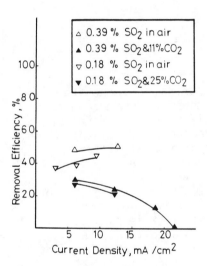

Fig.6 Effect of $CO_2$ on $SO_2$ removal efficiency.

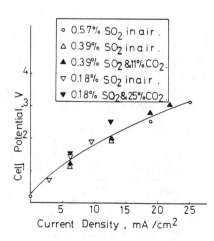

Fig.8 Effect of current density on cell potential.

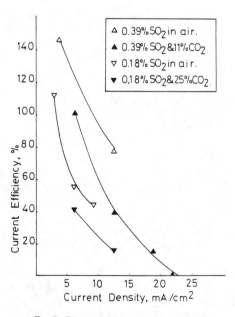

Fig.7 Effect of $CO_2$ on current efficiency.

# DOE ALTERNATIVE FUELS SOLICITATIONS

**Michael E. Card**
U.S. Department of Energy

We are mutually involved in a giant national experiment. A massive effort is being made to move the United States toward more independence in energy. As a Nation we have moved from merely seeking to develop "energy options" in this decade to the much more specific goals of actual production - to producing the equivalent of a half million barrels of oil per day by 1987 and two million barrels a day by 1990.

Congress has initiated these efforts by appropriating over $5.5 billion to the DOE to begin the implementation of a alternative fuels program for commercial scale production. The size of this task is hard to grasp. Economically, it has the scope of war. On a commercial scale, alternative fuels plants can cost from $1 to $6 billion each. Over the next twenty years, alternative fuels investment in the United States could total between 500 billion to a trillion dollars. A new industry is now in the gestation stage. Birth of this industry will create many challenges and new opportunities. The DOE alternative fuels solicitations have helped initiate this new industry.

It seems clear that DOE has been in transition since its creation in 1977. DOE's past approach to the domestic development of alternative fuels capacity has predominately been based on the philosophy that developing technologies to a preceived state of commercial readiness would serve as a necessary and sufficient condition for commercialization to occur under the influences of normal market forces.

This historical pattern changed dramatically in July 1979 when the President announced major new initiatives for the domestic development and production of alternative fuels. These included P.L. 96-126, the Department of Interior and Related Agencies Appropriation Act for FY 1980; P.L. 96-304, FY 1980 Supplemental Appropriations and Rescission Act; and P.L. 96-294, the Energy Security Act. I will review each of these in depth as they relate to DOE alternative fuels initiatives. I would also like to tell you how we are organizing for the management of the grants and agreements arising from the solicitations, and where we are going from here. I'll start with P.L. 96-126.

DOE's primary synthetic fuels production activities began with the signing of P.L. 96-126 on November 27, 1979. Under this authority, DOE has completed two significan actions:

1.  We have solicited and evaluated 971 proposals for alternative fuels feasibility studies and cooperative agreements, selecting 110 proposed alternative fuels projects for awards totaling approximately $200 million. These projects were divided into 99 Feasibility Studies and 11 Cooperative Agreements.

2.  We continued support for the first domestic commercial-scale plant to produce High Btu Gas from Coal, through a $22 million cooperative agreement to the Great Plains Gasification Project, headed by the American Natural Resource Corporation.

This initial activity heralded significant progress in the program toward the goal of stimulating domestic, commercial production of synthetic fuels equivalent to production of two million barrels of oil daily by 1992. Congress recognized the importance of continuing this interim DOE activity without interruption. In the FY 1980 Supplemental Appropriation and Rescission Act (P.L. 96-304), the Statement of Managers states that the interim authorities allow "a large, aggressive program" in DOE. The conferees stated:

1.  "$3 billion is immediately available to carry out interim Defense Production Act activities as authorized by S. 932. This interim activity along with the $2 billion previously provided in P.L. 96-126 . . . allows a large aggressive program. The managers expected the Department to issue solicitations as quickly as possible to use these funds and maintain current momentum in the program."

2.  "$100 million is provided for feasibility studies . . . to continue programs begun in the FY 1980 appropriation."

3.  "$200 million is provided for cooperative agreements . . . to continue programs begun in the FY 1980 appropriation."

The Defense Production Act (DPA) program ($3 billion) thus became a key part of an overall interim alternative fuels production program which totals $5.5 billion. The DPA authority of Section 305 will lapse when the President determines that the Synthetic Fuels Corporation is fully operational. At that time, the DPA's projects and associated resources will be eligible for transfer to the Corporation. The remainder of the interim program funds will be transferred to the Corporation on June 30, 1981.

On August 1, 1980, solicitations were issued for the second round of Feasibility Studies and Cooperative Agreements. Proposals were due September 30, 1980.

The solicitations differed from the first round in that Heavy Oil and Tar Sand projects were eligible if Congress enacted legislation to implement the intent of the conference committee. Direct combustion of municipal waste also became eligible. This was done by the Joint Continuing Resolution for FY 1981. Over 1000 proposals were received in response to these solicitations. A team of some 300 professionals evaluated these proposals.

Under P.L. 96-304, in addition to the $300 million for Feasibility Studies and Cooperative Agreements, solicitations were issued on October 15, for proposals for Financial Assistance under the Federal Nonnuclear Research and Development Act (P.L. 93-577), as amended, for the Development of Alternative Fuels and also for Financial Assistance under Title I, Part A of the Energy Security Act (P.L. 96-294) for the Development of Synthetic Fuels under the Defense Production Act. On November 15, eighteen proposals for projects were reviewed in response to these solicitations. Four projects were submitted under both solicitations.

SOLICITATIONS FOR PROPOSALS FOR FINANCIAL ASSISTANCE
UNDER TITLE I, PART A OF THE ENERGY SECURITY ACT
(PUBLIC LAW 96-294) FOR THE DEVELOPMENT OF SYNTHETIC FUELS
UNDER THE DEFENSE PRODUCTION ACT

This solicitation was for the purpose of inviting interested parties to submit proposals for financial assistance to expedite the production of synthetic fuels for national defense needs, as provided for by Title I, Part A of the Energy Security Act (P.L. 96-294). The types of financial assistance available through this solicitation consisted of:

o    Contracts for purchase of, or commitments to purchase, synthetic fuels for Government use for defense needs;

o    Price guarantees through purchase commitments in the event that the Government exercises the right to refuse delivery of a synthetic fuel and pays an amount by which the contract price exceeds the prevailing market price;

o    Loan Guanantees to finance the construction of synthetic fuel production facilities in conjunction with contracts to supply synthetic fuels for Government use for defense needs;

o    A minimum liquid production threshold of 10,000 barrels per day and a maximum level of crude oil equivalent of synthetic fuel of 100,000 barrels per day per project;

o    Up to $1.5 billion for price guarantees, purchase, and purchase commitments; and

o    $500 million as a reserve for loan guarantee defaults, and up to an additional $500 million to be used for additional loan guarantee reserves from the funds available for price guarantees, purchase, and purchase commitments (reducing funds available for these purposes). The amount of outstanding loan guarantees may not exceed $3 billion.

Advance payments may be made in conjunction with contracts to purchase, or commitments to purchase synthetic fuels, subject to the requirements as set forth in the solicitation.

The solicitation was issued in keeping with the Congressional declaration of policy as set forth in Section 102 of the Energy Security Act to:

    ". . . insure the national defense preparedness which is essential to national security, it is (also) necessary and appropriate to assure domestic energy supplies for national defense needs."

The solicitation was very flexible in terms of how proposers could respond. Proposers seeking financial support for synthetic fuel projects need not have proposed that all of the products to be produced by those facilities comprising the project be intended for sale to the Department of Defense. Synthetic fuel projects proposed should, in fact, be designed to suit the maximum potential of the technology used, within the context of sound economic and commercial practice. Proposals may, where technically and economically practicable, propose the production of fuels derived from blends of synthetic crude oils (or products) and conventional crudes and/or fuels with synthetic crudes and/or synthetic fuels may only receive financial assistance for the synthetic crude and/or synthetic fuels portion of the project, except for those portions or facilities for which separations of cost cannot be made.

The basic intent in structuring the scope of the DPA solicitation was to expedite the construction and operation of commercial scale facilities for the production of synthetic fuels for national defense needs at the earliest time practicable and in a manner consistent with commercial practices. It was recognized that we will be able to fund but a relatively few worthwhile projects. P.L. 96-304 provided that proposals received by DOE in response to the DPA solicitation will be considered as resposive to a solicitation of the U.S. Synthetic Fuels Corporation to the extent that the proposed projects meet the eligibility criteria for funding by the Corporation.

As part of its implementation of the law, the Department issued a conditional loan guarantee commitment of $250 million to Great Plains Gasification Associates for assisting in the financing of the first year construction costs of 125 million cubic feet per day high Btu coal gasification facility to be located in Mercer County, North Dakota. On Wednesday, November 19, 1980, this was raised to $1.5 billion, and DOE announced its first loan guarantee.

We are reviewing eight proposals in the following categories: shale oil, tar sands, and other.

SOLICITATION FOR PROPOSALS FOR FINANCIAL ASSISTANCE
UNDER THE FEDERAL NONNUCLEAR RESEARCH AND
DEVELOPMENT ACT (PUBLIC LAW 93-577)
FOR THE DEVELOPMENT OF ALTERNATIVE FUELS

This program solicitation was for the purpose of inviting interested parties to submit proposals for financial assistance to expedite the commercial production of alternative fuels. Two billion dollars of assistance is available. The types of financial assistance available through this Solicitation consists of:

o   Purchase commitments, or price guarantees, for alternative fuels, and

o   Loan guarantees for financing the demonstration of commercial size alternative fuels production facilities.

For purposes of determining eligibility for financial assistance under this solicitation, as defined in Public Law 96-126 (as amended by Public Law 96-304), alternative fuel is defined as gases, liquid, or solid fuels and chemical feedstocks derived from the following domestic resources: coal, oil shale, tar sands, unconventional natural gas, lignite, peat, and other minerals.

Direct burning of these resources is not within the definition of alternative fuels. A mixture of coal and petroleum liquids is within the definition of alternative fuels. Under Section 19 of the Federal Nonnuclear Energy Research and Development Act, unconventional natural gas is an eligible domestic resource for loan guarantees only if the project converts the gas to an alternative fuel.

This solicitation was issued in keeping with the Congressional declaration of policy as set forth in Public Law 96-126 to:

". . . expedite the domestic development and production of alternative fuels and to reduce dependence on foreign supplies of energy resources by establishing domestic production at maximum levels at the earliest time practicable. . ."

The goal supported by this solicitation is to provide financial assistance to those projects for which such assistance represents the only remaining determining factor to proceed with the construction and operation of commercial scale alternative fuels production facilities. Organizations interested in pursuing this goal and whose projects meet the criteria as set forth in the solicitation were invited to respond.

Proposers were encouraged to consider and to propose the utilization of any single incentive, or combination of incentives, available under this solicitation in a manner which best serves the financial needs of the proposed project and in a manner that effects the maximum production capacity consistent with minimizing the level of Government liability.

On November 15, thirteen proposals for projects were received under this solicitation in the areas of shale oil, tar sands, peat, coal-oil mixtures, and other. Evaluation is now underway at DOE headquarters.

Where do we go from here? Management of the assistance grants and negotiated cooperative agreements will be a major undertaking. The DOE Office of Procurement, programmatic offices, and national laboratories are all currently organizing themselves for the management of this work.

When the government sets such domestic energy production goals, it has de facto committed itself to involvement in the domestic fuel development and production marketplace.

Such involvement can be accomplished while still adhering to our American tradition that recognizes the competitive private marketplace as the principal mechanism for building industrial productive capacity, for serving the demand of the consuming public, and for sustaining and advancing our national economic strength. We view the marketplace as being best suited for the transfer of the products and services of industry to their ultimate consumer. We view deviations from competitive or conventional marketplace practices with concern.

However, we must also recognize that for the next several years few of these projects for the synthetic fuels will be economically competitive with imported oil. Private industry is not likely by itself to undertake major alternative fuel projects because of the following factors:

o    Many of the benefits of developing conventional and alternative resources, such as national security and downward pressure on world oil prices, do not occur directly to private firms.

o     Private companies face major uncertainties with large scale synthe-
      tic fuel plants.  These uncertainties include future OPEC price
      policies, future world oil and gas production levels, the risks of
      scaling up new technologies to large size plants, investing in
      facilities with uncertain life cycle costs, or the uncertainty of
      government policy.

o     The immediate financial costs for plants and projects are often
      extremely large.  Each plant can cost billions of dollars, and its
      exact operating design and economics are unknown.  Many private
      firms cannot put such large amounts at risk since they represent a
      majority of the firm's assets.

o     There are long and costly procedures that must be followed in order
      to satisfy all Federal, State and local regulatory requirements.
      These further increase the risk of project failure and large
      financial losses.

Because of these constraints, therefore, some intervention by the
Federal government is necessary, until we can get to the point where the
markets will be able to meet our needs.

In broad terms, the Department of Energy's responsibilities for supporting the private sector include the following:

o     Accelerating the rate at which new energy products and services
      find their way into the energy marketplace.  This means accelerating
      the process at a greater rate than otherwise would occur and doing
      this in a manner consistent with goals of economic efficiency and
      the need to reduce imports.

o     Hastening informed consumer participation in the decisions among
      alternatives so that the most cost effective technologies can enter
      the energy marketplace.

o     Employing identifiable marketplace behavior patterns, in conjunction
      with known national needs and energy resource scarcities, in order
      to establish priorities for the development of needed energy
      technologies.

o     Eliminating constraints through minimum government intervention and
      expenditures.

The Assistant Secretary for Resource Applications is the principal
programmatic agent for ensuring that the DOE market-centered activities
proceed at the pace required by national policy.  She has the responsibility for carrying out the intent of Congress for alternative fuel
development and production as stated in legislation.

Resource Applications will provide ongoing support to the development of
alternative fuels through the following activities:

o     Administer the solicitation and selection of alternative fuels
      projects for the $5.5 billion in feasibility study, cooperative
      agreement, loan guarantee, and price and purchase guarantee incen-
      tives provided in Public Laws 96-126, 96-304, and 96-294.

o     Take responsibility for leasing activities essential to allow the private sector to obtain the coal, oil shale, tar sands, and other Federally-owned resources necessary for projects.

o   Continue to identify the need for, and encourage DOE sponsorship of, research, development, and market-oriented demonstration activities that will bring new processes to readiness for full-scale commercial operation.

o   Continue to provide the principal DOE marketplace assistance for those alternative fuels and technologies not eligible for SFC support.

o   Continue to examine how regulatory, tax, and other off-budget incentives can best interact with the incentives provided to stimulate the maximum possible rate of industrialization at the lowest new cost to the government.

o   Develop overall energy marketplace policy, planning and analysis, and programming efforts consistent with the national energy policy.

Resource Applications is also the point of contact for cooperation between DOE and the Department of Interior in matters pertaining to the leasing of Federal resources. RA provides the staff support for the Leasing Liaison Committee.

We consider the following to be our major energy production and supply functions:

o   Provide financial incentives to the private sector to stimulate early industrial development and use of new energy sources.

o   Determine those non-financial actions necessary to facilitate wider-scale use of conventional and alternative energy sources.

o   Promote institutional arrangements and provide assistance to non-Federal entities to overcome near-term barriers to the production and use of these energy sources. These actions include attempts to mitigate environmental, permitting, licensing, financial and market barriers.

In many instances the construction and operation of an alternative fuels plant could have substantial impact on States and local communities. Water requirements, socioeconomic needs, and potential environmental effects must be carefully controlled to ensure that we achieve the desired benefits without unacceptable side effects.

Our efforts in forward planning and assistance are designed to let us achieve these results through community impact assistance, environmental impact assessments, transportation and water supply planning and coordination with other government agencies at the Federal, State and local levels.

In 1976, I initiated a "Study of Energy R&D in the Private Sector" utilizing participation of the private sector through the Industrial Research Institute. Over a three-year period, retired corporate executives interviewed senior management of 100 companies. Their major findings included:

o   A consistent national energy strategy is needed for private companies to formulate effective long range energy strategy.

o    There is a serious lack of industry-government cooperation and
     mutual trust.

o    DOE must play a positive role in improving government-industry
     cooperation in promoting energy RD&E and Commercialization.

The DOE Alternative Fuels Program which I have discussed this morning
takes giant steps in responding to these stated private sector concerns.
Much more remains to be done.  It is my intent to assure that we continue working positively in cooperation with industry to further our
mutual goals in the successful birth of the Alternative Fuels Industry.

# A PERSPECTIVE ON THE ECONOMIC READINESS
# OF METHODS FOR PRODUCING COAL LIQUIDS

Ronald L. Dickenson, Dale R. Simbeck and A. James Moll
Synthetic Fuels Associates, Inc.

INTRODUCTION

Two routes have been used historically to convert coal into liquid fuels: direct coal hydrogenation, and what is usually referred to as "indirect" liquefaction which is coal gasification followed by catalytic reaction of the produced synthesis gas. In Germany and Japan during World War II production reached approximately 75,000 barrels per day (b/d) of gasoline from coal via the Bergius direct liquefaction process and 25,000 b/d of gasoline/diesel fuels via gasification and Fischer-Tropsch reaction of synthesis gas. Lurgi and Winkler coal gasification technologies were used in the gasification step. South Africa currently produces gasoline and diesel fuel from coal with Lurgi gasifiers followed by Fischer-Tropsch synthesis, as well as methanol from coal via Koppers-Totzek gasification and ICI methanol synthesis. In addition, direct and indirect coal liquefaction processes are currently approaching large scale demonstration in the United States and elsewhere.

This paper outlines important factors influencing the choice between direct and indirect coal liquefaction and looks at the reality of the two routes to liquid fuels. Addressed are key aspects of technological and economic readiness such as:

        o  Technology background and status
        o  Process design alternatives
        o  Product and by-product market considerations
        o  Importance of coal type and site selection

OVERVIEW

Figure 1 is a simplified schematic diagram showing the production of a gasoline from bituminous coal via both direct and indirect coal lique-faction. There are some rather significant analytical differences between coal and gasoline. To convert coal to gasoline or other light liquid fuels, the following must be accomplished:

- o The hydrogen/carbon (H/C) ratio must be substantially increased
- o Sulfur, oxygen, and nitrogen must be removed
- o Ash, metals, and water must be removed.

Direct coal liquefaction processes normally accomplish the above tasks by addition of hydrogen to the coal at pressures of 2000-3000 psig and temperatures of 700-900° F for residence times of 30-60 minutes. The hetero atoms are removed in their hydrogenated forms as $H_2S$, $H_2O$, and $NH_3$. The importance of oxygen content in coals being considered for direct coal liquefaction will be discussed later in this paper. It is a fact of life that the raw products of the leading direct coal liquefaction processes (SRC-I, SRC-II, H-Coal and Exxon Donor Solvent) are not retail liquid products. The raw distillable liquids from these coal liquefaction processes require considerable "refinery type" hydrotreating to be converted into gasoline or other light transportation liquids. This additional catalytic hydrogenation is required to further increase the H/C ratio and remove hetero atoms remaining after the liquefaction step. The overall hydrogen consumption to convert bituminous coal into gasoline is about 8,000 standard cubic feet (scf) per barrel of gasoline. Hydrogen consumption for the coal liquefaction step is 4,000-6,000 scf/barrel and the required downstream refinery hydrotreating is therefore 2,000-4,000 scf/barrel. The lower the coal liquefaction hydrogen consumption the lower will be the quality of the coal liquefaction products, however, the higher the hydrogen consumption required in refining the coal liquids. Although coal liquefaction plants would produce their own hydrogen, it is reasonable to project the cost of this hydrogen at about $2.50/1,000 scf, therefore, the hydrogen related cost of gasoline from coal would be about $20/barrel. (Unless otherwise stated, all costs referred to in this discussion are mid-1980 dollars.) Our studies at Synthetic Fuels Associates (SFA) indicate that the cost of both raw coal liquids and refined retail products from the coal liquids is being strongly influenced by the hydrogen requirements. This may sound trivial, but it is not, as we hope the following discussion shows.

Indirect coal liquefaction processes accomplish the previously stated requirement for conversion of coal to gasoline by destruction of the coal into synthesis gas ($H_2$ and CO), then purification of that gas, and finally catalytic reaction of the synthesis gas into gasoline. Indirect coal liquefaction consumes large amounts of oxygen and converts much of feed coal carbon into $CO_2$ thus increasing the H/C ratio of the coal by subtracting carbon. On the other hand, direct liquefaction adds hydrogen. Synthesis gas reactions require a $H_2$ to CO molar ratio of at least 2 to 1. Depending on the type of coal gasification process, large amounts of steam are consumed in the gasifiers or in CO-$H_2O$ shift reactors to obtain the required $H_2$ to CO ratio. The overall synthesis gas reaction is:

$$\text{hydrogen} + \text{carbon monoxide} \longrightarrow \text{gasoline} + \text{water} + \text{heat}$$

$$16H_2 + 8CO \longrightarrow C_8H_{16} + 8H_2O$$

337

This reaction can be accomplished via Fischer-Tropsch (F-T) technology or by first converting the synthesis gas to methanol and then reacting the methanol to gasoline with Mobil's Methanol to Gasoline (MTG) technology. One of the primary products of both synthesis gas-to-gasoline reactions is water. The overall thermal efficiency of the coal to gasoline indirect processes are low (about 50%) due to the production of large amounts of water and $CO_2$.

## DIRECT COAL LIQUEFACTION

Direct coal liquefaction processes are all basically the same. Figure 2 shows a generalized direct liquefaction process flow diagram. Figure 2 could easily represent SRC-II, H-Coal or Exxon Donor Solvent with bottom recycle. The only major difference between the current developmental coal hydrogenation processes and the original Bergius process is the development of residue partial oxidation gasification. In the original Bergius plants, the distillation residue was centrifuged for maximum recovery of recycle solvent and the solids discarded.

The major design alternative in direct coal liquefaction processes is hydrogen production. The available feedstocks for hydrogen production are unconverted coal residue, methane, and coal. General plant utility requirements and the need to maximize coal conversion in hydrogenation reduce the choice to methane or coal residue. Our analysis indicates that hydrogen from residues is the best choice if the methane produced in the process can be sold at an energy price parity with the liquids. If methane markets dictate a lower price, it likely becomes more economical to reform the methane into hydrogen and use medium Btu gas (MBG) produced from gasification of the residues as the reformer furnace fuel.

The higher the pressure, temperature, and residence time (severity) in the hydrogenation reactor, the higher the hydrogen consumption and the higher the quality of products. The H-Coal process, in the syncrude mode of operation, produces the higher quality product, but at the expense of a catalyst and high hydrogen consumption and partial pressure. SRC-II produces lower quality products, but at a lower hydrogen partial pressure and with lower hydrogen consumption.

The projected capital investment for 50,000 b/d direct coal liquefaction systems range in the order of 2.0 billion dollars, or $40,000 per daily barrel. The cost of the raw coal liquids range from $40 to $60 per barrel depending mainly on capital charges and coal cost. The capital and product cost closely follow the product quality and hydrogen consumption. In summary, the highest cost direct coal liquefaction processes produce the highest quality liquids.

The raw products can be considered as a $C_5$-400° F naphtha and a 400-900°F fuel oil. Figure 3 gives general analysis of these raw coal liquids. The naphtha appears to be an excellent feedstock for high octane gasoline production using conventional hydrotreating followed by reforming. Based on recent published work,[1,2,3] the projected capital cost for refining the raw coal naphtha into gasoline ranges from $4,000 to $7,000 per daily barrel with an added cost to the products of $5 to $8 per barrel. The cost and difficulty of upgrading the raw coal derived fuel oil into gasoline or No. 2 fuel oil appears to be much greater than for the naphtha. The projected costs for fuel oil upgrading range from $8,000 to $13,000 per barrel. Upgrading the raw coal derived fuel oil involves relatively expensive refinery processing such as severe hydrotreating, hydrocracking, and fluid catalytic cracking. An important consideration is that the quality of the raw coal derived fuel oil is much poorer than conventional residual fuel oil because of the low hydrogen to carbon ratio and the high levels of nitrogen and oxygen.

SFA believes it would normally be more cost effective to directly burn the coal derived 400 to 900°F fuel oil in the existing residual fuel oil markets. The conventional residual fuel oil thus displaced could be upgraded into gasoline and No. 2 fuel oil at a lower cost than upgrading the heavy coal derived distillate fuel oil.

## INDIRECT COAL LIQUEFACTION

Figure 4 shows a block flow diagram of SASOL-II, the world's foremost indirect coal liquefaction plant. Important points in the SASOL-II design include:

- o Use of Lurgi gasifiers and Synthol F-T reactors
- o Partial oxidation of the methane into synthesis gas
- o High levels of lead required to raise the octane of the raw F-T gasoline produced
- o Use of significant amounts of local hydroelectric power
- o About 30 wt.% of the products used are chemicals, rather than liquid fuels.

The use of Lurgi gasifiers at SASOL reflects requirements dictated by the feed coal more than by the products desired. The coal used at SASOL favors the Lurgi gasification process because of its high ash content, high ash fusion temperature, and high reactivity. The main disadvantage of Lurgi gasification for this application is the high methane yield which is an inert in the F-T reaction. Approximately 55% of the net energy in the raw products leaving the F-T reactor is in methane. Most of this methane is from the gasifiers, however, some is also produced in the F-T reactor. The methane is converted to synthesis gas in Lurgi "Autotherm" reactors and then converted into liquids by recycling to the F-T reactors. Fluor Engineers and Constructors, Inc. has projected the capital investment of a 50,000 b/d SASOL-II type plant built in the United States to be about $2.5 to $3.2 billion or equivalent to $50,000 to $64,000 (1979 dollars) per daily barrel.[4] The cost of the resulting gasoline and diesel fuel would be $70 to $90 per barrel depending mainly on the capital charge rate and coal cost.

It is unlikely that a plant identical to SASOL would be considered in the United States. Likely modifications for a U.S. based plant would include:

- o Marketing of methane if Lurgi gasifiers are used
- o Consideration of additional coal gasification processes
- o Elimination of some of the chemical production[5]
- o More efficient utilities.

The use of Lurgi gasification for indirect coal liquefaction in the U.S. appears to be substantially influenced by the ability to market methane at a price equivalent to the coal liquids. If methane markets dictate a lower price, it will likely be more economical to use partial oxidation gasifiers which do not produce methane.

Many of the U.S. coals because of their low ash fusion temperature, low reactivity and caking characteristics favor entrained partial oxidation gasifiers over fixed-bed gasifiers. The projected investment for a 50,000 b/d (oil equivalent) Lurgi based co-product liquids and methane plant is about $2 billion or equivalent to $40,000 per daily barrel. The cost of these products would be $50 to $70 per barrel depending on the capital rate and coal cost, assuming the methane is sold at the equivalent energy price of about $10 to $14 per million Btu. Similar costs can be projected for Koppers-Totzek based indirect coal liquefaction plants with no co-product methane. Advanced gasifiers such as Texaco, Shell/Koppers, and British Gas Corporation/Lurgi slagger, have the potential of producing indirect coal liquids at $40 to $60 per barrel, depending mainly on capital charges and coal costs.

With respect to the basic products of indirect liquefaction, the use of F-T technology appears to be favored for diesel fuel production whereas the Mobil MTG process appears to be the favored technology for production of high octane unleaded gasoline. Methanol can be used directly as a fuel or for production of higher value products:

> o   Fuel grade
>     -direct firing
>     -transportation fuel
>
> o   Chemical grade methanol
> o   Methyl tertiary butyl ether (MTBE)

Current testing also indicates "neat" methanol could be used in spark-ignition internal combustion engines and combustion turbines. A mixture of methanol with 5 to 20% diesel may find use in diesel type I.C. engines. However, careful market analysis on a specific basis is required to assess the potential of methanol in existing gasoline and diesel fuel markets.


CRUDE OIL "DISPLACEMENT"

The most economical form of coal liquefaction is usually overlooked--this is the use of coal directly for heat in existing oil refineries. We sometimes refer to this as "phantom" coal liquefaction. Approximately 10% of the energy in a barrel of crude oil is consumed during conventional refining, about half of which is consumed for hydrogen and plant fuel gas, and the other half in the production of utility and process steam.

It is well known that coal can be directly burned in pulverized coal boilers to produce steam. Although the initial cost of steam from coal would be high because of the capital investment of the new boiler and coal handling facilities, deregulation of both oil and gas should soon make the use of steam from coal less expensive than steam from refinery gas or fuel oil. Environmental regulations represent the major uncertainty delaying the industrial conversion to coal for steam production. The U.S. Environmental Protection Agency (EPA) was scheduled to announce its proposed industrial coal combustion environmental standards in October 1980, however, these standards had not yet been published at the time this paper was prepared. These proposed industrial environmental standards will have significant influence on the future of direct coal use by industry. This uncertainty is delaying a large potential market for steam coals.

Coal can also be used indirectly in existing oil refining by gasification into a medium Btu gas (MBG) to replace refinery gas in existing furnaces.

340

Rather than being used as refinery fuel, the displaced refinery gas can be used more economically for hydrogen **pro**duction, sold over the fence as a chemical feedstock or purified and marketed as pipeline gas. The pipeline gas option does not appear economical without complete deregulation of natural gas.

Oxygen blown coal gasification to produce MBG is more economical than producing low Btu gas (LBG) from air blown gasification when the transportation and existing furnace retrofit costs are considered. Furthermore, large MBG plants supplying several customers appear to be cost effective in reducing the investment and increasing the annual load factor relative to small producers.

SELECTION OF SITE, COAL AND CONVERSION PROCESS

The selection of coal process, liquid product type, and plant site is complex. The coal selection is most important for direct coal liquefaction since hydrogen consumption is the dominant cost in direct coal liquefaction. As mentioned earlier, low rank coals contain relatively large amounts of oxygen, therefore, require larger amounts of hydrogen to liquefy. The hydrogen required for converting Powder River Subbituminous coal into raw coal liquids is about twice that of converting Illinois No. 6 Bituminous coal into the same raw coal liquids. Additional hydrogen is required to upgrade raw coal liquids into gasoline. The capital investment for direct liquefaction of Wyodak Subbituminous has been projected to be over 40% higher than for direct liquefaction of Illinois No. 6 Bituminous; moreover, the Wyodak coal liquids still contain large amounts of oxygen.[6] This much higher capital investment overpowers the lower coal cost advantages of lower rank Western U.S. coals.

The capital cost of indirect coal liquefaction appears to be about the same for all rank coals. Coal cost is important for indirect coal liquefaction, due to the inherent inefficiency. The less expensive low rank Western U.S. and the Gulf Coast coals appear to be favored for indirect coal liquefaction.

Figure 5 illustrates SFA's projection of coal use in liquid fuels markets. We believe that one of the most attractive uses of coal will be direct firing or gasification in existing refineries to increase liquid yield.

Oil refineries will favor direct coal liquids because they have the existing facilities for upgrading coal-derived naphtha, and have existing markets for coal-derived fuel oil to replace conventional residual oil. The displaced residual oil can then be upgraded to higher value liquids with reasonable modifications to existing refineries. The location of current residual markets and capital cost of direct coal liquefaction appear to favor construction of direct coal liquefaction plants in the Eastern U.S. near high rank Bituminous coals. Some direct coal liquefaction plants may also be built in the Midwest and use Illinois basin Bituminous coals, however, there is a weaker residual fuel demand in that area. Lower non-union construction costs and stronger residual markets could lead to construction of direct coal liquefaction in the Gulf Coast with barge delivery of Eastern or Midwest Bituminous coal.

Indirect coal liquefaction will be favored by "energy entrepreneurs", chemical companies, and oil refiners not having strong residual fuel markets available. Indirect coal liquefaction will produce the highest value liquid fuels, e.g., methanol, MTBE, gasoline, and diesel fuel.

Texas and Louisiana lignites have excellent potential as feedstocks for indirect coal liquefaction plants on the Gulf Coast where there is a high demand for chemicals. Economics and the lack of residual fuel markets should favor indirect liquefaction of low cost Western Subbituminous coal in the Western U.S. However, because of the high value of the products, indirect coal liquefaction can also be considered for Midwest and Eastern coals.

In conclusion, the U.S. liquid fuel market is large and complex. First generation conversion technology is available for production of liquids by indirect methods. However, direct systems are still in the pilot stage and, except for a limited number of demonstration systems, are not expected to be in commercial production until the late 1980's. There is not a clear winner between direct or indirect coal liquefaction. There is large potential for both. The choice is specific to an individual company's position in the marketplace, existing plant and fuel distribution system investment, and coal feedstock availability. As the price of oil continues to rise and its long term availability becomes more tenuous, the full potential of coal will be realized--shrewd and perceptive investors are already moving quickly to be in position for the next disruption in world crude oil supply or price.

REFERENCES

1.  R. E. Conber, G. M. Garrett, and J. A. Weiszman, "Crude Oil Versus Coal Oil Processing Study," UOP Process Division, AIChE Meeting, Boston, Massachusetts, August 19-22, 1979.

2.  H. A. Frumkin, R. F. Sullivan, B. E. Strangeland, "Converting SRC-II Process Products to Transportation Fuels," Chevron Research Company, AIChE Meeting, Boston, Massachusetts, August 19-22, 1979.

3.  T. R. Stein, et al., "Upgrading of Coal Liquids for Use as Power Generating Fuels," Prepared by Mobil Research and Development Corporation for Electric Power Research Institute, Report No. AF-444, October 1977.

4.  J. Mullowney, "A Fluor Perspective on Synthetic Liquids," Presented at the Congress of the Australian Institute of Petroleu, Sydney, Australia, September 1980.

5.  R. L. Dickenson and D. R. Simbeck, "SNG from Coal: By-Products Need Attention," The Oil and Gas Journal, pp. 65-68, March 12, 1979.

6.  P. A. Buckingham, et al., "Engineering Evaluation of Conceptual Coal Conversion Plant Using the H-Coal Liquefaction Process," prepared by Fluor Engineers and Constructors, Inc. for Electric Power Research Institute, Report No. AF-1297, December 1979.

## Figure 1
## Alternative Routes for Coal Liquefaction

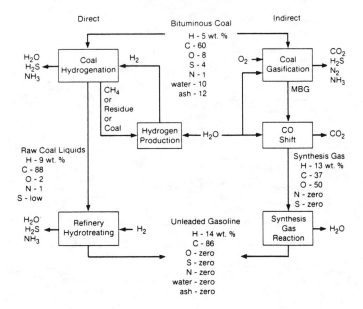

## Figure 2
## "John Doe" Direct Coal Liquefaction Process

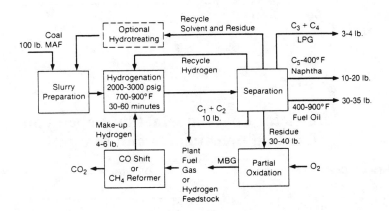

## Figure 3

### RAW COAL DERIVED LIQUIDS

| Naphtha (C$_5$ - 400° F) | Analysis (wt. %) |
|---|---|
| H | 10 - 13 |
| C | 86 - 88 |
| N | 0.2 - 0.5 |
| O | 0.5 - 2.0 |
| S | 0.05 - 0.2 |
| | |
| Fuel Oil (400 - 900° F) | |
| H | 8 - 10 |
| C | 87 - 89 |
| N | 0.5 - 1.0 |
| O | 1.3 - 3.5 |
| S | 0.1 - 0.3 |

343

## Figure 4
## SASOL-II Block Flow Diagram

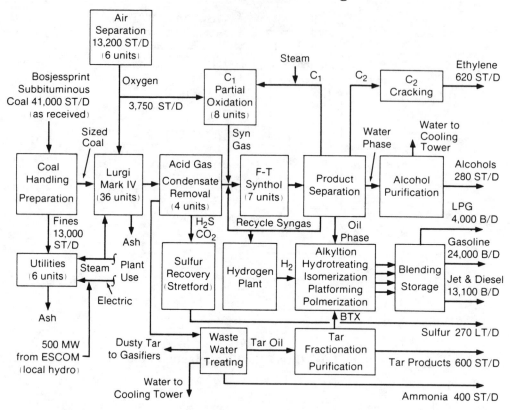

## Figure 5
## Routes for Coal Liquefaction Depend on Existing Investments and Markets

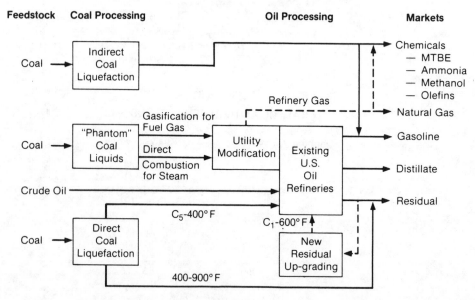

# SYNTHETIC FUELS AND THE ENVIRONMENT

**Minh–Triet Lethi**
U.S. Environmental Protection Agency

It is no secret that synthetic fuels will assume an expanding role over the next several decades in the country's energy supplies.

We at the Environmental Protection Agency recognize the importance of synthetic fuel development and are ready to take prompt, effective, and reasonable action to ensure that this development can take place in an environmentally sound fashion, and according to the timetable set by the President and Congress.

At the same time, while we believe that sound development is possible, we also believe that the environmental problems of synfuels are substantial, and in some cases, represent serious risks to human health and welfare. Large-scale synfuels plants have the potential for great environmental damage. EPA sees its job as influencing public and private investment toward fuels and technologies at the safer end of the investment spectrum, and requiring adequate controls on those synfuel facilities which do attract capital.

EPA is making research, regulatory and policy efforts to achieve this goal of environmentally sound development. I would like to summarize what we see as the environmental problems of synfuel production and outline the aproach we plan to take to investigate and mitigate them.

But let me say first that we can not do the job of protection alone with our major tools, permits and regulations. Achieving the positive goal of environmentally sound development requires a partnership among all who are involved - a partnership that we want very much.

Synthetic fuel production for the four major technologies -- oil shale, coal gasification, coal liquefaction, alcohol/gasohol -- creates environmental problems in all of the major media: air, water, and land.

The air problems are particularly acute. The synthetic fuel industry will emit a wide range of air pollutants with adverse environmental effects. Oil shale production for example will emit nitrogen oxides,

345

sulfur, hydrocarbons, and ammonia, as well as particulate matter. In addition, potentially hazardous trace elements, such as arsenic, boron, and mercury will also be discharged. Furthermore, much of the development, particularly in oil shale, is going to take place in and near areas of pristine air quality. The air problems created by the coal gasification and liquefaction processes are similar. Preserving air quality is going to be, thus, a major challenge.

Water supply and quality problems of Synthetic Fuel production can be just as complex. Adequate water supplies, particularly in the oil shale and coal areas of the West are very uncertain. While water supply may not end up being the absolute limiting factor in very many cases, that does not mean the problem will not be acute. Water quality problems are potentially even more serious in many cases. Process water, particularly for the coal technologies, will be highly contaminated by toxic and carcinogenic materials and could represent threats to wells, rivers, and other public drinking supplies if not properly controlled. Total recycling of process water for re-use in synthetic fuel production will be a big help in this regard, but it will not totally solve the water quality problem. Leaching or run-off from spent oil shale, or coal ash residue, for example, could possibly contaminate ground water supplies with a variety of organic and inorganic substances, including toxics and carcinogens.

Solid waste problems are related to water quality problems. Both oil shale and the coal conversion processes produce enormous amounts of solid waste, almost all of which may present a potential threat to human health and well-being. Oil shale production, for example produces spent shale that is greater in volume than what is removed from the ground, while coal technologies, both gasification and direct liquefaction, will produce vast quantities of ash. These waste materials will likely contain varying amounts of potentially hazardous materials. For example, the Department of Energy has estimated that a commercial scale coal gasification plant could produce up to 1 million tons of coal ash waste per year. This material will contain hazardous trace elements including arsenic, cadmium, lead, and mercury.

We are also concerned about product toxicity in the use of synthetic fuels by handlers and consumers, both industrial and the general public. Coal derived liquid fuels, particularly those produced by the direct liquefaction processes, are a particular case in point. These coal liquids are not the same as normal crude oil products and contain substances which are potential mutagens or carcinogens. Public exposure to such fuels through normal consumer use represents a potentially serious health problem. More research and study needs to be performed, both to determine the nature of these substances and to devise appropriate regulatory and safety measures.

It is fair to say, then, that much is unknown about the environmental effects of synthetic fuels. It is also clear that existing regulations will likely not be a guarantee of safeguarding human health and welfare. Much research remains to be done and new regulations will have to be written.

What EPA is Doing. Recognizing the full range of environmental problems and need to solve them promptly and effectively, EPA is undertaking a multi-media approach to the synthetic fuel industry.

The key word in EPA's future plans for synthetic fuels regulation is integration, both internal and external.

Internally, EPA is synchronizing relevant research concerning the environmental risks of synfuels, so that preregulatory documents and ultimately regulations from our air, water, solid waste and toxic programs will be produced concurrently for each technology. That is, EPA will develop a unified body of regulations applicable to the synthetic fuel industry. This effort is rather unique for the Agency. We always try to be aware of the many interrelated environmental problems caused by any industry. But in the case of synfuels we are making the most concerted effort we have ever made to develop coordinated standards.

To do this, we have established an Alternate Fuels Group charged with coordinating all Synthetic Fuel research and regulations development. This group includes division directors from all the major EPA labs and offices charged with research and regulatory development.

Because it will take at least several years before we can actually promulgate new standards, we are developing what we call Pollution Control Guidance Documents to assist developers and permit officials. They will not have the force of law, but they will indicate our best guess of the likely ranges of performance and costs of control technologies. We expect to publish them in 1981 and 1982. We are conscious of the fact that we have a special opportunity to help shape this new industry, and we want to take advantage of it.

We expect to publish Pollution Control Guidance Documents for each major technology: oil shale, direct liquefaction of coal, indirect liquefaction and high-, medium- and low- Btu gasificaition of coal production. We may also publish one for alcohol.

We are planning to circulate drafts of these documents to DOE, interested industry and environmental groups before they are published, and generally attempt to stimulate discussion of environmental concerns and EPA's role in synfuels.

Of the synfuels technologies, we are closest to promulgating standards for alcohol. Alcohol production, particularly, ethanol production, presents relatively conventional environmental problems, principally those characteristic of fossil-fuel stationary sources and of facilities producing bioorganic wastes. We expect that many of our current regulations will be satisfactory. We are now collecting field samples, and we welcome the cooperation of industry in this process and in the production of reasonable regulations for this pioneering synthetic fuel technology.

Finally, we are working to improve our performance in the area of permits and environmental review.

We have set up a Permits Coordination Group which will carefully track permits for all energy installations, including synfuels and identify potential problems early to enable appropriate action in a timely fashion.

I should mention that EPA has already begun to issue permits for synthetic fuel production facilities and apply the full range of existing regulations for air, water, and solid waste regulations.

EPA's Denver Regional office, has already promptly issued 5 air quality permits for oil shale development. Best Available Control Technology and prescribed monitoring requirements as determined by existing regulations were defined for these permits.

This early group of permits included the Colony Development Operation of ARCO and TOSCO Corporation, the first commercial scale oil shale facility to be permitted in the United States. EPA's permit will ultimately allow Colony to expand and produce 46,000 barrels per day of low sulfur distillates and other by-products.

This landmark permit will also allow the company to construct and operate:

1) a 66,000 ton per day underground oil shale mine,

2) a surface oil shale retorting facility capable of extracting oil from oil shale, and

3) extensive support facilities including a 194-mile long pipeline and loading terminal.

The EPA permit process for all of this was short and augurs well for the environmentally sound and timely development of synthetic fuels. The PSD permit process for the extensive Colony operation was completed in only 12 months, and is indicative of EPA's desire to take prompt action.

While we have been prompt in issuing these first permits, we still have a big job in front of us. Much research must be done so that guidance documents may be prepared and standards set. Many air, water and solid waste and toxic standards need to be devised as the synthetic fuels industry commercializes. This is quite a challenge.

We look forward to a mutual effort of all those who are interested in producing a viable and environmentally sound synthetic fuels industry. We hope that those who are in charge of financing and developing synfuels also go about their business with the environmental problems in mind and take every step to see that synfuels are developed in an environmentally sound manner. If we do not develop synfuel technology with these concerns in mind, we will not have fulfilled our responsibility to our generation and those of the future.

Thank you.

TABLE 1. IMPACT ELEMENTS USED IN SELECS METHODOLOGY

| No. | Category Title | Impact Elements | Application* |
|---|---|---|---|
| 1 | Employment & Population | Percent Population Increase | M,C |
| | | Percent Change in Unemployment | R |
| 2 | Services | Percent Increase in Police Officers | M,C |
| | | Percent Increase in Fire Fighters | M,C |
| | | Percent Increase in Doctors | M,C |
| | | Percent Increase in Hospital Beds | M,C |
| | | Percent Increase in Dentists | M,C |
| | | Percent Increase in Teachers | M,C |
| | | Percent of Water Supply Capacity Margin | M,C |
| | | Percent of Wastewater Treatment Capacity Margin | M,C |
| | | Percent Reduction in Life of Solid Waste Disposal Facility | M,C |
| | | Percent Increase in Park & Recreational Area | M,C |
| 3 | Bonding Capacity | Percent of Bonded Debt Margin | M,C |
| 4 | Housing | Percent Increase in Housing Units | M,C |
| 5 | Land Use | Percent Reduction of Undeveloped Municipal Land | M |
| | | Percent Reduction of Undeveloped County Land | C |
| | | Percent Reduction of Prime Farmland | C |
| 6 | Water Availability | Percent of Average Surface Water Flow | R |
| | | Percent of 7-Day, 10-Year Low Flow | R |
| 7 | Air Quality | Percent of Annual Particulate PSD Standard | R |
| | | Percent of 24-Hour Particulate PSD Standard | R |
| | | Percent of Annual $SO_2$ PSD Standard | R |
| | | Percent of 24-Hour $SO_2$ PSD Standard | R |
| | | Percent of $NO_x$ Annual Concentration Margin | R |
| 8 | Water Quality | Percent Reduction of Average Flow Assimilation Capacity | R |
| | | Percent Reduction of 7-Day, 10-Year Low Flow Assimilation Capacity | R |
| | | Percent of Allowable Water Pollutant Loads | R |
| 9 | Endangered Species | Species Habitat Proximity | R |
| 10 | Aesthetics & Important Sites | Visual Prominence | R |
| | | Important Site Proximity | R |
| 11 | Tax Rates | Percent Change in Average Tax Rate | M,C |

*M = Determined for Each Municipality
C = Determined for Each County
R = Determined for Overall Region

TABLE 2. HYPOTHETICAL ENERGY CONVERSION SYSTEMS EVALUATED WITH THE LEVEL 1 SELECS METHODOLOGY

| No. | Location | Technology |
|---|---|---|
| 1 | Pittsburgh, PA | Low Btu Coal Gasification |
| 2 | Charleston, WV | Coal Liquefaction |
| 3 | Jackson, OH | In Situ Coal Gasification |
| 4 | Kansas City, MO | Coal-Oil Mixture Power Generation |
| 5 | Bloomington, IL | Magnetohydrodynamic Power Generation |
| 6 | Vernal, UT | Tar Sands Conversion |
| 7 | Bismarck, ND | Lignite Liquefaction |
| 8 | Farmington, NM | High Btu Coal Gasification |
| 9 | Rawlins, WY | Oil Shale Conversion |

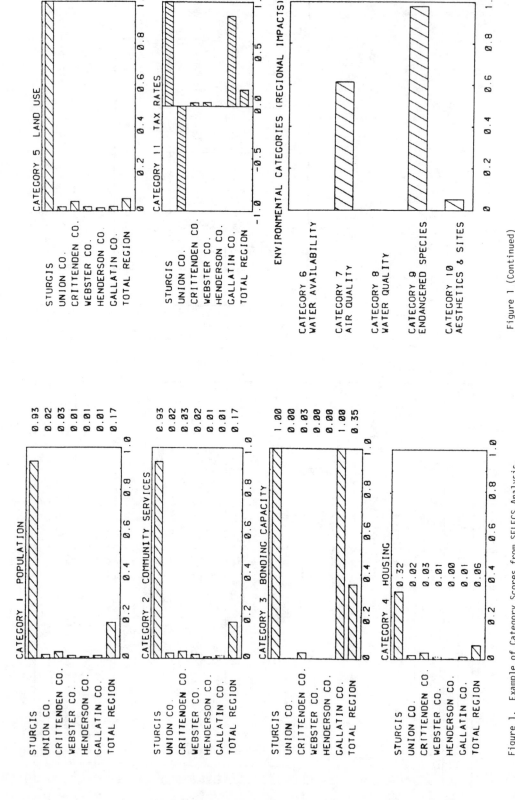

Figure 1. Example of Category Scores from SELECS Analysis.

350

## POTENTIAL ENVIRONMENTAL IMPACTS OF
## SYNTHETIC FUELS PRODUCTION

**Peter W. House**
U.S. Department of Energy

INTRODUCTION

This year, the Energy Security Act of 1980 (P.L. 96-294) became law, thereby establishing a Synthetic Fuels Corporation whose expressed purpose is to accelerate the Nation's efforts in the production of gaseous and liquid fuels from coal and other fossil energy sources.  The passage of the law took over a year of study and debate both by the Executive Branch and the Congress. The debate covered such topics as need, extent, strategy, financing, timing and impacts.  This discussion is about one of these areas, the potential environmental impacts of the synthetic fuel production levels originally proposed by the Administration.

During the weeks proceeding the Administration's submission to the Congress an interagency task force examined all facets of the proposals, prepared the background documents, and helped draft the final legislative package.  The environmental portion of this interagency effort consisted of a group from the Environmental Protection Agency, Council on Environmental Quality, and the Department of Interior, and was chaired by the Department of Energy.  The report of this group stated that there were no "fatal" problems with the Administration's proposal, but suggested a need for further study of environ- mental concerns in the technologies of modified in-situ oil shale and direct liquefaction of coal.  The report engendered considerable interest, especially a portion of the effort which was mininterpreted by some as suggesting that actual future sites for the synfuels plants were being specified.  As a member of the task force and environmental study group, let me state that there was no way that we actually could have performed such an analysis. But the suggestion that we might have tried assured widespread attention. On the basis of this interest we decided to provide a more in-depth analysis and made this report available in the early summer.

Primary Impacts

The follow-up study was intensive.  We tried to get in touch with everyone who might have done or be doing analysis of the environmental impacts of synfuels or the regional potential for siting plants.  The report examined

four principal areas; siting, regulations, permitting, and the individual
energy technologies per se.  In summary, the following findings were reported:

   o  <u>There were no "show stoppers" from a purely technological perspective</u>.

As long as one assumed that best engineering practices (BEP) were followed
as well as the use of best available control technologies (BACT), there
appeared to be no known environmental, health, or safety problems that could
not be adequately handled.  Some concern was noted for modified in-situ oil
shale processes in the area of retort contamination of aquifers and leaching,
and in the area of worker safety for direct liquefaction processes.  In these
cases, prudence and further study was advised.

Although some cause for optimism was justified, the caveat of BEP and
BACT should be noted.  The findings merely stated that the pollutants <u>could</u>
be handled, not that they would be.  Maintenance, enforcement, and good
will are all required both by the industry and by the local public officials
regulating the industry.  The success record of existing regulations on
existing plants might lead one who was cynical or cautious to question whether
the potential assumed by the study would ever come to pass in reality.  But
the fact remains, that, at present, there is no known problem that can't be
taken care of, if we want to.

   o  <u>Permitting is not a severe problem</u>.

After an exhaustive review of a large number of permitting laws and regulations from several Federal agencies and states, permitting did not appear,
in the main, to be as severe as many were making it out to be at the time.
The permitting process seemed to be operating well within the normal time
period that companies use for planning and developing large complexes such
as synfuels plants.

But again, a caution should be raised so that one does not get the idea that
all permitting proceeds as neatly as the regulations might suggest.  On the
one hand, litigation and public protest can hold up the issuance of a permit
for a considerable time period.  The number of such occurrences is small,
but there is no reason to suppose that they will not occur.  On the other
hand, there is apt to be some reluctance to embrace wholeheartedly such schemes
as "one-stop shopping" or Joint Review or other such fast tract measures on
the part of those companies who have not experienced serious permit delays.
Just as the enforcement of pollution laws is the prerogative of local officials,
so is the issuance of permits.  There is no reason to believe that these permits
are handled with equal professional skill or diligence in every jurisdiction.
In places where there are already informal means for securing permits, a
large, highly visible, complex mechanism for reviewing and issuing permits
may not be welcome or useful.

   o  <u>Siting of plants was not found to be much of a problem</u>.

The study tried to see if there were enough places in the Nation where a
50,000 barrel per day (BPD) plant could reasonably be sited.  The siting
question was examined from the perspective of potential air and water pollution, water quality, socioeconomic, and ecological impacts.  An abundance of
areas that could hold potential sites became evident.

This effort only addressed itself to the measurable technical/physical aspects
of the siting question.  Some attempt was made to take institutional features
into account, but political and public opinion could change drastically about
any issue at any time.  For example, the mere fact that there is physically
enough water available does not automatically mean that the water will be made

available for a synfuels plant.

In general, though, it is difficult to imagine how one could not find room
in almost any region for some type of synfuels plant.

o   Regulations, as they are presently promulgated, are not seen to be a
    problem for any of the known variants of the synthetic fuels proposal
    of 1,500,000 BPD by 1990.

Unfortunately, not the same level of assurance can be given for those laws
or regulations which are awaiting promulgation.  The proposed versions of
these regulations are normally submitted for public comment.  These public
comment documents are designed so that all relevant perspectives are taken
into consideration in formulating the final provisions of the regulations.
Some of these viewpoints are extreme and have the potential of causing
concern for one or the other adversarial groups engaged in the issue.  The
mere presentation of an extreme view during the comment period means that it
could influence the ultimate way in which the regulation will be promulgated;
but there are no guarantees.

## Secondary Impacts

The few findings I have noted so far could be considered only the beginning,
however.  As the synthetic fuels production program matures more and more
plants will come on line.  Their mere magnitude will begin to engender some
potential for the unanticipated problems.  Let us postulate three areas where
such difficulties may occur - health, land use, and agriculture.

Health:  The findings of the study referred to above were predicated on the
assumption that there would be some time to carry out the required research
and characterization assessments before commitment had to be made for the
full commercialization of given synfuel processes.  The normal progression
of a research project through pilot to demonstration to a commercial-sized
facility ordinarily allows the environmental scientist enough time to monitor
environmental insults caused by operation of the facility at various stages,
and to use the results of the monitoring and concomitant research to recommend
engineering design changes as the next phase of development is started.  Both
the original Executive proposal that went to the Congress and the legislation
that emerged had built into them a modified form of this research and redesign
idea in that both proposals were in two-stages.  By this I mean that the first
stage was more or less a proof of concept, in that promising technologies were
tested to see just how they actually performed.  Those that met expectations,
would, it may be assumed, be the prime candidates for the next round of design,
construction and operation.

But there is no guarantee that synthetic fuel developments will actually work
this way.  Because no unproven process was supposed to be a candidate for the
Synthetic Fuels Corporation, by definition, any project considered was ready
to be commercialized.  Clearly, however, all the required research has not been
done on all of the candidate processes from the perspective of environmental,
health, and safety concerns.  This rushing to market could mean, as one analyst
estimated, that a half-dozen or so plants could be in full-scale operation
before significant environmental findings were available.  Discovery of serious
health and safety problems could be problematical indeed, under such circumstances.

Land Use:  A second unanticipated result might come to pass in that the construction required could result in the growth and development of a portion of
the country heretofore considered relatively rural in nature.  Prime candidates
for this growth could come about in the Mountain States of the West.

The term "boomtown" has been used for some time to describe a situation where a relatively small community attempts to absorb the impacts of a relatively large perturbation to its economic base. There are several situations which could give rise to such upsets, but all result in the influx of large sums of money and numbers of people into a community over a very short time period. Military requirements often can cause several such situations, ranging from direct stimuli such as base locations, weapons making, or defense weapons placement, to indirect ones such as the building of an aircraft industry, as in California, or shipyards, or in North Carolina and San Diego. Private industry developments, as in the resource extraction areas leading to ghost towns, provide other examples. Recently the energy area has been targeted as a culprit causing "boomtown" effects. Siting a new, large utility plant or a coal mine have received national attention due to localized growth rate impacts.

The emergence of the synthetic fuel production goals on the part of the Federal Government and the acceptance of these goals on the part of industry has brought to the fore the potential for socioeconomic impacts in heretofore rural areas where energy, especially coal, resources are concentrated. As of this moment, no one knows where these plants will be, but examination of specific locales could give us some indication of what these impacts could ultimately be. For example, the present plan for the development of a syn-fuels industry anticipates some 400,000 BPD from oil shale by 1990 with all of this seemingly coming from the Piceance Basin in Colorado. By the year 2000, this number could be as high as 2,000,000 BPD.

This is considerable growth indeed. It is particularly problematical if one considers that the estimates of this study are only those related to energy and defense. But growth of this magnitude from a single source is bound to stimulate other growth by other primary industries attracted by the development. For example, the confluence of metallurgical coal and high-grade iron ore could presage a new steel industry. The existence of steel plants might attract some prefabrication industries, and so forth.

The growth, regardless of what form it takes, could be sufficiently large that today's assumptions made about the West and its future will have to undergo basic and fundamental change. Several examples of what the change might be come to mind. The conventional wisdom which cites scarce water in the West is normally based on an extrapolation of existing trends and conditions into the future. What happens if these trends don't come to pass? If one were to imagine such rapid growth from just energy-related industries, it is reasonable to envision that compared to cities such as Denver, Houston, and San Francisco which started as single-industry cities, the very size of these new "energy cities," and the population that they require, will attract other industries and begin a development cycle resulting in considerable growth in the West. This growth will mean a fundamental change in the character of the West, and makes much of our present planning projections irrelevant.

Let us examine a couple of these possibilities. The agricultural sector has considerable land under management and irrigation that could be squeezed by large influxes of people spilling out of sprawling industrial centers into present-day farmland. Associated with these people is the use of recreational vehicles and the presence of domestic animals that could constrain present farming methods. Recreational resources and wildlife will both feel con-siderably more pressure than at present, and the influx will certainly change the present complexion of the "outdoors" imprint of these areas. Finally, the real potential for urban sprawl, the need to travel long distances, and the increase in population and the number of vehicles per capita will lead to the release of large amounts of pollution into the atmosphere. These institutional and organizational changes added to the ecosystem in the West suggest some significant environmental difficulties, especially with regard to environmental resources such as air, water and land use.

In short, the impacts of energy-related growth will not be in the nature of a boom/bust cycle, but will more closely approximate the growth and development of a underdeveloped country. There is likely to be some short-term problems, related to initial growth rate, but at present there is little anticipation of the probable longer-run picture.

## Agriculture

A noticeable portion of the synthetic fuels production goal is to be made up by gasohol and other agriculturally-derived fuels. In essence, these sources, trees, grains, or other farm-related activities, are competitors to the normal agri- and silva-culture based industries in place today. The land that is used to grow food may not be available to grow energy and vice-versa. This is not a problem if: (1) a food surplus exists; (2) surplus land is available to accommodate both food and energy production; (3) productivity of the present land base could be substantially increased; or (4) if the energy is produced from processed food residues or vice-versa.

The food surplus may or may not be present through time although there are still many crops which are presently produced in greater quantities than are presently required by our Nation. At the same time, it should be remembered that many of these food surpluses are shipped abroad to meet foreign needs and represent a significant element in our balance-of-trade problems. There is no formal land set aside by the Federal program at this time. The better soil types are likely all under production, and some skepticism exists as to whether any significant production capacity can be expected from cultivation of unused agricultural lands.

Finally, productivity is not only a function of original soil productivity, farming method, and crop type, but of the amounts of fertilizer, herbicides, and insecticides applied. Current data indicates a leveling-off of agricultural productivity. One of the reasons is that the chemicals applied to the soil are petrochemical in nature. The higher cost of petroleum and other energy sources has pushed up the chemical feed-stock costs to such levels that their use becomes uneconomical. There is some question as to whether these higher costs can be borne by escalating energy uses better than the food-related ones. In any case, there is clearly some reason for concern about the inflation potential of introducing a challenge-response mechanism (energy-food) into this portion of our Nation's economic system.

## SUMMARY

The issues addressed in this paper are by no means exhaustive, either in terms of the potential impacts of a synthetic fuels industry on the environment or in terms of the depth with which the ones we have addressed are covered. The three we discussed are, to the extent they are judged to be real potential problems, "grist" for full-scale research projects. Until that time, although there are no obvious immediately serious environmental problems that are not able to be overcome by application of known technology, there may still be surprises, and there will obviously be difficulties in the areas of implementation and enforcement. If there are to be environmental issues in the synfuels arena, they are apt to be addressed and readily resolved.

## Reference

Synthetic Fuels and the Environment: An Environmental and Regulatory Impacts Analysis, DOE/EV-0087, U.S. Department of Energy, Washington, DC, May 1980.

## SYNFUEL DEVELOPMENT IN THE WEST:
## NATIONAL VERSUS LOCAL PERSPECTIVE

### Douglas C. Larson
Western Interstate Energy Board

I am Douglas Larson, Executive Director of the Western Interstate Energy
Board.    The Energy Board is an organization of 16 Western state
governments. That's state governments, not the federal government.

In its three years of existence, the Board has established the western
regional solar energy center, coordinated state actions during the 1979
fuel shortage, and developed a program to solve the West's low level
radioactive waste disposal problems.   However, the major emphasis of the
Board has been in resolving on-the-ground energy resource development
issues, including the development with the Interior Department of a new
federal coal leasing program, the establishment of a program for the
joint state-federal review of major energy transmission corridors in the
West and the resolution of coal mine reclamation issues.

It is from this on-the-ground vantage point that I would like to speak
about the difference in perspective regarding synthetic fuel development
in the West--differences between the national perspective as exemplified
by the actions of the federal government and the local or regional
perspective as exemplified by Western state governments.

Before I delve into the differences, there is one overriding point I
want to leave you with.   That is, the differences in perspective do not
preclude agreement among the parties on specific synthetic fuel plants
or synthetic fuel strategies.

National Perspective

Prior to November 4, the national or federal government perspective on
synthetic fuels development was extraordinarily clear in   comparison
with the uncertainty of prior years.  With the enactment of the Energy
Security Act, the Congress and the Carter Administration spoke
very clearly and made a massive commitment to subsidize the production
of synthetic fuels.

This commitment is in stark contrast to the earlier synthetic fuel
initiatives of the Ford Administration, which were killed in Congress in

1975 and 1976 by a unique coalition of environmentalists who were concerned about the impact of synthetic fuel development on the environment and conservatives who believed that the subsidy of synthetic fuel development is not the proper role of government. In 1979, that coalition was eroded by the gasoline lines in California and the East Coast and the subsequent overwhelming political need for Congress and the Administration to do something, anything to show the government was acting to get the country out of its supply shortage. While opposition to government subsidies of synfuels has eroded, it still exists. Indeed, with the conservative sweep led by President-Elect Reagan, it is unclear how the government's synfuel initiative begun in 1979 will fare in the next few years.

However, for the purpose of a comparison of national and local perspectives, I will assume that the federal government remains on the course set by the Energy Security Act.

What exactly is that course? First, the government is willing to spend upwards of $88 billion to subsidize the production of synthetic fuel from basically non-renewable sources. Furthermore, the initial push will be to produce middle distillate fuel for use by the military. However, Congress has mandated that the subsidies be used to help develop production facilities which utilize diverse technologies. And, indeed, the first project to receive any significant money is the Great Plains Coal Gasification Project in North Dakota.

We also know that the federal government's preference for the type of subsidies is purchase agreements or loan or price guarantees. If that fails, loans and outright government ownership or partnerships are a last resort. We also know that Congress wants a geographic distribution of these facilities.

Western States' Perspective

There is general agreement among the major coal and oil shale producing states that the nation needs to develop several different types of synthetic fuel facilities which will definitively determine the economic and technical feasibility of the technologies. There is recognition that the development of a synthetic fuels industry will be enormously expensive and will not produce significant quantities of fuel for more than a decade and that stringent conservation efforts should remain a top priority for the immediate future.

There is also agreement among the states that synthetic fuel facilities must meet environmental standards and that social and economic impacts from these developments must be mitigated.

Finally, there is a strong belief that synthetic fuel development should further the states' economic development desires--desires which vary from state to state.

Constraints

Let me take a few minutes to discuss the major constraints or potential constraints to synthetic fuel development in the West. It is over these constraints that the state and federal interests are most likely to conflict.

Air Regulation:    Contrary to the common wisdom, it is my belief that

357

air quality, not water availability, will be the limiting constraint on synthetic fuel development as well as all energy development in the major Western coal and oil shale states. For example, there have already been major emission cutbacks in current projects under construction in the coal producing states of Montana and North Dakota because of Class I increment limitations. In Western North Dakota, where the only commercial scale synfuel plant in this country is under construction, the $SO_2$ increment in Class I Teddy Roosevelt National Park has been entirely used. Barring a change in federal law or a modification of modeling techniques, no future major $SO_2$ emitting source can be constructed in that region unless there are offsetting reductions in current $SO_2$ emissions.

In Montana, the Northern Cheyenne have successfully petitioned the EPA for a reclassification of their reservation to Class I. As a result, a coal-fired powerplant in the area, Colstrip Units 3 and 4, was required to substantially upgrade its pollution control equipment in order to avoid violating the new Class I increments. Because of meteorological differences from one site to another, it is not clear how wide a geography is affected by the Colstrip powerplant and the available increment on the reservation.

Other parts of the West are also feeling the constraints from PSD and related visibility regulation. In the Powder River Basin in Wyoming, there are growing indications that the existing particulate increments have been entirely used by existing and permitted mines in the area. In Utah, coal mining and conversion in the southern part of the state has already been precluded by Class I designations in the numerous national parks and monuments in the area. Further north in Utah, in the oil shale and tar sands resource regions there are concerns about potential Class I limits as well as Class II constraints. In Colorado, government agencies are likely to be facing in the very near future critical decisions on which oil shale projects get air permits and which are left out because of an insufficient increment to accommodate all development plans.

What is interesting about these constraints is that in the immediate future they will typically not result in conflicts between a pro-development federal government and state governments charged with enforcing environmental laws. In most of these states, the EPA retains the responsibility for enforcing PSD rules, as the states have not yet assumed that responsibility. Thus we may be witnessing one division of the federal government fighting with another division. This, of course, assumes that the Congress does not radically alter the Clean Air Act Amendments in the next legislative session.

Water: Water availability is more likely to be an area of conflict between the feds and the states because the states generally exercise greater control over water supplies than over air regulation. However, the question of water availability is not going to be as widespread an area of conflict as air. In many potential synfuel producing areas water is available, at least for a first round of synfuel plants.

Indeed, some states in the West, such as Colorado, exercise very little control over the use of water supplies and thus water supplies will flow toward energy development because energy development can always outbid current agriculture uses. Other states, such as Utah and Montana have some state government check against the unsupervised transfer of water from one use to another use.

This is not to say that developers will have any easy time securing the necessary water. In some areas, such as the Powder River Basin in Wyoming, there just isn't sufficient water to support major coal conversion facilities. Indeed, the nation's only air cooled coal-fired powerplant is located just outside of Gillitte.

Additionally, there is a general consensus among the energy producing states that the diversion of water to energy development must not eliminate the current agricultural base in these states.

Socio-Economic Impacts: Depending on the responses of the federal government, the states and the developer in specific locations, the potential conflict between the feds and the states could come to a head first in the area of socio-economic impacts. Over the past five years, progress has been made in providing the necessary financial resources to impacted local governments by developers, states (e.g. severance taxes, impact funds, etc.) and to some extent the federal government (e.g. increase in the states' share of mineral leasing revenue). However the West is just beginning on the steep portion of the curve of accelerated energy development. Whether these measures will be sufficient is an open question.

As well, there are shortsighted forces at work in Eastern and Midwestern states to dismantle the cornerstone of the impact mitigation strategy-- state severance taxes. Should this primary impact mitigation tool of the states be dismantled, we are headed toward substantial conflict between the industry and a pro-development federal government and state governments rendered unable to cope with resultant socio-economic impacts. There will be no winner in this fight because the industry will find it is unable to keep a competent work force under poor living conditions and the energy consuming states and the federal government will find that development will be stymied by industry's problems and unwilling host states.

Projected Synthetic Fuel Development in the West

The number of firms expressing an interest in developing synthetic fuels in the West is rapidly expanding as a result of the new federal government subsidies for feasibility studies, cooperative agreements, price and loan guarantees, etc. It is difficult to keep an exact count of all the interested parties and even more difficult to determine how serious some of the interested parties are.

Given these caveats, it appears that there are approximately 20 commercial oil shale projects proposed, 15 located in Colorado and 5 in Utah. No commercial shale projects are currently proposed in the other shale resource areas of the West--Wyoming and California. Tar sands projects are limited to 3 in Utah. There are 24 coal liquid or gasification plants proposed--6 in Wyoming; 7 in Montana, 1 in Colorado; 1 in Utah; 1 in Alaska; 2 in California; 3 in North Dakota and 3 in New Mexico. In addition, the nation's first commercial coal gasification plant is under construction in North Dakota. Again, it is difficult to determine how many of these projects will actually be built and how many more may be proposed. Nonetheless, the impact from this number of facilities is substantial.

Other Energy Development in the Same Areas

It must be remembered that with the exception of California and

Alaska, all of these proposed synthetic fuel projects are proposed in the same areas which are currently experiencing a boom in the development of other energy resources--coal mining, coal-fired powerplants, uranium development, and in some cases booming conventional oil and gas development. Let me briefly describe some of this development in these six states. In Western North Dakota, where all of the state's synthetic fuel facilities are planned, we are also expecting a jump in coal production from 15 million tons in 1979 to as much as 72 million tons in 1990. As well, the Williston Basin located in the same area, is probably the hottest oil drilling area in the nation. An additional four coal-fired powerplants are also expected. In Montana, coal production is projected to soar from 32 million tons in 1979 to upwards of 130 million tons in 1990. In Wyoming, we have the largest expected increase in coal production of any state in the Union, from 72 million tons in 1979 to 175 million tons in 1990. At least two new powerplants, major uranium development, and several coal slurry pipelines are expected. Wyoming is also the center of the biggest new conventional oil and gas find in the lower 48 states in many years--the Overthrust Belt.

In northwestern Colorado, where the shale development is expected, we are also expecting a jump in coal production from 13 million tons in 1979 to 38 million tons in 1990. Additional coal-fired powerplants are likely to supply the enormous amount of electricity needed by coal mining and the synthetic fuel facilities.

Utah is also experiencing the same conventional oil and gas boom from the Overthrust Belt as is Wyoming. In addition, construction of the world's largest coal-fired powerplant (Intermountain Power Project) will begin shortly. The state is expected to see a jump in coal production from 12 million tons last year to 44 million tons in 1990. Some of this coal may be used to fire additional powerplants planned for the state.

Finally, in New Mexico the coal resource is again co-located with much of the conventional oil and gas development as well as uranium development. Coal production is forecasted to increase from 16 million tons in 1979 to 58 million tons in 1990. Two new powerplants are expected. While the uranium market is currently depressed, a rebound in the market could strain local facilities.

Add to this, the potential development of the massive MX missile system in the same general vicinity and you have a significant strain on environmental, economic and social resources in the areas where synthetic fuel development is expected.

Synthetic fuels will have to compete with other energy development and military projects for scarce resources. At the state and national levels hard decisions may have to be made on which of these types of developments have priority.

Innovative Means of Avoiding Potential Conflict

By this time, you have no doubt gathered that there is abundant potential for conflicts over synthetic fuel development--conflicts between environmental groups and developers; conflicts between states and the federal government; and conflicts among federal agencies. If as a nation we proceed on the course which we have too often followed--a course which has each party, the energy company, the state, the EPA,

DOE, environmental groups, etc., pursuing their objectives until we wind up in court--then I can guarantee a resurgence of suspicion between the states and the federal government and perhaps an unprecedented level of conflict.

Fortunately, in the recent history of energy development in the West we have several very hopeful models to follow to avoid that bleak scenario. These models of cooperation are still the exception, but they are appearing more frequently. Let me spend several minutes outlining some of these success stories because I believe they hold the key to both the federal government and the states meeting their objectives in synthetic fuels as well as conventional production.

<u>Intermountain Power Project:</u> In 1977 a consortium of utilities and REA's proposed development of the world's largest coal-fired powerplant next to Capitol Reef National Park in Utah. A waiver of air regulations would have been needed to permit operation. Interior Secretary Andrus refused to grant the needed waiver, but instead of ending DOI's actions there, the Secretary joined with the State of Utah and local governments in Utah to locate alternate sites for the project. Two other sites were found which did not require a waiver of law and subsequently the project sponsor chose one of the alternate sites and expeditiously received all necessary federal permits to construct at the alternative site.

<u>Federal Coal Leasing Program:</u> To gain a full understanding of the import of the cooperative effort on federal coal mangement program, you need to understand some of the history of federal coal leasing in the West. First, the federal government owns outright 60 percent of the coal in the West and through interspersed land ownership patterns controls approximately 80 percent of the reserves. Second, in the past two decades federal coal leasing has been characterized by vast speculation followed by a leasing moratorium in the early 1970's, two false starts in developing new leasing programs and debilitating lawsuits resulting in essentially no leasing for 10 years. Finally, there has historically been no love lost between the federal land mangement agencies and the states in the West.

In spite of these less than favorable conditions for the orderly development of a resource, the states and the Interior Department have been successful in designing and now implementing a new federal coal management program which meets the needs of both parties. The unexpected success in forging a partnership betwen the states and the federal government on coal leasing is the product of the following conditions.

(1) For one of the few times involving a major energy program, a federal agency has fully recognized that the affected states are not just another interest group, but in fact, must be partners in the program if it is to succeed. (There is a lesson here for the Energy Department and the Synthetic Fuels Corporation.)

(2) Both the Interior Department and the states recognize that each party has primary interests and responsibilities. For example, the states (often by default) have major responsibilities in coping with the social and economic consequences of coal development while the federal government has responsibility for insuring as fair return to the federal treasury for federal coal.

(3) The coal state governors and Secretary Andrus placed a high priority on development of a workable federal coal program and provided the necessary staff support through the Office of Coal Leasing in Interior and the Coal Committee of the Western Interstate Energy Board.

(4) All parties recognized that this was essentially the last chance to develop a workable coal program. If this program failed, Congress would probably resolve the issue itself in a hasty manner which would be unacceptable to all parties.

The heart of the new program is the establishment of state-federal regional coal teams. A regional coal team has been established for each of the five major coal basins in the West. The coal team consists of an appointee of the governor of each state in a basin, the affected state BLM directors, and a team chairman appointed by BLM Director Gregg. While the teams are advisory in nature, they have become the forum for resolution of nearly all coal leasing issues early in the process thereby avoiding the costly delays operators experience when different levels of government disagree on whether or under what conditions an area should be developed. Our experience thus far indicates that the teams' recommendations have in fact become Interior Department policy.

This unprecedented departure from the past actions by the federal government on coal leasing holds substantial promise for resolving exactly the type of conflicts we are likely to see in synthetic fuels development. Indeed, we are now in the process of expanding the program to oil shale leasing.

Additional Innovations: The Colorado Department of Natural Resources has devised a new program to promote the cooperative review of a proposed project by the multitude of state, federal, and local government permitting agencies. It is called the Joint Review Process. There are currently three major resource development projects participating in this joint review process. While the participation of the company and the government agencies is purely voluntary and none of the three projects has completed the review process, it is hoped that the process can institutionalize the type of cooperation experienced in projects such as IPP.

In another new development, the Energy Board is expanding the federal-state cooperation in the coal leasing program to energy corridors across public lands and facility siting. The Board has signed a cooperative agreement with the Bureau of Land Management and the Forest Service to further this objective. While this effort is in its infancy, it is hoped it will (1) result in greater state input into BLM and Forest Service decisions on the location of energy facilities on federal land, including interstate pipelines and transmission lines and (2) provide a forum for state and federal regulators to coordinate their actions on specific projects.

In addition to these government actions, there are growing examples of the industry fulfilling its responsibilities to mitigate the impacts from its development. In the Overthrust Belt, the major oil companies have formed the Overthrust Belt Industrial Assocation for the express purpose of helping impacted communities. Either through voluntary actions, or as a result of pressure from state siting authorities, the industry has also helped meet socio-economic impacts at the Laramie

River coal-fired powerplant and developed or proposed the development of new communities in Utah, Wyoming, and Colorado. This increased level of social responsibility by developers is also in their self-interest, as was shown by the delays and increased costs at the Jim Bridger powerplant in Wyoming where poor living conditions led to rapid turnover in the workforce and subsequent delays.

Conclusion

In conclusion, the different perspectives on synthetic fuel development within the federal government and between the federal government and the producing states have the potential for debilitating conflict with no party--not the federal government, nor the states, nor the industry, environmental groups, or local citizens--realizing their objectives. No question about it, the major push to develop synthetic fuels will further strain the already strained capability of the energy producing West.

My message today is that it will take a level of cooperation among all parties that we have rarely experienced in the past for the development of a viable and acceptable synthetic fuels industry. The cooperative efforts I mentioned earlier, must become the rule, not the exception.

That is not a small order. It will require restraint and common sense. Above all it will require a recognition that there is room for compromise, that we in the West are not facing the black and white decision where someone is the clear winner and someone else is the clear loser. There is room for creative solutions where all parties can gain, including the synthetic fuels industry.

## SELECTING SITES FOR SYNFUELS PLANTS

**Charles A. D'Ambra** and **Gaylord M. Northup**
The Center for the Environment and Man, Inc.

## 1.0 Introduction

With enactment of the Energy Security Act of 1980, the U.S. has taken major steps toward establishment of a viable synthetic fuels industry, as a means of greater energy independence. The Act establishes synthetic fuel production goals equivalent to 500,000 barrels of oil per day (bbl/day) by 1987 and 2,000,000 bbl/day by 1992. To put these numbers in perspective, the latter goal constitutes 29% of U.S. oil imports during the first eight months of 1980, 11% of all U.S. petroleum consumption, and 5% of the anticipated total U.S. demand for energy in 1980.[1] It will take 20 to 45 commercial scale facilities -- each producing 50,000 to 100,000 bbl/day of oil or 250 to 500 million standard cubic feet per day of high Btu gas -- to produce the energy equivalent of 2 million bbl/day of oil.

To achieve these production goals, commitments for and construction of synfuels plants must be expedited, and the Energy Security Act does provide strong financial incentives. However, a number of existing environmental statutes and implementing regulations will have a direct impact on how, where and when synfuels are produced. Pertinent Federal statutes include:

- National Environmental Policy Act (NEPA)
- Clean Air Act
- Clean Water Act
- Safe Drinking Water Act
- Resource Conservation and Recovery Act (RCRA)
- Toxic Substances Control Act (TSCA)
- Endangered Species Act
- Fish and Wildlife Coordination Act
- Surface Mining Control and Reclamation Act (SMCRA)
- Occupational Safety and Health Act (OSHA)
- Historic Sites Act.

Additional laws are applicable if federal resources are involved, and state and local requirements will also affect synfuels development. The major implications for organizations developing a synthetic fuel project include:

- Environmental Impact Statement
- Emission limits
- Performance standards
- Numerous permits
- Elimination of some resources and sites
- The possibility of litigation by groups opposed to the project.

## 2.0 Environmental and Socioeconomic Impacts and Site Selection

### 2.1 Site Selection as a Method to Expedite Project Implementation

While federal incentives and regulations applicable to synfuels development might be altered by the incoming Administration and Congress, it will nonetheless be prudent to go beyond standard engineering and economic criteria for siting synfuels plants by considering environmental and socioeconomic impacts. Aside from altruism, the motive for considering these impacts during the site selection process is to minimize objections to the project which could lead to expensive delays, denial of permits, litigation, and/or poor public image. Since these factors can directly affect the viability and profitability of a synfuel project, it makes good business sense to consider potential environmental and socioeconomic impacts before making a final decision on the plant location. If major impacts are foreseen, it may be in the developers' best interest to choose another location.

### 2.2 Potential Socioeconomic and Environmental Impacts

Employment at commercial-scale synthetic fuel plants using coal, oil shale or tar sands as the energy feedstock will be typically in the range of 700-1000 persons during the operational phase and as high as 5000 or more during periods of peak construction activity. The fraction of these positions filled by people already residing in the area will depend primarily on the size and skills available in the local labor force. Many positions will be filled by persons who do not presently reside in the area. Furthermore, the synfuels plant will create local opportunities for business expansion and establishment of new businesses. New workers and their dependents will increase the demand for a host of public and private services, thereby increasing job opportunities and the incentive for outsiders to open businesses or seek employment in the area. When these effects are aggregated, about 7.5 new residents can be expected for each employee at an operating synfuels plant who does not presently reside within commuting distance. This ratio will probably be somewhat smaller for the larger, temporary construction labor force.[2]

If the population increases rapidly -- generally, more than 5% per year-- "boom town" impacts could occur.[3] Even without extremely rapid growth, socioeconomic impacts could occur on a moderate to major level. Some or all of the following conditions might be experienced:

- Inadequate services, shopping facilities and entertainment.
- Housing shortage and price escalation.

- Declining industrial productivity and profitability due to high labor turnover and low employee morale.
- Inflation above national average.
- Higher tax rates, particularly in surrounding communities and counties which do not receive tax revenue from the major new facility.
- Poor traffic flow.
- Limitations on capital development for expanded services in nearby municipalities and counties (many states have statutory bonded debt limits).
- Transfer of control away from long-time residents, if they become a minority of a town's population.
- Increased rates of alcoholism and crime.

Loss of agricultural and undeveloped land may also be significant in areas which do not presently have large amounts of such areas.

Construction and operation of synthetic fuel plants may significantly degrade air quality, and Federal New Source Performance Standards (NSPS), Prevention of Significant Deterioration (PSD) standards, and visibility impairment regulations will be applicable. The quality and quantity of water resources will frequently be of concern on the local, regional, state and national levels. Public exposure to toxic substances from solid waste leachates, accidental releases from process streams, and possibly during product utilization may also be a concern.

Several endangered and threatened species are protected by law, and special interest groups may oppose projects which could harm other species or rare or important ecosystems. Objections will be raised if the proposed facility is perceived to have an unacceptable aesthetic impact due to visual, aural or olfactory characteristics. Finally, if sites with historical, scientific (e.g., archaeological) or special local significance (some of which are protected by laws and regulations) could be damaged, it is reasonable to expect that efforts will be made to block the project.

## 2.3 Beneficial Impacts

The choice of sites can also affect the magnitude and significance of beneficial impacts. Local unemployment will decline and many residents will acquire more desirable employment. A large inflow of dollars into the local economy can be expected. Tax revenues may increase, particularly in the town and/or county which will have its property tax base substantially increased by the large investment in the synfuel plant. The nation will be able to decrease its reliance on imported energy, thus helping to improve the balance of trade and enhancing national security.

## 3.0 General Siting Guidelines

Given the number of potential problem areas, few, if any, sites are likely to satisfy all engineering, economic and environmental ideals. However, the following general guidelines for avoiding major adverse environmental and socioeconomic impacts may be useful:

- Environmental and socioeconomic factors should be addressed before a final site is selected, perhaps when a small number of locations have been identified as acceptable on the basis

366

of engineering and economic criteria. This will facilitate identification of potential adverse impacts and obstacles to efficient project implementation. When evaluating potential socioeconomic impacts, other major commitments for development in the area should be considered, since conclusions on labor availability and the rate of growth may be altered.

- Desirable socioeconomic conditions include:
  - Adequate existing labor force for a high fraction of employment requirements, in order to avoid excessive growth rates.
  - Well established services (e.g. water and wastewater treatment capacity and health services) sufficient for an expanded population.
  - Towns and counties near the site that will not be experiencing more than moderate growth when construction of the synfuels project begins.
- Desirable environmental factors include:
  - Adequate to abundant water resources.
  - Absence of potential problems with endangered and threatened species.
  - Criteria air pollutant concentrations that are below standards, or relatively easy and inexpensive emissions offsets can be obtained.
  - Absence of detrimental impact on important sites.

## 4.0  SELECS Methodology:  An Impact Evaluation and Site Screening Tool

### 4.1  Purpose and Organization of SELECS

While the aforementioned general guidelines may provide some preliminary guidance, sophisticated techniques are needed to evaluate the complex relationships between synfuels plants and environmental and socioeconomic systems. Under sponsorship of the U.S. DOE Pittsburgh Energy Technology Center (PETC), The Center for the Environment and Man has developed the Site Evaluation for Energy Conversion Systems methodology (SELECS).[4,5] SELECS, which is specifically tailored to synfuels development, integrates a number of impact evaluation algorithms and models into a systematic, quantitative framework. The purpose of SELECS is not only to protect environmental and socioeconomic resources, but also to foresee and avoid problems that could delay or block synfuels projects.

Realizing both the advantages of an objective evaluation and that the significance of potential impacts requires subjective judgments about their importance, the SELECS framework defers subjective decisions until the probable impacts have been evaluated by objective methods. For example, the population increase of a given town or county is estimated objectively, then a relative measure of the significance of this impact is applied. (Subjective judgments are applied by use of Relative Importance Curves and Impact Weights; the mechanics of this procedure are discussed in References 4 and 5.) Different measures of impact significance can be applied and alternative results can then be discussed in a relatively objective manner. The analysis produces impact evaluation scores in several categories which highlight the most significant impacts and facilitate identification of useful ameliorating actions.

Since the initial stages of development three years ago, SELECS has been conceptualized in three levels of increasing sophistication:

367

- Level 1: a relatively simple technique which any interested party can apply to evaluate impacts during the operational phase of a synfuels plant. Because of this emphasis on workability, it was contractually required that Level 1 SELECS contain computations which can be performed with a simple calculator within one working day (data collection and analysis of the results bring total time requirements to two to three staff-weeks). A User's Manual for Level 1 SELECS was published in 1980.[5]

- Level 2: a more comprehensive, computer automated technique suitable for analyzing several site/process alternatives and for sensitivity analysis. Impacts during both the construction and operational phases would be addressed. Level 2 SELECS has been partially developed during the most recent phase of work for DOE/PETC in which the impacts of two commercial-scale coal liquefaction plants at six potential sites in Kentucky were evaluated.[6]

- Level 3: a sophisticated methodology suitable for rigorous analysis of finalist alternatives. It would incorporate advanced simulation models for air quality, water quality, employment, and socioeconomic impacts. In a Level 3 SELECS analysis a substantial portion of the analysis required for an environmental impact statement would be performed.

## 4.2  Evaluation of Impacts

In the most recent applications of SELECS (discussed below) 31 Impact Elements (measures of impact) distributed among 11 Categories in SELECS were utilized. The Impact Elements, which are listed in Table 1, can be supplemented or replaced within the SELECS framework by other measures. A Relative Importance, on a scale of 0 (no impact) to +1 (maximum adverse impact), is determined for each Impact Element. Beneficial impacts are determined on a scale of 0 to -1. Impact Elements within each Category are weighted and Category Scores, which also fall between 0 and 1, are computed. As a result, quantitative measures of probable impact and perceived significance are produced in each Category.

## 4.3  Applications of SELECS

During development of the Level 1 SELECS methodology, nine scenarios of energy conversion processes at various sites across the nation were postulated. These hypothetical examples, listed in Table 2, served to demonstrate the methodology using real data.

Recently, several of the impact evaluation algorithms in Level 1 were refined and the methodology was computerized and applied to the potential implementation of commercial-scale SRC 1 and Coal to Methanol to Gasoline facilities at each of six potential sites in Kentucky:

- Daviess County (two sites)
- Union County
- Muhlenberg County
- Marshall County
- Mason County.

For comparison, a coal-fired steam-electric power plant consuming a comparable amount of coal and a non-energy intensive manufacturing plant with employment requirements comparable to the synfuels plants were also evaluated at each site. A comparison of the site/process alternatives will be the subject of a future paper. To illustrate results from a SELECS analysis, Category Scores for one of the 24 site/process alternatives are shown in Figure 1. (Since most of the bars for the most favorable alternatives are very short, one of the less favorable alternatives has been chosen to illustrate the results.)

5.0 Conclusions

Several environmental statutes are applicable to synthetic fuels development in the U.S., and most of these will probably remain as the synfuels industry emerges.

Although the possibility of major adverse environmental impacts from synthetic fuel facilities exists, synfuels development and reasonable standards of environmental protection need not be mutually exclusive. Adverse impacts can often be ameliorated and beneficial impacts can be enhanced. Comparative site evaluation can identify locations for synfuels plants where impacts will generally be considered acceptable, thereby reducing delays and expenditures required for project implementation. In short, synfuels development can be expedited by responsible decision making.

The SELECS methodology is a logical, systematic framework for site screening and impact analysis. It also facilitates identification of measures which reduce adverse impacts. SELECS has been developed to an extent suitable for preliminary site screening and impact analysis. Further development to an advanced, comprehensive methodology suitable for any major synfuel project is planned.

6.0 References

1.  "Energy Indicators." Energy Magazine. Vol. 5: No. 4, Fall 1980.

2.  Synfuels Interagency Task Force. Synthetic Fuels Commercialization Program, Washington, DC. December 1975.

3.  Gilmore, John S. and Mary K. Duff. Boom Town Growth Management, Boulder, CO, Westview Press, 1975.

4.  Northrop, G.M., C.A. D'Ambra, and R.L. Scott. "SELECS: A Workable Methodology for Evaluating Socioeconomic and Environmental Impacts of Energy Conversion Facilities at Potential Sites," Proceedings of the Sixth Energy Technology Conference, Washington, DC, Government Institutes, February 1979.

5.  Northrop, G.M., C.A. D'Ambra, and R.L. Scott. Environmentally-Acceptable Fossil Energy Site Evaluation and Selection: Methodology and User's Guide (CEM Report 4231-625), Hartford, CT, The Center for the Environment and Man, Inc., February 1980.

6.  Northrop, G.M. and C.A. D'Ambra. Application of the SELECS Methodology to Evaluate Socioeconomic and Environmental Impacts of Commercial-Scale Coal Liquefaction Plants at Six Potential Sites in Kentucky (CEM Report 4231-704), Hartford, CT, The Center for the Environment and Man, Inc., November 1980.

TABLE 2. HYPOTHETICAL ENERGY CONVERSION SYSTEMS EVALUATED
WITH THE LEVEL 1 SELECS METHODOLOGY

| No. | Location | Technology |
|-----|----------|------------|
| 1 | Pittsburgh, PA | Low Btu Coal Gasification |
| 2 | Charleston, WV | Coal Liquefaction |
| 3 | Jackson, OH | In Situ Coal Gasification |
| 4 | Kansas City, MO | Coal-Oil Mixture Power Generation |
| 5 | Bloomington, IL | Magnetohydrodynamic Power Generation |
| 6 | Vernal, UT | Tar Sands Conversion |
| 7 | Bismarck, ND | Lignite Liquefaction |
| 8 | Farmington, NM | High Btu Coal Gasification |
| 9 | Rawlins, WY | Oil Shale Conversion |

TABLE 1. IMPACT ELEMENTS USED IN SELECS METHODOLOGY

| No. | Category Title | Impact Elements | Application* |
|-----|----------------|-----------------|--------------|
| 1 | Employment & Population | Percent Population Increase | M,C |
| | | Percent Change in Unemployment | R |
| 2 | Services | Percent Increase in Police Officers | M,C |
| | | Percent Increase in Fire Fighters | M,C |
| | | Percent Increase in Doctors | M,C |
| | | Percent Increase in Hospital Beds | M,C |
| | | Percent Increase in Dentists | M,C |
| | | Percent Increase in Teachers | M,C |
| | | Percent of Water Supply Capacity Margin | M,C |
| | | Percent of Wastewater Treatment Capacity Margin | M,C |
| | | Percent Reduction in Life of Solid Waste Disposal Facility | M,C |
| | | Percent Increase in Park & Recreational Area | M,C |
| 3 | Bonding Capacity | Percent of Bonded Debt Margin | M,C |
| 4 | Housing | Percent Increase in Housing Units | M,C |
| 5 | Land Use | Percent Reduction of Undeveloped Municipal Land | M |
| | | Percent Reduction of Undeveloped County Land | C |
| | | Percent Reduction of Prime Farmland | C |
| 6 | Water Availability | Percent of Average Surface Water Flow | R |
| | | Percent of 7-Day, 10-Year Low Flow | R |
| 7 | Air Quality | Percent of Annual Particulate PSD Standard | R |
| | | Percent of 24-Hour Particulate PSD Standard | R |
| | | Percent of Annual $SO_2$ PSD Standard | R |
| | | Percent of 24-Hour $SO_2$ PSD Standard | R |
| | | Percent of $NO_x$ Annual Concentration Margin | R |
| 8 | Water Quality | Percent Reduction of Average Flow Assimilation Capacity | R |
| | | Percent Reduction of 7-Day, 10-Year Low Flow Assimilation Capacity | R |
| | | Percent of Allowable Water Pollutant Loads | R |
| 9 | Endangered Species | Species Habitat Proximity | R |
| 10 | Aesthetics & Important Sites | Visual Prominence | R |
| | | Important Site Proximity | R |
| 11 | Tax Rates | Percent Change in Average Tax Rate | M,C |

*M = Determined for Each Municipality
C = Determined for Each County
R = Determined for Overall Region

370

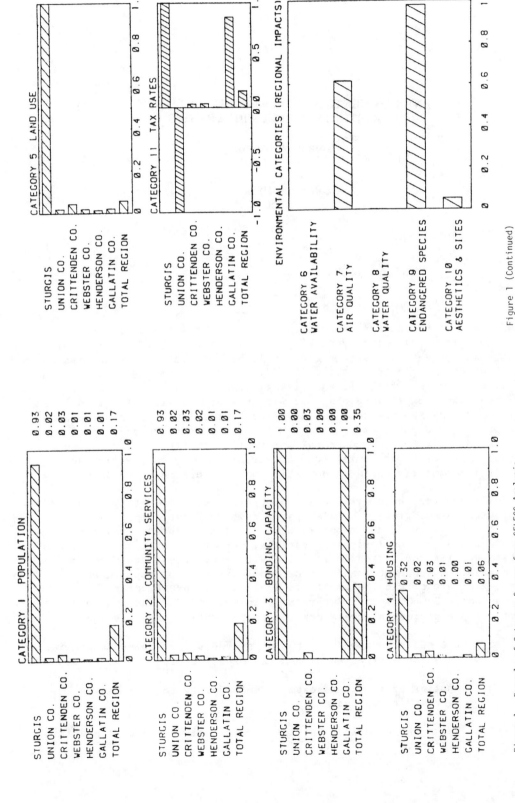

Figure 1 (Continued)

Figure 1. Example of Category Scores from SELECS Analysis.

371

# ENVIRONMENTAL DEVELOPMENT CAPACITY
# FOR SITING SYNFUELS INDUSTRIES

**Robert V. Steele**
SRI International

**F. Jerome Hinkle**
U.S. Department of Energy

## 1.0  Introduction

As Title I of the Energy Security Act of 1980 is implemented, a
framework for evaluating synthetic fuels plant siting strategies and
public and private investment risks will be critical to the success of
the Synthetic Fuels Corporation (SFC) in reaching its legislated goals.
Traditional market analyses and risk assessments may be adequate for
private investments alone, but the wide variety of financial assistance
offered by the SFC may lead to extensive cost and risk sharing between
the private and public sectors.  The Federal concern extends to
evaluating numerous environmental and public benefit issues arising from
a massive effort to encourage the construction of a completely new
industrial capacity based on domestic coal and oil shale resources.

To fully analyze the constraints and opportunities of various synfuel
deployment strategies, a comprehensive framework for comparing
alternatives is needed.  Along with market and financial analysis
(product slates, cost/prices, locations, market penetration and growth
rates) and transportation/logistics/distribution questions, the natural
resources, environmental and socioeconomic impacts of deployment
strategies are critical elements in any overall investment risk
assessment.  The Office of Environment in DOE and SRI International have
devised an experimental framework for comprehensive analysis that:

o    Creates a workable concept of synfuels development potential
     based on environmental and resource factors.

o    Applies an impact assessment methodology to four key
     geographic regions.

o    Ranks these regions (at the subregional level) on six critical
     factors.

o    Examines the intersection among all six factors to determine the overall development potential of subregional areas.

We suggest here how several planning and evaluation needs of a Federally-supported enterprise can be satisfied by the use of these tools. The Energy Security Act empowers the SFC to solicit* for feasibility studies, cooperative agreements, and financial assistance to encourage commercialization of a variety of synfuel technologies. The SFC also must prepare a comprehensive strategy to achieve the national synthetic fuel production goal and submit the proposed strategy to the Congress for approval:**

"The comprehensive strategy must set forth the recommendations of the Board on the goals of the United States Synthetic Fuels Corporation and schedules for their achievement. Such strategy shall include comprehensive reports on ...(a) the economic and technical feasibility of each facility including but not limited to information on product quality and cost per unit of production; (b) the environmental effects associated with each facility as well as projected environmental effects, and water requirements; and (c) recommendations concerning the specific mix of technologies and resource types to be supported...(and) shall directly address and give emphasis to private sector responsibilities in the efforts necessary to achieve such goal..."

Several important aspects of these functions could be met by an analysis framework proposed here:

o    Evaluate and compare proposals for specific site developments in different regions.

o    Provide guidance in evaluating feasibility studies and investment proposals.

o    Assist in preparing the terms of loan guarantees, purchase agreements, etc., to circumscribe the limits of risk to both the government and the developer.

o    Provide incentives to guide siting synfuels facilities in particularly desirable or overlooked areas.

o    Evaluate the growth-limiting aspects of resource depletions, where other developments (e.g., power plants) or multiple synfuels plants compete for water, coal, labor , PSD increments, etc.

o    Devise overall planning strategies by evaluating several alternatives that may optimize growth, yield lowest investment risk, "optimize" economic development opportunities and environmental quality, or prioritize the locations for particular sizes of incentives and public investments.

---

\* Title I, Part B, Subtitles D, E, and F.
\*\* Subtitle C, Section 126(b).

2.0 Method of Approach

2.1 Critical Factors

The critical factors used in developing the analytical framework were those derived in a previous impact assessments.1  The important components of each factor are discussed briefly below.

o  Air Quality--Of primary concern are Prevention of Significant Deterioration (PSD) areas, as defined in the Clean Air Act Amendments of 1977, and nonattainment areas.

o  Water Availability--Both physical factors (stream flows) and institutional factors (e.g., state allocation policies and Indian water rights) can affect the availability of water for synfuel production.  In addition, the quantity and quality of waters that are to receive plant discharges are other important components of water availability.

o  Socioeconomic Capacity--The major concern is whether communities are able to successfully absorb the rapid population growth associated with constructing and operating synfuel facilities.  Their ability to do so is reflected in such factors as their population size, infrastructure level, and growth history.

o  Ecological Sensitivity--Both the potential for reclamation of mined lands and the susceptibility of natural ecosystems to disturbance associated with large-scale industrial activity are important considerations.

o  Environmental Health--The potential for impairing human health in the work force and the population surrounding a synfuel plant is of concern.  The risk factors are still largely undefined because knowledge is lacking about the kinds and quantities of toxic materials to be released from actual synfuel plants.

o  Federal Land Management--This factor is principally of importance in the West where the federal government is a major land holder.  Traditional management practices on federal lands, as well as policies instituted under the Federal Land Policy and Management Act of 1976, may be in conflict with increased exploration and development of coal resources, as well as with the siting of synfuel facilities.

2.2 Assumptions

The analytical framework and the results derived from it depend on several critical assumptions:  (1) We did not consider factors that play a major role in siting decisions but that do not affect the environmental aspects of the problem--factors such as resource availability, transportation of resources and products, availability of markets, and choice of conversion technology;  (2) in the foreseeable future, environmental laws and regulations will not be relaxed;  (3) coal will not be transported long distances from the mine site.

374

The coal conversion plant sizes were chosen on the basis of current conceptual designs for commercial facilities and fall in the range of 40,000 to 60,000 barrels/day, oil equivalent. Oil shale plant sizes were based on announced plans by companies for commercial projects.

### 2.3 Selection of Study Areas

We chose study areas by looking for a sufficient concentration of resource to support at least one plant over its lifetime. To help define the areas that contain sufficient, accessible coal to support synfuels development, we used the SRI Coal Resources Model.[2] This model uses the USGS/USBM data base on coal resource characteristics, quantities, and locations (by county), and employs a sophisticated computes extraction cost curves for more than 10,000 coal seams in the coterminous United States.

The quantity of resource available at or below current coal prices was then used to define the study areas. The minimum quantity of resource required must support a single plant over a 30-year lifetime, or approximately 250 million tons.

Figures 1a and 1b show the study areas selected. Counties surrounded by or adjacent to coal resource counties are included in the study areas. For the sake of brevity, only the Four Corners/Rocky Mountain and Illinois Basin study areas are discussed here. However, study areas in the Northern Great Plains and Appalachia were analyzed also.[3]

### 2.4 Analytical Approach

Using the appropriate technology-specific and region-specific data, we analyzed the six constraining factors concurrently, seeking parameters that would tend to make the siting of synfuel facilities in a certain area more or less difficult than in other areas. Siting difficulty could result either from constraints imposed by the regulatory process or from the effort required to mitigate the impacts of development.

Once the relative degree of siting difficulty (or conversely, the relative ease of siting) had been determined for subareas within the study areas and for all six constraining factors, the results were combined in a composite picture showing relative ease of siting based on environmental considerations.

We used maps to display the results of all the analyses at a geographical scale that is appropriate for each constraining factor. A semiquantitative rating scale represented the degree of siting difficulty associated with each factor. The five rating categories for the relative ease of siting were: low, medium low, medium, medium high, and high; they indicate the potential for successfully siting synfuel plants in an area.

### 3.0 Air Quality

We performed the air quality analysis in two steps. First, we modeled air pollutant concentrations that will result from selected synfuel operations. We modeled the annual and worst case 24-hour average particulate (TSP), sulfur dioxide ($SO_2$) and nitrogen dioxide ($NO_2$) concentrations, and the 3-hour worst case $SO_2$ concentrations under meteorological conditions that might be encountered in the study areas. We also modeled peak 24-hour $SO_2$ concentrations under meteorological conditions that produce the highest downwind concentrations to determine how facilities might affect the air quality in a nearby Class I PSD region.

The second step of the analysis entailed examining current air quality regulations, discussing with state and federal air quality officials trends in state and federal attitudes regarding possible regulations, and documenting current legal challenges to the regulations. In particular, we examined current air quality, and the status of proposed and suggested Class I PSD regions.

The federal air quality legislation and state air quality regulations define the following categories within the study areas: Current and proposed Class I PSD areas and Class II PSD areas. The law specifically states how much the short- and long-term concentrations of $SO_2$ and particulates can increase because of industrial development. It will be extremely difficult to site synfuel plants in PSD Class I regions because emissions from the plants will generally result in concentrations that exceed the PSD increments for Class I regions. The CAAA also specifies Class I areas also must not be adversely affected by nearby development, and many areas with Class II designation are presently in violation of ambient air quality standards for one or more air pollutants. These factors lead to a clear overall ranking for the regions:

o    Class II PSD regions that are not in violation of air quality standards will be the easiest in which to site new facilities

o    Class I PSD regions will be the most difficult in which to site new facilities

Figure 2 shows the method for ranking the regions within the study areas. The results of the air quality ranking are shown in Figures 3a and 3b.


4.0  Water Availability

The availability of water resources depends not only on surface water flows, or physical water availability, but also on institutions and laws regulating the availability of water at the state or federal level and on the potential for water quality problems, primarily in the East. These three assessments combine to provide an overall ranking of water availability.

   4.1  Surface Water Flow

Our analysis of streamflows was based on several key assumptions:

o    Surface water will provide the primary water supply.

376

o    Some risk exists that surface water will not be available
     at all times.

o    Supplemental water supplies may be required.

Some risk exists that surface water will be unavailable for portions of
each year, and the more water demanded from a stream, the higher the
potential risk that the demand cannot be met.  The degree of this risk
depends on variability in precipitation and watershed characteristics
such as the size of the drainage basin, elevation, steepness of
topography, bedrock geology, and reservoir storage.  We selected
acceptable levels of risk based on flow duration curves.  These curves
represent historical streamflow at a particular stream gage, indicating
the percentages of time that certain flows took place in the stream.
Thus, the degree of risk is measured by the percentage of time that
upstream demands can be met.

Our method of approach consisted of four main steps:

1)  Obtaining Streamflow Records.  We selected 80 U.S. Geological Survey
(USGS) gaging stations whose drainage basins covered the entire area of
the four study areas and obtained the flow duration curve for each gage.

2)  Summing Water Demands.  We not only considered demands for
diversions of water, but also assessed demands for maintenance of fish
and wildlife habitat (i.e., "instream flow requirements").  For this
habitat in particular, the instantaneous flow of a stream is of prime
importance, and we chose to evaluate "withdrawal demands"--the amount of
water withdrawn from surface water to meet demands at any one instant.
We included water used for agriculture, industry, mining, thermal
electric generation, municipalities, and rural development, and water
lost by evaporation or export out of the drainage basin.  We projected
of competing water demands for 1990.

We formulated a hypothetical synfuel demand for each river basin by
assuming that one synfuel plant can be supplied by each coal resource
county within the basin in the West, or by groups of five coal resource
counties in the East.  In those basins that did not intercept coal
resource counties, we assumed one or two plants would be supplied,
depending on the size of the basin.  We based oil shale demands on
announced plans of companies that intend to use surface water supplies.

3)  Determining the Percent of Time the Demands Are Met.  We then
compared the estimated total upstream demand to the flow duration curve
to determine the percentage of time that the demand could be expected to
be met, given the historical streamflow pattern at the gage.

4)  In Marginal River Basins, Determining Whether a Stream Can Support
an Impoundment.  If demands could be met less than 60% of the time, we
determined whether the stream could support an impoundment.  The
impoundment had to meet three criteria:

o    Provision of water for synfuel facilities only

o    Release of water at a rate to meet the instream flow
     requirement of the gage

377

o    Size that could be feasibly constructed by an energy concern
      (50,000 acre-feet or less).

We ranked each river basin as high, medium, medium low, or low on the
basis of the percentage of time the estimated upstream demand can be
expected to be met (Table 1).

Table 1

CATEGORIES OF RANKING FOR SURFACE WATER FLOW

| | |
|---|---|
| High | Water demands met at least 80% of time. |
| Medium | Water demands met 60-80% of time. |
| Medium Low | Water demands met less than 60% of time, but streamflow can support impoundment of less than 50,000 acre-feet. |
| Low | Water demands met less than 60% of time. |

4.2  Institutional Factors

The institutional analysis was carried out on a state-by-state basis.
State water laws, water allocation policies, interstate compacts, and
general state attitudes toward water use were examined.

In the West, Colorado, New Mexico, South Dakota, and Utah received the
most favorable ratings for institutional water availability, with
Montana receiving the lowest rating.  To illustrate, Colorado has a
generally positive attitude toward developing its energy resources,
including a willingness to contribute to meeting national needs.
Moreover, Colorado's water laws will not particularly obstruct water use
for energy needs.  Transfers between basins and between uses are
permitted, and have successfully taken place.  Colorado does have some
interstate compact problems, however, in that the state contributes 73%
of the Colorado River's flow, but is allocated just 20% under the
relevant compacts.  Potential problems with Indian water rights may be
encountered in downstream claims that the Navajo Tribes may make over
and above those Indian rights already allocated by the U.S. Supreme
Court.  Federal reserved rights are also being adjudicated in Colorado,
but these may not prove to be injurious amounts.

Montana, by way of contrast, has indicated its resistance to having its
water used for meeting national energy needs.  Montana's statutes
reflect this position.  For example, the recent state allocations placed
industrial and energy use far down the preferred list for Yellowstone
River water.  Montana is a signatory of the Yellowstone River Compact,
which forbids transfers out of the Yellowstone River Basin without the
consent of all the signatory states--a significant institutional
constraint.  Montana is also the site of Indian tribes that have made
substantial claims to water flowing through the state, thus opening

water rights allocated under state law to question. Nonetheless, Montana does have several large federal reservoirs that might provide water for energy development, provided that several key issues are resolved.

It is virtually impossible to compare the humid eastern states with the semiarid western states. The institutional factors found in the West arise because of the meager precipitation found there. Although the East has not heretofore been subject to problems that characterize western water, evolving eastern issues are now reaching a point where comparisons are possible. In general, however, few distinctions can be made, except for a shift in water law to a permit system such as has occurred in Maryland and Kentucky; these distinctions may be augmented by consideration of general state policies.

### 4.3 Water Quality

We employed two basic assumptions in our assessment:

o    No discharge from synfuel facilities in the West.

o    Dilution effect in the receiving waters.

As most recent studies of water quality have done, we assumed that synfuel plants in the West will discharge no effluents. The cost of cleaning effluents sufficiently for discharge to receiving waters is much higher, given the limited nature of that resource in the West, than is extensive recycling of water with final discharge to evaporation ponds. Therefore, we have evaluated only the water quality effects of discharge from synfuel plants in the East.

We evaluated the effect on stream water quality by assuming that at least one synfuel plant was located in each river basin. We concentrated on evaluating the dilution effect of the discharge, and did not evaluate chemical kinetics in the receiving waters.

We assumed that each synfuel plant would be required to meet effluent standards established by the EPA that are similar to those established for refineries and steam electric power plants. On the basis of an analysis of those standards and assumptions about best available technology for synfuel plants, we estimated effluent rates for selected pollutants from a standard size plant. We restricted our analysis to pollutants for which state standards exist or for which some determination of acceptable water quality can be made.

Table 2 lists the categories of ranking used in this assessment.

### 4.4 Overall Water Availability

The final step in the assessment was to develop a composite ranking that combines the surface water flows, institutional factors, and water quality analyses into one overall assessment of water availability. The results of this composite ranking are shown in Figures 4a and 4b.

Table 2

CATEGORIES OF RANKING FOR WATER QUALITY

High            All standards met.

Medium          No more than one standard violated.

Low             More than one standard violated.

To arrive at this ranking we overlaid the maps for each part of the
water analysis. We then determined how the rankings were affected by
each other to arrive at the overall ranking. For example, in Western
river basins, a ranking of "low" for surface water flows could be raised
to an overall ranking of "medium," if the institutional factors were
rated "high." Because institutional considerations are extremely
important in the West, we weighted the rankings for institutional
factors higher than those for surface water flows in the western
regions. In the East, we allowed the lowest ranking for either water
quality or quantity to be definitive so that, for example, if a basin
was ranked medium for water quality, the overall ranking could be no
higher than medium. Institutional factors were weighted lower in the
East than in the West.


5.0  Federal Land Management

In this analysis we examined the restrictions to mineral entry,
prospecting, leasing, and development on federal lands in the western
United States (only) to determine how such restrictions may inhibit the
development of a commercial synfuel industry that employs coal or oil
shale as a feedstock.

We examined legislative and administrative actions that restrict coal
mining to determine how much they might constrain synfuel development.
The constraint levels were determined on the basis of the general
qualitative level of the analysis and the degree to which exploration,
mining, and coal conversion plant siting would be restricted.

The four general constraint categories in which 10 federal land
management units were placed appear in Table 3. Then, we mapped the
existing units. (The color maps resulting from the analysis were not
reproducible in this paper.)

Federal land use controls will not seriously constrain siting of coal
mines or synfuel plants within the Four Corners/Rocky Mountain area. In
only three counties (Hinsdale, Pitkin, and San Juan, Colorado) do
absolute constraints exceed 25% of the total acreage, but in no county
does the area so controlled exceed 50%. Two Colorado counties reach the
25-50% level of severely constrained land. Several coal resource
counties have more than 25% of their land in national forests--Garfield,
Utah, and Gunnison and Routt, Colorado. The only coal resource county
with more than 50% of its area under federal control is Gunnison,
Colorado.

380

Table 3

FEDERAL LAND MANAGEMENT CONSTRAINTS TO MINING
AND SYNFUEL PLANT SITING

| Constraint | Relative Ease Of Siting | Land Classification |
|---|---|---|
| Absolute | Low | National parks |
| | | National monuments |
| | | National Wildlife Refugees |
| | | National wilderness preservation system |
| Severe | Medium Low | RARE II wilderness |
| | | Administration endorsed wilderness |
| Moderate | Medium | RARE II further planning |
| | | National forests and BLM lands |
| Minor | Medium High | National forests and BLM lands with scattered ownership patterns |
| None | High | Non-federal lands |

6.0  Socioeconomic Capacity

Full-scale commercial synfuel production has potential for causing major
social and economic impacts in the areas surrounding the conversion
plants.  The impacts result from a combination of the rapid buildup of a
large construction work force followed by a sharp decline to a much
smaller operating work force, and from the location of the coal and oil
shale resources in largely unpopulated rural areas.  On the other hand,
when a siting opportunity exists in an area with (or with access to)
substantial urban resources, the potential for severe dislocation is
greatly lessened.

For each candidate county, we considered indicators of potential
capacity, choosing those indicators for which data were available, those
whose relationship to major areas of socioeconomic impact was known, and

those that allowed us to distinguish between the ability of different areas to absorb those impacts. For each indicator used, we defined a region of influence around each study county. Our approach thus considers a county and its supporting region.

We first identified counties with very high or very low potential absorptive capacity based on population because it is the strongest proxy for the available infrastructure in a county.

A county with a population of 250,000 or more would have extensive human and urban resources, and would more likely satisfy a large portion of the demands of synfuel siting, including the construction and operating labor force. We ranked such counties "high."

On the other hand, a sparsely populated, isolated county would not likely satisfy synfuel demands from local resources, and most of the work force would be in-migrants. In a region of 40,000 or less, a plant siting could result in more than doubling the population in 3 years, followed by a traumatic population decline in the operating phase. Thus, if the population of a county and its proximate areas was 40,000 or less, we judged its potential absorption capacity to be low.

The minority of counties that ranked either high or low were not treated in the second step, although data on them were collected. Counties that fell into the medium category were ranked on the basis of five indicators, summarized below.

- o   Population Centers. This indicator is designed to represent the urban resources (i.e., public and private infrastructure) available to each study county.

- o   Historical Growth Rate. This indicator was used to reflect an area's potential capacity to accommodate future growth on the basis of its recent growth history.

- o   Available Labor Force. This indicator is a rough measure of the in-migration that may be required to satisfy employment generated by the plant. The availability of secondary-sector labor is not considered.

- o   Percent of Income from Agriculture. This indicator measures the dominance of agriculture in the local economy.

- o   Median Age. The potential for conflict between "newcomers" and "old timers" is, in part, indicated by the differnces in the age profiles of the two groups.

The weighted scores for each indicator were summed to give a total score for each county in the "medium" category. The results of the socioeconomic ranking procedure are shown in Figures 5a and 5b.

The Four-Corners/Rocky Mountain study area has two counties with 250,000 or more residents, and the Illinois Basin has three such counties.

Twenty-one "low" counties (less than 40,000 population) were located in the Four Corners/ Rocky Mountain study area. There were no "low" counties in the Illinois Basin study area.

Of all Four Corners/Rocky Mountain counties in the medium range, 70% were classified as medium high. This reflects the influence of the larger cities in or proximate to this region (Albuquerque, New Mexico; Salt Lake City, Utah; and Denver, Colorado). "Medium high" counties are located near these population centers, tended to have a large construction and manufacturing labor force, and tended to have a moderately low median age. The "medium low" counties in this region generally had a smaller labor force and higher median age.

All the Illinois Basin counties rated medium high or above, reflecting the relatively greater supply of urban and human resources available in the region.

## 7.0 Ecological Sensitivity

The ecological sensitivity of an area may be defined in terms of a ratio between the quality of the ecological resources before synfuel development and that likely to exist after development. The term "ecological resources" includes both aquatic and terrestrial habitats in natural ecosystems, as well as agricultural ecosystems.

Although the terrestrial ecological impacts of synfuel operations may be the most apparent (e.g., land disturbed by mining or attendant urbanization in an area), aquatic ecosystems may also be significantly affected. Direct physical impacts include channelization and rerouting of existing streams, impoundment and reduction of stream flows, reduction of streamside and watershed vegetation cover, and disruption of underwater aquifers that supply year-round water to surface streams. Sediments resulting from surface disturbance and erosion may also increase, as may the toxicity potential from plant effluents, from settling ponds and from solid wastes.

We have included agricultural lands within ecological resources because those lands, as highly managed and productive ecosystems, form a functional whole with their environments, just as natural ecosystems do.

We have defined the ecological sensitivity of an ecosystem in terms of three interrelated factors:

(1) Ecosystem Quality--The inherent quality of natural habitats and agricultural lands (measured by such factors as productivity, diversity, and uniqueness) that may be directly or indirectly preempted or altered by synfuel development. The higher the ecosystem quality of an area, the more ecologically sensitive it is and the lower its relative ease of siting has been rated.

(2) Disturbance Susceptibility--The likely duration, extent, and degree of indirect habitat preemption or deterioration resulting from the human population increases anticipated to accompany synfuel development; the amount of land and the

extent of waterways already highly disturbed or polluted; and the physical resiliency of an ecosystem to land form alterations. The greater the disturbance susceptibility of an area and the less it has already been disturbed, the higher we rate its ecological sensitivity and the lower its relative ease of siting.

(3)  Rehabilitation Potential--The capacity of man to rehabilitate natural habitats and agricultural lands to ecologically stable, productive, and useful conditions after such areas have been altered by synfuel development operations, represented by revegetation potential and watershed management potential. An area with high rehabilitation potential is rated as having low ecological sensitivity and more developmental potential.

We consulted mapped and tabulated data sources to evaluate the three major ecological sensitivity factors. We ranked areas as having an overall high, medium, or low ecological sensitivity on the basis of our subjective evaluation of their composite ratings for the various components analyzed. For each component, high, medium, and low values were defined in relation to other values within a particular region. The high, medium, and low ecological sensitivity rankings were translated into low, medium, and high ease of siting, respectively, which are indicated in the maps shown in Figures 6a and 6b.

The overall ecological sensitivity of areas within the Four Corners/Rocky Mountain study area were either high or medium. Regions within this study areas were given high ecosystem quality rankings, largely because of their varied and often numerous native wildlife, good aquatic habitats, and large areas of undeveloped or unimproved land. The agricultural ecosystem rankings were generally fairly low, but the disturbance susceptibility rankings were fairly high, largely because in this region the reestablishment of the original vegetation diversity may require 40 years or more.

A large percentage (more than 85%) of this region's areas were ranked as highly ecologically significant for a variety of reasons related to the diversity of topographic, climatic, and soil conditions, as well as biotic communities. Included are southwestern Wyoming's and Utah's cold shrub deserts (e.g., sagebrush, saltbrush, and greasewood) that extend into Colorado; the pinyon-juniper-oak woodlands and pine-douglas fir forests and alpine meadows at upper elevations; and the desert grasslands of New Mexico. The vegetation types are an index of the drought, temperature extremes, and saline/alkaline soils that characterize the region.

In contrast to the western study regions, a number of areas in the Illinois basin have low ecological sensitivity, partly because the region has had a long history of surface mining.

Most of the natural ecosystem rankings for this area were fairly low. On the other hand, the ecological sensitivity of the agricultural ecosystems in large parts of the region was high. Approximately one-fourth of the counties have more than 75% of their land in cropland, much of it prime. Largely on the basis of existing human population

density, we ranked most of the counties as having medium disturbance
susceptibility.

The restoration of the lands in the Illinois Basin to productive pasture
land seems quite feasible.  However, we have noted the current Illinois
law that requires productive fertile soils to be returned to cropland
and have thus assigned prime farmland a low rehabilitation ranking.
Farmland other than prime has been given high rehabilitation ranking.
Forested areas and hilly areas are given medium rankings.  Most of the
areas in Kentucky, where the reclamation laws are not as strict as those
in Illinois, but are becoming more stringent, are given high
rehabilitation potential rankings, as are the farmland areas in Indiana.


8.0  Environmental Health

Among many possible health-affecting constituents and contaminants of
oil shale and coal conversion processes and their liquid products are:

o    Polycyclic aromatic hydrocarbons
o    Aromatic amines and other nitrogenous compounds
o    Trace metals, especially heavy metals
o    Radioactive nuclides.

The presence of these compounds in the process streams of commercial
synfuel plants is inferred from pilot-scale measurements as well as
theoretical treatments.

The list of possible health effects from synfuel generation includes:

o    Cancer
o    Central nervous system effects
o    Liver damage
o    Mutagenic effects

The number of occurrences of disease in a population exposed to
pollutants from synfuels depends on the number of people exposed to
various levels of pollutants and the probabilities that those exposures
will result in disease.

Within limits, neither the number of people associated with a plant
(workers and their families, and the induced population) nor the
distribution of residence location or mobility patterns are dependent on
plant location.  Thus, the preplant population variation largely
determines the population at risk and is represented by population
density, which is available at the county level.

Our reasoning for the locational dependence of the probability of
disease is more complicated and much more speculative.  First, we
observe that the effects cited above, cancer is the most likely to be
manifested at low exposures and is also probably the most serious type
of effect.  Cancer is also known to be a disease that often has multiple
causes.  Synergistic carcinogenesis produces more cancer occurrence for
the sum of two exposures than the sum of the occurrences for the two
exposures separately; in fact, any risk factor that increases the risk
for one carcinogen could also increase the risk for another similar
carcinogen.

We therefore decided to use the preexisting occurrence of cancers as an indication of the excess cancer rate, the Atlas of Cancer Mortality for U.S. Counties 1950-1969. The atlas presents maps of counties to indicate those counties where cancer mortality rates are significantly higher or lower than the U.S. average.

The county-level data on population density and cancer incidence by body site (skin, lung, and bladder) were combined to arrive at an overall assessment of the relative health risk posed by siting synfuel plants. Those results were mapped onto the four study areas, and are displayed in Figures 7a and 7b.

In the Four Corners/Rocky Mountain study area, it seems clear that health risk considerations have relatively little effect on synfuel plant siting. Only three counties in the Four Corners/Rocky Mountain Area received ratings as low as "medium low," and those were due primarily to high population densities.

In the East, the situation differs substantially. None of the counties in the Illinois Basin received "high" ratings, and some counties received "low" ratings. The "low" ratings are usually in counties that have high cancer rates, as well as high population densities. The absence of "high" ratings is generally due to the uniformly higher population density as compared with the West.

The overall rarity of low ratings in the health area reflects our judgment that the methodology for health assessment is of relatively low discrimination and is based on less than ideal data. Our purpose here is to suggest that when both population densities and pre-existing cancer rates (or, in fact, other rates of disease that relate to synfuels) are high, more care is justified in selecting the specific site and assuring the lowest feasible levels of population exposure.

9.0  Composite Results

Sections 3.0 through 8.0 summarized analyses of the important environmental considerations for siting synfuel plants for each of six constraining factors. The analyses were formulated in a way that allowed mapping of the relative ease of siting plants using a semiquantitative rating scale. Although each set of maps is useful in its own right, it is worthwhile to consider how the combination of some or all the constraining factors would affect the opportunities for synfuel plant siting in each of the four study areas.

We have combined the individual maps of constraining factors into composite maps. We first combined the maps for those factors that are perceived to be the most constraining because of strong regulatory support or physical resource requirements. These factors are air quality, water availability, and federal land management. We then separately combined the factors that represent important social and environmental concerns, but which are perceived to be less constraining because of the absence of regulatory requirements. Finally, we combined all the factors in composite maps that represent the total effect of environmental constraints to synfuel plant siting.

The composite maps were prepared by using the following combining rules:  (1) an area that has a "low" rating for any single factor receives a composite rating of "medium low;" (2) an area that has two or more "low" ratings receives a composite "low" rating; (3) other composite ratings are approximate averages of individual ratings. Although these combining rules emphasize the low end of the scale, they serve a very useful purpose in that they highlight those areas in which there are major barriers to plant siting.

The composite maps for the Four Corners/Rocky Mountain study area (Figures 8a, 9a, and 10a)  show major constraints to plant siting due to (1) Class I areas, federally owned land, and low availability of water, and (2) considerable extent and overlap of ecologically sensitive areas and non-urbanized areas.

Figure 10a shows the dominance of ecological and socioeconomic factors. Areas around the major urban centers (e.g., Salt Lake City, Albuquerque) that tended to receive higher socioeconomic and ecological ratings did not have higher composite ratings because of generally lower air quality ratings and lower environmental health ratings.

Only a few areas appear to be prime candidates for synfuel development on environmental grounds--that is, developable without extensive mitigation measures.  It is within these areas that, all factors considered, siting opportunities will occur that require the least extensive mitigation measures will pose the least difficult permit acquisition, and will result in the least environmental and social disruption.

The composite rating maps for the Illinois Basin study area (Figures 8b, 9b, and 10b)  appear much different from those of the western study areas.  First of all, there tend to be many "high" rated areas based on water availability and air quality (federal land management was not a factor here).  Second, because of the absence of "low" ratings for socioeconomic capacity, and only a few "low" ratings for environmental health, there are no composite "low" ratings in Figure 9b.

The result is that Figure 10b shows a more even distribution than Figure 10a, with a few "low" areas and even a few "high" areas. Overall, the Illinois Basin displays ample opportunity for synfuels plant siting without the extensive mitigation requirements needed in the west.  The most restrictive  areas are those with water quality problems, air quality problems due to existing industrial and urban development, and a high density of prime cropland.

10.0  References

1.   Dickson, E. M. et al., "Synthetic Liquid Fuels Development: Assessment  of Critical Factors," Vols. I and II, ERDA 76-129/2 (1976).

2.   Dickson, E. M., and I. W. Yabroff, "Coal Resources Model:  An Impact Analysis Tool," prepared by SRI International for the U.S. Department of Energy (January 1979).

3.   R. V. Steele et al., "Environmentally Based Siting Assessment for Synthetic Liquid Fuel Facilities," prepared by SRI International for the U.S. Department of Energy (April 1980).

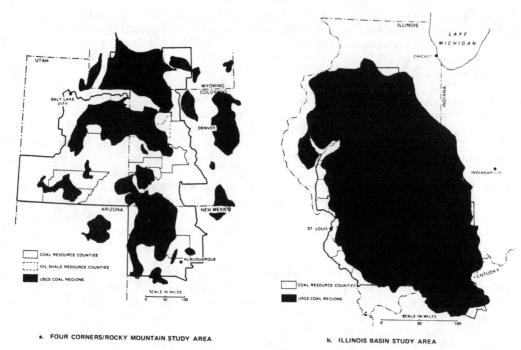

a. FOUR CORNERS/ROCKY MOUNTAIN STUDY AREA          b. ILLINOIS BASIN STUDY AREA

FIGURE 1.   STUDY AREAS USED IN SITING ASSESSMENT

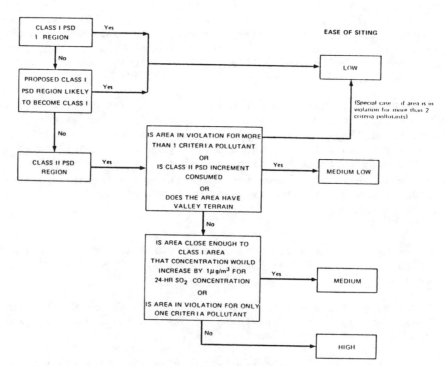

FIGURE 2.   METHOD FOR RANKING THE CONSTRAINTS ON EASE OF SITING

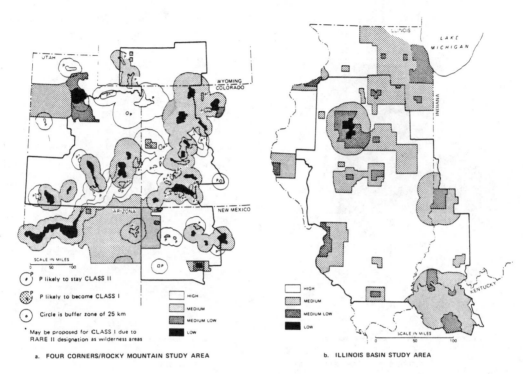

a.  FOUR CORNERS/ROCKY MOUNTAIN STUDY AREA

b.  ILLINOIS BASIN STUDY AREA

FIGURE 3.   RELATIVE EASE OF SITING BASED ON AIR QUALITY

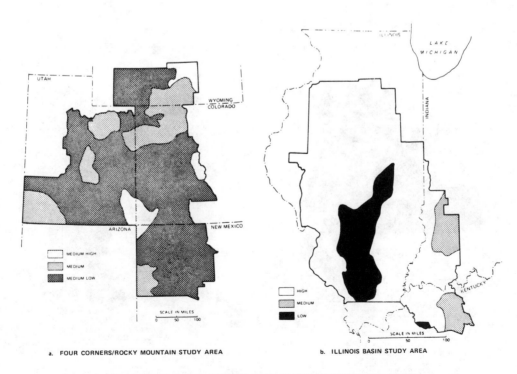

a.  FOUR CORNERS/ROCKY MOUNTAIN STUDY AREA

b.  ILLINOIS BASIN STUDY AREA

FIGURE 4.   RELATIVE EASE OF SITING BASED ON OVERALL WATER AVAILABILITY

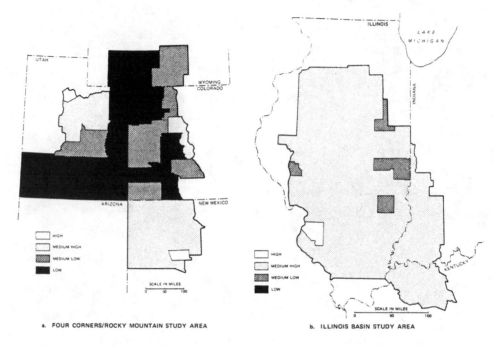

a. FOUR CORNERS/ROCKY MOUNTAIN STUDY AREA    b. ILLINOIS BASIN STUDY AREA

FIGURE 5. RELATIVE EASE OF SITING BASED ON SOCIOECONOMIC CAPACITY

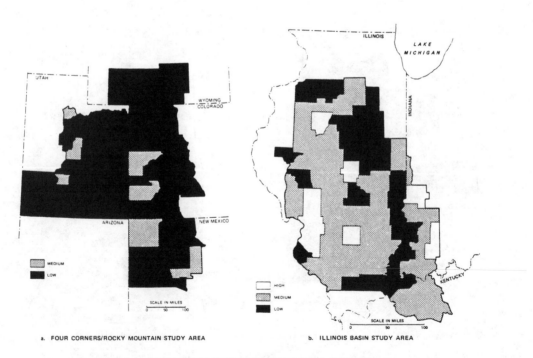

a. FOUR CORNERS/ROCKY MOUNTAIN STUDY AREA    b. ILLINOIS BASIN STUDY AREA

FIGURE 6. RELATIVE EASE OF SITING BASED ON ECOLOGICAL SENSITIVITY

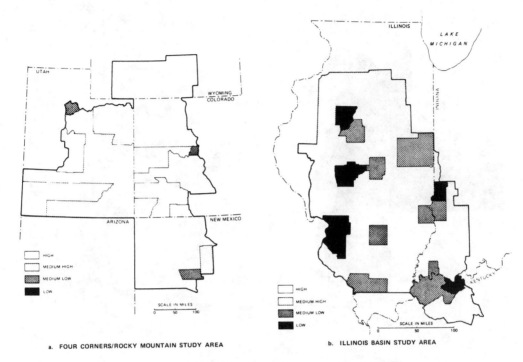

a. FOUR CORNERS/ROCKY MOUNTAIN STUDY AREA      b. ILLINOIS BASIN STUDY AREA

FIGURE 7.    RELATIVE EASE OF SITING BASED ON ENVIRONMENTAL HEALTH

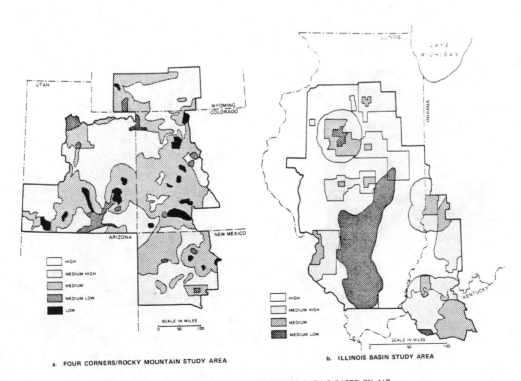

a. FOUR CORNERS/ROCKY MOUNTAIN STUDY AREA      b. ILLINOIS BASIN STUDY AREA

FIGURE 8.    RELATIVE EASE OF SITING: COMPOSITE RATING BASED ON AIR
QUALITY, WATER AVAILABILITY AND FEDERAL LAND MANAGEMENT

391

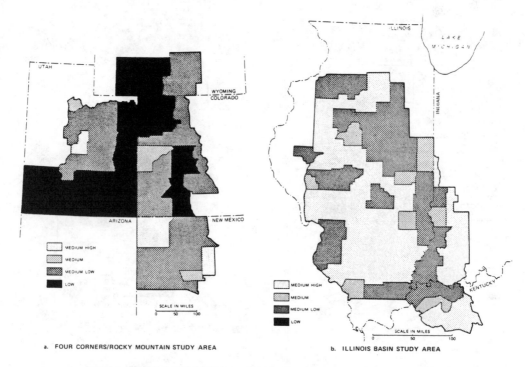

a. FOUR CORNERS/ROCKY MOUNTAIN STUDY AREA          b. ILLINOIS BASIN STUDY AREA

FIGURE 9.   RELATIVE EASE OF SITING:  COMPOSITE RATING BASED ON ECOLOGICAL
SENSITIVITY, SOCIOECONOMIC CAPACITY AND ENVIRONMENTAL HEALTH

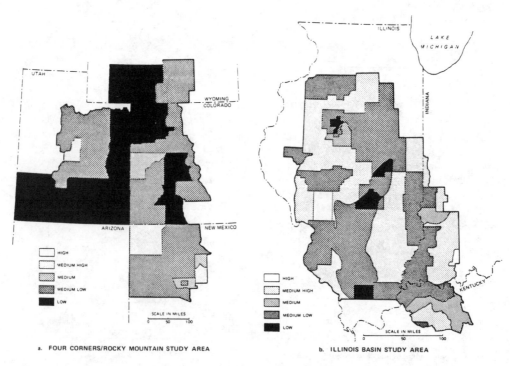

a. FOUR CORNERS/ROCKY MOUNTAIN STUDY AREA          b. ILLINOIS BASIN STUDY AREA

FIGURE 10.   RELATIVE EASE OF SITING:  COMPOSITE RATING BASED ON ALL FACTORS

392

## POTENTIAL HEALTH PROBLEMS
## FROM COAL CONVERSION TECHNOLOGIES

**R. A. Wadden** and **A. L. Trabert**
University of Illinois

Nature of the Problem

The conversion of coal to desirable solid, liquid, or gaseous products, or to usable energy, is an important technology for an industrialized society. Because of the demand for low-sulfur, environmentally acceptable fuels, coal liquefaction and gasification are of particular interest at the present time. However, there are several factors which make the potential for coal processing plants to discharge hazardous materials different from other fossil or nuclear fuel processing facilities. One of these is the nature of the raw material. Coal has a complex heterocyclic structure which must be broken down to produce desirable products. Cleavage of these cyclic linkages, and the chemical form of the resultant compounds are difficult to analyze and predict. In addition, raw coal has less homogeneity than gas or oil in terms of physical structure and the distribution of trace metals. Its structure also presents materials handling problems, with attendent environmental discharges, not associated with fluids' movements. For example, coal feeds usually require crushing and grinding into small size particles which are then often incorporated into oil or water slurries in order to be uniformly introduced into the liquefaction or gasification process.

Another factor of significant concern is the size of projected coal processing plants. Commercially feasible facilities typically will reduce in excess of 20 000 tonnes of coal per day to products and wastes. These conversions ordinarily require addition of hydrogen to the coal to produce saleable materials and can be generally classified as hydrogenation processes. Because of the large amounts of material being processed, the potential exists for environmental discharge of considerable quantities of hazardous materials. In this regard the environmental implications of *in situ* conversion, where even the quantities of coal converted are not well defined, are also of concern.

The characteristics of each particular coal conversion process are also a major factor in hazard evaluation. Figure 1 shows a generalized flow sheet combining a number of processing schemes. The intent of the flow sheet is to emphasize the elements common to most hydrogenation processes. Each particular process is considerably more complex and requires careful analysis of its products and wastes. However, some general observations can be made about the nature of the conversion process. Most hydrogenation processes require energy addition at elevated temperatures and pressures in order to break down the coal. The necessary energy is ordinarily derived from oxidizing a portion of the incoming coal. Hydrogen is supplied either through direct or indirect utilization of the hydrogen atoms in water or through use of a hydrogen donor solvent. Net water usage for process purposes has been estimated at $1-10 \times 10^6$ gal/day for a 20 000 tonne/day plant. This volume does not include cooling water requirements which may be ten times as large.

A final factor which needs consideration is that of expediency. With increasing international concern about the ready availability of petroleum for fuels and chemical feedstocks, there is great interest in developing coal-based alternatives. In the rush to construct and operate coal processing plants, there is a tendency to overlook or minimize the warning signs already received from available data.

Available Health Data

Only one study of the actual health hazards of coal hydrogenation is presently available.[1-4] A group of 359 workers was examined for skin cancer over a period of 5 years (1955-1960) at a liquefaction plant (West Virginia, U.S.A.) designed to process 273 tonne/day of incoming coal. Of this group, 10 men developed skin cancer. Exposure varied from several months to 23 yr, but all significant lesions were found in workers with less than 10 yr exposure. The reported incidence was 16-37 times the incidence of skin cancer expected in the chemical industry.[4] Precancerous lesions were also found on the skin of 40 additional workers.

Analyses of the liquid materials produced in the plant identified a number of carcinogenic materials. An increase in carcinogenicity with increasing liquid boiling point, usually with respect to liquids with boiling points $>260°C$, was observed in laboratory animals.[3] This relationship was particularly evident for the oil, produced in the plant, which was recycled to slurry the incoming coal. In addition, air measurements in working areas revealed benzo($\alpha$)pyrene concentrations often in excess of $0.05$ $\mu g/m^3$ with a peak value of $18$ $\mu g/m^3$.[2] Benzo($\alpha$)pyrene has been identified as being strongly carcinogenic.[5] These concentrations are in contrast to typical U.S. urban levels of $0.003-0.006$ $\mu g/m^3$.[5]

In 1977 a follow-up study was instituted to identify and trace the original 50 workers with skin lesions.[6] One worker was lost to follow-up, 5 had died from non-cancer causes, and there was one case each of lung and prostate cancer. After a latency period of 18-20 years, the cancer incidence did not appear to be marked. It is well to note that none of the deaths had been autopsied.[7] In addition, the author has pointed out that this review should not be the basis for drawing any conclusions about cancer incidence save for this highly selected group. He has recommended an epidemiological study which would identify the remainder of the workforce who were exposed to the coal hydrogenation plant discharges but who did not develop skin cancer in the 1955-1960 interval.  The possibility exists that some of the workers who had not developed skin lesions during that period may have developed more serious organ cancers after the surveillance program was discontinued in 1960.[6]  It has also been reported that since the end of December 1958, the workers have been followed and that at first there were one to two new cases of skin cancer every year, but after the plant stopped operating the incidence was about one new case every five years.[8]

Mortality and morbidity data at other coal hydrogenation facilities have not been systematically collected or analyzed, at least in the published literature.  Several deaths have been reported due to high CO exposures or inadequate ventilation of vessels during inspection or cleaning.[8,9]  Health statistics on occupational groups in other coal conversion operations, such as coke ovens and coal-tar processing, have shown significantly higher lung cancer rates relative to groups without such occupational exposures.[10-15]  In particular, coke oven workers appear to have 2.5 times the incidence of lung cancer as other steel workers.[12]  British workers heavily exposed to the products of coal carbonization in gasworks, where coal is heated in retorts with the primary purpose of producing flammable gas, were found to have greater incidence of death due to lung cancer (140%) and bladder cancer (208%) compared to workers exposed only to by-products.[10,11]  Analysis of gasworks air showed a general level of about 3 $\mu g/m^3$ of benzo($\alpha$)pyrene and over 200 $\mu g/m^3$ in the worst working area.[16]  The environment on top of a coke oven or in a gasworks is, of course, different from that likely to occur in hydrogenation plants which will eventually be designed to process raw materials continuously.  Continuous processing requires more complete containment (and, consequently, better control) of all substances than do batch operations such as coke ovens.

Potential Process Discharges

Although a considerable number of full-scale coal hydrogenation plants have been built outside the U.S., until recently little data have been available to evaluate the types of hazardous materials discharged to the environment.  An on-going study of the Lurgi Kosovo Kombinant plant, Pristina, Yugoslavia, represents the most complete stream characterization data set for a commercial coal gasification plant in the published literature at the present time.[17,18]  Although more data will be forthcoming from this joint U.S.-Yugoslav effort, the information reported here supplies a representative evaluation of a plant of Lurgi design, at steady state, with relatively few environmental controls.  Table 1 summarizes "best value" concentrations of vented gas streams (streams 3.6, 13.6 and 7.1 being subsequently being sent to flare) based on repeated sampling, and engineering and analytical judgement.  Streams 7.1 ($H_2S$ vent) and 7.2 ($CO_2$ vent) represent discharges from the Rectisol acid gas clean-up step and would ordinarily be processed further for $H_2S$ removal. Even so, the concentrations of CO, HCN, $NH_3$ and various mercaptans ($CH_3SH$, $C_2H_5SH$) are significant in most of the streams and the quantities emitted are notable.  Also of considerable concern are the relatively high concentrations of benzene and other aromatics.[17]

Table 2 shows the polycyclic aromatic hydrocarbon content of the Kosovo oil and tar and the raw and treated phenolic water.  Using the particulate oil and tar fractions of gaseous waste streams, B(a)P concentrations of 20-1500 $\mu g/m^3$ were estimated for flare and other exhaust gases.  If these estimates prove to be accurate (and confirming analysis is presently being carried out) the discharge of B(a)P is significant.

The Kosovo stream characterizations are very useful as they can be used in conjunction with measured flow rates, appropriate air quality models and meteorological conditions to predict human exposures. Methods are also available for estimating some of the health effects which might result from such dosages (e.g.,[19]).

Other data on coal hydrogenation plants are available (e.g.,[20-22]) but not in as complete form as the Kosovo information. One test of a commercial low-Btu gasifier revealed 0.01 and 0.004 ppm of nickel and iron carbonyl, respectively.[21] Waste aqueous streams often contain significant concentrations of phenols, HCN and ammonia.[17,22] In addition data are needed to describe start-up, shut-down and upset conditions.

Possible Exposed Populations

Workers and dependents represent those who are most likely to be exposed to the highest concentration of hazardous materials. Dependents are at risk to workplace hazards brought into the home on clothes and skin. Significant toxic hazards may include polynuclear aromatic hydrocarbons and other carcinogens such as $\beta$-naphthylamine and benzidine, nitrosoamines and possibly $Ni(CO)_4$. Compounds capable of potentiating carcinogenesis such as $SO_2$, acridines, dodecane and phenols are also present. Organs conspicuously at risk are the skin and lung, other parts of the respiratory tract and the genitourinary system, especially the bladder. Effects other than cancer such as dermatitis of the skin and liver, and central nervous system (CNS) disorders from phenols, $H_2S$, COS, $CS_2$, iron and nickel carbonyl and aromatic amines may be significant. Nonmalignant respiratory effects, such as chronic bronchitis, as well as cardiovascular effects due to CO and $CS_2$ are also possible problems. Acute lung irritation can be caused by a wide variety of pollutants such as HCN, $NH_3$, $SO_2$, $H_2S$, and phenols.[8] Benzene is myelotoxic and leukemogenic.[23]

Several groups of workers may be at high risk to coal hydrogenation emissions. Workers exposed for more than 5 years or who are over 45 are possibly more susceptible to skin cancer. Job categories which involve high exposure are at risk. Appearance of skin cancer may be a warning sign for subsequent cancer in the respiratory and digestive tracts. A significant association has been reported of these effects in men with scrotal cancer.[24]

Medical monitoring requires a pre-employment physical, including history and laboratory tests; regular checkups, more frequently and in more detail for high risk groups; long-term follow-up of high-risk individuals after transfer to other job categories or termination; and full record keeping, including work history and exposure data. Protective procedures include good plant housekeeping (e.g., prompt clean-up of spilled material, prevention of leaks, regular maintenance of pollution-control equipment); proper protective clothing and cleaning of clothes; specific and detailed operating instructions and work practices for preventing hazardous exposures, particularly during shut-down, startup and vessel and tank clean-out; thorough worker education on the potential hazards of the hydrogenation process; and regular and frequent monitoring of hazardous materials in work areas.[7]

Both normal and hypersusceptible populations in communities in the plant vicinity need to be considered in evaluating community exposure. Groups hypersusceptible to hydrogenation discharges include children, the elderly and those with illness or genetic dysfunction which would make them more vulnerable to the stress of exposure. Collection and periodic review of cancer, disease and birth defect registeries are necessary requirements of community health surveillance. Monitoring for pollutants in air and water in the plant vicinity would also be prudent. When pollutant quantities discharged from the plant can be determined, methods are available for estimating community exposure.[19]

The chemical composition of liquid and gaseous products from coal hydrogenation is much different from the composition of equivalent materials extracted from petroleum or natural gas. Consequently, product users may also be at risk. Product oils are likely to have high aromatic contents and significant amounts of polynuclear hydrocarbons, some of which are carcinogenic. Burning characteristics may also be different. The hydrogen content of product gas from coal gasification will be much higher than that in natural gas. Whether this higher hydrogen content will be significant in terms of hydrogen embrittlement of existing pipelines and gas holders or a hazard in industrial and domestic burners remains to be determined. Low-Btu product gas will also contain higher CO concentrations than in natural gas. Evaluation of the potential environmental health consequences to users of these and other hydrogenation products requires more detailed information about the composition and variability of such products and the chemical and physical changes they may undergo during combustion or further chemical processing.

Recommendations for Control and Planning

Most of the problems of control in hydrogenation plants are probably solvable, providing that the nature and source of the pollution are properly described. In addition, low-sulfur, refined fuels produced by such processes are much more acceptable in urban areas than raw coal or heavy, high-sulfur oils. However, because of the large quantities and complex nature of the coal and the design and operating characteristics of the conversion process, there must be a recognition that these new types of operations may produce kinds of pollutants with which we are unfamiliar.

Available process stream data for full-scale and pilot plants, although meagre, show significant concentrations of a number of pollutants which are recognized hazards to lung and skin, and some which are known carcinogens. A major effort is needed to further characterize the wastes and products of the hydrogenation process, particularly in terms of toxic materials which may appear in low concentrations. Complete quantitation of pollutant discharges requires long-term operation of large-scale pilot plants. (This recommendation is contrary to the way pilot plants are usually operated. Because of time and economic restrictions there is a tendency to change operating conditions frequently.)

Another important concern is one of educating plant and regulatory personnel to occupational and environmental problems which are caused by release of hazardous materials from coal processing. The most effective control device available still is turned on or off by a human being whose judgement is subject to error. Chemical plant managers and operators are used to periodic upset conditions with consequent discharges to the environment. Their major concern is to return the plant to productive operation, sometimes without sufficient concern for the types and quantities of emissions. There is a need to build into both the design and operation of such plants a consciousness of minimizing or eliminating the environmental release of any stream which has a reasonable potential for containing toxic substances.

Considerable effort should be expended to review, monitor and protect the occupational health of plant workers.  Retrospective epidemiological investigations on identified groups of workers at pilot or commercial-sized installations where medical records exist (e.g., pre- and post- employment physicals) would be extremely useful in trying to better delineate long-term health implications.  Continuous monitoring of pollution levels in worker, user and community settings as well as prospective epidemiological evaluations and the development and maintenance of pertinent disease registeries on these populations is a necessary concomitant to responsible development of coal resources.

Control of pollutants from an established technology is often difficult and costly.  It is much more desirable to define and quantitate hazards, and develop appropriate characterization and control methods, before a new technology is launched.  Designers, operators and regulators must be aware and alert to the potential problems of coal hydrogenation in order to properly protect the health of the worker and the public.

Acknowledgement

This paper was abstracted from a report, commissioned by the Beijer Institute for Energy and Human Ecology, Royal Swedish Academy of Sciences and the United Nations Environmental Program, which was part of the Institute's on-going program on environmental impacts of coal utilization.

References

1.  Sexton, R.J.  "The hazards to health in the hydrogenation of coal. I.  An introductory statement on general information, process description, and a definition of the problem."  Arch. Environ. Health, 1, 181-186 (1960).

2.  Ketcham, N.H. & Norton, R.W.  "The hazards to health in the hydrogenation of coal.  III.  The industrial hygiene studies."  Arch. Environ. Health, 1, 194-207 (1960).

3.  Weil, C.S. & Condra, N.I.  "The hazards to health in the hydrogenation of coal.  II.  Carcinogenic effect of materials on the skin of mice."  Arch. Environ. Health, 1, 187-193 (1960).

4.  Sexton, R.J.  "The hazards to health in the hydrogenation of coal. IV.  The control program and the clinical effects."  Arch. Environ. Health, 1, 208-231 (1960).

5.  NAS.  Particulate Polycyclic Organic Matter.  National Academy of Sciences, Washington, D.C. (1972).

6.  Palmer, A.  "Mortality experience of 50 workers with occupational exposures to the products of coal hydrogenation processes."  J. Occup. Med., 21, 41-44 (1979).

7.  NIOSH (National Institute for Occupational Safety and Health). Recommended Health and Safety Guidelines for Coal Gasification Pilot Plants.  NIOSH Publication 78-120, Cincinnati, Ohio (1978).

8.  NIOSH (National Institute for Occupational Safety and Health). Criteria for a Recommended Standard: Occupational Exposures in Coal Gasification Plants.  NIOSH Publication 78-191, Cincinnati, Ohio (1978).

9.  Partridge, L.J.  "Coal-based ammonia plant operation."  <u>Chem. Eng.</u>
    <u>Prog.</u>, <u>56</u>(no. 8), 57-61 (1976).

10. Doll, R., Fisher, R.E.W., Gammon, E.J., Gunn, W., Hughes, G.O.,
    Tyrer, F.H., & Wilson, W.  "Mortality of gasworkers with special
    reference to cancers of the lung and bladder."  <u>Brit. J. Ind. Med.</u>,
    <u>22</u>, 1-12  (1965).

11. Doll, R., Vessey, M.P., Beasley, R.W.R., Buckley, A.R., Fear, E.C.,
    Fisher, R.E.W., Gammon, E.J., Gunn, W., Hughes, G.O., Lee, K., &
    Norman-Smith, B.  "Mortality of gasworkers--Final report of a pro-
    spective study."  <u>Brit. J. Ind. Med.</u>, <u>29</u>, 394-406  (1972).

12. Lloyd, J.W.  "Long-term mortality study of steelworkers. V.  Respira-
    tory cancer in coke plant workers."  <u>J. Occup. Med.</u>, <u>13</u>, 53-67
    (1971).

13. Redmond, C.K., Ciocci, A., Lloyd, J.W., & Rush, H.W.  "Long-term
    mortality of steelworkers.  VI.  Mortality from malignant neoplasms
    among coke oven workers."  <u>J. Occup. Med.</u>, <u>14</u>, 621-629 (1972).

14. Mazumdar, S.C., Redmond, C., Sollecito, W., & Sussman, N.  "An
    epidemiological study of exposure to coal tar pitch volatiles among
    coke oven workers."  <u>J. Air Pol. Control Assoc.</u>, <u>25</u>, 382-389 (1975).

15. Kawai, M., Amamoto, H., & Harada, K.  "Epidemiological study of lung
    cancer."  <u>Arch. Environ. Health</u>, <u>14</u>, 854-864 (1967).

16. Lawther, P.J., Commins, B.T., & Waller, R.E.  "A study of the con-
    centrations of polycyclic aromatic hydrocarbons in gas work retort
    houses."  <u>Brit. J. Ind. Med.</u>, <u>22</u>, 13-20 (1965).

17. Bombaugh, K.J., Corbett, W.E., Lee, K.W., & Seames, W.S.  "An
    environmentally based evaluation of the multimedia discharges from
    the Lurgi coal gasification system at Kosovo."  Presented at the
    <u>Symposium on Environmental Aspects of Fuel Conversion Technology -</u>
    <u>V</u> - St. Louis, Mo., Sept. 16-19, 1980.

18. Patterson, R.K.  "Ambient air downwind of the Kosovo gasification
    complex: a compendium."  Presented at the <u>Symposium on Environmental</u>
    <u>Aspect of Fuel Conversion Technology - V</u> - St. Louis, Mo., Sept. 16-
    19, 1980.

19. Wadden, R.A. (Ed.).  <u>Energy Utilization and Environmental Health:</u>
    <u>Methods for Prediction and Evaluation of Impact on Human Health</u>.
    John Wiley & Sons, New York (1978).

20. Wadden, R.A. & Trabert, A.L.  <u>Potential health problems from coal</u>
    <u>conversion technologies</u>.  Position paper prepared for the United
    Nations Environmental Program and the Beijer Institute for Energy
    and Human Ecology, Stockholm, Sweden (1980).

21. Thomas, W.C., Trede, K.N., & Page, G.C.  <u>Environmental Assessment:</u>
    <u>Source Test and Evaluation Report-Wellman-Galusha (Glen Gery) Low-</u>
    <u>Btu Gasification</u>.  U.S. Environmental Protection Agency Report
    EPA-600/7-79-185, Washington, D.C. (1979).

22. Ghassemi, M., Crawford, K., & Quinlivan, S.  Environmental Assessment Report: Lurgi Coal Gasification Systems for SNG.  U.S. Environmental Protection Agency Report EPA-600/7-79-120, Washington, D.C. (1979).

23. NIOSH (National Institute for Occupational Safety and Health). Criteria for a Recommended Standard: Occupational Exposure to Benzene.  NIOSH, Washington, D.C. (1976).

24. Holmes, J.G., Kipling, M.D., & Waterhouse, J.A.H.  "Subsequent malignancies in men with scrotal epithelioma."  Lancet, July 25, 214-215 (1970).

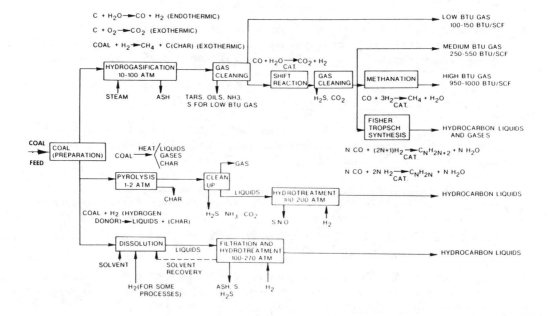

Figure 1.  General Elements of Coal Hydrogenation Processes[20]

Table 1. Vented gas streams for Kosovo Lurgi gasification process[a] (Reference 17)

| Stream | 3.2 | 3.6 | 13.1 | 13.3 | 13.6 | 13.7 | 14.5 | 7.1 | 7.2 |
|---|---|---|---|---|---|---|---|---|---|
| Compound | Lock Hopper Vent Gases Low Pressure | High Pressure | Tar Tank Vent | Medium Oil Tank Vent | Tar Separation Expansion Gases | Phenolic $H_2O$ Tank Vent | Stripper Vent | $H_2S$ Vent | $CO_2$ Vent |
| **Fixed Gases (Vol. %)** | | | | | | | | | |
| $H_2$ | 37. | 32.0 | TR[c] | TR[c] | 11.0 | TR[c] | NF[c] | 0.7 | TR[c] |
| $O_2$ | 0.27 | 0.24 | 19. | 0.45 | TR[c] | 13. | -- | TR[c] | TR[c] |
| $N_2$ | 0.18 | 0.14 | 77.5 | 1.1 | TR[c] | 39. | -- | TR[c] | TR[c] |
| $CH_4$ | 8.6 | 10.5 | 0.16 | 7.6 | 3.5 | 0.2 | TR[c] | 4.3 | 1.2 |
| CO | 14.6 | 12. | TR[c] | 5.9 | 1.1 | NF[c] | NF[c] | 1.1 | TR[c] |
| $CO_2$ | 36.5 | 42. | 0.86 | 69. | 77.5 | 35. | 55. | 88. | 94.0 |
| **Hydrocarbons (Vol. %)** | | | | | | | | | |
| $C_2$-$C_5$ | 0.41 | 0.79 | 0.01 | 0.98 | 1.24 | 0.066 | TR[c] | 1.81 | 1.88 |
| $C_6^+$ | 0.12 | 0.08 | 0.37 | 2.4 | 1.3 | 1.8 | NF[c] | 0.21 | NF[c] |
| **Aromatic Hydrocarbons (ppmv)** | | | | | | | | | |
| Benzene | 760 | 550 | 2,000 | 7,650 | 9,600 | 11,000 | -- | 110 | 1.0 |
| Toluene | 220 | 100 | 960 | 1,400 | 1,200 | 2,300 | -- | 8 | TR[c] |
| Xylene & Ethylbenzene | 75 | 38 | 220 | 140 | 1,500 | 280 | -- | -- | 1.4 |
| Phenols | 6 | 2 | 22 | 45 | 4.2 | TR[c] | 6,200 | NF[c] | NF[c] |
| **Other Compounds (ppmv)** | | | | | | | | | |
| $H_2S$ | 12,700 | 3,500 | 6,900 | 26,000 | 9,000 | 12,600 | 19,500 | 45,400 | 39 |
| COS | 110 | 120 | 110 | 96 | 120 | 41 | NF[c] | 420 | 62 |
| $CH_3SH$ | 420 | 460 | 390 | 5,200 | 2,500 | 2,100 | 290 | 2,100 | 8.5 |
| $C_2H_5SH$ | 220 | 210 | 240 | 2,100 | 1,600 | 7,200 | 100 | 780 | 4.4 |
| $NH_3$ | 2,400 | NF[c] | 2,600 | 11.4 | 19,300 | 12,000 | 418,000 | 2,200 | 46 |
| HCN | 600 | 170 | 130 | -- | 65 | 38 | 4,800 | 200 | 13 |
| Flow Rates $Nm^3$/gasifier-hour @ 25°C [b] | 21 | 230 | 0.51 | 1.7 | 28 | 5.5 | 260 | 3,600 | 3,600 |

a Nominal 384 tonnes/day coal feed per gasifier.

b Ordinarily 5 of 6 Kosovo gasifiers are in operation at one time.

c NF = not found; TR = trace.

Table 2

Polynuclear aromatic hydrocarbon analysis for
Kosovo product and wastewater streams[17]

| Component | Phenolic Water mg/ℓ | Phenosolvan Wastewater mg/ℓ | Light Tar mg/kg | Medium Oil mg/kg |
|---|---|---|---|---|
| Benz(a)anthracene | 0.92 | NF* | 490 | 156 |
| 7, 12-dimethylbenz(α)anthracene | 0.22 | NF | 1090 | 62 |
| Benzo(β)fluoroanthrene | 0.68 | NF | 306 | 115 |
| Benzo(α)pyrene | 0.18 | NF | 210 | 68 |
| 3-methylcholanthrene | <0.004 | NF | 26 | NF |
| Dibenz(a,h)anthracene | 0.02 | NF | 23 | 6.6 |
| 252 molecular weight group | 1.26 | 0.19 | 945 | 282 |

* NF = not found

# A HUMAN HEALTH ASSESSMENT MODEL FOR POLLUTANTS
# DISCHARGED FROM EMERGING ENERGY TECHNOLOGIES

**Amiram Roffman** and **Joseph A. Maser**
Energy Impact Associates

INTRODUCTION

Human exposure to pollutants discharged from new energy technologies
has recently gained attention due to possible massive utilization of
these technologies.  Of particular interest are the possible occupational and public health effects of synfuel facilities.

Emission sources, discharge rates and characteristics are not well quantified for synfuel plants.  Some preliminary data are available from:
(1) paper studies based on engineering design data, type of processes
involved, and chemistry, physics and mass balance concepts, (2) laboratory
tests and studies of bench scale units and process development units and
(3) field tests and studies at pilot plants.  A summary of the sources of
possible emissions and discharges from coal gasification and liquifaction
facilities is included in Tables 1 and 2 respectively.  The discharges
are in air, water and solid waste.  They could include major, minor and
trace pollutants of organic or inorganic nature.  Quantities of emissions
depend on the process employed, the size of the facility, the type of
control technologies employed, the maintenance and safety measures utilized, and the nature of the releases, routine or accidental.

In spite of the lack of sufficient source characterization data, an assessment of the possible environmental and health effects of pollutants
discharged from synfuel facilities is desirable.  Pollutant pathways and
fate are an important part of such an environmental and health effect
assessment.  This paper describes key elements of a Human Exposure and
Health Effect Model (HEHEM) that has been developed for assessing the
fate of pollutants discharged into air, water and soil (Table 3).  The
model is capable of estimating pollutant dose to a human accounting for

different physical, chemical and biological processes governing the rate of change in pollutant concentrations in the different media and organisms involved.  This model includes considerations of pollutant uptake by ingestion, inhalation and deposition.

The utility of the model is illustrated in examples that address the fate, human exposure and the health effects of mercury and benzo(a)pyrene.


METHODOLOGY

Pollutants discharged from synfuel plants are subject to transport and transformation, as well as accumulation and depletion in the physical and biological environments.  The modeling scheme describing the fate of a pollutant in media consists of four models:

Physical Environment:

(1)  Atmospheric model describing the fate of pollutants in air.  It accounts for diffusion, advection, wet and dry deposition and chemical and physical transformations.  Figure 1 is a description of the various components of this model.  The degree of sophistication of the atmospheric models utilized in this scheme ranges from simple Gaussion diffusion models to advanced models that are based on fluid dynamics equations, chemical kinetics reactions, and attachment and recombination of ions and aerosols.

(2)  Hydrospheric model describing the fate of pollutants in water. It accounts for diffusion and transport, hydrolysis, biolysis, volitalization, photolysis and exchange with sediment reservoirs. Figure 2 is a description of the various components of this model.  The degree of sophistication in this scheme ranges from simple two-dimensional analytical equations to advanced three-dimensional fluid dynamics models that account for chemical and biological kinetics.

(3)  Pedospheric model describing the fate of pollutants in soil. It accounts for deposition rate on ground surfaces, quality of irrigation water, quantity of rainfall and the soil characteristics (field irrigation efficiency).  Figure 3 is a description of the various components of this model.  In this scheme a steady state model is employed.

Biological Environment:

(4)  Biospheric model describing the fate of pollutants in biota. Figures 4 and 5 describe qualitatively the various components of this model.  Basically, the model describes pollutant uptake and concentrations in the trophic levels of terrestrial and aquatic ecosystems accounting for various processes such as ingestion, inhalation and external exposure (dry deposition and immersion in water).  Output of this model provides the total human body burden under steady state conditions or the daily dose of a pollutant.

The biospheric model described in Figures 4 and 5, has been quantified using methodologies similar to those outlined by Le Clare et al[1] and various U.S. Nuclear Regulatory Commission (NRC) publications.[2,3]  The various equations used to calculate pollutant concentrations in lower forms of biota and man are as follows (see nomenclature for terms):

I. Pollutant Concentrations in a Primary Terrestrial Organism

Primary terrestrial organisms considered in this analysis are vegetative. Other primary organisms can include animals and man. Pollutant uptake by vegetation consists of: (1) uptake due to dry deposition on the vegetation surfaces, (2) pollutant uptake by spray irrigation through foliar deposition and (3) pollutant uptake from soil via the roots. The expression for the pollutant concentration in vegetation is:

$$C_{pt} = C_d + C_i + C_r$$

or:

$$
\begin{aligned}
C_{pt} = {} & 8.64 \times 10^4 \ \frac{C_a V_d f_r f_{gs}}{P \lambda_{pt}} \ [1 - \exp(-\lambda_{pt} t_d)] \\
& + 1.44 \ T_w f_r C_w \ \frac{V}{P} \ [1 - \exp(-\lambda_{pt} t_g)] \\
& + 4.46 \times 10^{-6} f_t f_s C_w \ \frac{V}{\lambda_\ell} \ [1 - \exp(-\lambda_\ell t_i)]
\end{aligned}
\tag{1}
$$

II. Pollutant Concentration in a Secondary Terrestrial Organism Due to the Consumption of a Primary Terrestrial Organism

The concentration of a pollutant in a secondary organism (animals for example) consuming vegetation with a pollutant concentration of $C_{pt}$, μg/g, is given by the expression:

$$C_{st} = 1.44 \ T_{st} P_s f_{st} \ \frac{C_{pt}}{m_s} \ [1 - \exp(-\lambda_{st} t_{in})] \tag{2}$$

III. Pollutant Concentration in Primary Aquatic Organisms

Pollutant concentration in primary aquatic organisms, such as vegetation, is given by the expression:

$$C_{pa} = C_w B_p \tag{3}$$

IV. Pollutant Concentration in a Secondary Aquatic Organism Consuming a Primary Aquatic Organism

The internal concentration in a secondary organism (e.g., fish) consuming a primary aquatic organism is given by the expression:

$$C_{sa} = 1.4 \ T_{sa} P_s f_{sa} \ \frac{C_w B_p}{m_s} \ [1 - \exp(-\lambda_{sa} t_{in})] \tag{4}$$

V. Pollutant Concentration in a Tertiary Terrestrial Organism Due to the Consumption of a Secondary Organism

The pollutant concentration in a tertiary organism (e.g., a carnivore) consuming a secondary organism such as meat with a pollutant concentration $C_{st}$, μg/g, is given by the expression:

$$C_{tt} = 1.44 \ T_{tt} \ P_t \ f_{tt} \ \frac{C_{st}}{m_t} \ [1 - \exp(-\lambda_{tt} t_{in})] \tag{5}$$

VI.  Pollutant Concentration in an Organism Due to Drinking Water

The pollutant concentration in an organism (e.g., Man) consuming water is given by:

$$D_{dw}C_{dw} = 1.44 \ T_{tt}P_{w}f_{ww} \frac{C_{w}}{m_{t}} \ [1-\exp(-\lambda_{tt}t_{id})] \tag{6}$$

VII.  Pollutant Concentration in an Organism Due to Inhalation

The pollutant concentration due to inhalation by an organism (e.g., man) is given by:

$$D_{a}R_{a} = 1.44 \ T_{tt}b_{r}f_{bi} \frac{C_{a}}{m_{t}} \ [1-\exp(-\lambda_{tt}t_{bi})] \tag{7}$$

VIII.  Pollutant Concentration in an Organism Due to Dry Deposition and Immersion in Water

Pollutant concentration in an organism (e.g., man) due to dry deposition and immersion is given by:

$$D_{ex}R_{ex} = 8.64 \times 10^{6} \ \frac{C_{a}V_{d}f_{ra}f_{ghs}}{P \ \lambda_{tt}} \ [1-\exp(-\lambda_{e}t_{a})]$$

$$+ \ 1.44 \ T_{tt}f_{rw}C_{w}A_{p}[1-\exp(-\lambda_{tt}t_{w})] \tag{8}$$

Human exposure is described in two modes:

o    total daily dose, the total amount of pollutant uptake per day

o    total body burden, the steady state concentration of a pollutant retained in the body.

IX.  The total daily dose $D_{t}$ to man in µg/day is:

$$D_{t} = C_{pt}I_{pt} + C_{st}I_{st} + C_{st}'I_{st}' + C_{pa}I_{pa} + C_{sa}I_{sa}$$

$$+ \ C_{dw}I_{dw} + C_{a}I_{a} + C_{ex}U_{ex} \tag{9}$$

where, $C_{pt}$, $C_{st}$, $C_{st}'$, $C_{pa}$, $C_{sa}$, $C_{dw}$, $C_{a}$ and $C_{ex}$ respectively are pollutant concentrations in lower biota or in the physical environment, and:

$I_{pt}$, $I_{st}$, $I_{st}'$ $I_{pa}$, $I_{sa}$, $I_{dw}$, $I_{a}$ and $U_{ex}$ are the corresponding daily intake rates.

X.  The total body burden $C_{T}$ in man in µg/g is:

$$C_{T} = D_{pt}R_{pt} + D_{st}R_{st} + D_{st}'R_{st}' + D_{pa}R_{pa} + D_{sa}R_{sa}$$

$$+ \ D_{dw}R_{dw} + D_{a}R_{a} + D_{ex}R_{ex} \tag{10}$$

where, $D_{pt}$, $D_{st}$, $D_{st}'$, $D_{pa}$, $D_{sa}$, $D_{dw}$, $D_{a}$ and $D_{ex}$ respectively are the daily doses due to individual organisms, air and deposition and immersion with subscript nomenclature similar to that for the pollutant concentrations (for example pt stands for dose from a primary terrestrial organism). $R_{pt}$, $R_{st}$, $R_{st}'$, $R_{pa}$, $R_{sa}$, $R_{dw}$, $R_{a}$, $R_{ex}$

are the corresponding retention factors in man that are functions
of the biological half life, fraction of pollutant retained, fraction
remaining after biological decay and length of time of exposure.

APPLICATION OF THE HUMAN EXPOSURE MODEL

A summary of the fate of pollutants in the total environment and exposure
to man is given in Figure 6. The first step in the human exposure assessment involves estimates of pollutant concentration in the physical environment (air, water, soil) and the possible interactions between the media.
Results of calculations yield concentrations in the three media to which
the biological environment is exposed. These concentrations are used as
input to the biospheric model which includes an assessment of pollutant
concentrations in the food chain, as well as pollutant concentrations in
man due to inhalation, deposition and immersion.

## Fate of Mercury in the Environment and Human Exposure

Mercury is used as an example of trace metal behavior in the environment.
For the purpose of this example, a coal gasification facility located on
a major river in the United States is considered. Energy facility characteristics and maximum discharges, as well as a description of the environment are given in Table 4.

Also in this example, which represents a worst case situation, it is
assumed that:

1.  Humans are living 2 kilometers from the facility,
2.  Their primary sources of food (including beef, milk, fish and vege-
    tables) are of local origin,
3.  Their drinking water is obtained from the river at a distance 23
    meters from the outlet,
4.  Fish spend predominant residence time 23 meters from the outlet,
5.  Beef and milk cattle are grazing for 10 percent of their food, 2
    kilometers from the facility,
6.  Vegetables are grown 2 kilometers from the facility and are irri-
    gated to supplement natural precipitation,
7.  The human of concern is the "standard man."

## Physical Environment

Assuming maximum operation of the plant and emissions as indicated in
Table 5, the atmospheric model (Figure 1) predicts a 24-hour average
concentration of $3 \times 10^{-3}$ $\mu g/m^3$ incremental above ambient ($2 \times 10^{-2}$ $\mu g/m^3$)
2 kilometers downwind from the plant (Table 5). Although this concentra-
tion is high, this worst case will assume it to be the concentration to
which man is exposed every day. The concentration will diminish with
distance from the plant to a point of nearly ambient concentration 5
kilometers downwind.

The hydrospheric model (Figure 2) predicts the concentration of mercury
23 meters downstream from the output source to be 0.001 mg/l (Table 5).
This concentration includes an ambient concentration of 0.6 $\mu g/l$. At a
distance of 65 meters downstream the mercury concentration in the river
is projected to be nearly equivalent to the ambient concentration.

The pedosphere model (Figure 3) predicts the concentration of mercury 2
kilometers from the gasification facility to be 0.22 $\mu g/m^3$ (Table 5).
Five kilometers from the source the concentration in soil is 0.17 $\mu g/m^3$.

407

The values obtained from the physical environment models are used in determining the fate of mercury in the biosphere and finally into man.

## Biological Environment

Considering the terrestrial food chain (Figure 4), primary uptake by plants can occur through soil, water and air. These inputs have been taken into account in Equation 1. Solution of this equation results in a concentration of $3.81 \times 10^{-4}$ µg of mercury per gram of plant material. This vegetation will serve as a food source for herbivorous organisms.

An example of a herbivore consuming vegetation is a cow eating pasture grass. The concentration of mercury in cattle is $4.38 \times 10^{-5}$ µg/g and is obtained from Equation 2. This would be the concentration in the beef at the time of slaughter. It is assumed for this example that the concentration of mercury in a cow's milk would be approximately 7 percent of that in the muscle. Therefore, the mercury concentration in milk equals $2.92 \times 10^{-6}$ µg/g.

In the aquatic foodchain, uptake of mercury by phytoplankton is stated by Equation 4. This equation by-passes several considerations and deals directly with a bioconcentration factor. Using this equation, the concentration of mercury in phytoplankton is about 1.0 µg/g.

Consumption of phytoplankton by a herbivorous fish is considered in Equation 5 and results in a concentration of 6.8 µg of mercury per gram of fish.

## Human Exposure

Solving equation 9 using the mercury concentrations obtained previously for lower steps in the food chain results in a daily dose of mercury to a 70 kg standard man of 20.31 µg/day (Table 6).

The main sources of mercury uptake in man is from food, 17.85 µg/day. Fish products serve as the dominant contributor of mercury to man's daily food intake, 17.59 µg out of a total of 17.85 µg.

Equation 10 is used to calculate the total body burden, under steady state conditions, to man from different sources of mercury. The burden from each source can be calculated. For example, with a beef consumption rate of 213.7 g/day[3] and the half-life of mercury in man of 69 days,[10] the body burden from man consuming beef is $9.93 \times 10^{-6}$ µg/g. Table 7 indicates man's body burden for each major source of mercury. In this study, the total body burden of mercury to a standard man would be 1.37 mg at steady state (Table 7).

## Health Effects of Mercury

Both acute and chronic effects may occur in man as a result of exposure to mercury. Acute effects have been reported:[10]

1.  Inhalation of mercury vapor greater than 1 mg $Hg/m^3$ can damage lung tissue, causing acute mercurial pneumonitis

2.  Ingestion of $HgCl_2$ between 1 and 4 grams is lethal.

Most chronic effects are related to central nervous system dysfunction:

1. Chronic exposure to average air concentrations of 0.1 mg/m$^3$ have resulted in mental disturbances, tremors and gingivitis.

2. Chronic exposure as low as 0.06 mg/m$^3$ have resulted in non-specific symptoms such as loss of appetite, weight loss and shyness.

The Lowest Observable Effect Level (LOEL) of mercury occurs when blood levels are at 200 ng to 500 ng Hg/ml[10] (Table 8). This blood concentration corresponds to a total body burden of 30 to 50 mg/70Kg body weight and to long-term daily doses in the range of 200 to 500 µg Hg/70Kg. The LOEL can be used to establish an Acceptable Daily Intake (ADI). In this case, the ADI is 200 µg/day.[10]

According to the National Academy of Science (1977),[10] an uncertainty factor of ten can be applied to the ADI, because the LOEL were obtained from studies on prolonged ingestion by man and there was no indication of carcinogenicity. This would lower the daily uptake to 20 µg for a 70 Kg person to assure a margin of safety.

Estimated human exposure in this example is a total daily dose of 20.31 µg. This dose includes dietary intake of fish, meat, milk and vegetables, as well as intake due to drinking and inhalation. This total dose is ten-fold below the LOEL, and is equivalent to the ADI after the uncertainty factor of 10 has been applied. Thus, the estimated exposure level is within the margin of safety recommended by the National Academy of Science in spite of the use of worst case conditions.

## Fate of Benzo(a)pyrene in the Aquatic Environment, Human Exposure, Health Effects

Benzo(a)pyrene (BaP) is used as an example of a priority organic released into the environment from a coal gasification facility. Data for air releases of BaP could not be obtained, but water discharge data were found. For the purposes of this example a discharge rate of $1.382 \times 10^{-4}$ Kg/hr is assumed into the same environment described for mercury.

The hydrospheric model (Figure 2) estimates the highest BaP concentration in water to be $9.5 \times 10^{-10}$ µg/l at a distance 3 kilometers from the source. In this example the source of drinking water to man and the predominant residence of fish will occur at 3 kilometers from the source.

Using equation 3, the concentration of BaP in aquatic vegetation is $3.8 \times 10^{-8}$ µg/g, and using equation 4, the concentration in fish is $2.97 \times 10^{-8}$ µg/g. The total daily dose to man through the aquatic food chain and drinking water is $6.13 \times 10^{-8}$ µg/day. This daily dose results in a total body burden, at steady state, of $1.9 \times 10^{-11}$ µg/g or $1.3 \times 10^{-6}$ µg for a 70 Kg standard man.

BaP is a known carcinogen. Increased incidences of cancer have been correlated with increased exposures of BaP to man. An increase in cancer incidence of one in 100,000 ($10^{-5}$), is correlated with a concentration of 9.7 ng/l BaP in drinking water.[12] Table 9 presents the possible health risks of BaP and the exposures to BaP estimated from the model. The estimated exposures are considerably lower ($10^{-5}$) than the exposure which correlate to a cancer risk of one in 100,000.

NOMENCLATURE:  Abbreviations, Definition, Values (References)

$C_d$ = pollutant concentration in vegetation due to dry deposition of a pollutant on the vegetation surface, $\mu g/g$

$C_i$ = pollutant concentration in vegetation due to spray irrigation, $\mu g/g$

$C_r$ = pollutant concentration in vegetation from the soil via the roots, $\mu g/g$

$C_a$ = average ambient air concentration of a pollutant at the receptor, $3.0 \times 10^{-3}$ $\mu g/m^3$

$V_d$ = dry deposition velocity of a pollutant, $1.0 \times 10^{-2}$ m/sec

$f_r$ = the fractional retention of a pollutant on the vegetative surface, 0.25 (1)

$f_{gs}$ = the average fraction of the ground surface area that is covered with the vegetation, 0.75 (1)

$\lambda_{pt}$ = effective biological decay constant of a pollutant in vegetation, 0.053 days$^{-1}$ (3)

$t_d$ = the duration of deposition of a pollutant on the vegetation, 60 days (1)

$P$ = average vegetation density, 2700 $g/m^2$ (1)

$T_w$ = effective half-life of a pollutant due to weathering or removal, 13 days (3)

$C_w$ = concentration of a pollutant in water, $1 \times 10^{-3}$ $\mu g/cm^3$ (5)

$V$ = volume of irrigation water applied to vegetation, 905 $cm^3/m^2$-day (1)

$t_g$ = growth time of vegetation, 60 days (1)

$f_t$ = transfer coefficient from soil to vegetation via roots, 0.56 $\mu g/kg$ fresh weight per $\mu g/kg$ dry soil (4)

$f_s$ = the fraction of a pollutant that reaches soil, 0.75 (3)

$\lambda_\ell$ = the fractional loss of a pollutant in soil due to leaching, approaching 0 days$^{-1}$ (6)

$t_i$ = the time interval over which irrigation has been assumed to occur, 30 days (1)

$T_{st}$ = effective biological half-time of a pollutant in the secondary terrestrial organism, 10 days (assumed same as man, 7)

$P_s$ = consumption rate of vegetation by the secondary terrestrial organism, 5000 g/day (3,8)

$f_{st}$ = the fraction of ingested pollutant that is retained in the body of the secondary terrestrial organism, 0.75 (assumed same as man, 7)

$\lambda_{st}$ = effective biological decay constant of a pollutant in the secondary terrestrial organism, 0.0693 days$^{-1}$ (assumed same as man, 7)

$t_{in}$ = period of ingestion, 730 days (8)

$m_s$ = mass of the secondary terrestrial organism, $4.54 \times 10^5$ g (9)

$T_{tt}$ = effective biological half-life of a pollutant in the tertiary organism, 69 days (10)

$f_{tt}$ = the fraction of ingested pollutant that is retained in the body of the tertiary organism, 0.75 (7)

$\lambda_{tt}$ = effective biological decay constant of a pollutant in the tertiary organism, 0.010 days$^{-1}$ (10)

$m_t$ = mass of the tertiary organism, 70,000 g (7)

$P_{tt}$ = consumption rate of secondary organism by the tertiary organism, 213.7 g/day (3)

$B_p$ = the bioaccumulation factor for a pollutant in a primary aquatic organism, 1000 µg/g / µg/m$^3$ (11)

$T_{sa}$ = effective biological half-life of a pollutant in a secondary aquatic organism, 112 days (10)

$P_s$ = consumption rate of primary aquatic organism by secondary organism, 25 g/day (assumed average)

$f_{sa}$ = the fraction of ingested pollutant (in the aquatic organism) that is retained in the body of the secondary organism, 0.75 (assumed)

$\lambda_{st}$ = effective biological decay constant of a pollutant in the secondary organism under consideration, 0.0062 days$^{-1}$ (10)

$m_s$ = mass of the secondary organism under consideration, 500 g (assumed)

$P_w$ = consumption rate of water by the organism under consideration, 2000 g/day (7)

$f_{ww}$ = the fraction of a pollutant in the drinking water that is retained in the organism under consideration, 0.80 (10)

$t_{id}$ = period of drinking, 25,550 days (3)

$b_r$ = breathing rate of air by a human, 20 m$^3$/day (3)

$f_{bi}$ = the fraction of inhaled pollutant that is retained in the body of a human, 0.75 (7)

$t_{bi}$ = period of inhalation, 25,550 days (3)

411

$f_{ra}$ = the fractional retention of a pollutant on skin due to ambient air exposure, o (10)

$f_{ghs}$ = the average fraction of ground surface area occupied by human

$t_a$ = exposure time to ambient outdoor air, days

$f_{rw}$ = the fractional retention of a pollutant on skin due to immersion in water

$A_p$ = absorption and adsorption rate of a pollutant onto the skin due to immersion in water, 0 $\mu g/g$ / $\mu g/m^3$ (10)

$t_w$ = exposure time during immersion, days

$I_b$ = daily ingestion of beef, 213.7 g/day (1)

$I_m$ = daily ingestion of milk, 798.4 g/day (1)

$I_f$ = daily ingestion of fish, 18.7 g/day (1), 2.2 g/day fresh water fish (10)

$I_v$ = daily ingestion of vegetables, 645.2 g/day (1)

$I_w$ = daily ingestion of water, 2000 g/day (1)

$I_a$ = daily ingestion of air, 20 $m^3$/day (1)

REFERENCES

1.  LeClare, P., et al, Standard Methodology for Calculating Radiation Dose to Lower Form Biota, Report AIF/NESD-006, Washington, DC, Atomic Industrial Forum Inc., 1975.

2.  U.S. Nuclear Regulatory Commission, "Calculation of Annual Doses to Man from Routine Releases of Reactor Effluents for the Purpose of Evaluating Compliance with 10 CFR Part 50 Appendix I," Regulating Guide 1.109, October 1977.

3.  U.S. Nuclear Regulatory Commission, "Calculational Models for Estimating Radiation Doses to Man from Airborne Radioactive Materials Resulting from Uranium Milling Operations, Draft Regulatory Guide and Value Impact Statement, "Task RH 802-4, May 1979.

4.  Lindberg, S.E., et al, "Atmospheric Emission and Plant Uptake of Mercury from Agricultural Soils near the Almaden Mercury Mine," Journal of Environmental Quality, 8 (1979), 572-578.

5.  Memphis Light, Gas & Water Division, "Industrial Flue Gas Demonstration Plant Environmental Permitting Overview," Memphis, Tennessee, 1980.

6.  Maser, J.A., Energy Impact Associates, Inc., personal communication to Roffman, H.K., Energy Impact Associates, Inc., November 23, 1980.

7.  International Commission on Radiological Protection, Report of Committee II on Permissible Dose for Internal Radiation, ICRP Publication 2, London, Pergamon Press, 1959.

8.  U.S. Department of Agriculture, "Consumption of Feed by Livestock, 1909-56; Relation Between Feed, Livestock, and Food at the National Level," Production Research Report No. 21, Washington, DC, Government Printing Office, 1957.

9.  U.S. Department of Agriculture, Economics, Statistics, and Cooperation Service, "Livestock and Meat Statistics, Supplement for 1978," Statistical Bulletin No. 522, Washington, DC, Government Printing Office, 1978.

10. U.S. Environmental Protection Agency, Office of Water Planning and Standards, "Mercury, Ambient Water Quality Criteria, #297925, Washington, DC, 1979.

11. Thompson, S.E., et al, "Concentration Factors of Chemical Elements in Edible Aquatic Organisms," UCRL-50564, Rev. 1, Livermore, California, Lawrence Livermore Laboratory, University of California, 1972.

12. Environmental Protection Agency, "Water Quality Criteria; Availability," Federal Register, 44, Part XI, (October 1, 1979), 56655-56657.

413

TABLE 1

A SUMMARY OF SOURCES AND TYPE OF POSSIBLE EMISSIONS AND DISCHARGES
FROM COAL GASIFICATION FACILITIES

| Module | Air | Waste Water | Solid Waste |
|---|---|---|---|
| Coal Pretreatment | Particulates, organic compounds | TDS, TSS, organics, inorganics | Particulate coal, trace elements |
| Coal Gasification | Particulates, oil tar aerosols, sulfur cyanide, organic species and trace elements | TDS, TSS, $NH_3$, sulfides, trace elements, cyanides, phenols, oil, grease and organics | Trace metals and organics, inorganics |
| Gas Purification | Particulates, sulfur species, organic compounds, trace elements, cyanides, $NH_3$ | TDS, TSS, organics, sulfur and cyanide compounds, phenols, $NH_3$, trace elements | Carbon, sulfur compounds, organics, inorganics, trace elements, solid sludges |

TABLE 2

A SUMMARY OF SOURCES AND TYPE OF POSSIBLE EMISSIONS AND
DISCHARGES FROM COAL LIQUIFACTION FACILITIES

| Module | Air | Waste Water | Solid Waste |
|---|---|---|---|
| Coal Pretreatment | Particulates, hydro-carbon vapors | TDS, TSS, organics, inorganics | Particulate coal, trace elements |
| Coal Liquifaction | CO, $NO_x$, organics, trace elements, sulfur compounds | Phenols, tars, $NH_3$, thiocynates, sulfides, chlorides, spent catalyst, trace elements and condensate | Particulate coal, ash, slag, mineral matter, char, spent catalyst, trace elements |
| Separation | Organics, sulfur compounds, $NH_3$, particulates, trace elements | Oils, hydrocarbons, phenols, $NH_3$, sulfides, trace elements | Particulate coal, ash, slag, mineral matter, trace elements |
| Purification and Upgrading | $H_2S$, $CO_2$, CO, $NO_x$, $NH_3$, organics, particulates, trace elements | Light hydrocarbons, dissolved salts, phenols, $NH_3$, sulfides, trace elements | Particulate coal, spent catalyst, ash, slag, mineral matter, char, trace elements |
| Auxiliary Processes such as: Water Cooling Oxygen Generation Sulfur Recovery Acid Gas Removal Control Equipment Land Fill | $CO_2$, CO, organics, sulfur compounds, $NO_x$, $NH_3$, trace elements, spray bio-cides, anticorrosives, particulates, fly ash | Phenols, tars, $NH_3$, sulfur and cyanide compounds, chlorides, trace elements, TDS, TSS, organics, oils | Particulate coal, spent catalyst, ash, slag, mineral matter, char, spent absor-bent, sludges, spent regen-erants, organic residues, sulfur compounds |

## TABLE 3

### HUMAN EXPOSURE AND HEALTH EFFECTS MODEL

| ELEMENTS | | CONSIDERATIONS |
|---|---|---|
| SOURCE CHARACTERIZATION: | o AIR<br>o WATER<br>o SOLID WASTE | ROUTINE AND ACCIDENTAL RELEASES |
| FATE OF POLLUTANTS: | o ATMOSPHERE<br>o HYDROSPHERE<br>o PEDOSPHERE<br>o BIOSPHERE | PHYSICAL AND BIOLOGICAL INTERACTIONS |
| HUMAN EXPOSURE:<br>(DOSE) | o INGESTION<br>o INHALATION<br>o DEPOSITION | PUBLIC AND OCCUPATIONAL EXPOSURES |
| HEALTH DAMAGE FUNCTION: | o ACCUTE<br>o CHRONIC<br>o CARCINOGENIC | PUBLIC AND OCCUPATIONAL EXPOSURES |

## TABLE 4

### SITE DESCRIPTION SUMMARY

**Energy Facility**

| | |
|---|---|
| Type: | Coal Gasification Plant |
| Input Coal: | Mixed Eastern |
| Input Rate: | 3200 tons/day |
| Mercury Content: | 0.19 ppm |
| Mercury Emitted as Gaseous Form: | 95% |
| Mercury Emitted in Wastewater: | 0.002 mg/l |
| Wastewater Flow Rate: | 144 m$^3$/hr |

**Environment**

| | |
|---|---|
| Type: | River Valley |
| Topography: | Nearly Flat |
| Soils: | Alluvial, Dredged Material, Sand and Gravel Over Very Dense Silty Clay |
| Hydrology: | Flow Rate of River = 1.7 x 10$^{10}$ m$^3$/hr |
| Meteorology: | Average Annual Precipitation = 50 Inches<br>Average Annual Wind = 9 MPH |
| Terrestrial Ecology: | Vegetation - Agricultural Crops, Forested Area, Herbaceous Meadow<br>Wildlife - Typical Floodplain Mammals<br>Birds, Amphibians and Reptiles |
| Aquatic Ecology: | Vegetation - Diatoms, Green and Blue-green Algae<br>Fish - Shad, Carp, Catfish |
| Land Use: | Rural within 3 Kilometers<br>Suburban within 10 Kilometers |

## TABLE 5

### CONCENTRATIONS OF MERCURY IN THE PHYSICAL ENVIRONMENT

ATMOSPHERE

| Distance from Source | Concentration |
|---|---|
| 2 Kilometers | 23 ng/m³ |
| 5 Kilometers | ~20 ng/m³ (Background) |

HYDROSPHERE

| Distance from Source | Concentration |
|---|---|
| 23 Meters | 0.001 mg/l |
| 65 Meters | ~0.0006 mg/l (Background) |

PEDOSPHERE

| Distance from Source | Concentration |
|---|---|
| 2 Kilometers | 0.22 µg/m³ |
| 5 Kilometers | 0.17 µg/m³ |

## TABLE 6

### DAILY DOSE OF MERCURY TO MAN

| Source of Mercury | Daily Dose | Percent of Total |
|---|---|---|
| Beef | 0.009 µg/day | 0.05 |
| Milk | 0.002 µg/day | 0.01 |
| Vegetables | 0.246 µg/day | 1.21 |
| Fish | 17.60 µg/day | 86.63 |
| Water | 2.00 µg/day | 9.84 |
| Air | 0.460 µg/day | 2.26 |
| TOTAL | 20.32 µg/day | 100.00 |

## TABLE 7

### MERCURY BODY BURDEN TO MAN

| Source of Mercury | Body Burden |
|---|---|
| Beef | $9.93 \times 10^{-6}$ µg/g |
| Milk | $2.48 \times 10^{-6}$ µg/g |
| Vegetables | $2.55 \times 10^{-4}$ µg/g |
| Fish | $1.87 \times 10^{-2}$ µg/g |
| Water | $2.77 \times 10^{-4}$ µg/g |
| Air | $3.41 \times 10^{-4}$ µg/g |
| Total | $1.96 \times 10^{-2}$ µg/g |
| For "Standard Adult" 70 kg | 1.37 mm/70 kg |

## TABLE 8

### HEALTH EFFECTS OF MERCURY

| Measure | Recommended | Estimated for the Example Under Consideration |
|---|---|---|
| Lowest Observable Effect Level (LOEL) Corresponds to: | | |
| 1. Total Body Burden | 30 to 50 mg/70 kg | 1.37 mg/70 kg |
| 2. Long-term Daily Dose | 200 to 500 µg/70 kg | 20 µg/70 kg |
| Acceptable Daily Intake (ADI) | 200 µg/day | 20 µg/day |

## TABLE 9

### HEALTH EFFECTS OF BaP

| Cancer Risk | Recommended | | Estimated |
|---|---|---|---|
| $10^{-7}$ | Total Body Burden | $1.3 \times 10^{-1}$ µg/70 kg | $1.2 \times 10^{-6}$ µg/70 kg |
| | Daily Dose | $6 \times 10^{-3}$ µg/day | $5.8 \times 10^{-8}$ µg/day |
| $10^{-5}$ | Total Body Burden | 13 µg/70 kg | |
| | Daily Dose | $6 \times 10^{-1}$ µg/day | |

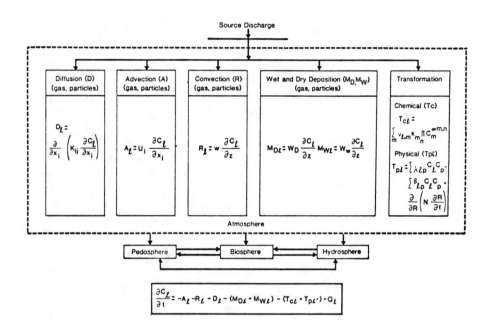

The nomenclature of parameters, variables and terms used in the equation describing the fate of pollutants in the atmosphere is as follows:

where:

$D_\ell$ = Diffusion term

$A_\ell$ = Advection

$R_\ell$ = Convection

$M_{W\ell}$ = Wet deposition

$M_{D\ell}$ = Dry deposition

$T_{c\ell}$ = Chemical transformation

$T_{p\ell}$ = Physical transformation

$Q_\ell$ = Loading of substance into the system

$N$ = Number density of particles

$C_\ell$ = Concentration of species

$X_i$ = Cartesian coordinates, x,y,z

$K_{ij}$ = Eddy diffusion coefficients

$U_i$ = Horizontal wind speed components u,v

$R$ = Particle radius

$w$ = Vertical wind speed

$W_D$ = Dry deposition velocity

$W_W$ = Wet deposition velocity

$\nu_{\ell,m}$ = Stoichiometric coefficient for components p in the m th reaction

$k_m$ = Reaction rate constant for the m th reaction

$\alpha_{m,n}$ = Order of reaction of species n in the m th chemical reaction

$C_p$ = Concentration of particles p

$\beta_{\ell,p}$ = Attachment coefficient between particles $\ell$, and p

$\lambda_{\ell,p}$ = Recombination coefficient between between particles $\ell$, and p

FIGURE 1 - FATE OF POLLUTANTS IN THE ATMOSPHERE

417

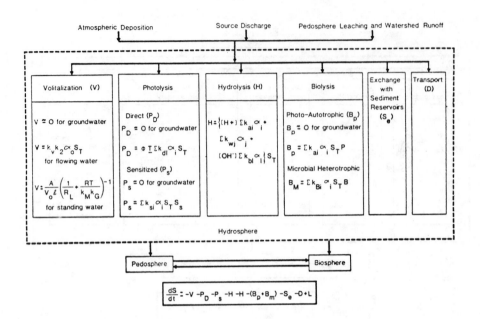

The nomenclature of parameters, variables and terms used in the equation describing the fate of pollutants in the hydrosphere is as follows:

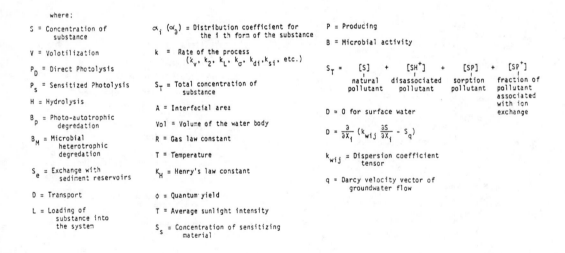

where:

S = Concentration of substance

V = Volotilization

$P_D$ = Direct Photolysis

$P_s$ = Sensitized Photolysis

H = Hydrolysis

$B_p$ = Photo-autotrophic degredation

$B_M$ = Microbial heterotrophic degredation

$S_e$ = Exchange with sediment reservoirs

D = Transport

L = Loading of substance into the system

$\alpha_i$ ($\alpha_g$) = Distribution coefficient for the i th form of the substance

k = Rate of the process ($k_v$, $k_2$, $k_L$, $k_G$, $k_{di}$, $k_{si}$, etc.)

$S_T$ = Total concentration of substance

A = Interfacial area

Vol = Volume of the water body

R = Gas law constant

T = Temperature

$K_H$ = Henry's law constant

$\phi$ = Quantum yield

T = Average sunlight intensity

$S_s$ = Concentration of sensitizing material

P = Producing

B = Microbial activity

$S_T = [S] + [SH^+] + [SP] + [SP^+]$
  natural pollutant  disassociated pollutant  sorption pollutant  fraction of pollutant associated with ion exchange

D = 0 for surface water

$D = \frac{\partial}{\partial X_i}\left(k_{wij}\frac{\partial S}{\partial X_i} - S_q\right)$

$k_{wij}$ = Dispersion coefficient tensor

q = Darcy velocity vector of groundwater flow

FIGURE 2 - FATE OF POLLUTANTS IN THE HYDROSPHERE

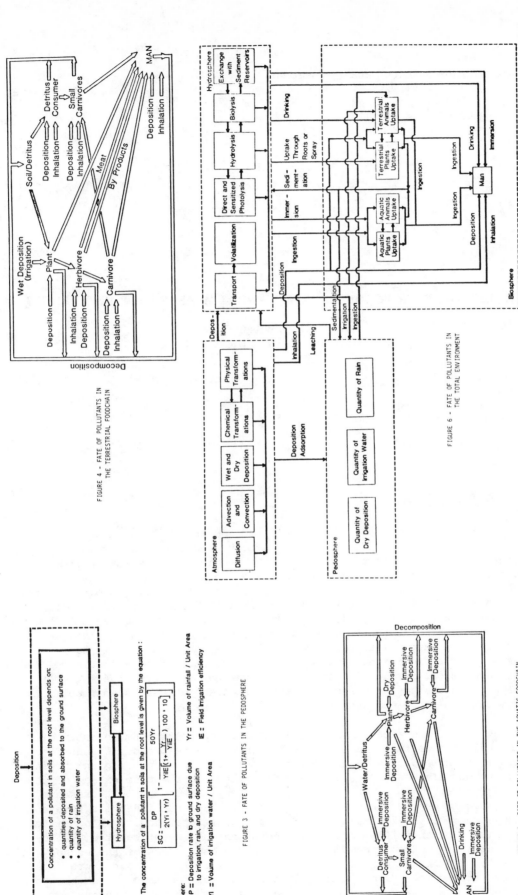

FIGURE 4 - FATE OF POLLUTANTS IN
THE TERRESTRIAL FOODCHAIN

FIGURE 6 - FATE OF POLLUTANTS IN
THE TOTAL ENVIRONMENT

The concentration of a pollutant in soils at the root level is given by the equation:

$$SC = \frac{DP}{2(Yi + Yr)} \left[ 1 - \frac{50\,Yr}{YiIE \left(1 + \frac{Yr}{YiIE}\right) 100 \cdot 10} \right]$$

Where:

DP = Deposition rate to ground surface due
to irrigation, rain, and dry deposition

Yi = Volume of irrigation water / Unit Area

Yr = Volume of rainfall / Unit Area

IE = Field Irrigation efficiency

FIGURE 3 - FATE OF POLLUTANTS IN THE PEDOSPHERE

FIGURE 5 - FATE OF POLLUTANTS IN THE AQUATIC FOODCHAIN

419

## INDUSTRIAL ENERGY CONSERVATION—WHAT MEASURE FOR EFFICIENCY?

**D. B. Wilson**
New Mexico State University

Introduction

The current National Energy Plan is first and foremost an energy conservation strategy.  Whether this continues to be federal policy remains to be
determined.  What is certain is that the chemical and petroleum industry
have accepted and are accomplishing energy conservation.  Impetus is
found in fuel costs but also in acceptance that the U.S. economy does not
use energy efficiently.  For every one million dollars of gross domestic
product, the U.S. uses 1,513 tons  of oil or its energy equivalent; by
contrast looking at countries with equivalent standards of living, West
Germany uses just 997 tons;  Sweden, 865 tons; and France, 817 tons.[1]
These figures are further analyzed to indicate that two categories of use
are major areas for improvement - industry and transportation (automobiles).  This study concerns only industrial energy use, specifically the
chemical and petroleum industry.

As mentioned, the acceptance by the chemical and petroleum industry of
the challenge of improved energy efficiency is widespread.  Witness the
papers at this conference and the many publications appearing in the industry journals.  What has not been determined is the true extent of the
energy conservation effort that can be accomplished.  While there have
been suggested guidelines for energy conservation published i.e. Project
Independence at 15-25% reduction, as part of federal policy, a consistent
definition of the efficient use of energy has not been accepted.  Allocations and allotments have been assigned in many areas of the economy
based on historical use or 'lobbied' compromise, rather than a pre-deter-
mined efficiency.  Mainly this exists because of the many approaches to
measuring the energy efficiency of petroleum and chemical processes.
Figure 1 lists six thermodynamic approaches or thermodynamic functions used to measure energy efficiency and gives their definitions.[2-7]

This paper analyzes these six approaches as applied to ammonia production as a means for their comparison and evaluation.

Ammonia Process

The specific ammonia production process that will be examined is based on methane both as the primary raw feed material (plus air) and methane as fuel for all energy requirements in the process.[3] Actual operating experience (1975) for this type of process shows that $40.4*10^6$ BTU's (methane equivalent) were being used to produce one ton of ammonia ($11.3*10^3$ kcal/kg). The basic question is whether this is an efficient operation or an inefficient operation.

Figure 2 is a representation of the energy used (Sankey diagram) for the ammonia process. This figure is also representative of the mass-energy balance method of energy-use analysis. Strictly speaking this analysis, i.e., Sankey diagram, should account for all energy flows for the process. Since this 'First Law' application accounts for all energy it implies a 100% efficient operation. More appropriately this form of analysis would result in a process efficiency based on the following definition:

$$\eta = \frac{\text{Theorectical}}{\text{Actual}} \qquad (1)$$

For the overall ammonia operations which were included in the 1972 survey, this definition gives a value ranging from 42% to 57% efficiency. This range of value is represented in the Sankey diagram, Figure 2 also. In addition the energy balance approach allows a person to focus on the individual steps of a process and identify operations where potential energy conservation measures could be applied.[8]

While Figure 2 summarizes the mass - energy balance method, the remaining thermodynamic based approaches are compared by examining three specific process changes in the ammonia process: (1) preheating primary reformer air, (2) revision to compressor surface condenser, and (3) separation of hydrogen from the convertor purge gas stream. These steps have been shown to lead to improved energy efficiency.[9]

Results and Discussion

A.  Preheating Primary Reformer Air

It is a well-established fact that the energy efficiency of any furnace can be improved by recovering waste heat in the flue gases to preheat combustion air. Most reforming furnaces already use some measure of heat exchange between the flue gases and the combustion air. This system is used to contrast the various thermodynamic functions shown in Figure 1. Figure 3 summarizes the evaluation of each of the thermodynamic variables for this operation. All appropriate thermodynamic property values are taken from Lewis and Randall.[10]

This sample calculation concerns the change in sensible heat of a process or utility stream. Its analysis should reflect that energy availability as BTU's (KJ) and hence energy efficiency is dependent on the value of the absolute temperature of the system. The values used are representative of the flow schematic for a former 600 ton per day ammonia plant near Carlsbad, New Mexico.

## B. Revision to Surface Condensor

This specific improvement (or alternative) is applicable to systems using gas compressors that are steam turbine driven. This type of compressor has a condensing-type steam driver which exhausts steam into the surface condenser. The essential feature of this operation from the standpoint of energy efficiency concerns vacuum losses occurred in operating above design during the summer months. Figure 4 summarizes the analysis.

This steam-drive turbine is operated as a cyclic process, i.e. the maximum thermodynamic efficiency is the Carnot efficiency. Operational efficiencies for the turbine are well represented by the ratio of the actual enthalpy change in expansion across the turbine compared to the theoretical (adiabatic, reversible). The specific example pertains to operation at above design capacity.

## C. Hydrogen Recovery from Purge Gas

The purge gas stream from the ammonia convertor loop is used to control the buildup of inerts due to recycle. Most operations return this purge stream to the head of the process as fuel for the primary reformer. Figure 5 summarizes the analysis.

This example calculation compares the energy efficiency of hydrogen as water vapor, the result of combustion, or as ammonia, the result of recovery and recycle through the convertor. Standard temperature and pressure are assumed, although the high pressure of the hydrogen (purge stream) facilitates its separation through membrane permeation.

Although Equation (1) is the approximate definition for efficiency that is applied in practice, i.e. the theoretical value is the 'design' value, it is in need of modification since most processes are not operating at 'design' conditions. A better working definition of efficiency is a ratio of some "yield" to some "expense".[6] Since it has been shown that exergy and availability are effectively calculated the same, the appropriate efficiency definition is

$$\eta = \frac{\text{Exergy Yield}}{\text{Exergy Expense}}$$

or

$$= \frac{\text{Availability Yield}}{\text{Availability Expense.}} \qquad (2)$$

## Conclusion

It should not be surprising that each of the thermodynamic measures of energy efficiency gives an indication of process changes that can result in energy conservation. It is further not surprising that while computer design and computer control are extremely useful tools for energy efficient operation of chemical and petroleum processes, it is still the engineer who initiates energy conservation measures.

As in all human ventures subjective judgements will result in the engineer making that measure for efficiency that he or she is most comfortable using. This paper has shown that while First Law evaluations alone are not sufficient, any of the approaches combining First and Second Law measures do provide the necessary and sufficient information for energy conservation decisions.

## Acknowledgement

This work was supported in part by the Energy and Minerals Department of the State of New Mexico through funds administered by the Energy Institute of UNM.  The author is solely responsible for its content.

## Nomenclature

| | |
|---|---|
| $E_s$, | Essergy, Kcal/kg |
| $E_x$, | Exergy, Kcal/kg |
| $\Phi$, | Availability, Kcal/kg |
| $H$, | Enthalpy, Kcal/kg |
| $U$, | Internal Energy, Kcal/kg |
| $S$, | Entropy, Kcal/kg $^\circ$K |
| $W$, | Work, kilocalories |
| $Q$, | Heat, kilocalories |
| $p$, | Pressure, absolute, atm |
| $T_o$, | Absolute Temperature (Surroundings), $^\circ$K |
| $T$, | Absolute Temperature (System), $^\circ$K |
| $\eta$, | efficiency |

## References

1.  Boretsky, M., "Opportunities and Strategies for Energy Conservation," Tech Review, 57, July/August 1977.

2.  Austin, L.G., "Fuel Efficiency via the Mass-Energy Balance," Chem Tech, 631, October 1974.

3.  Pinto, A. and P.L. Rogerson, "Impact of High Fuel Cost on Plant Design," CEP, 95, July 1977.

4.  Riekert, L., "The Efficiency of Energy-Utilization in Chemical Processes," CES, 29, 1613, December 1974.

5.  Sussman, M.V., "Standard Chemical Availability," CEP, 37, January 1980.

6.  Voight, H., "Evaluation of Energy Processes Through Entropy and Energy," RM-78-60, Int. Institute for Applied Systems Analysis, Austria, November 1978.

7.  Evans, R., "A Proof That Essergy is the Only Consistent Measure of Potential Work," PhD Dissertation, Dartmouth University, University Microfilms International, Ann Arbor, Michigan, 1970.

8.  Wilson, D.B., "Energy Conservation in the Petro/Chemical Industry," Energy and Minerals Department, Report #681234, State of New Mexico, December 1980.

9.  Maclean, D.L. and V.C. Chae, "Energy-Saving Modifications in Ammonia Plants," CEP, 98, March 1980.

10. Lewis, G.N. and M. Randall, "Thermodynamics," revised by K.S. Pitzer and L. Brewer, McGraw-Hill Book Company, New York, 1961.

Method          Reference

1. FIRST LAW          Austin, 2

$$\Delta U = \delta Q - \delta W$$

2. SECOND LAW        Pinto & Rogerson, 3

$$W_{LOST} = T_o \Delta S_{PROCESS} + Q$$

3. CHEMICAL POTENTIAL     Riekert, 4

$$\varepsilon = H - H_o - T_o (S-S_o)$$

4. AVAILABILITY        Sussman, 5

$$\Delta \phi = \Delta H - T_o.\Delta S$$

5. EXERGY          Voigt 6

$$Ex = H - T_o S$$

6. ESSERGY (essence of energy)    Evans, 7

$$E_s = U + P_o V - T_o S - \Sigma \mu_o N$$

Figure 1
Thermodynamic Measures for Energy Evaluation

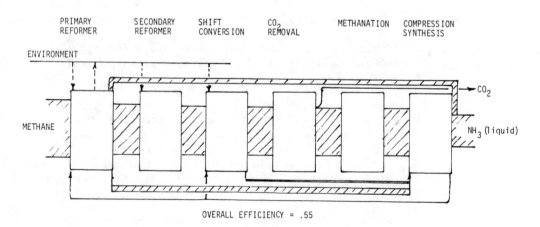

OVERALL EFFICIENCY = .55

Figure 2
Sankey Diagram for the Ammonia Process

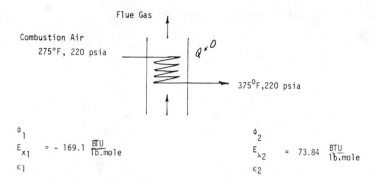

Flue Gas

Combustion Air
275°F, 220 psia

Q = 0

375°F, 220 psia

$\phi_1$
$E_{x_1}$ = - 169.1 $\frac{BTU}{lb.mole}$
$\epsilon_1$

$\phi_2$
$E_{x_2}$ = 73.84 $\frac{BTU}{lb.mole}$
$\epsilon_2$

First Law: $H_{AIR,2} - H_{AIR,1} = H_{GAS,1} - H_{GAS,2}$

Second Law: $W_{LOST} = T_0 \Delta S_{PROCESS} + Q$

Figure 3

Preheating Reformer Combustion Air

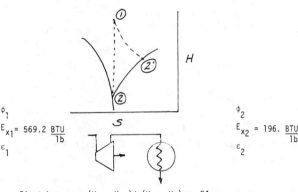

$\phi_1$
$E_{x_1}$= 569.2 $\frac{BTU}{lb}$
$\epsilon_1$

$\phi_2$
$E_{x_2}$ = 196. $\frac{BTU}{lb}$
$\epsilon_2$

First Law: $\eta = (H_1 - H_{2'})/ (H_1 - H_2) = .84$

Second Law: $W_{LOST} = T_0 (S_{2'} - S_2) = 49.2$ BTU/lb

Figure 4

Operation of Surface Condenser on Gas Compressor

Purge Gas

Hydrogen

Water Vapor

$\epsilon$
$\phi$    - 54.64 $\frac{Kcal}{gmole}$
$E_x$

$\epsilon$    $H_2$: -8.52 $\frac{Kcal}{mole}$
$\phi$    $O_2$L -13.4 $\frac{Kcal}{mole}$
$E_x$    $N_2$: -12.5 $\frac{Kcal}{mole}$

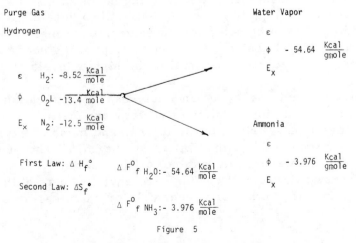

Ammonia

$\epsilon$
$\phi$    - 3.976 $\frac{Kcal}{gmole}$
$E_x$

First Law: $\Delta H_f^\circ$    $\Delta F^0_f$ $H_2O$:- 54.64 $\frac{Kcal}{mole}$

Second Law: $\Delta S_f^\circ$

$\Delta F^0_f$ $NH_3$:- 3.976 $\frac{Kcal}{mole}$

Figure 5

Separation of Hydrogen from the Purge Gas

# THE ENVIRONMENTAL DESIRABILITY OF NUCLEAR PROCESS HEAT

**Leon Green, Jr.**
Consultant

## Introduction

> Nuclear energy is unusable for any purpose other than electric power production.
>
> -James R. Schlesinger [1]

Realization that heat delivered to a chemical process"...can be worth several times as much as if it were merely supplied to a heat engine to generate electricity" was first enunciated in 1947 by John J. Grebe of the Dow Chemical Company, who argued that such an application of high-temperature nuclear heat would be "...economically more attractive than more-or-less marginal competition with coal for power production."[2] That the substitution of nuclear heat for fossil combustion heat could also reconcile the conflicting needs of energy production versus environmental quality was pointed out two decades later.[3] As indicated by the opening quotation, however, awareness of the economic and environmental desirability of nuclear process heat has not, until (perhaps) very recently, penetrated the higher levels of government.

Given our present energy dependence, exacerbated during the past four years by a national policy of nuclear abnegation, it is clear that coal is indeed our "bridge to the future" and must be exploited as rapidly as environmental compatibility will permit. The bridging or transition period, however, will extend for the next several decades, during which interval the role of coal is here envisioned as shifting from that of a primary fuel to that of a feedstock for chemical processes.

Nuclear Heat Source

The ultimate future here foreseen for coal derives from the possibility of substituting for combustion heat the nuclear fission heat released in a high-temperature, gas-cooled reactor (HTGR). Reactors of this type are being developed in the Federal Republic of Germany, Japan and the USSR as well as in the United States. The FRG and Japanese programs are the most advanced and are proceeding at a level of effort several times that available to the U.S. program. The level of the Soviet program is not known, but their publications in the open literature suggest substantial activity. Low-level research programs are continuing in France and Switzerland.

Although different reactor core designs are employed by these efforts, all employ helium-cooled, graphite-moderated, thermal-spectrum reactors and their operation can in general be represented schematically as indicated in the flow diagram of Figure 1, which depicts the parameters of a current U.S. design. [4-6] Details of the flow sheet need not be elaborated for purposes of this discussion, but it should be ndted that the pacing development item is the intermediable heat exchanger (IHX) which couples the hot, primary helium flow (shown here at 850°C) to the secondary helium loop, which in turn "fires" the process heat exchanger. Perhaps exploiting some metallugical advancements, the FRG and Japanese programs demand even higher temperature performance from the IHX (950 and 1000°C respectively). The relative status of the several ongoing programs has been recently reported. [7-12]

Process Heat Applications

Some of the processes to which high-temperature nuclear heat can be applied are indicated in Figure 2. Coal gasification is the objective of the German program, [7] first using nuclear-generated steam and later substituting nuclear heat for the combustion heat released in situ during the conventional "autothermal" reaction. The Japanese program [8] is aimed principally at using the hot synthesis gas or "syngas" ($H_2$ + CO) for direct reduction of iron ore in nuclear steelmaking. An alternative use for syngas would be as feedstock for synthesis of methanol (and hence gasoline via the Mobil-M dehydration process) or of Fischer-Tropsch liquids, the two so-called "indirect" liquefaction methods.

Perhaps the nearest important application of nuclear process heat, however, will be the steam reforming of methane to produce hydrogen for "direct" liquefaction of coal (e.g., SRC-II), for synthesis of ammonia, [13] for refining or hydrotreating of crude oil or for a variety of other industrial uses. One of these is recovery of oil from shale via the hydroretorting process, [14] an alternative to conventional steam retorting particularly well suited for use with low-grade shales. [5]

A combination of industrial applications can be accomodated by a nuclear-chemical energy center [5] as conceptually depicted in Figure 3.

By using coal liquids as feedstock, the center can be decoupled from the synthetic liquids plant (which may be geographically remote) and located close to industrial demand centers, thus gaining the advantage of short transportation distances for its output products.

Reactor Safety

The use of high-temperature nuclear process heat as here envisioned requires that the reactor be part of the industrial complex involved, for which it would also cogenerate local power and process steam. Fortunately, the outstanding safety characteristics unique to the HTGR can permit such close-in siting. Unlike that of a conventional light-water reactor (LWR), the helium-cooled graphite core of an HTGR can neither melt nor react chemically with its coolant. As a result of this inherent safety advantage, the consequences of the worst imaginable HTGR accident (i.e., depressurization and loss of circulation) are lower by two to three orders of magnitude (15) than those of the equivalent LWR accident (i.e., LOCA and core meltdown) as estimated by the Rasmussen report. Even prior to the Three Mile Island accident this safety advantage had already commended the HTGR to reactor critics in the ongoing nuclear safety controversy[16] and subsequently general recognition that the HTGR permits no "China Syndrome" is growing. [17]

Emission Control

In contrast to the situation involving fossil combustion, where the carbon dioxide produced is diffused into the atmosphere, the $CO_2$ produced in processing fossil material as feedstock using externally-generated nuclear heat is under control in concentrated form. Scrubbed from the process stream, it is conventially discharged as waste, but this need not be the case. Beneficial uses are possible and economic as well. [3] Aside from the direct use of $CO_2$ in speeding plant growth, its potential for indirect use as a constituent in synthesis of valuable compounds is considerable.

A good example is urea. Synthesized by reaction with ammonia, urea is a principal ingredient of solid nitrogen-base fertilizers. It is also a feed supplement for ruminant livestock, as well as the base for production of plastics, adhesives and coating materials. Soda ash and its many derivatives constitute another route to useful products.

In addition to providing carbon dioxide control, the avoidance of fossil combustion in processing fossil feedstocks obviates the problem of $NO_x$ and $SO_x$ emissions. Sulfur is routinely recovered from such processes in elemental form. In cases where the coal contains significant amounts of uranium or thorium (as do some lignites) the process ash can be mined to recover these valuable materials.

Conclusion

The foregoing brief discussion has only touched upon the points which together constitute a strong environmental case for the use of nuclear fission heat instead of fossil combustion heat in fuel processing and other process industries. Not included was the subject of nuclear waste management, since this is primarily a political, not a technical, problem and hence inappropriate for discussion on this occasion.

References

1. J.R. Schlesinger, Secretary of Energy, Remarks before the Princeton University, Princeton, New Jersey, April 14, 1978.

2. L. Green, Jr. and T.D. Anderson, "History of Nuclear Process Heat Applications in the United States", Proc. First National Topical Meeting on Nuclear Process Heat Applications, Los Alamos, N.M., October 1-3, 1974, LA-5795-C, (CONF-741032), USAEC, November, 1974, pp. 62-69.

3. L. Green, Jr., "Energy Needs versus Environmental Pollution: A Reconciliation?," Science, Vol. 156, pp. 1448-1450, 1967.

4. R.N. Quade, D.L. Vrable and L. Green, Jr., "Production of Liquid Fuels with a High-Temperature Gas-Cooled Reactor", 2nd Miami International Conference on Alternative Energy Sources, December 10, 1979, Miami, Florida.

5. R.N. Quade and D.L. Vrable, "High-Temperature Process Heat Application with an HTGR", Conference on the Utilization of Small and Medium Size Power Reactors in Latin America, May 12-15, 1980, Montevideo, Uruguay.

6. D.L. Vrable and R.N. Quade, "Design of the HTGR for Process Heat Applications", Proceedings of the 15th Intersociety Energy Conversion Engineering Conference, American Institute of Aeronautics and Astronautics, 1980, pp. 1107 - 1112.

7. R. Pruschek, E. Arndt and R. Harth, "Status of Nuclear High Temperature Process Heat Development in the Federal Republic of Germany (Coal Gasification and Long Distance Energy Transport)", Ibid, 11. 1074-1079.

8. H. Murata, H. Nakamura, M. Hirata, S. Ikari and S. Yasukawa, "A Challenge to Create New, Manageable and Possible Energy Supply System (JAERI's Projection on Nuclear Process Heat Application)", Ibid, pp. 1080-1090.

9. S.R. Penfield, Jr., "The High Temperature Gas-Cooled Reactor for Nuclear Process Heat Applications: Status of the U.S. Program", Ibid, pp. 1091-1096.

10. J. Makherbe, "Feasibility of a Steam Reforming Plant Heated by an HTGR", Ibid, pp. 1097-1100

11. P. R. Kasten, "High-Temperature Gas-Cooled Reactors and Process Heat", Ibid, pp. 1101-1106

12. O. F. Kimball, "Structural Materials Requirements for Nuclear Process Heat Gas Cooled Reactor Systems Applications", Ibid, pp. 1113-1119

13. L. Green, Jr., "An Ammonia Energy Vector for the Hydrogen Economy", Hydrogen Energy Progress (T. N. Veziroglu, K. Fueka and T. Ohta, Editors), Pergamon Press, Oxford and New York, 1980, pp. 1265-1272.

14. L. Green, Jr., "Hydroretorting of Oil Shale with Nuclear Process Heat", ASME Paper No. 76-WA/Ener-3, December 1976.

15. R. C. Dahlberg, S. L. Koutz and H. M. Agnew, "Inherent Safety in Nuclear Power Plants: The Case for Gas-Cooled Reactors", to be published.

16. J. Primak, "The Nuclear Safety Controversy", Chem. Engineering Progress, 70, 26 (1974)

17. H. M. Agnew, "Nuclear Power in Perspective", presented at the Commonwealth Club, San Francisco, California, June 29, 1979.

## HTGR-PH FLOW DIAGRAM

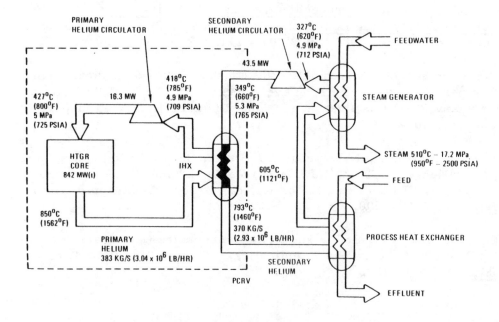

Fig. 1 Flow diagram for process heat HTGR

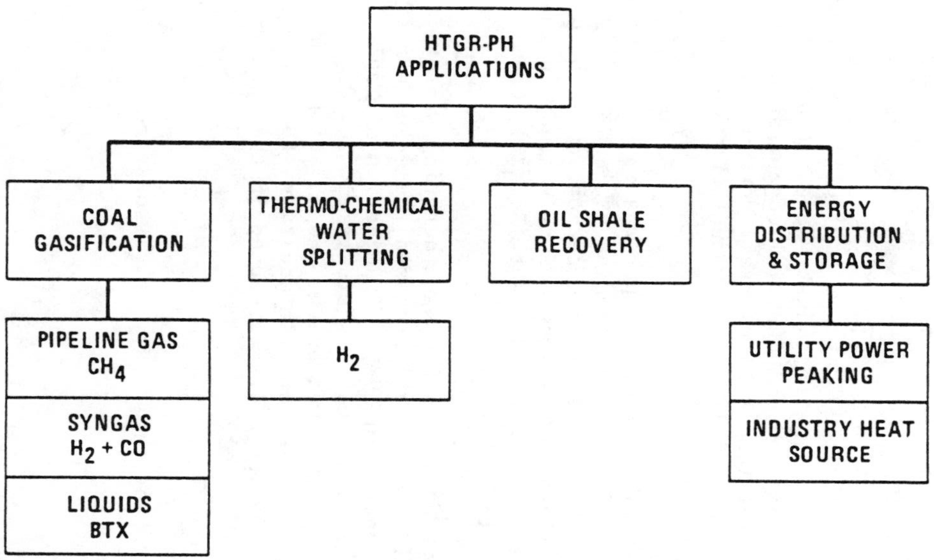

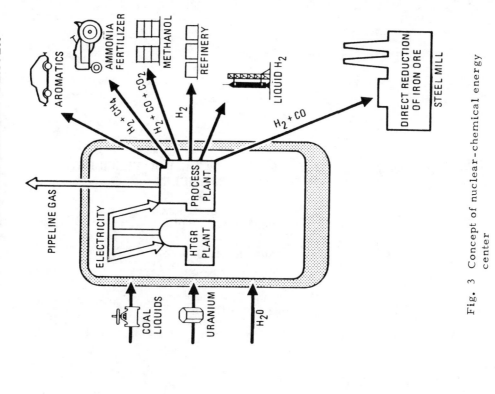

**NUCLEAR-CHEMICAL ENERGY CENTER**

Fig. 3 Concept of nuclear-chemical energy
center

Fig. 2 Nuclear process heat applications

431

# INDUSTRIAL COGENERATION: ECONOMIC FEASIBILITY
## OF SELECTED EXAMPLES

**C. L. Kusik** and **L. K. Fox**
Arthur D. Little, Inc.

**J. R. Hamm** and **K. D. Weeks**
Westinghouse Electric Corporation

**I. King** and **V. P. Buscemi**
Gibbs and Hills, Inc.

Introduction

For its energy needs, the United States today relies heavily on natural
gas and petroleum-based fuels, with industry accounting for almost 40% of
the nation's energy consumption. Industrial cogeneration represents one
way of reducing industrial (and agricultural) consumption of energy.
Such a system involves the simultaneous production of electricity or shaft
power, and useful thermal energy.

The program of work described in this paper was supported by DOE's Office
of Industrial Programs so that the economics of industrial cogeneration
and the barriers to its more general applications could be understood.
Industrial sectors examined in the scope of this study include: Food
(SIC 20), Textiles (SIC 22), Pulp and Paper (SIC 26), Chemicals (SIC 28),
and Petroleum Refining (SIC 29).

Industrial energy requirements in the manufacturing sector (SIC 20 through
39) called for about 10.4 quads of purchased fuels and $635 \times 10^9$ billion kWh
of purchased electric energy in 1976. Examination of Table 1 indicates
that those sectors of interest to this study (SIC 20, 22, 26, 28 and 29)
account for about 56% of purchased energy and 45% of purchased electric
energy in the manufacturing sector. In addition, the table shows that an
estimated 2.66 quads of waste fuels were used in the same categories.

Approach

Our approach in undertaking this study involved seven major elements in
which we:

1. Selected 10 plants representing good near-term opportunities for cogeneration in five major energy-consuming industrial sectors mentioned above. Two plants were chosen from each industry sector.

2. Developed energy demand profiles for selected plants. The energy demand profiles are shown in Figure 1. Annual energy use is summarized in Table 2. Except in certain unusual situations, cogeneration systems were sized to meet the steam requirements of the plant.

3. Examined various cogeneration systems for each plant, considering fuel availability/costs and state-of-the-art technology for cogeneration systems which included: a) coal- and oil-fired power steam generators; b) oil-, natural gas-, and coal-fired process steam boilers; c) slow-speed, two-stroke cycle marine diesel engines; d) high-speed, four-stroke cycle industrial diesel engines; e) heat recovery steam generators (HRSG's); f) waste heat boilers (generating saturated steam); g) steam turbine/generators; h) gas turbine/generators; i) combined cycle subsystems with various steam turbine configurations; j) coal-fired gas cycles; k) particulate removal and flue gas desulfurization systems, including electrostatic precipitators, scrubbers, and baghouses; l) Rankine bottoming cycles, including fluorinol-85, methanol and ammonia as working fluids; and m) energy storage devices such as steam accumulators, packed beds, and moving beds in various configurations.

4. Selected appropriate return on equity (ROE) hurdle rates for industry, utility, or third-party ownership. These ROE hurdle rates ranged from about 16 to 25%, depending on industry sector. In the utility sector, the ROE's assumed typically ranged from 9 to 13%.

5. Selected, characterized, and optimized cogeneration systems for each of these plants to achieve maximum energy savings for given minimum values of return on investment (ROI) or return on equity (ROE). Cogeneration systems were optimized to achieve maximum ROE until the hurdle rate was achieved. Once the hurdle rate was achieved, the ROE was then kept constant (at the hurdle rate) and energy savings were maximized. Calculations for optimizing the cogeneration system were done on the basis of debt-equity ratios reflecting the company or industry sector.

   Costs (e.g., for labor, electricity, capital equipment, etc.) reflect those at the plant site in mid to late 1978. For example, labor rates for plant construction were those found in a nearby urban area; electricity and fuel costs were those paid by the plant being analyzed for incorporation of a cogeneration facility. The optimization was done by considering pollution control standards and by using prices for energy forms likely to be used in the industrial sectors of interest in the time period of 1980 to 2000; energy costs and escalation rates are summarized in Table 3.

6. Developed conceptual designs and capital cost estimates for each plant. The designs contain all the major components necessary for a cogeneration facility with the exception of mechanical and electrical distribution systems. Site preparations for these facilities were assumed to be minimal requiring only clearing and grubbing. Soils in each facility were assumed to have sufficient bearing capacity so that spread footings could be used for the foundations of all equipment and structures. Capacities and sizes of systems and components such as cranes, pumps, drivers, deaerators, ductwork, breeching, stacks, piping,

433

electrical switchgear and wiring, insulation, alarms and controls were
determined by using the principles and engineering guidelines common
to accepted industrial and utility practice. Each component specified
is of standard design and is available from at least two manufacturers.
Depending on regulatory requirements, particulate control was achieved
either by electrostatic precipitators, baghouses, or scrubbers.
Desulfurization of flue gas was obtained through the use of venturi
scrubbers or tray towers.

Standard estimating techniques were used in developing the capital
cost of each facility. Installed equipment costs were derived starting with major component costs acquired through direct solicitation of
manufacturers, published book prices, and in-house data. Information
consisting of a site plan, plot plan, flow diagram, one-line diagram,
brief preliminary specifications of major components, cost estimate,
and schedule were developed for each cogeneration facility.

7. Developed capital and operating cost estimates to calculate cash flow
   projections and determine return on investment for three potential
   ownership options of the cogeneration facility: a) industry owner-
   ship--which would consider a cogeneration venture which would result
   in savings of purchased electricity: if process steam raising boilers
   were old and had to be replaced, industry would consider cogeneration
   as an incremental investment to a steam raising boiler; b) utility
   ownership--which would look at incremental investments required over
   that of a central electric power generating boiler and, as a result,
   would generate increased revenues from steam sales; and c) third-
   party ownership--with fixed take-or-pay contracts for steam and
   electric power looking at the cogeneration system as a new venture.

   For each ownership option, we used appropriate debt-to-equity ratios
   and took into account various tax credits allowed for the facility.
   Return on equity investment calculations include escalation and
   interest during construction.

Findings and Conclusions

Results are summarized in Table 4 which shows that the 10 plants selected
for detailed analysis were drawn from DOE Regions I, III, IV, V, VI, and
IX (as illustrated in Figure 3). These six regions account for 84% of
industrial energy use in the manufacturing sectors (SIC 20-39). In
addition, Table 4 shows that the cogeneration systems considered ranged
from small units needing under 100 million Btu/hr to large units with fuel
demands of just under 3 billion Btu/hr. Our findings are presented from
three viewpoints: 1) technical-economic, 2) financial/return-on-invest-
ment, and 3) energy conservation.

## Technical-Economic Considerations

In general, we found that most of the plants operated 24 hours per day with
some facilities closing down on weekends. Seasonal variations were noted
in certain situations. Nevertheless, the energy demand profiles were
found to be relatively constant, especially over the course of a 24-hour
day. Because of this, heat storage devices were usually inappropriate,
except for the brewery where a steam accumulator proved to be economic to
meet short-duration peak demands (for this batch process). We believe
that the plants considered here are fairly typical ones in the industrial
sectors covered in this study. However, the results and findings discussed
here should not be extrapolated to facilities operating less than 24 hours

a day, operating only part of the year, or exhibiting a large fluctuation in steam or electrical power demand.

Table 4 shows that the optimized cogeneration system involved steam topping turbines (back pressure or extraction) in 7 out of the 10 cases. In these seven optimized cogeneration systems, coal was the most economic fuel choice in six cases; wood, in plentiful local supply, was the fuel choice for the seventh plant (a textile mill). The optimized steam topping cycles typically exhibited relatively high thermal to electric ratios $(kW_t/kW_e)$ of 7.9 to 14.9, as shown in Table 4.

Cogeneration systems for the remaining three cases were selected largely as a response to local energy availability and fuel costs:

- A gas turbine and waste heat boiler were found to offer the optimal economics and energy savings for a plant in DOE Region VI (Gulf Coast).

- A combined cycle with a single automatic extraction/back pressure steam turbine was the optimum system for the large, complex petroleum refinery located on the West Coast. The gas turbine combustor was fired with naphtha and the heat recovery steam generator was fired with low-sulfur residual oil.

- Inclusion of a Rankine bottoming cycle has the potential of being the best choice for the medium-size, medium-complexity petroleum refinery in DOE Region VI. The cogeneration system for this plant recovers energy from the 55 psia gas discharge of a high-temperature catalyst regenerator. Major system components include: 1) a turbo-expander, 2) heat exchangers, 3) a double automatic extraction/condensing generator drive steam turbine, and 4) a methanol Rankine bottoming cycle for completing the recovery of waste heat from the turbo-expander exhaust.

The above three cogeneration systems exhibited relatively low thermal to electric ratios of 1.2 to 2.3. Energy savings are summarized in Table 5.

With the exception of the petroleum refining industry, we generally found that the operation of an optimized cogeneration system would still result in net demand for electrical power, largely because of the high thermal to electric ratios characterized by use of steam turbines. However, where economically priced natural gas or petroleum fuels were available, the optimized cogeneration facilities (typically involving gas turbines) exhibited low thermal to electric ratios, resulting in electric energy generation that could more than meet plant requirements. An overview of the applicability of various cogeneration systems as a function of thermal to electric ratios and fuel use efficiency is presented in Figure 2.

### Financial and Return-on-Investment Considerations

Table 4 shows investments in cogeneration systems considered in this study ranged from $12.9 million for the smallest cogeneration facility to slightly more than $100 million for the largest, and that all 10 plants met return-on-investment criteria under at least one or more of the three ownership options evaluated. When cogeneration offered an attractive return on investment, it did so for a variety of reasons. Broadly, we categorized these situations that involved:

1. Low incremental investment costs achieved by the necessity of installing a new boiler to replace obsolete process steam boilers;
2. Low unit investment costs achieved by large-size cogeneration units;
3. High electric energy costs; and/or
4. Low cogeneration fuel costs

Moreover, the economics of cogeneration appear much more favorable when viewed against the investments required in supplying additional central electric utility capacity; with the utility owning and financing a cogeneration facility to supply a plant's utility needs (steam and electricity), return on investment and energy savings are simultaneously maximized. Results reported here are based on information and data available in the first quarter of 1979. Since then, the economics of industrial cogeneration have been potentially impacted by Federal Energy Regulatory Commission activities such as incremental gas pricing, value of cogenerated electric energy, and so on. Thus, returns on investment reported here for each specific situation should be used with caution and possibly re-evaluated based on current developments and seemingly ever higher energy price projections.

Energy Conservation Considerations

Examination of Table 5 shows a 15% average energy savings from cogeneration compared to separate generation of process steam and electric power. Extrapolation of these results to a national basis for the five industry sectors considered indicates potential energy savings equivalent to 630,000 to 750,000 barrels of oil per day or about 1.3 to 1.6 quads annually ($10^{15}$ Btu/yr). Even if a small fraction of such potential savings were realized, it would have a significant impact on energy conservation.

Acknowledgements

This paper is based on a two-volume report entitled, "Industrial Cogeneration Optimization Program," which was prepared by Arthur D. Little, Inc., Westinghouse Electric Corporation, and Gibbs & Hill, Inc. and is available from NTIS (DOE/CS/05310-01, January 1980). The report was prepared for the U.S. Department of Energy with Mr. John Eustis acting as the Government Project Officer whose suggestions and guidance of this effort is gratefully acknowledged. We wish to extend our appreciation to a large number of individuals who contributed to the report, including 1) from Arthur D. Little, Inc.: G.R. DeSouza, W.V. Keary, D.E. Johnson, R.F. Machacek, J.K. O'Neill, D. Shooter, R.P. Stickles, C.R. Canty, and L. Goddu; 2) from Westinghouse: S.L. Fletcher, C. Wood, D. Bachovchin, P. Kolody, R. Grimble, L. Miller, L. Jenkins, and N. Carpenter; and 3) from Gibbs & Hill: W.B. Spaab, J. Galperin, V.J. Pierce, A.M. Baird, S.N. Goldman, J. Perrin, L. Gettler, and H.E. Pieper.

We wish to thank the following for their overview and guidance of this effort: Paul P. de Rienzo, Vice President, Consulting Engineering, Gibbs & Hill; Richard M. Chamberlin, Westinghouse Project Manager; and Frederick G. Perry, Senior Staff Consultant, Arthur D. Little, Inc.

Finally, we wish to express our appreciation to the many industrial companies that cooperated with us in supplying data needed to complete this study.

TABLE 1

ANNUAL INDUSTRIAL ENERGY USE IN SECTORS OF INTEREST

| SIC | Sector | Purchased Fuels[a] (trillion Btu) | Waste Fuels[b] (trillion Btu) | Purchased Electric Energy[a] (billion kwh) |
|---|---|---|---|---|
| 20 | Food and Kindred Products | 804 | 20 | 39 |
| 22 | Textile Mill Products | 233 | -- | 28 |
| 26 | Paper and Allied Products | 1,146 | 999 | 43 |
| 28 | Chemicals and Allied Products | 2,521 | 120 | 145 |
| 29 | Petroleum and Coal Products | 1,197 | 1,530 | 28 |
| | Subtotal of 5 Sectors | 5,901 | 2,660 | 283 |
| | Other Manufacturing Sectors (in SIC 20-39) | 4,558 | n.a. | 352 |
| | Total Industrial Manufacturing* | 10,459 | n.a. | 635 |

*Industrial Manufacturing accounts for about 60% of industry's energy use. The difference is accounted for by non-manufacturing sectors such as mining, agriculture, fishing, etc. found in SIC 1-19.

[a]Source:  U.S. Bureau of Census, Annual Survey of Manufacturers, 1976.

[b]Source:  Guidelines for Developing State Cogeneration Policies, Resource Planning Associates, April 1979 (NTIS HCP/M8688-01; UC-95f).

n.a.  Not available.

TABLE 2

## SUMMARY OF PLANT ENERGY USE CHARACTERISTICS

| Industry | Plant | Average Electric Power Requirements (kW) | Annual Electric Energy Consumption (kWhe/YR) | Annual Process Steam Consumption (kWht/YR) | Annual Direct Process Heat Fuel Consumption (kWhc/YR) | Total Annual Energy Consumption (1) (kWheq/YR) | Rankings (2) |
|---|---|---|---|---|---|---|---|
| Food | Soybean Oil Mill | 1600 | $14.2 \times 10^6$ | $128.2 \times 10^6$ | $20.5 \times 10^6$ | $211.9 \times 10^6$ | (1) \|1\| <1> |
| Food | Brewery | 4500 | $38.9 \times 10^6$ | $172.1 \times 10^6$ | – | $313.6 \times 10^6$ | (2) \|2\| <2> |
| Textiles | Finishing Mill | 9000 | $65.0 \times 10^6$ | $412.6 \times 10^6$ | $184 \times 10^6$ | $855.1 \times 10^6$ | (4) \|5\| <7> |
| Textiles | Integrated Mill | 16000 | $116.7 \times 10^6$ | $279.7 \times 10^6$ | $25.4 \times 10^6$ | $687.9 \times 10^6$ | (8) \|3\| <4> |
| Pulp and Paper | Writing Paper Mill | 14000 | $120.9 \times 10^6$ | $782.4 \times 10^{6\,(4)}$ | – | $1265.9 \times 10^6$ | (7) \|8\| <8> |
| Pulp and Paper | Kraft Paper Mill | 10000 | $88.0 \times 10^6$ | $498.0 \times 10^{6\,(4)}$ | – | $837.2 \times 10^6$ | (6) \|6\| <6> |
| Chemicals | Agricultural Chemicals | 5100 (5800)[3] | $44.5 \times 10^6 (50.5 \times 10^6)^{(3)}$ | $310.2 \times 10^6 (361.9 \times 10^6)^{(3)}$ | – | $492 \times 10^6 (570 \times 10^6)^{(3)}$ | (3) \|4\| <3> |
| Chemicals | Air Separation | 67000 | $530.4 \times 10^6$ | $(1308-2615) \times 10^6$ | – | $(3054-4593) \times 10^6$ | (9) \|9\| <9> |
| Petroleum Refining | Medium Size and Complexity | 9415 | $82.5 \times 10^6$ | $498.1 \times 10^6$ | – | $821.7 \times 10^6$ | (5) \|7\| <5> |
| Petroleum Refining | Large/Complex | 120000 | $963.6 \times 10^6$ | $4720 \times 10^6$ | – | $8306 \times 10^6$ | (10) \|10\| <10> |

(1) Electricity generation efficiency assumed to be 35% and process steam boiler efficiency assumed to be 85%
(2) (Power), | process heat|, and <total energy consumption> in ascending order
(3) Projected to 1982
(4) Net after credit for black liquor recovery boiler

438

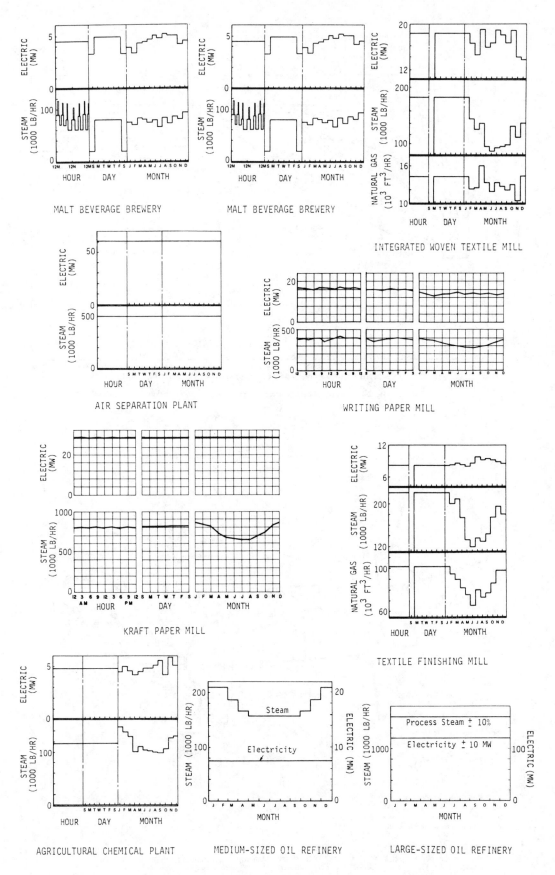

Figure 1: Energy Demand Profiles for Selected Plants

439

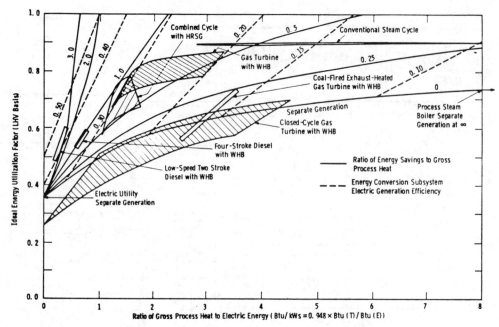

Fig. 2    —Energy utilization characteristics of candidate cogeneration systems

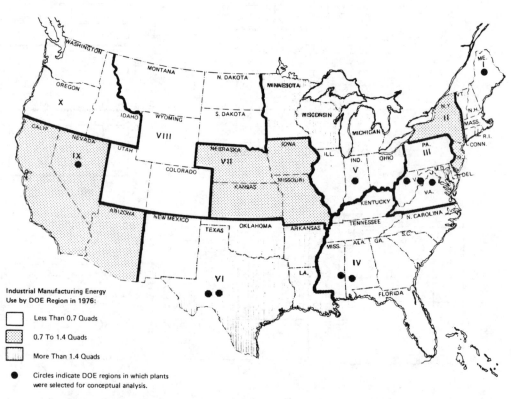

Industrial Manufacturing Energy
Use by DOE Region in 1976:

☐  Less Than 0.7 Quads

▨  0.7 To 1.4 Quads

▥  More Than 1.4 Quads

●  Circles indicate DOE regions in which plants
    were selected for conceptual analysis.

**Source:** Annual Survey of Manufacturers, 1976.

FIGURE 3        ENERGY USE IN INDUSTRIAL MANUFACTURING SECTOR BY DOE REGION

TABLE 3

SUMMARY OF FUEL AND ELECTRICAL ENERGY COST AND ESCALATION RATES USED FOR TEN PLANTS

(Time base - Dec. 1978)

| Industry | Type of Plant | Location State/DOE Region | Fuel Cost($/10⁶ Btu(HHV))/Escalation Rate (%/Year) | | | Electrical Energy | | |
|---|---|---|---|---|---|---|---|---|
| | | | Coal | Natural Gas | Fuel Oil | Cost (¢/kWh) | Escalation Rate(%/year) | |
| Chemical | Agricultural Chemicals | Maryland/$_{III}$ | 1.13/5.6 | — | 3.42$^{(8)}$/4.5 | 3.15$^{(7)}$ | 2.1 | (1) |
| | Air Separation | Texas/$_{VI}$ | 1.95/-1.0 | 1.61/4.4 | 2.28/5.8 | 2.03 | 6.0 | |
| | Brewery | Penna./$_{III}$ | 1.03/5.6 | 1.63/7.7 | 2.77/4.5 | 2.45$^{(7)}$ | 2.1 | |
| Food | Soybean Oil Mill | Indiana/$_{V}$ | 1.20/5.0 | 1.61/8.5 | 2.76/4.2 | 2.67$^{(7)}$ | 2.9 | |
| Petroleum Refining | Large/Complex | Calif./$_{IX}$ | 1.92/-0.3 | 2.40$^{(7)}$/6.8 | 2.06/7.6$^{(3)}$ | 4.0$^{(7)}$ | 2.6 | (4) |
| | Medium/Less Complex | Okla./$_{VI}$ | — | — | — | 2.1 | 6.0 | (6) |
| Pulp and Paper | Writing Paper Mill | Penna./$_{III}$ | 0.97$^{(7)}$/5.6 | 2.63$^{(5)(7)}$/7.7 | 5.25$^{(5)(7)}$/4.5 | 2.4$^{(7)}$ | 2.1 | |
| | Kraft Paper Mill | N. Hamp./$_{I}$ | 1.51/3.3 | — | 3.26$^{(8)}$/3.6 | 3.5$^{(7)}$ | 0.8 | |
| Textile | Finishing Mill | S. Carol./$_{IV}$ | 1.20/4.8 | 2.08$^{(7)}$/9.5 | 2.03$^{(7)}$/4.6 | 2.36 | 3.6 | |
| | Integrated Mill | Alabama/$_{IV}$ | 1.20/4.8 | 1.29/9.5 | 2.55/4.6 | 2.36 | 3.6 | (2) |

(1) Includes 10% local tax on energy purchased

(2) Wood was the preferred fuel for this plant. The cost of wood was assumed to be $1.84/10⁶ Btu (HHV) and the escalation rate for wood was assumed to be 4.8%

(3) Compliance low sulfur residual

(4) By-product light naphtha was an alternate fuel for this plant. The value of light naptha was estimated to be $2.58/10⁶ Btu (HHV) and the escalation rate was estimated to be 7.3%

(5) Updated to mid-1979 after oil price increases occurred

(6) Bottoming cycle

(7) Plant specific values - not DOE region

(8) Includes 12 1/2% Premium for Low-Sulfur Residual

TABLE 4.

SUMMARY OF COGENERATION SYSTEM INVESTMENTS AND ROE FOR PLANTS CONSIDERED

| SIC Sector | Industry | Type of Plant | DOE Region | Cogeneration System | | | | | Return on Equity[f] (ROE) % | | |
|---|---|---|---|---|---|---|---|---|---|---|---|
| | | | | System Type* | Fuel Type* | Fuel Use (10^9 Btu/hr) | $\frac{Kw_t}{Kw_e}$[b] | Total Capital Costs[c] ($ Millions) | Industry Ownership | Utility Ownership | Third Party Ownership |
| 20 | Food | Brewery | III | S.T. | Coal | 0.12 | 13.0 | 15.6[d] | 12.2 | 18.2 | < 0 |
| | | Soybean Oil | V | S.T. | Coal | 0.082 | 14.9 | 12.9[d] | 12.2 | 13.7 | < 0 |
| 22 | Textiles | Finishing Mill | IV | S.T. | Coal | 0.43 | 8.3 | 45.7 | < 0 | 26.6 | 15.1 |
| | | Integrated Mill | IV | S.T. | Wood | 0.30 | 10.6 | 26.7 | < 0 | 30.1 | < 0 |
| 26 | Pulp and Paper | Writing Paper | III | S.T. | Coal | 0.60 | 8.3 | 52.2[d] | < 0 | 23.7 | < 0 |
| | | Kraft | I | S.T. | Coal | 0.27 | 7.9 | 31.4 | < 0 | 34.8 | 26.7 |
| 28 | Chemicals | Agricultural Chemicals[a] | III | S.T. | Coal | 0.23 | 8.3 | 25.9 | 15.2 | 39.2 | 44.5 |
| | | Air Separation[a] | VI | G.T. | Nat. Gas | 0.34 | 1.4 | 24.8 | 18.0 | >1000[g] | 28.4 |
| 29 | Petroleum Refining | Large Sized Refinery | IX | C.C. | Naphtha/Resid. | 2.9 | 2.3 | 100.3 | < 0 | >1000[g] | < 0 |
| | | Medium Sized Refinery | VI | TE/OR | E | E | 1.2 | 20.0[e] | 88.3 | n.a.* | n.a.* |

[a] With cluster chemical complex.

[b] Thermal ($kwh_t$) to electric ($kwh_e$) energy ratio produced by cogeneration system.

[c] Capital cost estimates include necessary pollution control equipment; estimates exclude interest during construction and escalation.

[d] Initial capital costs shown here were used to calculate ROE; capital cost estimates based on final conceptual design were within ±4% of values shown here.

[e] Preliminary estimate.

[f] Based on incremental investment in cogeneration system; incremental investment is equal to total investment for third party ownership (see Section 6.1); interest during construction and escalation are considered for all ownership options.

[g] Incremental investments were found to be negative; thus ROE proved very large ($\infty$) and was set equal to 1000.

*Abbreviations used:
C.C. = combined cycle
G.T. = gas turbine with waste heat boiler
S.T. = steam turbine
TE/OR = turbo-expander with steam and organic rankine
E = energy recovery from high temperature exhaust gases
n.a. = not applicable

TABLE 5

SUMMARY OF ENERGY SAVINGS IF COGENERATION WERE IMPLEMENTED
IN SPECIFIC PLANTS CONSIDERED IN THIS STUDY

| Sector | Plant Considered in this Study | Energy Use, Separate Generation[c] (Billion Btu/Yr) | Savings if Cogeneration Were Implemented (Billion Btu/yr) | | | | Energy Savings As % of "Energy Use, Separate Generation" | |
|---|---|---|---|---|---|---|---|---|
| | | | Coal | Oil | Natural Gas | Total[a,e] | Oil and/or Natural Gas Saved | Total Energy Savings[a] |
| Food | Soybean | 688 | -498 | -- | 544 | 46 | 79 | 6.7 |
| | Brewery | 811 | -559 | 645 | -- | 86 | 80 | 10.6 |
| Textiles | Finishing Mill | 3,549 | -2,062 | 763 | 1,890 | 591 | 75 | 16.7 |
| | Integrated Mill | 2,606 | -837 | -- | 1,134 | 297 | 44 | 11.4 |
| Pulp and Paper | Writing Paper | 3,237 | -45 | -- | 501 | 456 | 15 | 14.1 |
| | Kraft Paper | 2,377 | -1,402 | 1,728 | -- | 326 | 73 | 13.7 |
| Chemicals | Ag Chemical | 1,890 | -343 | 584 | -- | 240[e] | 31 | 12.2 |
| | Air Separation | 8,315 | 5,602 | -- | -3,113 | 2,489 | -37 | 29.9 |
| Petroleum Refining | Medium | 3,511 | -- | -- | -- | 4,814 | -- | Footnote d |
| | Large/Complex | 33,971 | -- | 7,903[b] | -- | 7,903 | 23 | 23.3 |
| AVERAGE | | | | | | | | 15 |

[a]Savings in coal, oil, gas and other fuels (e.g., wood) (rounded values).
[b]Includes naphtha.
[c]Based on electric energy converted at about 10,500 Btu/kWh and thermal energy at 3,413 Btu/kWh,

[d]Excluded from average savings calculation since it represents using part of feedstock for generating fuels.
[e]For example, savings of 240 billion Btu/yr out of 1,890 billion Btu/yr are achieved by implementing cogeneration in the Agricultural Chemical plant.

443

# PRODUCTIVITY AND THERMAL ENERGY STORAGE TECHNOLOGY

Ben-Chieh Liu, Joe Asbury, Bob Giese,
Jarilaos Stavrou and Chuck Maslowski
Argonne National Laboratory

Introduction:  Background and Objective

Soaring fuel prices and the accelerating costs of power plant construction, coupled with scarcity of investment capital and pressing financial
and environmental problems, have made system planning in the utility
industry much harder.  System planning is especially difficult when
the criteria of energy conservation, efficient power generation, reliable
power supply, and equitable power distribution are to be met under peak-
load conditions because of changes in sectoral power demand over different times of the day, days of the season, and seasons of the year.

Peak-load generating capacity is not only expensive to operate but is
also utilized much less than its full capacity all year long.  Furthermore, peaking plants, which generally use oil, are started in response
to sudden rises in power demand, and oil prices have been skyrocketing
since the 1973 oil embargo.  As a result the cost for electric services
have risen substantially recently.  For example, the average price of
residential electricity surged from 2.09¢ per kWh in 1969 to 3.78¢ per
kWh in 1977; an increase by 81.0%.  (See H.S. Hauthakker [17].)

To improve the productivity (or efficiency) of the industry, various
cost-reduction approaches have been studied.  Time-of-use rate design,
marginal-cost pricing, and peak-load management are well known strategies recommended for coping with the peak-load problem in electrical
power generation, transmission, and consumption.  To alleviate the peak-
load problem from the demand sectors while reliably providing needed
electrical services at the lowest possible price, various storage technologies that reduce peak loads have been evaluated.

The on-site thermal energy storage (TES) system is one of the technologies that can improve the productivity of the industry.  Its major con-

444

tribution to load management is to smooth the demand curve between the peak- and the off-peak periods by storing and supplying power generated from non-peak generating capacity during the off-peak period. Thermal storage technology is thus expected to produce fuel cost savings in the short-run by using less oil or gas and more coal or nuclear fuels, and capacity-investment savings in the long-run without detracting from the reliability of the electrical industry. (For a detailed description on TES system, see Asbury, et al [4])

The primary objective of this paper is to summarize, from a social cost-benefit point of view, some micro- and macro-economic measures of the impact caused by the introduction of TES technology in U.S. residential areas. Although the results presented below should be considered pre-liminary, they do tend to support the conclusions reached by Miller and Coleman [24], Asbury et al. [4-7], Renshaw [26] and others [30] concerning the storage system, i.e., promotion of TES in residential areas is likely to be a feasible and cost-effective solution to the peak-load problem, and hence a justified technology for productivity improvement in the electrical industry. In what follows, a benefit-cost cross-impact approach model for technological impact assessment will be introduced. The micro-economic, technological impacts of TES for the residential customers will be analyzed in Section 3, the macro-economic impacts will be discussed in Section 4. Summary and concluding remarks are presented in the last section.

A BCCIPA Model for Technological Impact Assessment

Miller and Coleman [24], Asbury and Mueller [4-7], and Renshaw [26] have concluded that the change in electric capacity utilization brought about by efficient energy storage technologies would improve productivity in the utility industry and hence affect our social welfare. Since the demand curves reflect the utility functions of the individuals consuming the electric services, many technological impact studies have focused their social welfare analyses on a rather narrowly defined micro-economic analytical framework. These studies then assessed the costs and benefits of new technology by measuring the consumer's surplus, estimating the producer's surplus, or quantifying both simultaneously.

However, it is also well recognized that any technological change will have general effects on society in addition to the specific effects on the consumers and producers who are directly affected by the new technology. Moreover, any single direct effect is likely to create ripple effects over time and across regions. For a more general assessment of the TES technological impacts, an integrated model, namely, the benefit/cost, cross-impact assessment (BCCIPA) may be used. (See Liu [22,23] and Kneese and Bower [20] for applications.) While the benefit/cost measures may be quantitatively estimated for many micro-economic variables based on marked values and statistics, the cross-impacts of various macro-economic variables should generally be evaluated under uncertainty conditions, and more often than not, only qualitative results are produced.

The demand for TES from electricity is actually the derived demand for heating and cooling. People do not consume TES or electricity directly. Therefore, the demand functions for energy-related services must be constructed. While the shape of the demand curves may be directly related to the price or cost of the services per unit of time, the factors that shift the curves are exogenously determined by population and income growth, housing structure characteristics, land utilization

445

patterns, etc. Once the demand functions are developed, the consumer surplus can be measured under the demand curves. Then, changes in the consumer surplus (with and without TES) can be used to assess the gain or loss in social welfare. The supply of TES units and of electricity at least cost also requires an in-depth study of the cost curves of TES manufacturing and of electricity generation, transmission, and distribution.

In the micro-analysis in this paper, the opportunity costs of resource reallocation and improved efficiency in electricity generation, transmission, storage, and distribution are evaluated. The increase in net social welfare, with and without TES was measured through the SIMSTOR model (see Giese [15]) developed at Argonne National Laboratory. The estimated cost reduction in electric power provision may be employed as an equivalent measure for consumer's welfare.

Concern over energy independence in general and oil import substitution in particular requires study of TES impacts at aggregate levels as well as over time. Industrial structure in both the manufacturing and utility sectors will be affected by TES, and changes in performance and/or productivity brought about in these sectors by TES should affect those parameters in the overall economy. Impacts on total U.S. employment, national income, domestic price stability, balance of payments, resource conservation, and environmental protection are focal points for assessment and evaluation. Potential conservation policies that would be associated with TES when the technology is fully developed and implemented will undoubtedly produce an array of indirect cross-impacts.

Some cross-sectoral impacts of TES were evaluated with the assistance of the Wharton Annual Industrial Model [13] -- an econometric model with simultaneously determined industrial structural equations driven by endogenously determined final-demand forces derived from an input-output model. These cross-impacts were also estimated under the assumptions that investment reduction in the electric industry would be automatically transferred to the electrical equipment industry manufacturing TES systems, and that the increased capacity utilization rate would meet the increased demand for electricity because of a lower rate brought about by a more efficient peak-load management with TES. Those hypothetical events were assumed to occur instantaneously and the probability of their occurrences was also assumed to be unity, with little or no uncertainty at all.

Micro-Benefit/Cost Analysis of TES

The commercial feasibility of TES for utility load leveling has already been studied by Asbury, et al [4], who examined electric storage heating, storage air conditioning, and storage water heating applications. For each TES system, utility savings were estimated and compared with TES costs for two winter-peaking and two summer-peaking utility service areas. Total net savings from the use of TES for peak-load leveling ranged from $26 to $80 per customer per year in water heating. The net savings in space heating ranged from $310 to $568 per customer and, in air conditioning, from $252 to $305.

When the mix of generating capacity in the utility system is optimized, with the maximum number of TES residential customers, total net savings in the four service areas would amount of $7 million to $17 million per year. The total net savings do not seem to be very large when compared to total expenditures on electricity in these service areas. However, the figures were derived from 1976 data. With future population growth and residential development in these areas, the net savings are expected to grow at a higher than proportional rate. Table 1 shows the individual benefit/cost figures for the residential TES system for the four service areas.

Giese [15] has recently recomputed the potential benefit/cost figures for TES for space conditioning of residential buildings at the multi-regional level. Conventional space heating and cooling systems were compared to a central ceramic brick storage unit for heating and an ice storage unit coupled with a central air conditioner for cooling. In each case, the TES and conventional systems were matched to the requirements of a 1,500 square-foot, detached, centrally heating and cooled single-family unit characterized by a thermal load of 4 kWh heating degree-day and 4.25 kWh cooling degree-day, including duct losses. Current manufacturing data were used to determine the fixed capital and the control costs of both conventional and TES heating systems. Mature market costs were estimated for TES cooling systems. Each TES system was sized to provide 10 hours of storage capacity. Operating costs for each system were based on the Energy Information Administration (EIA) Series B price data from March 3, 1979, runs with imported oil prices fixed at \$21.50/bbl, \$23.50/bbl, and \$31.50/bbl for 1985, 1990, and 1995, respectively.

The total space conditioning market for any given year was defined to comprise new houses and the replacement market (about 5% of the existing 1980 housing stock). The data base, although originally centered on the 65 largest standard metropolitan statistical areas (SMSAs), was further disaggregated into intrastate regions served by summer- and winter-peaking utilities. Census data were used to define the potential TES market as well as the market shares for oil, gas, and electricity for space conditioning. Projections of population growth and migration from the U.S. Department of Agriculture were employed to project the potential TES market.

"The Weather Almanac" prepared by the Gale Research Company was used to obtain long-term heating and cooling degree-day data (65°F base) for SMSAs, upon which the rest of each state's data was estimated. The "Electrical World Directory of Electric Utilities, 1977-78 Edition was used to classify winter- and summer-peaking regions when determining system peaks. A constant 4% annual growth rate was assumed to project those peaks to the future.

It was projected that 228,000 of the TES heating systems would be sold in 1985 and 285,000 in each year from 1990 to 1995, but sales would decrease to 209,000 units in the year 2000. A total of 4.5 million systems may be sold during the next 20 years. More than 40.0% of these TES heating units (18 million units) are expected to be sold in DOE Region IV and 21.7% in DOE Region V. About 182,000 of the TES cooling systems were projected to be sold in 1995, and the cumulative total of new systems for the next 20 years was projected to be 2.1 million, of which DOE Regions V and IX would be the primary demanders, accounting for 23.8% and 22.5%, respectively. (See Table 2.)

The discounted cost savings of TES systems accrued to electric utility companies in winter-peaking residential service areas alone would amount to \$128 million in 1985 and \$1.48 billion in 2000. The total welfare gain attributed to the TES heating system for the next two decades may reach $14.2 billion. Most of the cost savings and welfare gains will occur in DOE Regions IV and V, which could benefit from the TES heating systems from 1980 to 2000. The comparable figures for the summer-peaking service areas are much smaller, however, the discounted cost savings for TES cooling systems in residential market would vary from $36 million in 1990 to $233 million in 2000. The total cost savings for the two decades was estimated to be $1.58 billion. (See Table 3.)

The cost-savings for TES systems described previously include both utility capital investment savings (in peak-load capacity expansion) and fuel cost savings (represented by fuel switching from oil or gas to coal or nuclear). They represent an estimated net producer's surplus resulting from more efficient operation and load management for communities that adopt new technology for power generation and consumption.

The consumer's surplus are not measured here because of data deficiency and the problems involved in average electricity rate calculation and price elasticity estimations which vary considerably among utility service areas within the DOE region. Assuming that the aggregate supply and demand curves are symmetrical with respect to the electricity rates, we may simply approximate consumer's surplus with that of the producer's. In other words, the estimated net consumer's benefit would be equal to the estimated producer's surplus minus the fixed and operating costs of installing the TES system. The net social benefit gained from improved productivity in the electric power industry brought about by TES may range from $15.8 billion to $31.6 billion for the next 20 years.

Macro-Economic Implications of TES

Some of the industrial and regional implications of TES relating to cost-savings in power generation and consumption have been illustrated and forecasted in the preceding section. Presented in this section are results of the Wharton Annual Industrial Model's simulation of macro-economic effects. Specifically, this section discusses the national income and employment effects of the proposed TES systems and a few strategic variables pertinent to national energy concerns such as oil imports and fuel switching.

As stated previously, the success of TES as a load management techno-logical improvement hinges on the reduction in peak-load capacity, in plant investment and fuel import costs to electric utilities. A reduc-tion in the rate of investment in peak-load capacity, coupled with a higher capacity utilization rate plus power generated through lower-cost fuels, obviously will increase the efficiency and hence the productivity of the entire industry. In other words, the productivity of each dollar's input in this industry is improved through the follow-ing effects: (1) a lower average cost curve for electricity generation resulting from the reduction in use of peak-load capacity, plus (2) re-duced fuel costs resulting from fuel switching from imported oil to domestically produced coal or nuclear and (3) other direct and indirect effects associated with TES technology that may generate additional external benefits through technological innovation, operating economies of scale and a varying mix of capital to labor ratios in the industries directly affected by the adoption of TES.

Reduced investment in peak-load capacity also lowers the average unit price of electricity, which alters the family budget for power consump-tion. The substitution and income effects of the lowered electricity price together with the industrial resource reallocation effect --

switching from peak-load capacity to TES manufacturing plant – will generate different equilibrium levels of national employment and gross national product. As a result of the income or employment changes, the demand for electricity will also change. Although each occurrence of these changes may be ascertained, their impacts over time would have to be approximated with a distribution function embedded with predefined probabilities of a continuing occurrence.

It is thought that the increased demand for electricity will be met by an increased capacity utilization rate in both base- and cycling-load plants. With TES, it is expected that more coal will be burned and less oil will be imported than otherwise would have been the case.

The peak-load capacity investment reduction estimated in the last section was used as a starting point for the macro-economic simulation in this section. According to the micro analysis, the peak-load capacity investment savings deflated to 1972 dollars, would range from $5 million in 1981 to $178 million in 1990. With these projected fixed cost reductions, the overall price reductions for electricity are calculated to range from 0.04% in 1981 to 0.81% in 1990. Given that the demand elasticity of electricity is about -0.87, as many empirical studies have estimated [27,28,29], the corresponding increases in total electricity expenditures in this country may be estimated at $16.5 million in 1981 and $326.0 million in 1990, respectively, again in 1972 dollars. In real terms, the increased expenditures would increase the gross output of electricity by 0.035% in 1981 and 0.70% in 1990. Assuming that all reductions in peak-load capacity investment by the electric companies is reinvested in the electrical equipment industries that manufacture TES systems, the data presented in Table 4 were derived and taken as exogeneously determined. These results were finally injected into the Wharton Annual Industrial Model [13] for macro-economic impact simulation.

In summary, six interdependent structural functions in the Wharton Annual Industrial Model were altered in order to assess the macro-economic impacts of the implementation of the proposed residential TES systems. To repeat, the changes made are the following:

  a. Reduction in Capacity Investment in Electric Utilities,
  b. Reduction in Average (gross output) Electricity Price for Residential Customers,
  c. Increase in Electricity Power Generation and Consumption,
  d. Increase in Overall Capacity Utilization Rate in Power Plants,
  e. Increase in the Coal Share of Fossil Generated Electricity,
  f. Increase in Investment in the Electrical Equipment Industries, with the amount assumed to be identical to that reduction in (a) for a conservative, lower bound estimation.

The Wharton Annual Industrial Model is a large-scale macro-econometric model of the U.S. economy designed for long-term forecasting and economic policy analysis. It contains over 2,200 variables and covers 56 industrial categories. A fully integrated input-output model provides the necessary bridge for connecting the production side and the demand side of the model. Specifically, the final demand sectors that determine national income are solved simultaneously with the input-output (I/O) system to generate estimates of industrial output that will meet the requirements of the final demand sectors.

The I/O bridge enables not only the conversion of final expenditures into effective demand for the output of individual producing sectors but also the conversion of sectoral prices to implicit deflators for the final expenditures. Industry value-added prices are translated into domestic gross output prices and the composite gross output prices are translated into final demand deflators systematically to assure the existence of an intertemporal market equilibrium. This approach and another slightly different approach that combines the input-output system with econometric models have become accepted procedures in policy analysis and scenario simulations. For example, Glassey [16] and Jorgenson and Hudson [18] employed the latter approach in tackling energy problems and policies.

To simplify the simulation procedures, the exogenously determined changes shown in Table 4 were injected into the Wharton model by adjusting the intercept terms in the six equations being studied. The behavioral relationships, captured in the slopes of those related equations, were not changed. In other words, only the constant terms estimated from the baseline data set in these six equations were adjusted in accordance with the new estimates produced by TES systems. It should be noted that the Wharton model is an econometric model that combines input-output techniques with regression methods for parameter estimation. Conceptually it should produce the material balance in engineering terms. However, it is simply impossible to expect such a material balance to be produced because technical coefficients in the input-output model are fixed from historical data. In addition, the relative price variables for each industry or sector, which are vital to I/O coefficient determination, were fixed in the model when the model was used for simulation. They are in fact very unstable. This undoubtedly complicates the simulation procedures and undermines the credibility of the results. Moreover, the ordinary least-squares-regression estimators for the parameters are best, unbiased linear estimators only if all necessary assumptions on model specification, independency, and normal behavior relationships among the variables and the error terms are satisfied. It is unlikely, if not totally impossible, for such a large-scale econometric model to be well specified and well behaved. Recognition of these econometric problems do not preclude us from using the model. Rather, it suggests that we must interpret the simulated results with care and use them with caution.

The simulation results, adhered to and from the Wharton model between 1980 and 1989 are presented in Table 5. Since models generally become less reliable the farther into the future they project, only ten years' simulation results with and without TES systems are presented. The benchmark data were taken from the model's April, 1980 premeeting solution. As Table 5 reveals, the impacts of TES on national employment and gross national product are positive, albeit relatively small. Should TES systems be introduced in 1981, the changes in GNP will not be observed until 1983 or later; the net increase in GNP brought about by TES may reach $500 million (in 1972 dollars) in 1990, a 0.03% increase over the control solution without TES. Along with GNP changes, there will be additional growth in employment, ranging from 1,000 in 1981 to 9,000 in 1987. It should be noted that these were derived from the most conservative estimate of investment switching from electric utility to electrical industries, and only this switching was considered.

With the multiplier effect, net growth in electricity sales resulting from TES may range from very little to 0.63% or $300 million (constant 1972 dollars) in 1990. This is accomplished with a lower growth rate in real electricity prices than otherwise would be. The simulated growth in price with TES differs more and more from the control case as time goes by, implying that consumer welfare expressed by the consumer's surplus increases at an increasing rate with TES. The real welfare increase in electricity consumption may also be reflected by the higher personal expenditures on electricity for the simulated case over the control case.

The fuel-switching effect of TES in terms of oil backout is somewhat significant. According to the simulation, TES can reduce the crude oil imports by 16.6 million barrels or 0.1 quad Btu-equivalent in 1990, a reduction of 0.67% from the control situation domestic coal production and consumption. The model results indicat that about 0.17 quad Btu-equivalent more of coal will be produced and burned. Furthermore, it should be noted that the oil import reduction estimated from the macro-simulation is not directly comparable to that estimated from the micro-analysis because of different model specification and assumptions.

Concluding Remarks

Electric power demand varies from hour to hour and region to region. A proper reduction of variation in load could reduce requirements for generating capacity (especially the peak-load capacity) as well as the amount of energy provided by imported oil. The result would be a reduction in the total cost of delivered energy and an improvement in industrial productivity. Thermal energy storage, which has been employed in European countries for years, has demonstrated its contribution to more efficient load management and hence improved productivity of the electric utility industry abroad.

This paper examined the problems of productivity and residential TES technology in the electric industry. It investigated the potential benefits and costs of the TES from a much broader industrial and national viewpoint. Not only the micro-economic efficiency arguments concerning power generation and electric service distribution were discussed, but the measurement problems involved in a benefit/cost assessment of the technological impacts of TES were delineated. Furthermore, a benefit/cost, cross-impact, probabilistic approach (BCCIPA) model for technological impact assessments, a model conceptually quite similar to the Environmental Quality and Residuals Management (EQRM) model developed by Kneese and Bower at Resources for the Future [20], was briefly described and a portion of the model was tested with the TES system being considered.

This paper was centered more on the benefit/cost quantification of the economic variables than on the cross-impact probabilistic evaluation of the social and environmental concerns. On a region-by-region basis, TES was found to be a highly efficient technology in residential space heating and cooling, especially in Federal Regions IV, V, and IX. The efficiency measures (or the benefit/cost ratios) naturally increase as imported oil prices increase. Even with oil price fixed at levels lower than they would likely be in the future, it is estimated that residential TES systems would save billions of dollars for utilities alone. The accumulated savings by electric utilities for the next two decades was estimated to be $14.2 billion for heating and $1.6 billion for cooling, if TES were to be substituted for the conventional system in some new single-family dwellings. When the peak-load capacity capital investments that are saved by TES are considered as the would-be expenditures necessary to meet the demand for electricity, this study found that almost all of the macro-economic impacts were positive and in favor of TES. A preliminary simulation over the next ten years showed (using the Wharton Annual Industrial Model) that the gross national

451

product and total national employment would all be greater if TES were
marketed than if not. Real prices for electricity would be lower and
the total output of elctricity would be higher; total oil imports would
be less and total domestic coal production would be more in the case
with TES. However, the relative differences between the two simulations
are not very large. For instance, the fuel-switching effect indicates
that TES would reduce annual oil imports by some 16.6 million barrels
and stimulate annual coal production by some 0.17 quad Btu-equivalent
in 1990. This is because we started with the most conservative, lower-
bound technological impact, i.e., only an equal amount of investment
would be switched from the utility to the electrical equipment manufacturing industries as a result of the introduction of TES to load
management. Resource reallocation effects would be much greater had we
also considered additional investment in the manufacturing industries
that are required by the TES system.

The impact assessment models in general and the BCCIPA model in particular, by their very nature, at best only approximate the working of our
society. The immense complexity of the real world makes it impossible
to fully understand and hence identify all technological impacts, let
alone accurately assess the impacts across the regions and over periods
of time with certainty. Thus, the quantitative results presented in
this study should be considered tentative and preliminary for many
reasons, including the highly possible model specification errors, data
deficiencies, and vulnerable assumptions. There remains much room for
model improvement and data refinement.

References

1.  Argonne National Laboratory, Benefits and Costs of Load Management:
    A Technical and Resource Material Handbook, ANL/SPG-12 (June, 1980).

2.  J. Asbury, et al., Assessment of Energy Storage Technologies and
    Systems Phase I: Electric Storage Heating, Storage Air Condition-
    ing, and Storage Hot Water Heaters, Argonne National Laboratory
    Report ANL/ES-54 (Aug. 1976).

3.  J.G. Asbury and A. Kouvalis, Electric Storage Heating: The
    Experience in England and Wales and in the Federal Republic of
    Germany, Argonne National Laboratory Report ANL/ES-50 (April 1976).

4.  J.G. Asbury, et al., Commercial Feasibility of Thermal Storage in
    Buildings for Utility Load Leveling, Proc. American Power Conf.,
    Chicago (April 1977).

5.  J.G. Asbury, et al., Assessment of Electric and Electric-Assisted
    Technologies in Residential Heating and Cooling Applications, Proc.
    Energy Use Management Conf., Tucson, Arizona (Oct. 1977).

6.  J. Asbury, et al., Assessment of Energy Storage Technologies and
    Systems, Phase II: Heat Pump and Solar Energy Application, ANL/
    SPG-3 (1978).

7.  J.G. Asbury, R.F. Giese and R.O. Mueller, Residential Electric
    Heating and Cooling: Total Cost of Service, paper presented at 16th
    Annual Illinois Energy Conference on Electric Utilities in Illinois,
    (Sept. 27-29, 1978).

8.  W.E. Balson and S.M. Banager, <u>Integrated Analysis of Load Shapes and Energy Storage</u>, Palo Alto: Electric Power Research Institute, EA-970, Research Project 1108-2 (March 1979).

9.  M.K. Berkowitz and F.C. Jen, "A Note on Production Inefficiency in the Peak-Load Pricing Model," <u>Southern Economic Journal</u>, 44 (2), (October 1977).

10. M. Boiteux, The Choice of Plant and Equipment for the Production of Electric Energy, J. Nelson, ed., <u>Marginal Cost Pricing in Practice</u>, Prentice-Hall, Englewood Cliffs, NJ (1964).

11. Chemical and Electrical System Program, Argonne National Laboratory, <u>Peak-Load Pricing and Thermal Energy Storage</u>, Symposium Proceedings, Chicago (July 15-17, 1979).

12. M.A. Crew and P.R. Kleindorfer, "Peak-Load Pricing with a Diverse Technology," <u>Bell Journal of Economics</u> 7, No. 1, pp. 207-231 (1976).

13. Electric Power Research Institute, <u>The Wharton Annual Energy Model: Development and Simulation Results</u>, EPRI. EA-1115 (July 1979).

14. _____, <u>Integrated Analysis of Load Shapes and Energy Storage</u>, EPRI, EA-970 (March 1979).

15. R.F. Giese, <u>Assessment of Regional Market Potential for Distributed Thermal Energy Storage</u>, Argonne National Laboratory, unpublished manuscript (1980).

16. C.R. Glassey, "Price Sensitive Consumer Demands in Energy Modeling - A Quadratic Programming Approach to the Analysis of Some Federal Energy Agency Policies," <u>Management Science</u>, 24(9), pp. 877-886 (May 1978).

17. H. Houthakker, "Residential Electricity Revisited," <u>The Energy Journal</u> 1(1), pp. 29-41 (1980).

18. E.A. Hudson and D.W. Jorgenson, "U.S. Energy Policy and Economic Growth, 1975-2000," <u>The Bell Journal of Economics and Management Sciences</u> 5, pp. 461-514 (1979).

19. P.L. Joskow, "Contributions to the Theory of Marginal Cost Pricing," <u>The Bell Journal of Economics,</u> Spring 1976, p. 202.

20. A.V. Kneese and B.T. Bower, <u>Environmental Quality and Residuals Management</u> Baltimore: The Johns Hopkins University Press (1979).

21. W.T. Lawrence, <u>Capital Cost Estimates of Selected Advanced Thermal Energy Storage Technologies</u> Chicago: Argonne National Laboratory, ANL/SPG-11 (June 1980).

22. B.C. Liu, "A BCCIPA Model for Water Resource Project Evaluation," E. Quano (ed.) <u>Water Pollution Control in Developing Countries,</u> Bangkok: AIT press (1978).

23. B.C. Liu, et al., "Measurement of the Socioeconomic Impact of Lake Restoration," <u>American Journal of Economics and Sociology</u>, 39(3) pp. 227-236 (July 1980).

24. D.M. Miller and W.R. Coleman, Residential Energy Storage - A Field Program to Measure Utility and Consumer Benefits from Using Electric Thermal Storage Devices, Proceedings of the American Power Conference, Vol. 40 (forthcoming) (1978).

25. I. Pressman, "A Mathematical Formulation of the Peak-Load Pricing Problem," The Bell Journal of Economics, (Autumn 1970) pp. 304-326.

26. E.F. Renshaw, "Expected Welfare Gains from Peak-Load Electricity Charges," Energy Economics, pp. 37-45 (January 1980).

27. V.K. Smith and C.J. Cicchetti, "Measuring the Price Elasticity of Demand for Electric Power: The U.S. Experience," Energy Systems: Planning, Forecasting and Pricing, (Ed. C.J. Cicchetti and W. Foell). Institute for Environmental Studies, University of Wisconsin, Madison (1975).

28. L.D. Taylor, "The Demand for Electricity: A Survey," Bell Journal of Economics, Vol. 6, (1975).

29. R. Turvey, "How to Judge When Price Changes Will Improve a Resource Allocation," The Economic Journal, pp. 825-832 (December 1974).

30. U.S. Dept. of Energy, Division of Energy Storage Systems, Peak-Load Pricing and Thermal Energy Storage Symposium Proceeding (July 15-17, 1979).

Table 1. Utility Savings and TES Costs

| Service Area | TES System | Discharge Period (hr) | No. of TES Customers | Generation Peak Load Reduction ($MW_e$) | Generation Peak Load Reduction (% of Peak) | Capital Gen. | Capital Tran. | Capital Dist. | Variable Fuel | Variable O&M | Variable Cycle | Total | TES Incremental Cost ($/yr/Customer) | Total Net Savings ($10^6$ $/yr) |
|---|---|---|---|---|---|---|---|---|---|---|---|---|---|---|
| A | Water Htg. | 4 | 50,000[b] | 24 | 3.3 | 37 | 15 | 11 | -7 | -1 | 4 | 58 | 26 | 1.61 |
| A | Water Htg. | 16 | 50,000[b] | 40 | 5.5 | 93 | 24 | -14 | 16 | 2 | 11 | 132 | 80 | 2.60 |
| A | Space Htg. | 8 | 2,800[a] | 40 | 5.5 | 841 | 441 | -36 | 127 | 19 | 30 | 1,421 | 568[c] 338[d] | 2.37[c] 3.00[d] |
| B | Water Htg. | 4 | 34,400[b] | 41 | 4.3 | 68 | 36 | 12 | -5 | -1 | 4 | 114 | 26 | 3.02 |
| B | Water Htg. | 16 | 37,500[b] | 46 | 4.8 | 161 | 37 | -9 | -8 | -1 | 10 | 190 | 80 | 4.13 |
| B | Space Htg. | 8 | 5,800[a] | 41 | 4.3 | 686 | 217 | -94 | -46 | -7 | 38 | 794 | 552[c] 310[d] | 1.40[c] 2.80[d] |
| C | Water Htg. | 4 | 145,000[b] | 114 | 0.9 | 40 | 23 | -10 | -9 | -2 | 5 | 48 | 26 | 3.08 |
| C | Water Htg. | 16 | 145,000[b] | 114 | 0.9 | 108 | 23 | 20 | -24 | -4 | 8 | 131 | 80 | 7.4 |
| C | Space Htg. | 8 | 54,000[b] | 0 | 0 | 220 | 0 | 0 | -136 | -23 | 40 | 152 | 552[c] 310[d] | -21.5[c] -8.5[d] |
| C | Air Cond. | 8 | 219,000[a] | 945 | 7.7 | 163 | 127 | 35 | 32 | 6 | 1 | 365 | 252 | 24.7 |
| D | Water Htg. | 4 | 28,100[b] | 29 | 1.5 | 47 | 31 | 22 | 2 | 0 | 8 | 110 | 26 | 2.36 |
| D | Water Htg. | 16 | 28,100[b] | 29 | 1.5 | 72 | 32 | 22 | 31 | 6 | 17 | 180 | 80 | 2.81 |
| D | Air Cond. | 8 | 18,300[a] | 141 | 7.5 | 355 | 234 | 198 | 121 | 22 | 21 | 950 | 305 | 11.8 |

[a] Number of TES customers at which total net benefits are maximized

[b] Number of TES customers at which the aggregate TES design-day kilowatt-hour load equals the aggregate design-day kilowatt-hour load of existing conventional-system customers

[c] Refers to 10-unit dispersed storage space heating system.

[d] Refers to central furnace system.

Winter Peaking Service

| YEAR | 1 | 2 | 3 | 4 | 5 | 6 | 7 | 8 | 9 | 10 | TOTAL U.S. |
|------|---|---|---|---|---|---|---|---|---|----|-----------|
| 1985 | 24 | 19 | 33 | 92 | 50 | - | 3 | 8 | - | - | 228 |
|      | (46) | (37) | (63) | (179) | (96) | - | (6) | (15) | - | - | (442) |
| 1990 | 30 | 24 | 41 | 115 | 62 | - | 3 | 10 | - | - | 285 |
|      | (194) | (158) | (267) | (755) | (406) | - | (23) | (64) | - | - | (1870) |
| 1995 | 30 | 24 | 41 | 115 | 62 | - | 3 | 10 | - | - | 285 |
|      | (342) | (278) | (471) | (1330) | (716) | - | (41) | (112) | - | - | (3290) |
| 2000 | 21 | 18 | 30 | 84 | 45 | - | 2 | 7 | - | - | 209 |
|      | (466) | (380) | (642) | (1820) | (976) | - | (56) | (153) | - | - | (4490) |

Summer Peaking Service

| YEAR | 1 | 2 | 3 | 4 | 5 | 6 | 7 | 8 | 9 | 10 | TOTAL U.S. |
|------|---|---|---|---|---|---|---|---|---|----|-----------|
| 1985 | - | - | - | - | - | - | - | - | - | - | - |
|      | - | - | - | - | - | - | - | - | - | - | - |
| 1990 | - | 25 | 23 | 12 | 36 | 14 | 7 | - | 35 | - | 152 |
|      | - | (48) | (45) | (24) | (69) | (27) | (14) | - | (67) | - | (294) |
| 1995 | - | 30 | 28 | 15 | 43 | 17 | 8 | - | 41 | - | 182 |
|      | - | (201) | (183) | (96) | (286) | (110) | (55) | - | (273) | - | (1200) |
| 2000 | - | 30 | 26 | 13 | 42 | 15 | 8 | - | 38 | - | 172 |
|      | - | (349) | (316) | (163) | (493) | (188) | (95) | - | (465) | - | (2070) |

NOTE: Figures without parentheses are annual quantities; figures in parentheses represent cumulative quantities.

Table 3. Discounted Savings for TES Systems, Residential Market,
by DOE Regions, 1985-2000
($10^6$)

Winter Peaking Service

| YEAR | 1 | 2 | 3 | 4 | 5 | 6 | 7 | 8 | 9 | 10 | TOTAL U.S. |
|------|---|---|---|---|---|---|---|---|---|----|-----------|
| 1985 | 22.6 | 0.6 | 7.0 | 56.3 | 37.6 | - | 1.5 | 2.7 | - | - | 128 |
|      | (42.2) | (1.1) | (13.1) | (105) | (70.2) | - | (2.8) | (5.0) | - | - | (239) |
| 1990 | 89.5 | 9.9 | 32.6 | 231 | 179 | - | 6.4 | 16.1 | - | - | 565 |
|      | (363) | (37) | (130) | (934) | (712) | - | (26) | (63) | - | - | (2,260) |
| 1995 | 190 | 40.4 | 83.8 | 435 | 393 | - | 12.9 | 35.6 | - | - | 1,190 |
|      | (1,180) | (210) | (489) | (2,800) | (2,390) | - | (81) | (216) | - | - | (7,370) |
| 2000 | 235 | 50.2 | 103.9 | 539 | 487 | - | 16.1 | 44.2 | - | - | 1,480 |
|      | (2,270) | (443) | (971) | (5,300) | (4,650) | - | (156) | (421) | - | - | (14,200) |

Summer Peaking Service

| YEAR | 1 | 2 | 3 | 4 | 5 | 6 | 7 | 8 | 9 | 10 | TOTAL U.S. |
|------|---|---|---|---|---|---|---|---|---|----|-----------|
| 1985 | - | - | - | - | - | - | - | - | - | - | - |
|      | - | - | - | - | - | - | - | - | - | - | - |
| 1990 | - | 4.9 | 4.8 | 3.4 | 8.2 | 5.1 | 2.2 | - | 7.4 | - | 36.0 |
|      | - | (9.1) | (9.0) | (6.3) | (15.3) | (9.5) | (4.1) | - | (13.8) | - | (67.2) |
| 1995 | - | 19.4 | 18.4 | 13.2 | 31.9 | 20.0 | 8.2 | - | 27.4 | - | 138 |
|      | - | (79.6) | (75.8) | (54.3) | (131) | (82.2) | (33.9) | - | (113) | - | (569) |
| 2000 | - | 32.9 | 30.8 | 23.0 | 53.1 | 30.4 | 17.3 | - | 45.4 | - | 233 |
|      | - | (222) | (209) | (154) | (361) | (214) | (109) | - | (310) | - | (1,580) |

NOTE: Figures without parentheses are annual quantities; figures in parentheses represent cumulative quantities.

Table 4. Projected Changes Directly Resulted from Peak-Load Capacity Investment Reduction

| Year | Peak-Load Capacity Investment Reduction ($M) | Average Electricity Cost (Price) Reduction (%) | Electricity Consumption and Output Increase ($10^6$ kWh) | Capacity Utiliz. Rate Change (%) | Base-line Data Without Storage | Storage Capacity Total Fuel-Switching (in billion/kWh) | Peak Capacity Replacement | % of Total Fossil Generated Elec. by Coal w/Stor. |
|------|------|------|------|------|------|------|------|------|
| 1981 | 5.3 | 0.040 | 840 | 0.035 | 1129.1 | 0.27 | 0.06 | 63.5 |
| 1982 | 10.6 | 0.053 | 1,180 | 0.047 | 1200.8 | 0.89 | 0.18 | 65.1 |
| 1983 | 20.6 | 0.079 | 1,760 | 0.069 | 1247.8 | 2.06 | 0.41 | 66.6 |
| 1984 | 41.3 | 0.106 | 2,400 | 0.092 | 1285.8 | 4.46 | 0.86 | 68.1 |
| 1985 | 81.0 | 0.204 | 4,690 | 0.177 | 1320.4 | 9.24 | 1.77 | 69.7 |
| 1986 | 105.6 | 0.325 | 7,610 | 0.283 | 1355.1 | 15.30 | 2.96 | 71.3 |
| 1987 | 109.9 | 0.444 | 10,590 | 0.386 | 1393.4 | 21.36 | 4.19 | 73.0 |
| 1988 | 118.4 | 0.561 | 13,620 | 0.488 | 1430.8 | 27.57 | 5.53 | 74.6 |
| 1989 | 135.3 | 0.675 | 16,660 | 0.587 | 1467.3 | 33.76 | 6.93 | 76.2 |
| 1990 | 178.4 | 0.808 | 20,290 | 0.703 | 1505.6 | 40.68 | 8.97 | 77.9 |

Note: Individual items may not add to total due to rounding.

Table 5. Assessed Macro-economic Impacts of TES

| Year | Control Solution | Simulation | Percent Difference |
|------|------------------|------------|--------------------|
| Gross National Product (billions of 1972 $) | | | |
| 1980 | 1441.4 | 1441.4 | 0.00 |
| 1981 | 1461.3 | 1461.4 | 0.00 |
| 1982 | 1506.7 | 1506.7 | 0.00 |
| 1983 | 1555.1 | 1555.3 | 0.01 |
| 1984 | 1604.2 | 1604.3 | 0.00 |
| 1985 | 1644.3 | 1644.5 | 0.01 |
| 1986 | 1690.3 | 1690.6 | 0.02 |
| 1987 | 1737.6 | 1738.0 | 0.02 |
| 1988 | 1790.4 | 1790.8 | 0.02 |
| 1989 | 1646.6 | 1847.1 | 0.03 |
| Employment, All Industries (millions) | | | |
| 1980 | 98.096 | 98.096 | 0.000 |
| 1981 | 98.946 | 98.947 | 0.001 |
| 1982 | 100.432 | 100.434 | 0.002 |
| 1983 | 101.850 | 101.854 | 0.004 |
| 1984 | 103.301 | 103.304 | 0.003 |
| 1985 | 104.585 | 104.591 | 0.006 |
| 1986 | 105.900 | 105.909 | 0.008 |
| 1987 | 107.217 | 107.227 | 0.009 |
| 1988 | 108.426 | 108.434 | 0.007 |
| 1989 | 109.682 | 109.687 | 0.004 |
| Electricity Gross Output (billions of 1972 $) | | | |
| 1980 | 45.3 | 45.3 | 0.00 |
| 1981 | 47.1 | 47.1 | 0.00 |
| 1982 | 49.2 | 49.2 | 0.02 |
| 1983 | 50.3 | 50.4 | 0.17 |
| 1984 | 51.1 | 51.1 | 0.11 |
| 1985 | 51.8 | 51.9 | 0.32 |
| 1986 | 52.7 | 52.9 | 0.32 |
| 1987 | 53.4 | 53.8 | 0.36 |
| 1988 | 54.6 | 54.9 | 0.53 |
| 1989 | 55.6 | 55.9 | 0.63 |
| Gross Output Price Index, Electric Utilities (1972 = 100) | | | |
| 1980 | 240.0 | 240.0 | 0.01 |
| 1981 | 266.1 | 266.0 | -0.04 |
| 1982 | 293.7 | 293.6 | -0.05 |
| 1983 | 316.5 | 316.3 | -0.05 |
| 1984 | 339.6 | 339.2 | -0.11 |
| 1985 | 369.1 | 368.4 | -0.20 |
| 1986 | 403.6 | 402.4 | -0.30 |
| 1987 | 443.7 | 442.0 | -0.39 |
| 1988 | 480.1 | 477.8 | -0.50 |
| 1989 | 518.7 | 515.6 | -0.60 |

Table 5 (Cont'd.)

| Year | Control Solution | Simulation | Percent Difference |
|------|------------------|------------|--------------------|
| Personal Consumption Expenditures, Electric Utilities (billions of 1972 $) | | | |
| 1980 | 17.1 | 17.1 | 0.00 |
| 1981 | 17.5 | 17.5 | 0.00 |
| 1982 | 18.1 | 18.1 | 0.04 |
| 1983 | 18.5 | 18.5 | 0.45 |
| 1984 | 18.9 | 18.9 | 0.29 |
| 1985 | 19.2 | 19.3 | 0.34 |
| 1986 | 19.6 | 19.7 | 0.80 |
| 1987 | 19.9 | 20.1 | 0.86 |
| 1988 | 20.3 | 20.6 | 1.29 |
| 1989 | 20.7 | 21.0 | 1.51 |
| Imports, Crude Petroleum (million barrels) | | | |
| 1980 | 2193.2 | 2193.4 | 0.01 |
| 1981 | 2120.3 | 2118.6 | -0.08 |
| 1982 | 2187.5 | 2188.2 | -0.03 |
| 1983 | 2732.4 | 2231.0 | -0.06 |
| 1984 | 2238.0 | 2237.1 | -0.04 |
| 1985 | 2233.5 | 2229.0 | -0.20 |
| 1986 | 2285.1 | 2279.5 | -0.24 |
| 1987 | 2334.0 | 2324.7 | -0.40 |
| 1988 | 2433.5 | 2418.7 | -0.61 |
| 1989 | 2491.5 | 2474.9 | -0.67 |
| Coal Production (Quadrillion Btu) | | | |
| 1980 | 16.72 | 16.66 | -0.32 |
| 1981 | 17.15 | 17.15 | -0.03 |
| 1982 | 18.29 | 18.30 | -0.05 |
| 1983 | 18.46 | 18.43 | -0.16 |
| 1984 | 18.85 | 18.83 | -0.15 |
| 1985 | 19.24 | 19.24 | 0.00 |
| 1986 | 19.59 | 19.64 | 0.24 |
| 1987 | 19.97 | 20.04 | 0.37 |
| 1988 | 20.37 | 20.49 | 0.60 |
| 1989 | 21.02 | 21.19 | 0.81 |
| Non-Residential Fixed Investment (billions of 1972 $) | | | |
| 1980 | 149.08 | 149.08 | 0.00 |
| 1981 | 148.95 | 148.95 | 0.00 |
| 1982 | 154.80 | 154.80 | 0.00 |
| 1983 | 162.83 | 162.84 | 0.01 |
| 1984 | 170.54 | 170.55 | 0.01 |
| 1985 | 177.57 | 177.59 | 0.01 |
| 1986 | 186.25 | 186.29 | 0.02 |
| 1987 | 194.62 | 194.67 | 0.02 |
| 1988 | 203.62 | 203.66 | 0.02 |
| 1989 | 212.71 | 212.75 | 0.02 |

# PROCESS MODELS: ANALYTICAL TOOLS FOR
# MANAGING INDUSTRIAL ENERGY SYSTEMS

**Stephen O. Howe, David A. Pilati** and **Chip Balzer**
Brookhaven National Laboratory

**F. T. Sparrow**
Purdue University

## Introduction

Mathematical programming models of energy-intensive industries in the United States are being developed at the National Center for Analysis of Energy Systems at Brookhaven National Laboratory (BNL). These industry-specific models are based on data from economic and engineering analyses of manufacturing processes and are intended to represent energy flows within an industry in considerable detail. Given a projection of the demand for an industry's manufactured goods and a scenario for the basic raw materials and energy prices, the models are designed to select those processes and levels of production that minimize the cost of meeting the projected demands. To the extent that an industry operates to minimize costs and maximize profits, these process models are an analytical aid to understanding the consequences of alternative scenarios. They are useful management tools which can be used to project energy demands, assess new technologies and analyze energy policy.

The purpose of this paper is to describe and illustrate how the process models developed at BNL are used to analyze industrial energy systems. Following a brief overview of the industry modeling program, the general methodology of process modeling is discussed. The discussion highlights the important concepts, contents, inputs and outputs of a typical process model. A model of the U.S. pulp and paper industry is then discussed as a specific application of process modeling methodology. Case study results from the pulp and paper model illustrate how process models can be used to analyze a variety of issues. Applications addressed with the case study results include projections of energy demand, conservation technology assessment, energy-related tax policies, and sensitivity analysis. A subsequent discussion of these results supports the conclusion that industry process models are versatile and powerful tools for managing industrial energy systems.

Table 1 gives the current status of industry models at BNL. The final step in the developmental sequence outlined in the table is the completion of a user-oriented version of each process model. These user-oriented versions are based on data obtained from a variety of public sources as documented in References 1 and 2. More highly developed versions of the models with data from other sources are, however, possible. Because relationships and data in the models are represented in terms and units conventionally understood within an industry, the models are amenable to improvement through criticism and review by industrial engineers and other outside experts. A goal of the industry modeling program is to establish relationships with private industry in order to enhance the validity and improve the utility of the models. Dialogues have begun with representatives of the paper and steel industries, as well as other sources of technical expertise.

Methodology

The industry process modeling approach is well developed in theory and in applications. An early, but comprehensive discussion of the approach is given in Reference 3. Central to the process modeling approach is the concept of an activity which transforms inputs of goods and services into outputs. Activities and their relationships in process models are defined to reflect actual and possible alternative structures of technologies at all stages of the production process. According to Figure 1 which describes process relationships in a simplified model, an initial activity in the production sequence could require a variety of raw material, energy, labor, and dollar inputs to create a unit of output which, in turn, could become one of several inputs for subsequent activities.

Process models are not only a catalog of alternative functional processes; they also encompass a mathematical description of the geographical structure of the industry and its internal and external environmental constraints. The scope of relationships considered is suggested by the following categories of equations from a typical process model:[4]

a) Process Description: These equations represent the chemical or physical transformation of inputs into outputs by activities. They include the concepts of material balance and process yield.
b) Supply and Distribution Balance: These equations represent links between intermediate processes in the model. They are transportation balances relating supply sources to demand sinks.
c) Capacity Constraints: These equations limit production activities to the capacity of the facilities designed for that production process. Since the facilities' capacity include the capacity "built" by the model, these equations are responsible for dynamism in the model in that investments in one time period increase the operating capacity in subsequent time periods.
d) Demand Stipulations: These equations drive the model. They link the demand for different products with the processes or external sources which supply these products. The representation of an optimal way of meeting these inelastic demands is the objective of the model.

e) Externally Imposed Restrictions: These may be important in the solution of the model. Political, historical, geographical and financial
considerations, that are beyond the system's domain, influence the
system's behavior. For example: federal regulations may demand a
certain level of performance from certain activities. These are constraints that the analyst manipulates to simulate different scenarios.

f) Inventory Balance Rows: These equations represent the inventory
balance between consecutive time periods.

g) Technological Change Balance Rows: These equations represent changes
in technology of existing facilities. The capacity changed from,
say, type A to type B, is deleted from the existing type A capacity
and added to the existing type B capacity.

Implicit in the preceeding equation categories are requirements for substantial quantities of input data. Figure 2 outlines the characteristics
of a typical process model and identifies some important types of input
data. The capacities of existing technologies, operating requirements,
energy prices and capital costs are representative of input data required
in addition to an engineering characterization of input-output transformations. Depending on the industry, certain additional constraints may
be needed to limit the availability of a particular technology or resource.

Generally, the models are designed to find the after-tax cost-minimizing
mix of activities which satisfy a specified set of demands for the final
products of an industry. This problem can be stated mathematically as
follows:[5]

$$\min_{x} \sum_{t} \frac{(1 - \tau)E_t(x) + K_t(x) - \tau D_t(x)}{(1 + r)^t}, \text{ subject to } Ax \le b, \quad x \ge 0$$

where $x$ = set of processes, $r$ = discount rate, $E_t$ = current expenses in
$t$, $K_t$ = capital expenses in $t$, $D_t$ = depreciation allowance in $t$, $\tau$ = profits tax rate; $Ax \le b$, the constraint set which relates the process to resource demand and supply. The constraint set includes the industry's
product demands, specification of material flow paths, equipment capacity, and other industry-specific behavior. Given the investment and
operating characteristics of specified technological options, such a
model selects investments and operating modes that result in minimizing
the present value of the industry's after-tax production costs. Simultaneously, this solution gives the predicted operating requirements
(energy, labor, and materials) to meet the specified industry product
demands. Also, studies have been done where the cost-minimization function, described above, has been replaced by energy-maximization or minimization functions and by functions which maximize or minimize the use of
particular technologies. In all cases, model results include industry
investment patterns, technology use and resource consumption as outlined
in Figure 2.

An Example of an Industry Process Model

The process model of the U.S. pulp and paper industry is both temporally
dynamic and regional. It considers the production of pulp and paper in
four regions of the United States over a 30 year time horizon divided
into three time periods (a five period, five years per period version is
available and is documented in Reference 2). Each ten year period is

459

modeled as a single representative year within the period; the base year or first representative year being 1975. Periods are linked in that production capacity existing or purchased in earlier periods is available in later periods after appropriate adjustments for physical depreciation. Outputs or inputs other than production capacity, however, may not accumulate in current periods for use in future periods. That is, no inventories of raw materials, semi-finished goods, or final outputs are allowed.

Five general types of activities are included in the model: purchases of raw materials, sales of by-product energy, material/energy conversion, capacity expansion, and transportation. Figure 3 gives a flow diagram of the major materials conversion processes within a single region. It shows the sequential processing relationships between the three types of fiber, eight types of pulp, and eight types of final paper products. The structure of the model, the sources of data, and results of validation case studies are discussed in detail in Reference 2.

## Scenario Description

Figures 4 and 5 give the scenario description assumed for the cases to be discussed in this analysis. Energy prices are based on the CONAES Demand Panel's Scenario B.[6] In this scenario, energy prices increase linearly in time and regional prices equalize in the year 2010. The demands for specific paper products are based on results in Reference 7. The rate of increase in aggregate demand averages about 3% per year.

In the cases presented here a discount factor or after-tax rate of return on investments of 10% is assumed. (For the eight paper companies in Reference 8, after-tax returns on investment ranged from 6.3 to 12.1% in 1977). Projected levels of investments are somewhat dependent upon the discount rate selected. Figure 6 shows that total investment in energy conserving equipment declines as the required rate of return increases. It varies by a factor of 2 over a reasonable range of rates which could have been selected.

## Case Study Results

Results selected from case studies done with the pulp and paper industry model are presented in Figures 7 through 11. These results are discussed briefly here to illustrate how a process model can be applied to a variety of energy-related issues.

Purchased Energy Projections and Technology Assessment. Figure 7 shows projections of purchased energy expressed in units of $10^{15}$ Btu for the industry from 1975 to 1995. Purchased energy is considered in these analyses to include coal, oil, gas and electricity and does not include purchased bark, hogged fuel or fuel derived from feedstocks. Case A is a simple hand-calculated extrapolation made for the purposes of comparison with model results. It is based on the demand projection in Figure 5 and the assumption that no changes in technology occur after 1975 (purchased energy per unit product is assumed constant). Cases B and C are results from cost-minimization solutions which consider the increases in energy prices described in Figure 4. Case B is a projection showing how energy conserving technologies available in 1975 could reduce purchased energy requirements in the future. Case C shows how a combination of all the energy conserving technologies discussed in detail in Reference 2 could reduce purchased energy requirements still more. (Case C in Figure 7 is labelled as case A, the base case in all subsequent figures.) A comparison of cases similar to B and C, one without a particular technology and one with a particular technology is a method of evaluating the energy conserving potential of the technology with an industry process model.

Btu Taxes and Investment Tax Credits. Figure 8 shows the effect of various Btu taxes on the amounts of energy purchased between 1975 and 1995. In case D, where the purchase of fuels and electricity was not allowed as an operating expense for income tax purposes, the price of purchased energy is effectively doubled relative to base case A with no Btu tax on energy.

Figure 9 illustrates how investment tax credits for energy conserving technologies can be evaluated with process models. In the base case a 10% selective investment tax credit is assumed. In cases with investment tax credits larger than 10%, less energy is purchased and higher levels of new technologies appear in the model solution. As tax credits approach 60%, new conservation technologies are favored so strongly in the model that premature retirement of old capacity occurs widely.

Sensitivity Studies and Alternative Objective Functions. Generally the objective function of a process model is a cost minimization function. The results discussed so far have all been from such models. It is possible, however, to define other objective functions and explore system behavior when the constraint on cost minimization is relaxed. Figures 10 and 11 show results from cases where the costminimizing objective function was replaced by an alternative function.

The projection of purchased energy labelled A in Figure 10 is the standard cost minimization base case. Case B is a case in which purchased energy is maximized subject to the constraint that costs in future periods do not exceed by more than 1% the costs in the cost minimization case. Similarly, in case C, purchased energy is minimized subject to the constraint that costs in future periods do not exceed by 1% the total production costs in the cost-minimization case. The spread between B and C represents the range of purchased energy projections of all industry operating modes within 1% of absolute minimum cost. About 60% of the difference is related to activities for the burning of feedstocks and self-generated sources of energy in case C, the energy minimization case, that did not enter the solution in case B, the energy maximization case. Nevertheless, the results do show that long run energy demands may vary significantly and yet have little impact on total production costs.

Results are summarized in Figure 11 from cases in which the utilization of particular technologies was either maximized or minimized subject to the constraint that costs do not exceed by more than 1% the total production costs of the cost-minimization case. In all cases except one, it was possible for a particular technology to be eliminated from the solution. In cases maximizing particular technologies the use of one technology increased by a factor of 9.6 relative to the base case. These results show that the market penetration of any of the energy conserving technologies in the pulp and paper model may vary considerably and yet have little effect on total production costs.

461

Discussion

The case study results from the pulp and paper model illustrate the range and power of the process modeling approach. The cases focussing on energy-demand projections show how future energy use is dependent on both the availability of particular energy conserving technologies and on the price of energy. The models are designed so that specific technological options available to an industry can be readily assessed and compared. Because technologies are explicitly represented, the effects of investment tax credits on specific technologies can also be readily evaluated. However, one limitation of process models is that they cannot foresee technological innovation; technologies considered in the models must be introduced explicitly. This fact could lead to underestimates of factor price sensitivity, especially in cases of large changes in prices over long periods.

Industry process models, by design, offer a systems approach to analysis of energy demand and conservation issues. Results from other applications of the pulp and paper model have identified technological interdependencies within the industry that could significantly affect the estimates of the potential of conservation programs.[9] In one instance, the model was used to investigate the conservation potential of three alternatives to a conventional technology. Given the model results for the use of each option, a separate calculation was made to find the engineering energy savings (alternative technology units in use multiplied by the energy savings per unit). The energy savings in the model results were about one to three times the engineering result, depending on the which alternative technology was being considered.

It is also possible that the opposite result could occur; introduction of an energy conserving process could cause greater overall energy use. Analysis of the effect of a selective tax credit for cogeneration led to the conclusion that such a policy could actually result in increased energy consumption.[9] In this case, the tax credit effectively subsidized the cost of steam and electricity, and investments in energy conserving options other than cogeneration were reduced.

Finally, the case studies with alternative objective functions presented in this analysis illustrate why the accuracy of projections based solely on the minimization of production costs should be questioned. Large variations in projected energy demands and market penetration of individual technologies occurred with relatively small effects on total production costs. This conclusion has important implications for energy modeling methodologies. It suggests that objective functions minimizing costs alone should be replaced by objective functions which reflect the unique realities in which each industry operates. This analysis has shown that process models are extremely flexible, both structurally and functionally. They are well suited to deal with alternative objective functions and to meet the evolving challenges of managing industrial energy systems.

Acknowledgements

The authors thank Arlean Vanslyke and Kathleen Melton for typing this manuscript. This work was done under contract No. RP-1762-1 for the Electric Power Research Institute and Contract No. DE-AC02-76CH00016 with the U.S. Department of Energy.

References

1.  F. T. Sparrow, D. Pilati, T. Dougherty, E. McBreen, L. L. Juang, "The iron and steel industry process model", Brookhaven National Laboratory Report No. 51073, Brookhaven National Laboratory, Upton, New York 11973, January 1980.

2.  D. A. Pilati, J. Chang, F. T. Sparrow, S. O. Howe, C. Balzer, B. McBreen, "A process model of the U.S. pulp and paper industry", Brookhaven National Laboratory Report No. 51142, Brookhaven National Laboratory, Upton, New York 11973, April 1980.

3.  A. S. Manne, H. M. Markowitz, (eds.), Studies in Process Analysis, John Wiley and Sons, New York, 1963.

4.  M. A. Kelly, "A model of the aluminum industry in the United States", M. S. Thesis, Virginia Polytechnic Institute and State University, Blacksburg, Virginia, 1978, pp. 46–47.

5.  D. A. Pilati and F. T. Sparrow, The Brookhaven process optimization models", Energy 5:417–428.

6.  Demand and Conservation Panel of the Committee on Nuclear and Alternative Energy Systems," U. S. energy demand: some low energy futures", Science, April 14, 1978.

7.  H. D. Nguyen. "A study of demand for paper and board products", Oak Ridge National Laboratory draft report.

8.  Ernst and Ernst, "Costs of capital and rates of return for industrial firms and class A & B electric utility firms", Report prepared for U. S. Department of Energy, Economic Regulatory Administration, June, 1979.

9.  D. A. Pilati and R. Rosen. "The use of the pulp and paper industry process and model for R & D decision making," Brookhaven National Laboratory Report No. 50839, Brookhaven National Laboratory, Upton, New York 11973, March 1978.

TABLE 1

| | | STATUS REPORT OF BNL INDUSTRY MODELING PROGRAM STAGE OF PROGRESS (11/80) | | | |
|---|---|---|---|---|---|
| INDUSTRIAL SECTOR | IDENTIFIED EXISTING MODEL | DATA BASE DEVELOPMENT | MODEL DEVELOPMENT | USER-ORIENTED VERSION DEVELOPMENT | USER-ORIENTED VERSION COMPLETE |
| IRON AND STEEL* | | | | | X |
| PULP AND PAPER* | | | | | X |
| ALUMINUM* | | | | X | |
| CEMENT* | | | X | | |
| PETROCHEMICALS* | X | | | | |
| PLASTICS AND POLYESTERS* | X | | | | |
| PETROLEUM REFINING* | X | | | | |
| COPPER | | | X | | |
| TEXTILES | | X | | | |
| DAIRY | | | X | | |
| MEAT | | | X | | |
| GLASS | | | X | | |

*CURRENT VERSION IN-HOUSE AT BNL.

463

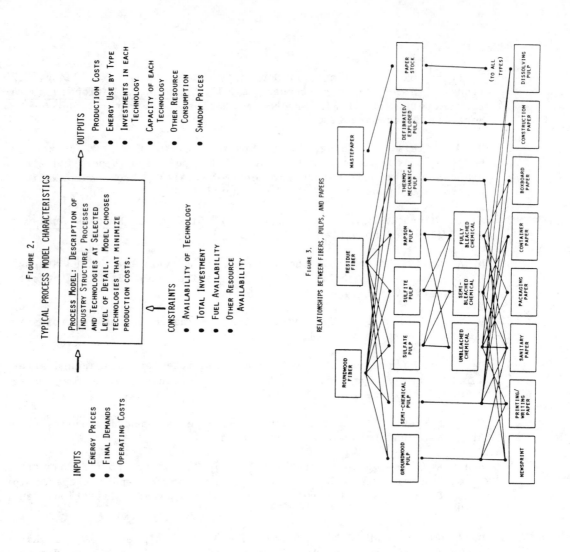

Figure 2.
Typical Process Model Characteristics

Inputs
• Energy Prices
• Final Demands
• Operating Costs

Process Model: Description of Industry Structure, Processes and Technologies at Selected Level of Detail. Model chooses technologies that minimize production costs.

Constraints
• Availability of Technology
• Total Investment
• Fuel Availability
• Other Resource Availability

Outputs
• Production Costs
• Energy Use by Type
• Investments in each Technology
• Capacity of each Technology
• Other Resource Consumption
• Shadow Prices

Figure 3.
Relationships Between Fibers, Pulps, and Papers

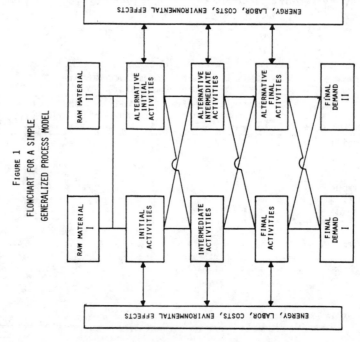

Figure 1
Flowchart for a Simple Generalized Process Model

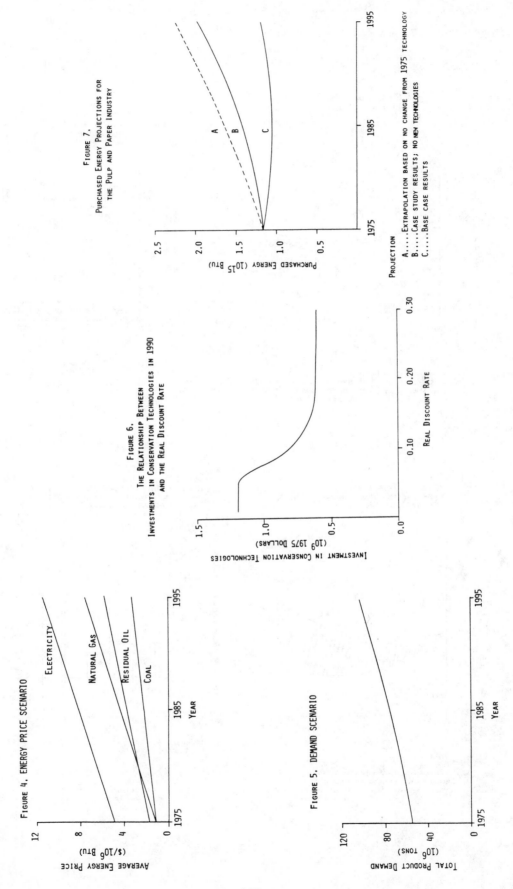

FIGURE 7.

PURCHASED ENERGY PROJECTIONS FOR
THE PULP AND PAPER INDUSTRY

PROJECTION

A.....EXTRAPOLATION BASED ON NO CHANGE FROM 1975 TECHNOLOGY
B.....CASE STUDY RESULTS; NO NEW TECHNOLOGIES
C.....BASE CASE RESULTS

FIGURE 6.

THE RELATIONSHIP BETWEEN
INVESTMENTS IN CONSERVATION TECHNOLOGIES IN 1990
AND THE REAL DISCOUNT RATE

FIGURE 4. ENERGY PRICE SCENARIO

FIGURE 5. DEMAND SCENARIO

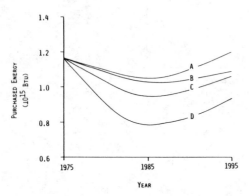

FIGURE 8.
THE EFFECTS OF ENERGY TAXES
ON PURCHASED FUELS AND ELECTRICITY
IN THE PULP AND PAPER INDUSTRY

POLICY

A.....BASE CASE; NO ENERGY TAX
B.....$0.25 TAX PER $10^6$ BTU PURCHASED FUELS AND ELECTRICITY
C.....$1.00 TAX PER $10^6$ BTU PURCHASED FUELS AND ELECTRICITY
D.....ELIMINATE TAX DEDUCTION ON PURCHASED FUELS

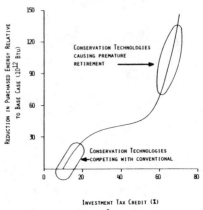

FIGURE 9.
THE EFFECTS OF SELECTIVE INVESTMENT TAX CREDITS ON THE AMOUNT OF
ENERGY PURCHASED BY THE PULP AND PAPER
INDUSTRY IN 1990

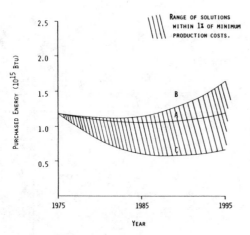

FIGURE 10.
PURCHASED ENERGY PROJECTIONS FOR
THE PULP AND PAPER INDUSTRY

PROJECTION

A.....BASE CASE RESULTS - MINIMUM COST SOLUTION
B.....MAXIMIZE PURCHASED ENERGY ⎤ ALLOWING THE COST OF
C.....MINIMIZE PURCHASED ENERGY ⎦ PRODUCTION TO INCREASE
                                    1% ABOVE THE MINIMUM.

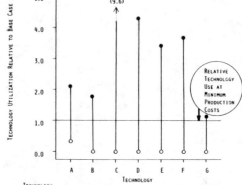

FIGURE 11. RANGE OF MARKET PENETRATION RESULTS WITH PRODUCTION
COSTS WITHIN 1% OF THE MINIMUM (FOR 1990)

COMPARISONS OF NEW TECHNOLOGY UTILIZATION

ALLOWING A 1%    ○  RELATIVE UTILIZATION LEVEL IN TECHNOLOGY MINIMIZATION CASE.
INCREASE IN
PRODUCTION COSTS ●  RELATIVE UTILIZATION LEVEL IN TECHNOLOGY MAXIMIZATION CASE.
ABOVE THE
MINIMUM

TECHNOLOGY

A.....RAPSON CLOSED CYCLE
B.....HEADBOX
C.....SLUDGE DRYING
D.....MACH NOZZLES
E.....VAPOR RECOMPRESSION
F.....FREEZE CRYSTALLIZATION
G.....HIGH EFFICIENCY TURBINES

466

# EFFECTS OF URANIUM MINING AND MILLING
# ON SURFACE WATER IN NEW MEXICO

**Lynn L. Brandvold**
New Mexico Bureau of Mines and Mineral Resources

**Donald K. Brandvold** and **Carl J. Popp**
New Mexico Institute of Mining and Technology

## Introduction

The overall energy situation with its increased demand for fuel for
nuclear reactors has had a stimulatory effect on uranium exploration,
mining, milling, and related activities in the area of New Mexico called
the Grants Mineral Belt.  In August of 1977, when the study reported here
began there were 21 active uranium mines, 36 exploration drilling rigs
and 38 development drilling rigs in the area.  This extensive activity
had and has the potential to introduce toxic materials and other substances into the Rio Grande, a major southwestern waterway.

The Grants Mineral Belt runs roughly from Albuquerque west to the
Arizona-New Mexico border (Figure 1).  The climate is semiarid to
arid.  Precipitation increases with increasing elevation.  The average
annual precipitation at Grants airport is 8.3 inches.  With the exception of the volcanically formed Mt. Taylor area most of the region is
plateau underlain by sedimentary rocks.  Except for high elevations
vegetation is generally sparse.

The area studied was that portion of the belt which drains east from the
Continental Divide to the Rio Grande.  The two main rivers are the Rio
San Jose and the Rio Puerco.  Two potentially important tributaries
because of their location are San Mateo Creek and the Rio Paguate.  San
Mateo Creek enters the San Jose only under flood conditions, while the
Rio Paguate is now a small permanent stream.  The Rio San Jose joins the
Rio Puerco which enters the Rio Grande at Bernardo about 50 miles south
of Albuquerque.  The lower reaches of the San Jose and Puerco rivers are
intermittent with detectable flow less than 50% of the time (1, 2).
When flow does occur it is often quite high, especially in the Puerco.
Ten percent of the time flow is higher than 50 cfs and one percent of

the time it is equal to or greater than 1000 cfs (3). Because of this extreme variability in flow, the arid climate, the land use, and the geologic history of the region, the Rio Puerco has a very high sediment load. The average annual sediment load of the Puerco near Bernardo is 5,584 acre-feet. Above its meeting with the Puerco, the yearly average for the Rio Grande is only 1,583 acre-feet (3).

Little information has been available on the effects of uranium mining in the Grants area on water quality. Unfortunately, there is essentially no record of the water quality in the streams before mining began. The United States Geological Survey maintains 9 gaging stations on the San Jose-Puerco system. With the exception of two very recently installed stations on Arroyo del Puerto and San Mateo Creek, these stations are used to measure only water quantity. Such measurements that have been done have only measured dissolved constituents and not substances carried by the suspended sediments. Material transported by the suspended particles may be especially significant for the Rio Puerco with its extremely high sediment load. The Puerco has been estimated to contribute only 16% of the water and over half the sediment to the Rio Grande below Bernardo (4).

Currently, there are 35 active mines, 5 mills and 4 ion exchange plants in the Grants area (Figure 2). Uranium occurs most frequently in the Westwater Canyon Member of the Morrison Formation. This formation is also an aquifer. In order to mine uranium, this water must constantly be pumped from the mines. This water generally passes through settling basins and then is either discharged, used as makeup water in the mills or for backfilling operations. As mining progresses, the uranium content of the water increases. Ion exchange plants are used to recover this uranium and these lower the uranium content to about 0.5-1 ppm. Radium contained in the water is removed by co-precipitation with barium sulfate.

After treatment, the water is discharged to arroyos, etc. During the time samples were taken for this project the following amounts of water were being discharged (Figure 3) (5): 200 gpm ion - exchange treated water to Arroyo del Puerto, 400 gpm from mine dewatering to Arroyo del Puerto, 1500 gpm from mine dewatering to San Mateo Creek, 4000 gpm from mine dewatering to San Lucas Arroyo, 1200 gpm to Juan Tafoya Creek, 800 gpm to Salado Creek and 20 gpm to Arroyo del Valle.

Although water from San Mateo Creek and Arroyo del Puerto does not normally enter the San Jose, it can do so under flood conditions remobilizing and transporting sorbed or precipitated constituents. Such a flood occurred in February of 1979 (6). Very high water west of Grants in the spring of 1980 may also have mobilized some material. The Rio Paguate runs through the Jackpile open pit mine draining this area to the San Jose. Besides uranium mining, other factors also affect the surface water quality. The city of Grants discharges treated sewage into the San Jose. An old copper mining area, the Sierra Nacimiento, is drained by the Rio Puerco. There are also many possible non-point sources.

The project reported here involved determining physical and chemical parameters of the water in the San Jose-Puerco system in New Mexico between March 1978 and September 1980.

Experimental

Because of the intermittent flow throughout the system, sampling times were dependent on flow conditions. Nine sampling trips were taken. The sampling sites were located as shown in Figure 4. Site 1 was on the San Jose above the intersection with San Mateo Creek. This site was dry every time except once. Site 2 was located in the lava beds just east of Grants. Site 3 was on the San Jose below the inflow from the Rio Paguate. Site 4 was also on the San Jose just before the confluence with the Puerco. Site 5 was on the Puerco where it is crossed by Interstate 40. Site 6 was on the Puerco below the confluence with the San Jose. Two samples were collected from Bluewater Lake. The following sites were added after the first two sampling trips and were sampled when water was flowing: site 7 on the Rio Puerco just before it enters the Rio Grande, site 8 on San Mateo Creek 3.2 miles below the point where it is joined by Arroyo del Puerto, site 9 on Arroyo del Puerto before it joins San Mateo Creek, and site 10 on an arroyo which carries a mine dewatering discharge to San Mateo Creek.

Field measurements were carried out for pH, conductivity and alkalinity. Samples for laboratory analyses were collected in acid washed poly-ethylene bottles. Samples used for heavy metal analyses were filtered as quickly as possible through 0.45 micron filters and acidified to pH 2 with redistilled nitric acid.

General water chemistries were determined by standard methods (7, 8). Arsenic, barium, beryllium, cadmium, cobalt, chromium, copper, lead, nickel, and vanadium were determined by flameless atomic absorption using a carbon furnace. Uranium (9) and molybdenum (10) were determined colorimetrically. Selenium was analyzed by a fluorometric method (11). Selenium was also done by the carbon furnace attachment to the atomic absorption using an EDL source. Mercury was determined by the cold vapor method of Hatch and OTT(12). Metal analyses were performed on both filtered and unfiltered samples. The unfiltered samples were digested according to the EPA method (8).

Results

Figure 5 lists the average values for ph and total dissolved solids (TDS). pH values at each site exhibited no general trends and were average for New Mexico surface waters. The San Jose had a higher average pH than the Puerco. The pH of the Puerco where it enters the Rio Grande is about the same as that of the Rio Grande (2). TDS increased as the system flows towards the Rio Grande. Part of this could be caused by evaporation but much can be attributed to redissolving previously deposited material since calcium and sulfate increase to a greater extent than do sodium and chloride ions. A substantial part of the increase appears between sites 2 and 3 and is probably caused by input from the Rio Paguate which flows through the Jackpile mine. In September 1980 it was announced that this mine was suspending operations. Future monitoring of the San Jose would show if the input from the Rio Paguate decreases. The TDS of the Rio Puerco at site 5 is about the same as the San Jose at site 4. The TDS of the Rio Puerco at site 7 is 2442 ppm as compared to the average of 360 ppm for the Rio Grande at Bernardo (14).

Suspended sediment data is shown in Figure 6. It is obvious that most of the sediment is contributed by the Rio Puerco (compare site 4 and 5). The average sediment load in the Rio Grande at Bernardo is 2.2 g/1 (14).

The dissolved trace element data was compiled and averaged for each site. Standard deviations were calculated for 4 or more samples. These averages were compared to those obtained by Popp and co-workers (14) for the Rio Grande just above the confluence with the Rio Puerco. Arsenic, barium, chromium, lead and zinc were found in concentrations similar to the Rio Grande. Copper and mercury (Figure 7) were somewhat elevated. Nickel and selenium (Figure 8) were similar except for high selenium at sites 8, 9 and 10 which are located below mining discharges. Molybdenum was elevated in both the San Jose and Puerco with the exception of Bluewater Lake (Figure 9). These values varied considerably. Again high values were found at sites 8, 9 and 10. Vanadium levels were high and variable. Uranium concentrations were elevated throughout the system, except for Bluewater Lake (Figure 10) although the values fluctuated. Here again, the highest concentrations were found at sites 8, 9, and 10. Gross alpha values were also high at sites 8, 9, and 10. Total trace element data differed from that for dissolved forms at the sites where large amounts of sediments were found, namely sites 5, 6 and 7. These sites on the basis of liquid volume contained 10 times greater total trace element over dissolved element amounts of mercury, molybdenum and vanadium, 100 times greater amounts of chromium, copper, lead, nickel, and zinc and over 1000 times greater amounts of uranium.

Because uranium is so greatly elevated and also because it is of course a "tracer" element for uranium ores it is interesting to look closer at its distribution. Figure 11 shows the dissolved and total uranium values for each sample trip for sites 2, 4, 5 and 6. These sites were chosen for the following reasons: it was not possible to display values for all the sites; these sites were all on the main system; site 2 was a perennial upstream site; site 4 was downstream on the San Jose just before the confluence with the Puerco; site 5 was on the Puerco and site 6 was below the confluence of the Puerco and San Jose. As previously stated, most of the uranium is associated with the suspended sediment. The amounts of uranium moving down the system can be seen to increase with time. Also apparent is the increased total uranium entering the system from the Puerco. It is interesting to note that there is not an increase in uranium between sites 2 and 4, which would reflect input from the Jackpile open pit mine. If the dissolved uranium is subtracted from the total uranium and this amount divided by the sediment load for each site, the uranium in the suspended sediments on a weight basis can be calculated (Figure 12). On this basis it is apparent that the percent of uranium is higher in the suspended sediments in the San Jose. This could be due to the fact that the area is naturally higher in uranium. Although then one would expect a consistently high value at site 2 for example and this is not the case. The percent uranium varies. This could indicate a far-removed source with the uranium mobilized to the San Jose by an unusual event such as flooding. There is not an increase in percent uranium in the suspended sediment between site 2 and site 3, which reflects input from the Jackpile open pit mine, nor would an increase be expected because of the reservoir on the Rio Paguate just below the Jackpile. The large input of uranium is, however, from the Rio Puerco. This is not only due to the very large sediment load of the Puerco but also to the much higher flow rate.

Vanadium occurs frequently with uranium. In the unmineralized sandstones of the Colorado Plateau vanadium concentrations average between

10 and 30 ppm (13). In the uranium mineralized areas the vanadium averages 4900 ppm. Although these values are not from the Grants Mineral Belt, the vanadium could be expected to be present in a similar distribution as uranium. The vanadium (Figure 13) is in fact, very similar. Large increases in vanadium are seen at sites 5 and 6 associated with the suspended sediment. Vanadium also increases with time. Somewhat different from the uranium data there appears to be an increase in total vanadium at site 4. This may be due to the greater solubility of vanadium. The largest amounts of vanadium in the suspended sediment (Figure 14) occur in the Rio San Jose, also consistent with the uranium data.

Conclusions

If the proposed Hidden Mountain Dam is built on the Puerco just below its confluence with the San Jose, the sediments containing uranium and vanadium and the other previously mentioned elements would be trapped behind the dam. This would be initially effective in removing these elements from the system, but depending on the form they are in and the depositional conditions, they may later be solubilized.

Mining discharges are resulting in elevated metal and gross alpha content in San Mateo Creek and the Arroyo del Puerto. Although this is not directly flowing into the San Jose the contaminants are going somewhere, perhaps being adsorbed on sediments, perhaps precipitating, perhaps reaching ground water. Snow melt, rain storms and flooding could mobilize those contaminants associated with sediments. Occasional high uranium in sediments in the San Jose suggest this could be occurring.

Increased TDS between site 2 and site 3 is most likely due to the Jackpile open pit mine. Elevated total vanadium values were detected between site 2 and site 3 and occasionally elevated uranium.

There was a general increase in uranium and vanadium with time over the San Jose and Puerco system. This doesn't appear to be related to any individual discharge but most likely reflects the general increase in activity in the area. As mining continues, this increase is expected to continue.

References

1. U.S. Geological Survey, Water Resources Data for New Mexico Water Year 1977, Water Data Report NM-77-1 (1977).

2. U.S. Geological Survey, Water Resources Data for New Mexico, Water Data REport NM-78-1 (1978).

3. New Mexico Water Quality Control Commission, Middle Rio Grande Basin Plan, Santa Fe, NM (1976).

4. Waite, D. A., Rio Puerco Special Project Evaluation Report, 1242, 3, Department of the Interior, Bureau of Land Management, Santa Fe, NM (1972).

5. Goad, Maxine. Personal Communication, NM Environmental Improvement Division, Santa Fe, NM (1980).

6.  NM Environmental Improvement Division, _Focus on the Environment_, vol. 2, no. 2, p. 3 (1979).

7.  American Public Health Association, _Standard Methods for the Examination of Water and Wastewater_, 13th Ed., Washington, D.C. (1971).

8.  U.S. Environmental Protection Agency, _Methods for Chemical Analysis of Water and Wastes_, EPA-600/4-79-020, Cincinnati, OH (1979).

9.  Francois, C. A., "Rapid spectrometric determination of submilligram quantities of uranium", _Anal. Chem._ 30, 50 (1958).

10. Perrin, D. D., "Improvements in the stannous chloride-thiocycanate method for molybdenum", _New Zealand J. Sci Tech_, 27A, 396 (1946).

11. Chan, Chris C.Y., "Improvement in the fluorometric determination of selenium in plant materials with 2,3- diaminonaphthalene", _Anal. Chim. Acta._, 82, 213 (1976).

12. Hatch, W. R., and Ott, W. L., "Determination of submicrogram quantities of mercury by atomic absorption spectrometry", _Anal. Chem._, 40, 2085 (1968).

13. Miesch, A. T., "Distribution of elements in Colorado Plateau uranium deposits--A preliminary report", U.S.G.S., Bulletin 1147-E (1963).

14. Popp, C. J., Brandvold, D. K., Brierley, J. A., Scott, N.J., and Gloss, S.P., "Heavy metals and pesticides in water sediments and selected tissue samples of aquatic life in the middle Rio Grande valley in New Mexico", Project EPQ-E3W7-06-072, 32 (1980).

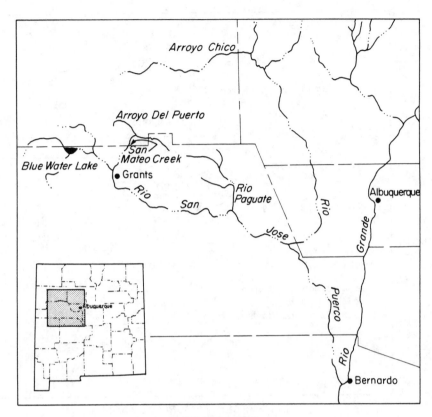

STUDY AREA

Figure 1

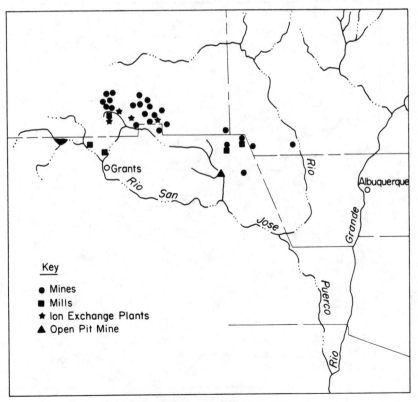

Key

● Mines
■ Mills
★ Ion Exchange Plants
▲ Open Pit Mine

MINING AND MILLING ACTIVITY

Figure 2

473

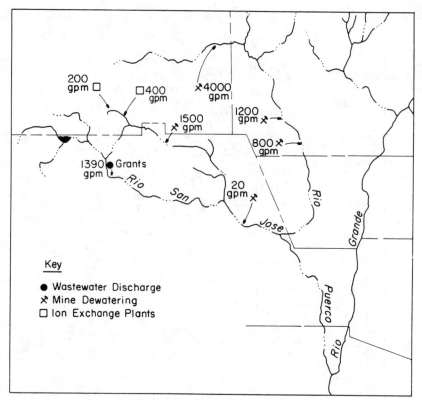

**DISCHARGES TO SURFACE WATER**
Figure 3

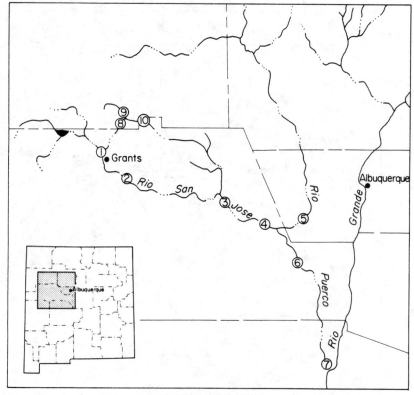

**SAMPLING SITES**
Figure 4

474

| Site No. | pH | | TDS (ppm) | |
|---|---|---|---|---|
| Bluewater | 8.06 | (2) | 254 | (2) |
| 1 | 7.85 | (1) | 185 | (1) |
| 2 | 7.51 ± .51 | (9) | 726 ± 130 | (9) |
| 3 | 7.95 ± .47 | (6) | 1595 ± 650 | (6) |
| 4 | 7.99 ± .4 | (7) | 1422 ± 503 | (7) |
| 5 | 7.46 ± .62 | (7) | 1470 ± 466 | (7) |
| 6 | 7.58 ± .54 | (9) | 1503 ± 856 | (9) |
| 7 | 7.54 | (3) | 2442 | (3) |
| 8 | 7.88 | (3) | 662 | (3) |
| 9 | 7.26 | (3) | 1333 | (3) |
| 10 | 6.74 | (3) | 557 | (3) |

Numbers in parentheses indicate number of samples

Standard deviations were not calculated for 3 or less samples.

AVERAGE VALUES FOR pH AND TDS

Figure 5

| Site No. | Suspended Sediment g/ℓ | |
|---|---|---|
| Bluewater | 0.03 | (1) |
| 1 | 1.057 | (1) |
| 2 | 0.356 ± .13 | (9) |
| 3 | 0.709 ± .816 | (9) |
| 4 | 1.251 ± 1.50 | (7) |
| 5 | 55.938 ± 11.67 | (7) |
| 6 | 61.513 ± 27.13 | (9) |
| 7 | 73.35 | (2) |
| 8 | 7.03 | (1) |
| 9 | 3.25 | (2) |
| 10 | 0 | (3) |

Numbers in parentheses indicate the number of samples.

Standard deviation not calculated for 3 or less samples.

SUSPENDED SEDIMENT

Figure 6

| Site No. | Cu | | Hg | |
|---|---|---|---|---|
| Bluewater | 22.0 | | 1.0 | |
| 1 | 25.4 | (1) | <.2 | (1) |
| 2 | 4.8 ± 6.22 | (6) | 0.4 ± .2 | (6) |
| 3 | 16.8 ± 2.8 | (4) | .3 | (1) |
| 4 | 17.2 ± 5.4 | (4) | .2 | |
| 5 | 19.4 ± 4 | (5) | .45 ± .25 | (4) |
| 6 | 28.6 ± 22.5 | (6) | .71 ± .56 | (5) |
| 7 | 24.1 ± 15.9 | (4) | .86 | (3) |
| 8 | 19 | (2) | .2 | (3) |
| 9 | 10 | (3) | .1 | (3) |
| 10 | | | .35 | |
| Rio Grande[1] | 10 | | <.2 | |

Values are given in ppb

Numbers in parenthesis are number of samples

[1]Upstream from confluence at Rio Puerco and Rio Grande (Popp, 1980).

AVERAGE VALUES FOR DISSOLVED COPPER AND MERCURY

Figure 7

| Site No. | Ni | | Se | |
|---|---|---|---|---|
| 1 | 41 | (1) | 5 | (1) |
| 2 | 44 ± 18 | (6) | 4 ± 2 | (6) |
| 3 | 71 | (3) | 5 ± 2 | (6) |
| 4 | 61 | (3) | <3 | (7) |
| 5 | 56 ± 47 | (5) | 6 ± 4 | (5) |
| 6 | 58 ± 43.5 | (5) | 10 ± 6 | (7) |
| 7 | 38.3 | (3) | 8 ± 5 | (4) |
| 8 | 45 | (2) | 52 | (2) |
| 9 | 46.8 | (3) | 133 | (3) |
| 10 | 28 ± 17 | (4) | 63.5 | (3) |
| Rio Grande[1] | 18 | | <5 | (2) |

Values are given in ppb

Numbers in parentheses are number of samples

[1]Upstream from confluence of Rio Puerco and Rio Grande (Popp, 1980)

AVERAGE VALUES FOR DISSOLVED NICKEL AND SELENIUM

Figure 8

| Site No. | Mo | | V | |
|---|---|---|---|---|
| Bluewater | <20 | (2) | <50 | (2) |
| 1 | 10 | (1) | 120 | |
| 2 | 64 ± 70.8 | (6) | 50 ± 40 | (4) |
| 3 | 16.8 ± 9.9 | (5) | 52.7 ± 42 | (4) |
| 4 | 40 ± 60 | (4) | 55 ± 45 | (4) |
| 5 | 16 ± 15 | (5) | 174 ± 232 | (6) |
| 6 | 25.7 ± 17 | (7) | 104 ± 71 | (9) |
| 7 | 40 | (3) | 173 | (3) |
| 8 | 320 | (2) | 89 | (2) |
| 9 | 878 | (3) | 108 | (3) |
| 10 | 305 ± 138 | (4) | 62 ± 50 | (4) |
| Rio Grande[1] | <20 | | 29 | |

Values are given in ppb

Numbers in parentheses are numbers of samples

[1]Upstream from confluence of Rio Puerco and Rio Grande (Popp, 1980)

AVERAGE VALUES FOR DISSOLVED MOLYBDENUM AND VANADIUM

Figure 9

| Site No. | $U_3O_8$ (ppb) | | Gross Alpha (pci/ℓ) | |
|---|---|---|---|---|
| Bluewater | 35 | (2) | | |
| 1 | 80 | (1) | | |
| 2 | 90 ± 54 | (4) | 7.7 ± 4.4 | (4) |
| 3 | 64 ± 54 | (4) | 9.3 | (3) |
| 4 | 38 ± 37 | (4) | 10 ± 2 | (4) |
| 5 | 66 ± 20 | (5) | 15 ± 6 | (4) |
| 6 | 48 ± 30 | (9) | 9 ± 4.4 | (5) |
| 7 | 56 | (3) | 18 | (2) |
| 8 | 1020 | (2) | 679 | (2) |
| 9 | 1501 | (3) | 748 | (2) |
| 10 | 636 ± 254 | (4) | 981 | (3) |
| Rio Grande[1] | 33 | | 10 | |

Numbers in parentheses are numbers of samples

[1]Upstream from confluence of Rio Puerco and Rio Grande (Popp, 1980)

AVERAGE VALUES FOR URANIUM AND GROSS ALPHA

Figure 10

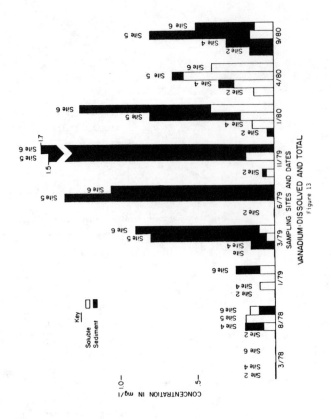

Key
☐ Soluble
■ Sediment

SAMPLING SITES AND DATES
VANADIUM: DISSOLVED AND TOTAL
Figure 13

CONCENTRATION IN mg/l

|  |  | Sampling Dates |  |  |  |  |  |  |  |
|---|---|---|---|---|---|---|---|---|---|
| Site No. | 3/78 | 8/78 | 1/79 | 3/79 | 6/79 | 11/79 | 1/80 | 4/80 | 9/80 |
| 1 |  |  |  |  |  |  |  | 37.8 |  |
| 2 | ND* | ND | ND | ND | ND | 81 | 160.6 | ND | 710 |
| 3 |  | ND | ND |  |  |  | ND | ND | 101 |
| 4 |  | 136 |  | 122 |  |  | 303 | ND | 62 |
| 5 |  | ND |  | 12.4 | 18.7 | 29.9 | 11.56 | 1.91 | 10.7 |
| 6 |  | 1.1 | 1.3 | 9.7 | 14.1 | 34.8 | 12.82 | 0.3 | 18.9 |

* Not detectable

VANADIUM IN SUSPENDED SEDIMENTS (ug/g)

Figure 14

Key
☐ Soluble
■ Sediment

SAMPLING SITES AND DATES
URANIUM: DISSOLVED AND TOTAL
Figure 11

CONCENTRATION IN mg/l

|  |  | Sampling Dates |  |  |  |  |  |  |  |
|---|---|---|---|---|---|---|---|---|---|
| Site No. | 3/78 | 8/78 | 1/79 | 3/79 | 6/79 | 11/79 | 1/80 | 4/80 | 9/80 |
| 1 |  |  |  |  |  |  | 199 |  | 142 |
| 2 | ND* | 19 | 347 | 42 | ND | 163 | 904 | ND | 64 |
| 3 |  | ND | 164 | 126 |  |  | 341 | 615 | 8 |
| 4 | ND | ND | 82 | 98 |  |  | 163 | 636 | 68 |
| 5 |  | ND | ND |  | 42 | 50 | 62 | 68 | 68 |
| 6 |  | ND | 1.2 | 4.8 | 74 | 85 | 78 | 184 | 28 |

*Not detectable

URANIUM IN SUSPENDED SEDIMENTS (ug/g)

Figure 12

# EVALUATION AND CONTROL OF PERMEABILITY DAMAGE DURING CARBONATE SOLUTION MINING OF URANIUM

**Terry R. Guilinger, I. H. Silberberg** and **Robert S. Schechter**
University of Texas at Austin

## Introduction

The sandstone uranium deposits of South Texas represent a possible major energy source. It is thought that these deposits were formed by the downdip migration of ground water carrying oxidized uranium leached from the host rock. When uranium-bearing waters reached a reducing zone, the uranium was precipitated. Much of the uranium ore in the area is low grade (<0.05% $U_3O_8$) and is at depths of 100-1500 ft.

Since 1960, various companies have been mining some of the higher grade deposits to depths of up to 200 feet, using conventional strip mining techniques. The concomitant surface disruption is extensive and the costs of mining and transporting to a mill such large amounts of material prohibit the utilization of low grade ore.

A mining technique that may overcome these difficulties to some extent and ultimately make more of the South Texas uranium deposits amenable to recovery is in-situ solution mining. This technique consists of pumping through the ore body a chemical solution that will dissolve the uranium minerals so that they may be leached from the ore and recovered from the solution. For this process to be economically feasible, a low-cost solution must be available that will dissolve a large portion of the uranium present, the uranium must be easily recoverable from the leach solution, the physical attributes of the ore body must be such that the leach solution can be pumped through the ore without great loss to the surroundings, and environmental hazards must be avoided.

The leaching process and its chemistry can be visualized as a two-step process. First is oxidation of the uranium mineral from a +4 valence state to a +6 state. Second is the complexing of the +6 uranium with an anion for greater solubility. Both acidic and basic solutions have been used with success in uranium leaching. Sulfuric acid is the most commonly used acidic leachant, with ferric iron as the oxidizing agent and sulfate ion as the complexing agent. In basic leach solutions, both sodium and ammonium carbonate have been used, with hydrogen peroxide as the oxidizing species.

The composition of the ore body dictates the type of lixiviant which should be used. In formations where the concentration of acid consumers, such as calcite, is low, sulfuric acid leaching is the preferred method of in-situ extraction of uranium. When calcite concentration is high and clay swelling due to sodium cation is not a problem, sodium carbonate is used as the lixiviant. For ore bodies where swelling clays and significant acid consumers are both present, ammonium carbonate is the preferred leaching medium.

After the completion of leaching, the ore body and resident aquifer must be restored. This means that the post-leached clays must be in equilibrium with a ground water of quality approximately equal to that present before leaching. For example, if ammonium carbonate were used as the lixiviant, regulatory agencies now require that most of the ammonium cations must be removed from the ore body before the site is abandoned, since there is normally very little ammonia in ground waters. A survey[1] revealed that this was the case in South Texas where a number of mines are now operating. However, removal of ammonia from a leached ore body is difficult because of the negatively charged clays which are present. These clays attach ammonium cations to themselves by a process of ion exchange with the contacting fluid. Thus, to completely restore an in-situ mining site, the ammonium must not only be pumped out with the lixiviant but it also must be exchanged on the clays with cations that are normally present in ground water.

Studies have been conducted which illustrate the difficulties encountered in completely removing ammonia from a leached ore body.[2] To circumvent this problem, a new process has been patented which utilizes chlorine gas to oxidize ammonia to nitrogen thus removing the environmental hazards of ammonia.[3] However, the high cost of chlorine may preclude any practical use of this process. Because of these difficulties, potassium carbonate is being examined as a potential substitute for ammonium carbonate. Potassium cations are present in many naturally occurring ground waters and do not present the potential hazard of being converted to nitrates as might be the case with ammonium cations. However, to be a viable substitute for ammonium carbonate, potassium carbonate should have similar properties during the leaching phase of an in-situ uranium mining operation.

One important property of a leaching solution is the effect of that solution on the permeability of the host rock. This paper considers the adverse effects on ore body permeability of the basic leach solutions. The solutions which have been examined experimentally are the carbonates of ammonium, sodium, and potassium. Also, the effect of clay stabilizing by "Cla-Lok" is considered.

Experimental
(A) Apparatus
The main features of the apparatus shown in Figure 1 are:

478

(1)  Ruska proportioning pump - This is a piston/cylinder type pump which is capable of delivering a constant flow rate as low as 1 ml/hour at pressures up to 10,000 psia.  The advantage of this pump is that its capacity is 1000 ml per stroke.  Thus, there is no pulsing of the fluid caused by continual reciprocation of the piston in the cylinder.  This feature is important in permeability studies since pressure pulsing can cause mechanical rearrangement of the particles in the porous medium.

(2)  Hassler cell - the cell consists of a cylinder with end caps containing pressure taps which lead to a differential pressure transducer. Inside the cylinder is a rubber sleeve which holds the ore sample in place.  External pressure is supplied in the annulus between the sleeve and cylinder by a nitrogen tank.  This pressure simulates overburden as found in underground formations and prevents channeling of the fluid by keeping the ore sample intact.  Screens are inserted at each end of the sample to prevent migration of the ore from the sample area.

(3)  Backpressure regulator - This regulator prevents fluid in the line from passing through until the fluid pressure exceeds an external pressure maintained by a nitrogen tank.  Thus, underground pressures can be simulated in this system by adjusting the external pressure.  Since permeability does not depend on absolute pressure, this item is not crucial to permeability studies.

(B)  Ore Preparation
   1. Consolidated Ore A
A drill press with a diamond coring barrel was utilized, with kerosene as the cutting fluid, to cut cores of one inch diameter and four inches in length.  The cores were then soaked in acetone to remove the kerosene and air dried to remove the acetone.  All cores were weighed.

The cores were then inserted into the Hassler cell and the rubber interior sleeve was clamped onto the sample by external pressure from the nitrogen tank.

   2. Unconsolidated Ore
Both ores obtained for this study were sifted through a U. S. standard sieve with a nominal opening of 0.59 mm, to remove large particles and then blended until uniform.  Ore preparation for the two batches differed from this point.

For Ore B, a sample of approximately 60 grams was weighed out and packed dry into the Hassler cell with external pressure applied by the nitrogen tank.  The total volume of the sample was measured by determining its length and diameter.

Ore A was prepared by weighing out a sample of approximately 90 grams and packing it into a one inch inside diameter rubber tube.  Water was also added and mixing was performed to eliminate air from the sample. The tube and ore were then frozen.  After removing the tube, the frozen core was cut to a length of four inches and inserted into the Hassler cell in the same manner as the consolidated cores.  After thawing, the four inch long samples were ready for testing.

(C)  Operating Procedure
The system was evacuated.  Simulated groundwater was allowed to fill the system.

The pump was then started and the fluid rate adjusted to the desired value. The pressure drop was continuously monitored by the pressure transducer and chart recorder.

After completion of the groundwater exposure, the system was again evacuated. Depending on the experiment to be performed, another solution was added to the feed tank. After evacuation, this solution was allowed to fill the system. The remainder of the procedure described for groundwater was again conducted. If other tests were to be performed on the ore sample, the procedure above was repeated as many times as necessary.

In all tests, the simulated groundwater contained 200 ppm sodium chloride and 150 ppm calcium chloride. The concentration of "Cla-Lok" used was 0.3 molar in 1% potassium chloride water. The concentrations of all other aqueous solutions were 0.1 molar unless otherwise specified. The experiments were conducted at a temperature of $22\pm2^{\circ}C$ with an external sleeve pressure on the ore samples of between 60 and 100 psig.

(D)  Porosity and Permeability Calculations

The grain density method was used to estimate the porosity and hence the pore volume of each sample. The porosity is given by

$$\phi = 1 - \frac{M_s}{V_s \rho_{ore}}$$

where:   $\phi$ = porosity

$V_s$ = total volume of ore sample

$M_s$ = mass of ore sample

$\rho_{ore}$ = ore density

The sample volume is calculated by measuring the length and diameter of each cylindrical sample.

Permeability was calculated using Darcy's Law for linear, horizontal flow in a porous medium. For these conditions, Darcy's Law assumes the following form:

$$k = - \frac{q\mu L}{A\Delta P}$$

where: q = volumetric fluid flow rate

L = sample length

A = cross-sectional area of the sample

$\mu$ = fluid viscosity

$\Delta P$ = pressure drop across the sample

This form of Darcy's Law was used to calculate the permeability of each sample. It should be noted that permeability depends only on the pressure difference applied, not on the absolute system pressure.

Results and Discussion

The characteristics of the blended ores are shown in Table 1. .The cation exchange capacity (CEC) is a measure of the amount and type of

480

clay contained in the ore samples. A higher CEC is indicative of
higher clay content when comparing two ores containing the same type
of clay. Thomas (5) has shown that montmorillonite is by far the most
commonly encountered clay in South Texas ores from the Catahoula
formation. Thus, Table 1 illustrates that Ore A is higher in clay
content than Ore B.

Permeability responses of each ore to the various lixiviants and control methods used in this study are shown in Figures 2 to 23.

All response curves are shown as continuous since the chart recorder
monitors pressure drop, and hence permeability, continuously.

(A) Groundwater/Carbonate Tests
  1. Unconsolidated versus Consolidated Ore
It was desired to use unconsolidated ore in the permeability tests for
reasons of ore characterization and reproducibility among tests. In
order to justify the use of unconsolidated ore, tests were conducted
with consolidated cores cut from aggregated ore available from an
Ore A strip mine. Experiments were conducted with ammonium carbonate
lixiviant preceded by a groundwater flush. Results of the permeability tests are compared to permeability responses with unconsolidated Ore A at similar conditions in Figure 2. The extents of permeability decline are nearly identical for the two aggregation states
of the ore, and both exhibit a slight recovery following the point of
maximum permeability decline. Although the pore volumes at which
minimum permeability occurred are different for the unconsolidated and
consolidated ores, the permeability responses are remarkably similar
and the use of unconsolidated cores is felt to be justified.

  2. Effect of Flow Rate
Lixiviant in a typical solution mining operation is injected at a flux
of approximately 10 feet/day. In order to accommodate this very slow
rate, permeability measurements would have to have been made over a
period of three to four days. Furthermore, the flow rates near the
wellbore are very much larger than those farther out in the formation.
Therefore, a study of the effect of flow rate on permeability response
was undertaken. Figure 3 shows the permeability response of unconsolidated Ore A to ammonium carbonate injection at rates comparable to
field rates. A comparison of a low flow rate test with an experiment
performed at a rate nearly three times as large is also shown. The
similarity of the permeability response curves shows that within the
range of flow rates studied there is no effect of rate on permeability.
Moore (6) reports that the cation exchange process is very rapid and
proceeds almost immediately to equilibrium when a clay is contacted by
a cation-containing water. The absence of a rate effect on permeability tends to confirm this local equilibrium and would indicate
that the permeability damage shown in Figure 3 is not caused by the
hydrodynamic shearing forces. We shall have more to say about the
reasons for this observed damage.

  3. Permeability Responses of Ores A and B to Carbonate Solutions
The three carbonates of this study, ammonium, potassium, and sodium,
were each injected into an unconsolidated ore sample and preceeded
by a groundwater flush. Permeability responses for ammonium carbonate
on Ore A are shown in Figure 4. The two curves shown are the results
of duplicate experiments and provide some measure of the reproducibility of an experiment. In both cases a minimum permeability is
encountered within the first ten pore volumes of leachant introduction.

481

This decline is to approximately 20% to 40% of the original groundwater permeability. In each test, a recovery in permeability is noted after continued leachant injection. Figure 5 shows the permeability response of Ore B to the same tests. Again, a minimum permeability in the first ten pore volumes of leachant is noted along with a recovery after continued injection. The maximum permeability decline in Ore B is not so severe as in Ore A. A comparison of the responses of each ore is shown in Figure 6. As shown in Table 1, Ore B has the smallest cation exchange capacity. As will be seen, this fact may very well account for the difference in the depth of the minimum between the two ores. Ores having the highest value of cation exchange can be expected to experience the greatest damage using the sequence of fluid injections followed here.

Results from tests with potassium carbonate as the leachant are shown in Figure 7 for Ore A and in Figure 8 for Ore B. Again, the same patterns of permeability response are exhibited by each ore as were found with ammonium carbonate. With potassium, however, the extent of the maximum permeability decline is slightly greater. Permeability declines with Ore B are again less severe than with Ore A. Figure 9 shows a comparison of the permeability responses of unconsolidated Ore A to potassium and ammonium carbonates.

When sodium carbonate is used as the leachant, a much greater loss in permeability results compared to either ammonium or potassium carbonate. This would be expected because of the differences in clay binding between adsorbed sodium and adsorbed ammonium or potassium cations. The test results for Ore A are shown in Figure 10 and for Ore B in Figure 11. Similarities among responses of Ore A with each of the carbonates are still evident in that the minimum permeability still occurs within the first ten pore volumes of leachant injection and a slight recovery is exhibited with continued injection of sodium carbonate. A comparison of the responses of Ore A to ammonium and sodium carbonates is shown in Figure 12. Ore B permeability responses to sodium carbonate are much different from those encountered previously. The permeability never reached a minimum even after declining to nearly 6% of its original groundwater value. These results illustrate that the Ore B is more sensitive to sodium than Ore A. This increased sensitivity is not yet understood, but the difference is felt to be significant.

(B) Permeability Responses Following Clay Stabilization
    1. Sodium/Potassium Mixtures
Since potassium exhibits less formation damage than does sodium, mixtures of sodium and potassium carbonates as the leachant may improve the permeability response of ore bodies as compared to a sodium leachant used alone. Results from tests utilizing sodium-potassium mixtures with molar ratios of sodium to potassium of from 1 to 9 up to 9 to 1 are summarized in Figure 13. From these tests it is evident that sodium/potassium leachant mixtures are no better than a sodium carbonate leachant alone until the molar ratio of potassium to sodium is nearly 2 to 1. This method of improving sodium permeability response is therefore not viable.

    2. Stepwise Salinity Increases
As suggested by Jones (7), sharp salinity contrasts may cause formation damage by increasing the local pressure within a clay aggregate and causing clay disruption and dispersion. This phenomenon was noticed when a water of high salinity was displaced by a low salinity

water. By analogy, increasing the salinity by small increments may reduce formation damage as compared to sharp salinity increases. In the previous tests with sodium carbonate and groundwater, the salinity was increased in one step from 200 ppm (i.e. $3.4 \times 10^{-3}$ molar) sodium to 0.1 molar sodium in solution. Figures 14 and 15 are results from two tests on Ore A with different strategies of increasing the sodium concentration slowly. It may be concluded that small salinity contrasts slow the permeability decline compared to large contrasts but the eventual decline is similar. This conclusion is not in contradiction to the results of Jones, since the direction of salinity change is opposite, and because the Berea cores used by Jones do not contain swelling clays. In the presence of montmorillonites it appears to be difficult to stabilize the permeability using sodium carbonates.

### 3. Multivalent Cation Clay Stabilizers

An inorganic polymer known as "Cla-Lok" was obtained from Halliburton Company. This polymer is used in oil field operations to stabilize clays. The mechanism of clay stabilizing by a polyvalent cation is by electrostatic adsorption of the cation onto several negatively charged sites on the clays. The cation not only binds the clay together by adsorbing between layers, but also reduces exchange by other cations. Because of the polymeric nature of "Cla-Lok," a curing period to allow for in-situ polymerization is necessary. Halliburton recommends a curing time of 24 hours.

The permeability response of unconsolidated Ore A to clay stabilizing by "Cla-Lok" with subsequent sodium carbonate injection is shown in Figure 16. The minimum permeability during exposure to sodium carbonate is approximately 35% of the original groundwater value. This compares to a minimum permeability of about 8% of original when no "Cla-Lok" is used. Thus, substantial improvement in permeability response to sodium has been effected by clay pretreatment with a multivalent cation clay stabilizer. The sodium carbonate with "Cla-Lok" permeability responses compare favorably with responses of Ore A to the carbonates of ammonium and potassium.

### (C) Permeability Responses to Formation Pretreatment

In the preceding sections, the minimum permeabilities in each case were less than 50% of their original groundwater values. A proposed mechanism for this drastic decline is the precipitation of calcium carbonate in the formation and subsequent plugging or bridging of pores by particle transport and attachment. As reported by Breston (8) and Torrey (9), formation damage by calcium carbonate precipitation at the interface between formation and injection waters has been observed. In the experiments performed in this work, the carbonate is available from the injection water. The calcium is available from the groundwater present and from adsorbed calcium on the clays. As leachant is injected, a calcium carbonate precipitate may form, not only at the boundary between formation and injection waters, but also may form as cation exchange takes place between adsorbed calcium and the leachant cation.

If precipitation were the main cause of permeability decline, then the permeability response curves would be expected to exhibit an immediate and drastic loss in permeability due to calcium carbonate solids formation. When all the calcium is consumed, no more loss in permeability would be expected and, in fact, a recovery in permeability may be predicted as some of the solids are flushed or dissolved by the leachant. These trends in permeability were noted previously

483

when groundwater was displaced by the carbonates of ammonium, potassium and, at times, sodium.

There are two options in preventing precipitation. First is the removal of carbonate ion from the injected lixiviant, so it will not be available for solids formation. This option is not practical, since without carbonate very little oxidized uranium would be brought into solution and recovered from the formation. The alternative to carbonate removal is calcium purging of the formation previous to carbonate introduction. This step may be accomplished by preceding lixiviant injection with formation exposure to a chloride solution of the cation being used in the lixiviant. Chloride is the anion of choice because its compound with calcium is very soluble in water.

Chloride preflushing was attempted with each of the cations used in this study. Results from these tests with ammonium as the cation are shown in Figure 17. A comparison of the results with each ore is shown. The minimum permeability in each case was between 70% and 80% of the original groundwater value. This compares to a minimum permeability of between 20% and 40% of the original when no chloride preflush is used. Chloride preflushing has substantially improved the permeability response of both ores to ammonium carbonate. Thus, calcium carbonate precipitation was a main cause for permeability decline. This conclusion is also substantiated by the absence of permeability recovery after the minimum permeability has been reached.

Results from tests with potassium carbonate with a chloride preflush exhibit the same dramatic improvements in permeability response as compared to responses without a chloride preflush. The permeability curves are shown in Figure 18 for Ore A and in Figure 19 for Ore B. A comparison of the permeability responses of Ore A to ammonium and potassium carbonates with chloride preflushes is in Figure 20.

Chloride preflushing was also attempted with sodium carbonate as the leachant. Results are shown in Figure 21 for Ore A and in Figure 22 for Ore B. Because of the clay swelling properties of sodium, there is a continuous permeability loss in both ores when exposed to sodium chloride solutions. The permeability response of Ore A to sodium carbonate with chloride preflush is compared to the response to ammonium carbonate with chloride preflush in Figure 23.

## Conclusions

1. Unconsolidated and homogeneous ore is used so that the composition of an ore sample is known. Results with unconsolidated ore show similar patterns of permeability response compared to experiments with consolidated, inhomogeneous ore.

2. In the range of flow rates studied, there is no effect of rate on permeability response. Thus, the cation exchange process is an equilibrium phenomenon for fluid rates between 9 feet/day and 30 feet/day.

3. Drastic permeability loss occurs in the ore samples when groundwater is displaced by carbonate solutions containing ammonium, potassium or sodium. The drop is immediate and reaches its maximum value within the first ten pore volumes of leachant flood.

4. Similar losses in permeability are exhibited by ammonium and potassium carbonate leachants. Losses with sodium carbonate are at

least more than twice as extensive.

5. Mixtures of sodium and potassium carbonates as the leachant are ineffective in preventing permeability decline due to sodium clay swelling until the mixture is nearly 2 to 1 potassium-to-sodium.

6. Small step increases in sodium concentration slow, but do not reduce, permeability decline.

7. The use of multivalent cation clay stabilizers such as "Cla-Lok" to prevent sodium clay swelling reduces permeability loss, but sodium leach solutions are still substantially more damaging to permeability than either ammonium or potassium leachants using chloride preflushes.

8. An important mechanism responsibile for permeability decline when carbonate leachant solutions displace a calcium-bearing ground-water is the precipitation of calcium carbonate and subsequent pore blockage by the solid. Thus, ore bodies which are exposed to a calcium (or magnesium) bearing groundwater should be preflushed prior to carbonate introduction to prevent permeability loss due to calcium carbonate precipitation.

9. A chloride preflush of the ore sample prior to leachant flood dramatically reduces and, in some cases, prevents permeability decline when either ammonium or potassium is used.

10. Preflushes utilizing sodium chloride are ineffective due to the clay swelling properties of sodium. Permeability decline for these cases is continuous throughout exposure of the ore sample to sodium ion.

11. The conclusions reached from the permeability studies are not ore-specific to Ore A. The same patterns of permeability response are shown with Ore B.

Nomenclature

| | |
|---|---|
| A | Cross-sectional area of ore sample |
| k | Permeability of ore sample |
| $k_o$ | Initial permeability of ore sample |
| L | Length of ore sample |
| $M_s$ | Mass of ore sample |
| $\Delta P$ | Pressure drop across ore sample |
| q | Volumetric flow rate |
| $V_s$ | Total volume of ore sample |
| $\mu$ | Fluid Viscosity |
| $\rho_{ore}$ | Ore sample density |
| $\phi$ | Porosity |

Acknowledgement

This research was supported by the U. S. Bureau of Mines, Department of the Interior and by the Texas Petroleum Research Committee.

References

1. Braswell, J., Breland, M., Chang, M., Hill, D., Johnson, D., Schechter, R., Turk, L., and Humenick, M., "Literature Review and Preliminary Analysis of Inorganic Ammonia Pertinent to South Texas Uranium In Situ Leach," Center for Water Resources Report #CRWR-ISS, EHE 78-01, The University of Texas at Austin, Austin, Texas (1978).

2. Hill, A. D., Walsh, M. P., Breland, W. M., Silberberg, I. H., Humenick, M. J., and Schechter, R. S., "Restoration of Uranium In Situ Leaching Sites," Paper presented at 53rd Annual Fall Technical Conference of the Society of Petroleum Engineers of AIME, SPE Paper No. 7534, (1978).

3. Yan, T. Y., "A Process for Removing Ammonium Ions from a Subterranean Formation After In Situ Uranium Leaching," Paper presented at the 55th Annual Fall Technical Conference of the Society of Petroleum Engineers of AIME, SPE Paper No. 9491, (1980).

4. Garwacka, A., Johnson, D., Walsh, M., Breland, M., Schechter, R. S. and Humenick, M. J., "Investigation of the Fate of Ammonia from In-Situ Uranium Solution Mining," Technical Report, Bureau of Engineering Research, The University of Texas at Austin, Austin, Texas, (February, 1979).

5. Thomas, G., Petrography of the Catahoula Formation in Texas, M.A. Thesis, The University of Texas at Austin, Austin, Texas, (June, 1960).

6. Moore, J. E., "Clay Mineralogy Problems in Oil Recovery," The Petroleum Engineer, 13-40, (February, 1960).

7. Jones, F. O., "Influence of Chemical Composition of Water on Clay Blocking of Permeability," Journal of Petroleum Technology, 441, (April, 1964).

8. Breston, J. N., "Conditioning Water for Secondary Oil Recovery in Appalachian Fields," The Oil and Gas Journal, 150, (August 24, 1950).

9. Torrey, P. D., "Preparation of Water for Injection into Oil Reservoirs," Journal of Petroleum Technology, 9, (April, 1955).

TABLE I

Comparison of Two Ores Used in This Study

| Ore | CEC[4] (meq/100 g ore) | Location |
|-----|------------------------|----------|
| A   | 11.3                   | Karnes County Texas |
| B   | 9.8                    | Live Oak County Texas |

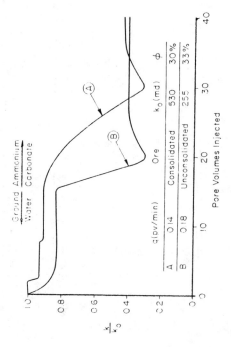

FIGURE 1. SCHEMATIC OF PUMPING SYSTEM

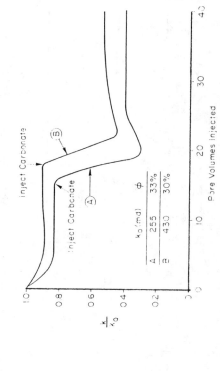

FIGURE 2. COMPARISON OF PERMEABILITY RESPONSES FOR CONSOLIDATED AND UNCONSOLIDATED ORE A

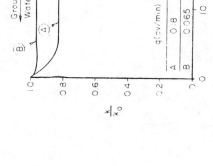

FIGURE 3. EFFECT OF FLOW RATE ON PERMEABILITY RESPONSE WITH AMMONIUM CARBONATE LEACHANT INJECTED INTO UNCONSOLIDATED ORE A

FIGURE 4. PERMEABILITY RESPONSE FOR UNCONSOLIDATED ORE A WITH AMMONIUM CARBONATE LEACHANT INJECTED AT RATE OF 0.18 PORE VOL/M

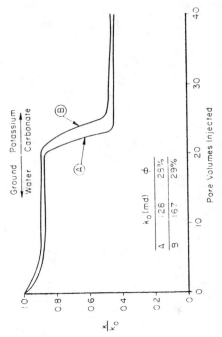

FIGURE 6. COMPARISON OF PERMEABILITY RESPONSES FOR UNCONSOLIDATED ORES A AND B

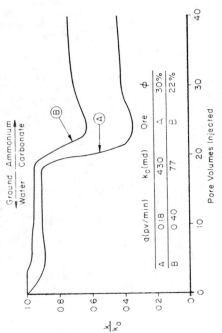

FIGURE 5. PERMEABILITY RESPONSE FOR UNCONSOLIDATED ORE B WITH AMMONIUM CARBONATE LEACHING

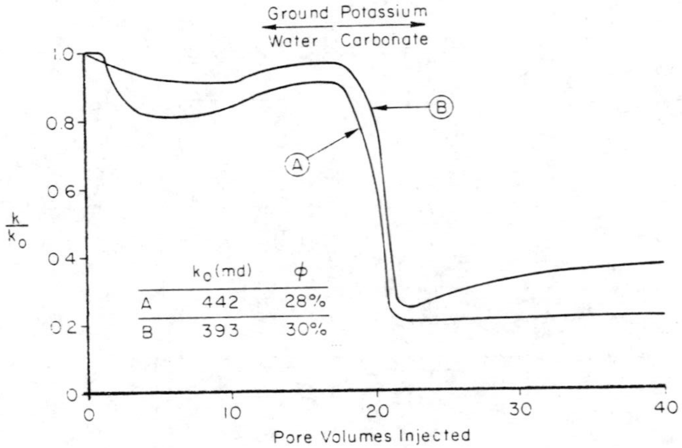

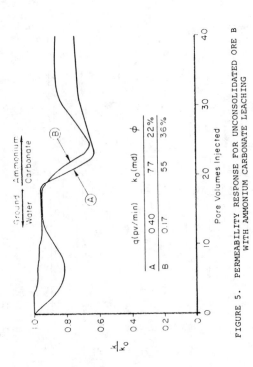

FIGURE 8. PERMEABILITY RESPONSE FOR UNCONSOLIDATED ORE B WITH POTASSIUM CARBONATE LEACHANT INJECTED AT A RATE OF 0.24 PORE VOL/M

FIGURE 7. PERMEABILITY RESPONSE FOR UNCONSOLIDATED ORE A WITH POTASSIUM CARBONATE LEACHANT INJECTED AT A RATE OF 0.18 PORE VOL/M

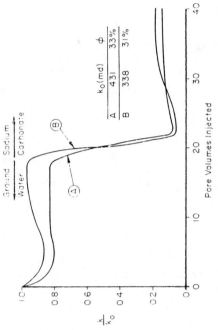

FIGURE 9. COMPARISON OF PERMEABILITY RESPONSES FOR AMMONIUM AND POTASSIUM CARBONATES WHEN INJECTED AT RATE OF 0.18 PORE VOL/M INTO UNCONSOLIDATED ORE A

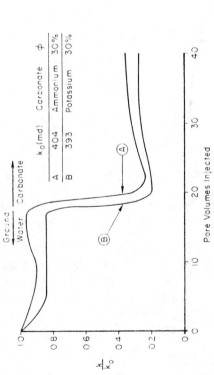

FIGURE 11. PERMEABILITY RESPONSE FOR UNCONSOLIDATED ORE B WITH SODIUM CARBONATE LEACHANT

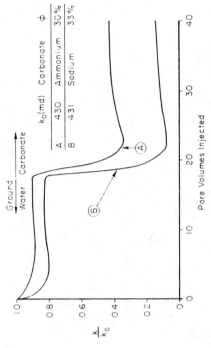

FIGURE 10. PERMEABILITY RESPONSE FOR UNCONSOLIDATED ORE A WITH SODIUM CARBONATE LEACHANT INJECTED AT RATE OF 0.19 PORE VOL/M

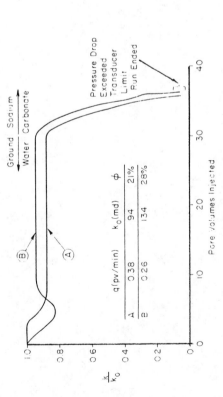

FIGURE 12. COMPARISON OF PERMEABILITY RESPONSES FOR AMMONIUM AND SODIUM CARBONATES WHEN INJECTED AT A RATE OF 0.18 PORE VOL/M INTO UNCONSOLIDATED ORE A

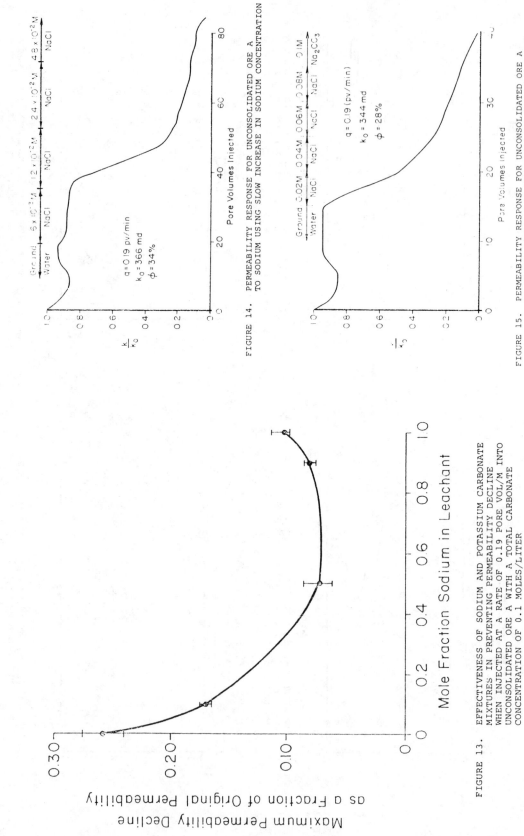

FIGURE 14. PERMEABILITY RESPONSE FOR UNCONSOLIDATED ORE A TO SODIUM USING SLOW INCREASE IN SODIUM CONCENTRATION

FIGURE 15. PERMEABILITY RESPONSE FOR UNCONSOLIDATED ORE A TO SODIUM USING SLOW INCREASES IN SODIUM CONCENTRATION

FIGURE 13. EFFECTIVENESS OF SODIUM AND POTASSIUM CARBONATE MIXTURES IN PREVENTING PERMEABILITY DECLINE WHEN INJECTED AT A RATE OF 0.19 PORE VOL/M INTO UNCONSOLIDATED ORE A WITH A TOTAL CARBONATE CONCENTRATION OF 0.1 MOLES/LITER

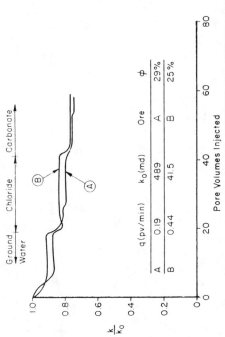

FIGURE 16. PERMEABILITY RESPONSE FOR UNCONSOLIDATED ORE A WITH SODIUM CARBONATE LEACHANT AND CLAY STABILIZERS INJECTED AT A RATE OF 0.22 PORE VOL/M

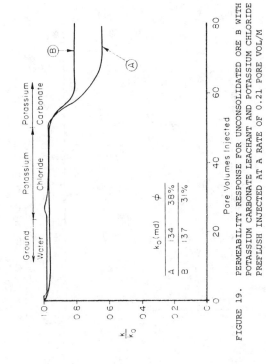

FIGURE 17. COMPARISON OF PERMEABILITY RESPONSES FOR UNCONSOLIDATED ORES A AND B WITH AMMONIUM CARBONATE LEACHANT AND AMMONIUM CHLORIDE PRE-FLUSH

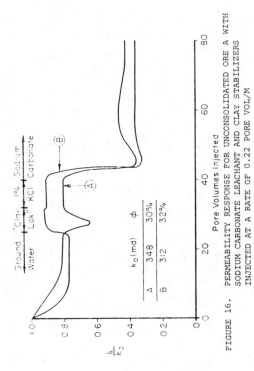

FIGURE 18. PERMEABILITY RESPONSE FOR UNCONSOLIDATED ORE A WITH POTASSIUM CARBONATE LEACHANT AND POTASSIUM CHLORIDE PREFLUSH INJECTED AT A RATE OF 0.19 PORE VOL/M

FIGURE 19. PERMEABILITY RESPONSE FOR UNCONSOLIDATED ORE B WITH POTASSIUM CARBONATE LEACHANT AND POTASSIUM CHLORIDE PREFLUSH INJECTED AT A RATE OF 0.21 PORE VOL/M

491

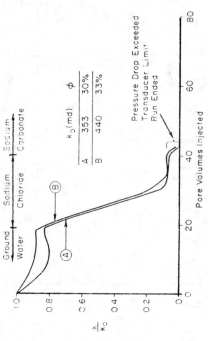

FIGURE 20. COMPARISON OF PERMEABILITY RESPONSES FOR AMMONIUM AND POTASSIUM CARBONATES WITH CHLORIDE PREFLUSHES ON UNCONSOLIDATED ORE A

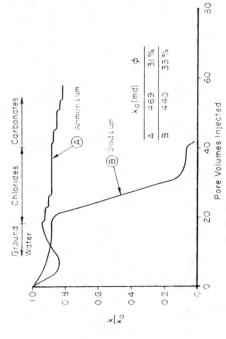

FIGURE 21. PERMEABILITY RESPONSE FOR UNCONSOLIDATED ORE A WITH SODIUM CARBONATE LEACHANT AND SODIUM CHLORIDE PREFLUSH INJECTED AT A RATE OF 0.24 PORE VOL/M

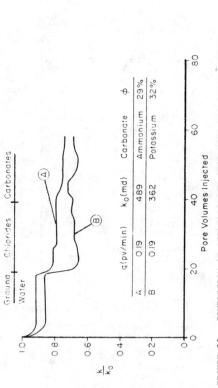

FIGURE 22. PERMEABILITY RESPONSE FOR UNCONSOLIDATED ORE B WITH SODIUM CARBONATE LEACHANT AND SODIUM CHLORIDE PREFLUSH INJECTED AT A RATE OF 0.22 PORE VOL/M

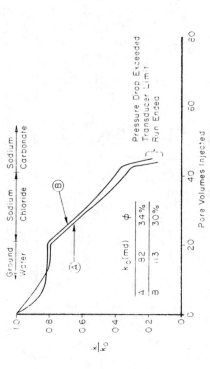

FIGURE 23. COMPARISON OF PERMEABILITY RESPONSES FOR AMMONIUM AND SODIUM CARBONATES WITH CHLORIDE PREFLUSHES WHEN INJECTED AT A RATE OF 0.19 PORE VOL/M INTO UNCONSOLIDATED ORE A

492

# ENVIRONMENTAL AND ENERGY CONSIDERATIONS OF FOREIGN NONFERROUS SMELTER TECHNOLOGY

**A. Christian Worrell III, Thomas K. Corwin** and **Mary A. Taft**
PEDCo Environmental, Inc.

**John O. Burckle**
U.S. Environmental Protection Agency

Introduction

This paper briefly addresses the environmental and energy problems
facing the domestic nonferrous smelting industry and presents an overview
of some alternative pyrometallurgical processes currently being used or
developed overseas.  Available data on environmental and energy aspects
of these foreign processes are summarized and compared with similar data
for commonly used domestic processes.  Because of the size and age of
the industry, copper production technology is emphasized.

Status of Domestic Industry

The domestic copper industry is relatively old.  A number of plants
found in the West were built over 50 years ago.  These plants were
constructed at a time when the availability of inexpensive fossil fuels
and high-grade ore made energy efficiency a minor design consideration;
environmental considerations were either not understood or thought of as
problems.  As a result of these factors, most operations of this industry
are energy intensive and produce significant quantities of pollutants.

Compliance with regulations promulgated under the Clean Air Act has
presented the industry with the bulk of its problems in the environmental
area.  Sulfur oxides and particulate matter that contains metal contaminants are the industry's most copious pollutants and often the most
difficult to control.

In addition to the considerable problems associated with the containment
of sulfur oxide and other air pollutants, the domestic nonferrous smelting
industry is faced with ever increasing energy costs and limited availability of various forms of fuel.  As ore grades have declined over the
years, energy requirements for mining, beneficiation, and smelting have
risen.

In 1976 the primary copper industry consumed approximately $70 \times 10^9$ Btu's of purchased fuels and electricity at a cost of over $110 million.[1] Unfortunately, environmental and energy considerations are generally in conflict with each other, given the technology now in place. Pollution control equipment is not only expensive to purchase and install, but its operation requires considerable amounts of energy, usually in the form of electricity for gas movement and cleaning devices.

Control of air pollutant emissions is often focused on $SO_2$ control. Because technologies for control of $SO_2$ require the removal of particulate contaminants, effective control of total particulate and trace element process emissions is accomplished as a "byproduct" of $SO_2$ control. Fugitive emissions are also a significant problem, however.

The most desirable approaches to emissions control are those which will offer not only environmental protection, but some financial benefit to the industry in return for the risk of the capital investment. Because the market is highly competive and the price of metal is determined on a worldwide basis, the cost of pollution control cannot simply be passed along to the consumer.

Retrofit pollution control measures that are capable of increasing the profitability of a smelting operation are somewhat difficult to identify. A technology capable of reducing emissions while simultaneously lowering energy consumption would constitute a highly significant means of increasing smelter profitability. Alternative prometallurgical processes are of interest because they can provide a strong $SO_2$ gas stream suitable for control by a conventional acid plant while simultaneously reducing gas stream volumes, energy consumption, and operating and capital costs. Certain pyrometallurgical technologies currently being used or developed overseas may, in the future, play an important role in solving domestic $SO_2$ and fugitives control problems while requiring a smaller amount of energy.

Foreign Technology

Some foreign smelters, especially those in Japan as well as a few European countries, have apparently had rather remarkable success in the control of polluting emissions to the atmosphere. These results have been achieved not only by installation of venting systems and pollution control devices, but also through application of improved process technologies, which often have reduced energy consumption.

Influences on Foreign Technology

There appear to be several reasons why some foreign smelters have installed improved technologies. Unlike the U.S. industry, European and Japanese smelters have faced a restrictive energy picture for many years, and there has been much greater incentive to develop more energy-efficient processes.

As long as ten years ago, many foreign smelters operated under tougher emission standards than those faced by their U.S. counterparts even today. They are located closer to population centers where there have been increased pressures for pollution control from government agencies and local citizenry. These pressures have led, in some cases, to very restrictive emission standards based upon potential health effects rather than emission levels achievable through the use of existing control technology. Such standards have been technology-forcing in nature and have spurred the development of new equipment. The rapid

economic growth and reconstruction of the industrial base of some
European countries and Japan in the years since World War II have provided further impetus toward the adoption of new technologies.  Many new
smelters have been built, and naturally the most modern equipment has
been selected.

Programs of direct and indirect government financial assistance have
also played a significant part in the development of improved foreign
technologies.  In Japan, for example, installation of pollution control
equipment has been encouraged through a range of government fiscal
policies, including accelerated depreciation allowances and reduced
fixed assets taxes.  Pollution control devices at smelters in that
country have also been subsidized through loans at favorable rates from
government agencies.[2]

## Process Descriptions

All of the factors discussed above have had some role in encouraging the
development of innovative technology and the resulting processes may
have some potential for application domestically.  Six of these processes
will be briefly described with emphasis on energy consumption and pollution control.

Mitsubishi Process.    In 1961, Mitsubishi Metals Corporation began
development of a process for continuous, pollution-free copper smelting.
After favorable pilot-scale operations, Mitsubishi built a commercial-
scale plant at the Naoshima smelter.[3]  A second commercial-scale plant,
owned by Texas Gulf, is located at Timmins in Ontario, Canada.[4]

The Mitsubishi process is composed of three metallurgical stages, each
of which is carried out in a separate furnace.  Concentrates are smelted
in the first, slag is cleaned in the second, and matte is converted to
blister copper in the third.[5]  Intermediate products in the molten state
move continuously between the furnaces.

The average sulfur dioxide content of the combined off-gases from the
three furnaces is greater than 10 percent.  When the smelting furnace is
operated with 25 percent oxygen-enriched air, $SO_2$ concentration is
fairly constant and permits economic recovery of sulfur as sulfuric acid
or elemental sulfur.[6,7]

A plant using the Mitsubishi process is estimated to require only 70 to
80 percent of the capital investment needed to build a conventional
smelter.  This low capital investment is attributed to relatively simple
engineering design involving the continuous gravity flow of molten
intermediate and final products and to higher output per volume of
smelting unit because of an increased smelting rate compared to reverberatory smelters.

The Mitsubishi continous smelting process has been estimated to require
from 14.0 to 20.4 x $10^6$ Btu per ton of anode copper.  This constitutes
fuel requirements of 78 to 90 percent of that required in hot calcine
reverb smelting.  Fuel oil requirements are very low - approximately 34
to 38 percent of that for the reverb.  Use of oxygen enrichment, maximum
usage of the heat of reaction of iron and sulfur, and the compactness of
the design all contribute to this economy.  Gas cleaning is also more
economical as a result of the continous design.  There are other aspects
of the process which somewhat offset these savings, however.  Continuous

smelting processes often either produce slag high in product or a product high in impurities. A high copper slag requires treatment for recovery of copper. Mitsubishi makes use of an electric slag-cleaning furnace which is estimated to consume from 1.3 to $3.8 \times 10^6$ Btu per ton of anode copper. In addition, production of oxygen requires another 1.3 to $1.5 \times 10^6$ Btu/ton.[8,9]

WORCRA Process. The WORCRA process, developed by Conzinc Riotinto of Australia, Ltd., is also a continuous direct smelting process for copper concentrates. Experimental evaluation of WORCRA began in early 1963 and encouraging results lead to the construction of a larger test furnace.[10,11]

Copper concentrates are smelted to matte, matte is converted to metal, and slag is cleaned in separate zones of an elongated hearth-type fur-nace. Concentrates and flux are added in the mildly oxidizing smelting zone and slag generally moves countercurrently to the matte; iron and other unwanted materials are continuously transferred to the slag after oxidation.[10,11] Moving slowly through the smelting and converting zones, matte is lanced with air (or oxygen-enriched air), causing conversion to white metal and then to copper. In the converting zone, the hearth slopes downward, and copper passes continuously to the "copper well," which overflows with low-grade blister copper. Extensive fire refining or converter processing may be required before casting.

The WORCRA furnace yields a constant stream of gas with an average $SO_2$ content of 10 percent. Use of oxygen-enriched air increases the $SO_2$ content.[5]

Capital costs for a WORCRA plant are expected to be 20 to 30 percent less than for a conventional reverberatory plant of a similar capacity. Lower operating costs are also predicted. Energy requirements for the WORCRA process should be between 70 and 80 percent of that required in a calcine feed reverb. Use of oxygen enrichment and the fuel equivalents in the sulfur and iron help reduce fuel usage. However, oxygen produc-tion will require additional energy. The production of lower grade blister may require additional converting or fire refining, but elim-inates the need for treatment of slag in a furnace or concentrating circuit.

Britcosmaco Process. The Britcosmaco process is also under development in Australia for pyrometallurgical treatment of sulfide ore concen-trates. Dry concentrate and flux are fed into a main smelting shaft with sufficient preheated or oxygen-enriched air for autogenous smelt-ing. An enriched white metal and a slag are produced and collect on a hearth in two layers.[5]

The slag flows along the hearth and contact with low-grade matte causes more copper to be rejected. As it falls down the main shaft, copper in the concentrate is oxidized and works through the slag layer. It dissolves at the top of the enriched metal layer and precipitates from the bottom as metal. The copper is removed by a bottom tapping siphon.[5]

According to its designers, the Britcosmaco process is advantageous because copper recovery is higher than for conventional pyrometallurgical treatment, energy requirements are lower, total gas volume is smaller, and $SO_2$ concentration in the off-gas is greater.[5]

INCO Oxygen Flash Smelting. The International Nickel Company (INCO) of Canada developed the oxygen flash smelting process in the 1940's to treat copper and nickel concentrates. By the early 1950's, INCO had successfully demonstrated the method on a commercial scale at its Copper Cliff plant.[12,13]

The INCO furnace is essentially a reverberatory furnace with an uptake shaft extending the length of the furnace roof. Concentrate is dried in a fluidized-bed drier; the solids are withdrawn from feed bins and transferred to two burners located at each end of the furance. Suspended in a horizontal flow of oxygen, the feed is injected into the furnace. Oxygen combines with some of the sulfur and iron to form $SO_2$ and iron oxides. The iron oxides in turn, combine with silica contained in the flux and concentrate to form a slag. The remaining sulfur and metal fractions collect in the matte, which is transferred to one of three Pierce-Smith converters for production of blister copper.[12,13]

Furnace off-gases contain about 80 percent $SO_2$ and about 2 to 3 percent of the feed which exits as dust during smelting operations. After the gas stream is cooled and cleaned of particulates, the $SO_2$-laden gas is piped to an adjacent liquefaction plant. The INCO oxygen flash furnace reportedly emits less $SO_2$ and fewer impurities than a reverberatory furnace handling a typical concentrate for North America. The process, however, is used with concentrates having impurity levels lower than those usually found in domestic United States concentrates. Behavior with high impurity concentrates has not been demonstrated.[13]

The INCO flash furnace has been estimated to offer 33 to 36 percent energy savings compared with a hot calcine feed reverb. Estimated requirements are 10 to 17.7 x $10^6$ Btu/ton anode copper. No fuel is used in the INCO flash furnace - smelting temperatures are achieved by the oxidation of the iron sulfides in the concentrate. Although over 3 x $10^6$ Btu/ton are required for oxygen production, the low volume and the high $SO_2$ content of the offgas offers some energy savings in gas handling and cleaning. Energy estimates differ in the area of waste heat recovery with one study allowing over 1.5 x $10^6$ Btu credit and another allowing no credit in this area.[8,9]

Top Blown Rotary Converter. The International Nickel Co. of Canada has adapted steel process technology (the Kaldo process) to nonferrous metallurgy in developing the top blown rotary converter (TBRC), which can smelt a variety of concentrates or byproducts. Boliden AB of Sweden has carried out additional development of the TBRC, and several commercial-scale plants are in use, including two in Sweden and one each in Ontario and British Columbia, Canada.[14,15] The process is currently licensed to Dravo in the United States.

After complex copper concentrates are dried and agglomerated in a briquetter, they are fed to the TBRC, which contains a molten sulfide bath. The vessel atmosphere is controlled by injecting natural gas and oxygen-enriched air into the molten mixture through a water-cooled lance. White metal is then sent to a second TBRC for further oxidation and elimination of impurities.[14,15]

The furnace rotates constantly, providing thorough contact between the gas and furnace contents and ensuring even distribution of heat. The molten metal is removed from the furnace periodically and cast into anodes. Slag is left in the vessel for recovery of valuable metals upon addition of new concentrate.

A close-fitting exhaust hood fits over the mouth of the converter during loading, smelting, and converting operations.[16,17]

The advantages claimed for the TBRC include lower particulate loadings, decreased fugitive emissions, and greater $SO_2$ concentrations (up to 50 percent with oxygen enrichment). The batch-type operation, however, does not produce a continuous stream of gas.

Data indicate that the capital cost of the TBRC process is less than that of conventional smelting techniques. The operating cost is also less because of lower maintenance, labor, and energy requirements.[14,15]

The TBRC has been estimated to require 38 percent less energy than hot calcine feed reverberatory smelting. The calculated energy requirement is $16.4 \times 10^6$ Btu/ton. The major area of savings is in fuel for smelting, which requires only $0.2 \times 10^6$ Btu/ton of direct fuel consumption. This savings is somewhat offset by $3.4 \times 10^6$ Btu for oxygen production, absence of waste heat recovery, and the electricity requirements for converter rotation.[12]

Kivcet Process.    The Kivcet process involves continuous smelting of complex sulfide concentrates with simultaneous production of lead, zinc, copper, nickel, and minor metals such as cadmium and tin. The process has been under research and development since 1963. The first commercial-scale plant was built in the Soviet Union.[18,19]

Sulfide concentrate, containing at least 25 percent sulfur, and oxygen are fed into either a cyclone or flash furnace where the concentrate is autogenously roasted and flash-smelted in a suspended state. The products leaving the bottom of the furnace pass into a distribution chamber, where off-gases and melt are separated. The off-gases are cleaned of particulates and sent to the acid plant; the melt is routed to the electric furnace.[18,19]

Matte containing copper, nickel, cobalt, and precious metals is formed and periodically tapped and further processed by conventional methods. Zinc is volatilized in the electric furnace and may be condensed to metal or oxidized to produce high-grade zinc oxide.[18-20]

Off-gases from the distribution chamber are very small in volume and high in $SO_2$ content. Approximately 80 percent of the sulfur in the concentrate passes into the off-gases, and may be converted to elemental sulfur, sulfuric acid, or liquefied sulfur dioxide. Slag from the electric furnace contains very little matte and can be discarded immediately without cleaning.

Overall capital costs of a smelter using the Kivcet process are claimed to compare favorably with other high-technology processes. Elimination of agglomerating, sintering, and slag cleaning installations and simplification of gas cleaning and $SO_2$ capture systems help reduce costs.[18,19]

Conclusions

This overview has incorporated information from many sources. Pitt and Wadsworth[8] and Kellogg and Henderson[9] were the sources of many of the energy data. Although there is a general agreement among these studies as to the relative energy requirements of various processes, they differ rather significantly in terms of absolute energy requirements.

The available data on costs of alternative technology are even more inconsistent than those for energy consumption. In general, the estimates that are published show most of the alternate technologies to be of lower capital cost than a greenfield smelter of conventional design. While cost estimates for technologies that are only in preliminary phases of development are often inaccurate, usually on the optimistic side, the nature of the design of many of the new technologies is such that the capital and operating savings claimed appear to be feasible.

Although this overview has focused on foreign innovative technologies, it is not intended to imply an absence of innovation in this country. Several promising pyrometallurgical processes such as the Q-S, Oxygen Sprinkle, and Amax Dead Roast processes are currently being developed domestically.

Many of the same factors that have encouraged research and development overseas are now coming into play in the United States; rising energy costs, strict environmental regulations and growing public awareness are major factors in this regard. These influences, coupled with the advanced age of many smelters, make the development and use of innovative technology highly attractive. However, industry must weigh these factors against uncertain markets and the huge capital outlays required for the construction of new plants based on technologies that have yet to be proven effective for domestic ores. The results of decisions made in this decade will have long term effects on the domestic nonferrous industry.

References

1.  Annual Survey of Manufacturers 1976, Fuels and Electric Energy Consumed. U.S. Department of Commerce, Bureau of the Census. U.S. Government Printing Office, Washington D.C.

2.  Corwin, T.K. "The Economics of Pollution Control in Japan." Environmental Science and Technology, Vol. 14, No 2, p. 154, February 1980.

3.  Suzuki, T. and T. Nagano. "Development of a New Continuous Copper Smelting Process." Presented at the Joint Meeting of the Mining and Metal Institute of Japan and the American Institute of Mining, Metallurgical, and Petroleum Engineers, Tokyo, Japan, May 24-27, 1972.

4.  Amsden M.P., et al. "Selection and Design of Texas Gulf Canada's Copper Smelter and Refinery." Journal of Metals, 30(7):16-26, July 1978.

5.  Price, F.C. "Copper Technology on the Move." Engineering and Mining Journal, pp. RR-WW, April 1973.

6.  "Mitsubishi's Continuous Copper Smelting Process Goes on the Stream." Engineering and Mining Journal, 173(8):66-68, August 1972.

7.  Nagano, T., et al. "Commercial Operations of Mitsubishi Continuous Copper Smelting and Converting Process." Extractive Metallurgy of Copper. Vol. 1. Metallurgical Society of AIME, 1976, New York. pp. 439-457.

8. Pitt, C.H., and M.E. Wadsworth. <u>An Assessment of Energy Require-ments in New Copper Processes</u>. U.S. Department of Energy. February 1980, Rough draft final report.

9. Kellogg, J.H. and J.M. Henderson. "Energy Use in Sulfide Smelting of Copper." <u>Extractive Metallurgy of Copper</u>. Symposium sponsored by the Metallurgical Society of American Institute of Mining, Metal-lurgical, and Petroleum Engineering. New York, 1976.

10. Reynolds, J.O. "WORCRA Copper and Nickel Smelting." Presented at the Joint meeting of the Mining and Metal Institute of Japan and the American Institute of Mining, Metallurgical and Petroleum Engineers, Tokyo, Japan, May 24-27, 1972.

11. Semrau, K.T. "Control of Sulfur Oxide Emissions from Primary Copper, Lead, and Zinc Smelters: A Review." Presented at the 63rd Annual Meeting of the Air Pollution Control Association, St. Louis, Missouri, June 1970.

12. Solar, M.Y., et al. "Smelting Nickel Concentrates in INCO's Oxygen Flash Furnace." <u>Journal of Metals</u>, January 1979.

13. Coleman, R.T., Jr. <u>Emerging Technology in the Primary Copper Industry</u>. Prepared by Radian Corporation for the U.S. Environ-mental Protection Agency under Contract No. 68-02-2608. Austin, Texas, August 1978.

14. Daniele, R.A., et al. <u>TBRC - A New Smelting Technique</u>. TMS No. A72-101. The Metallurgical Society of AIME, New York, 1972.

15. "Afton, New Canadian Copper Mine on Stream." <u>World Mining</u>, pp. 42-44, April 1978.

16. "Afton's Copper Smelter Proves Economic at 27,000 Tons Yearly." <u>Canadian Chemical Processing</u>, 63(2):22-24, March 21, 1979.

17. Thoburn, W.J., and P.M. Tyroler. "Optimization of TBRC Operation and Control at INCO's Copper Cliff Nickel Refinery." Presented at the 18th Annual CIM Conference of Metallurgists, Sudbury, Ontario, August 19-23, 1979.

18. "Soviet Continuous Smelting Process Licenses in West." <u>Engineering and Mining Journal</u>, 175(7):25, July 1974.

19. Kivcet pamphlet. Licensintorg, Moscow, Soviet Union. Distributed by Southwire Company, Carrolton, Georgia.

20. "Humboldt Wedag's Cyclone - Furnace Smelting Recovers Nonferrous Metals." <u>Engineering and Mining Journal</u>, 178(10):45,49, October 1977.

21. Dolezal, H., et al. U.S. Bureau of Mines. <u>Environmental Con-siderations for Emerging Nonferrous Metal-Winning Processes</u>. U.S. EPA draft report. March 1980.

| Technology | Capital cost percent of base case | Energy requirements percent of base case, (quantity $10^6$ Btu/ton of anode) | | |
| --- | --- | --- | --- | --- |
| | | Pitt and Wadsworth[8] | Kellogg[9] and Henderson | Other[21] |
| Calcine feed reverb (base case) | 100 | 100 (26.2) | 100 (15.6) | 100 (25.1) |
| Electric furnace | 100[E] | 144 (37.8) | 156 (24.3) | 112 (28.2) |
| Noranda | 80[E] | 74 (19.3) | 79 (12.3) | 53 (13.3) |
| Mitsubishi | 70-80 | 78 (20.4) | 90 (14.0) | - |
| INCO Flash Furnace | 70-80[E] | 67 (17.7) | 64 (9.9) | - |
| TBRC | - | 62 (16.4) | - | - |
| WORCRA | 70-80 | - | - | 75 (19.7)[E] |
| KIVCET | 70-80 | - | - | 75 (19.7)[E] |
| BRITCOSMACO | 70-80[E] | - | - | 75 (19.7)[E] |

[E] Estimated.

ESTIMATED CAPITAL COST AND ENERGY REQUIREMENTS FOR SELECTED TECHNOLOGIES

FIGURE 1

# ENERGY CONSERVATION AND POLLUTION ABATEMENT
## AT PHOSPHORUS FURNACES

**James C. Barber**
James C. Barber and Associates

The proposals made in this paper can result in large reductions in the
amount of energy consumed at phosphorus furnaces. Also, solutions are
offered for some major environmental problems facing phosphorus producers. The bases for proposals are ideas which came from operating
experience and from special studies conducted on production units.
Experiments were made recently at a university laboratory to make preliminary assessments of some of the proposals. However, much additional
developmental work is needed to reduce the ideas to practice.

## Phosphate Smelting

Elemental phosphorus is produced by smelting a mixture of phosphate ore,
coke, and silica rock in a submerged-arc electric furnace. Phosphorus
in the phosphate ore is combined as mineral fluorapatite. Coke is provided to supply carbon to react with oxygen and form carbon monoxide gas,
thereby reducing the combined phosphorus to the element. Silica rock
provides $SiO_2$ to combine with CaO in the phosphate ore and form slag.

Figure 1 gives equations for smelting phosphates, reduction of iron
oxide, and formation of ferrophosphorus. The ferrophosphorus is a by-
product of the smelting process. The phosphate ore is represented in
the smelting equation by tricalcium phosphate, but in practice the mineral contains impurities such as $SiO_2$, $Al_2O_3$, F, $Fe_2O_3$, and various
metal oxides. In some cases the phosphate may contain enough $SiO_2$ to
combine with CaO, and no silica rock is needed.

Metallurgical coke is the common source of carbon. This material contains about 12 percent moisture as received.

The silica rock should contain as much $SiO_2$ as possible, and this is usually about 95 percent. Impurities in this mineral are CaO, $CO_2$, $Fe_2O_3$, and $Al_2O_3$.

Elemental phosphorus vapor and a fuel gas consisting mainly of carbon monoxide come off the furnace through a duct. Calcium silicate slag is tapped from the furnace. In practice the slag is a mixture of calcium silicate and calcium aluminate, because the feed materials contain $Al_2O_3$ as an impurity. Heat loss in the slag amounts to about 20 percent of the electrical energy required for smelting, or about 8.2 million Btu per ton of phosphorus produced. (This is net heat loss in slag, or sensible heat loss less heat of slag formation.)

Equations 2 and 3 in figure 1 show that $Fe_2O_3$ is reduced to elemental iron and this element combines with phosphorus to form iron phosphide, a byproduct called ferrophosphorus. It is a metallic-like material consisting of a mixture of iron phosphides. Energy loss in the ferrophosphorus is relatively small--about 2 percent of the electrical energy input. Loss of phosphorus combined in the ferrophosphorus is of greater importance than energy loss, although ferrophosphorus is marketed for use in the metallurgical industry.

Figure 2 is a table showing the amounts of feed materials needed to make a ton of phosphorus. Seventy-five percent of the material is phosphate, 12 percent coke, and 13 percent silica rock.

Figure 3 gives the quantity of byproducts made per ton of phosphorus produced. Fluorides are shown as a byproduct, although phosphorus producers discharge these compounds as wastes (either aqueous waste or solid waste). About 91 percent of the fluorine put in the furnace comes out in the slag. The quantity of fluorides in the aqueous wastes is not sufficient to justify the cost of byproduct recovery. Precipitator dust is usually discarded or recycled to the process. At least one phosphorus producer distributes precipitator dust into agricultural markets to recover micronutrient values.

Preparation of Phosphates, Coke, and Silica Rock

for Feeding in Phosphorus Furnaces

## Use of Uncalcined, Unagglomerated Phosphates

## in Phosphorus Furnaces

Phosphates generally occur in nature as particles that are too small to use in furnaces. Small particles become entrained in the furnace gases and are carried into the phosphorus condensing system.

Most phosphates occur as sedimentary deposits, but some are igneous. Phosphates in sedimentary deposits are microcrystalline apatites that may be different in composition from pure mineral fluorapatite, $Ca_{10}(PO_4)_6F_6$. Carbonate may replace part of the phosphate, and other metals may replace part of the calcium.

In isolated cases phosphate particles are large enough to use in furnaces without agglomeration. A publication[1] gives some information on the use of various phosphates in the TVA furnaces. One unagglomerated phosphate tested was Florida hard rock having an average particle size of 0.67-inch and $P_2O_5$ content of 36.2 percent. This material was fed into a 15-MW furnace without calcination. It contained no clay as an impurity.

Performance of the furnace was superior to operation with nodules--an agglomerated and calcined material being used as the regular phosphate feed material. The overall conclusion drawn from this test is that calcination of high-grade phosphate is unnecessary.

Uncalcined Florida pebble was used to augment the supply of nodulized phosphate in the TVA furnaces. The pebble had an average particle size of 0.20-inch, and its $P_2O_5$ content was 31.4 percent. The uncalcined, unagglomerated Florida pebble was generally unsatisfactory as a feed material except in relatively small proportions at furnaces operating at powerloads less than about 20 MW. Poor operation of the furnaces was attributed to the small size of the Florida pebble.

Briquetted phosphate containing large amounts of clay is entirely unsatisfactory as furnace feed material unless the agglomerates are calcined.

## Use of Calcined, Agglomerated Phosphates in

## Phosphorus Furnaces

Phosphorus producers generally calcine and agglomerate the phosphate fed to the furnaces. Unbeneficiated ores or low-grade fractions from the beneficiation process are often used.

Clay in the phosphate serves as a binder to make compacts, pellets, or briquets. The green agglomerates have little strength and are unsuited for furnace feed until they are indurated by calcining. However, clay in the phosphate adversely affects the furnace operation, although the reason for this is not entirely known.

Phosphates may be agglomerated when clay is not present. In this case the material is ground to provide large surface areas, water is added, and the mixture is tumbled or rolled to form pellets. Discrete particles in the agglomerates are held together by surface tension forces, but they are easily deformed by pressure. The pellets are indurated by calcination, making a good furnace feed material.

At TVA, phosphate was agglomerated by nodulizing. The material was heated to the point of incipient fusion in a rotary kiln in order to obtain enough liquid phase for agglomeration. When clay is in the mixture the liquid phase can be obtained at lower temperature than is required with high-grade phosphates. Consequently, kiln maintenance cost is lower and energy requirement is decreased with the low-grade phosphates. At TVA, beneficiated Florida phosphate was added to the kiln feed to upgrade the furnace burden. The high-grade phosphate had no adverse effect on the nodulizing, but furnace operation was improved by incorporating this material in the mixture.

A TVA publication[2] describes the nodulizing process and gives results of studies to improve phosphate agglomeration.

## Compacting and Calcining in a Grate-Kiln

In 1970, TVA changed from nodulizing to the grate-kiln method of preparing phosphate for furnace feed. Phosphate was compacted and the resulting agglomerates were fed to a grate-kiln for calcining. Figure 4 illustrates this method of preparing the phosphate.

The phosphate mixture, consisting of Tennessee matrix and beneficiated Florida phosphate, was fed to a rotary dryer where the moisture content was reduced from 20 percent to about 12 percent. Partial drying improved the physical properties of the material. The partially dried mixture was agglomerated by compacting to a thickness of about 3/4-inch. The flakes broke into lumps. These agglomerates were fed to a grate-kiln calciner to complete the drying and to indurate the material by calcining to a temperature of 2200°-2300°F.

Effluent gases were treated in cyclonic dust collectors followed by a scrubber to remove the remaining dust and fluorine. Figure 5 shows the air pollution abatement equipment provided at the grate-kiln, and a description of the method of treating the effluent to remove fluorine is given in a published article.[3] Gas flow from the grate-kiln was about 685 cfm per ton of phosphorus produced per day. The dry dust collectors were supposed to reduce the dust concentration in the offgas to 1.0 grain per cubic foot, but this was not achieved in practice. The gases were scrubbed in a spray tower with a water:gas ratio of 0.018-gallon per cubic foot of gas. Fluorine removal was adequate but particulate emissions exceeded the standards. A higher water:gas ratio might have reduced particulates to the desired level, but this would require so much water that the effluent treating system would be overloaded.

Once-through water was used in the scrubber because recirculation caused pumps and piping to corrode. Too, the recirculated water contained so much suspended solids that spray nozzles became plugged and solids accumulated in the sump.

Figure 6 shows a pond for treating the effluent. Flourine was removed from the water by filtering through a large pile of granulated furnace slag, as shown. From 90 to 95 percent of the fluorine was removed in this manner, giving effluent containing about 30 ppm of fluorine.

## Preparation of Coke for Use in Phosphorus Furnaces

Much of the moisture in coke is adsorbed and will enter the reduction zone of the furnace where it will be reduced by carbon and will consume electric energy. Coke is dried before it is fed to the furnace. No special drying problems are involved and phosphorus producers have a choice of drying equipment. A rotary dryer was used at TVA.

Particulates are emitted during drying. Wet scrubbers may be preferred because of fire hazards when particulates are collected dry,

## Preparation of Silica Rock for Use in

## Phosphorus Furnaces

Silica rock is normally washed to remove clay before it is received. However, clay may remain in the washed material and cause handling problems. Clean silica rock contains only surface moisture and drying is unnecessary, but inadequately washed silica rock should be dried.

No special environmental problems are encountered in handling and drying clean silica rock. Particulate emissions may present a problem when the material contains clay.

Phosphorus Furnace Process

## Overall Description

Figure 7 is a diagram of a phosphorus furnace and condenser. The
mixture of phosphate, coke, and silica is fed to a bin and then flows by
gravity through feed chutes to the furnace. Gases come off the top of
the furnace, and in this diagram they are treated in an electrostatic
precipitator to remove dust. Some producers do not use electrostatic
precipitators. The gas mixture then flows to an open chamber where it
is contacted with water for adiabatic cooling. Phosphorus condenses and
collects as a liquid in a sump under the condenser. The gases are further cooled in a tubular unit and additional phosphorus is condensed.
Various combinations of spray chambers and tubular units are used in the
industry since condensing arrangements have not been standardized.
Liquid phosphorus is pumped from the sump to storage tanks where it is
stored under water. The remaining noncondensable gases, consisting
mainly of CO, are used as fuel.

## Operation of the Furnace

Gases are prevented from escaping from the furnace by resistance offered
by solids in the chutes and feed bin. Operators endeavor to maintain
atmospheric pressure inside the furnace, and under ideal conditions the
solids will prevent any gas from escaping from the feed bin. Ideal conditions are not realized in practice. Pressures less than atmospheric
cause air to be drawn into the furnace, and some of the elemental phosphorus is oxidized to $P_2O_5$, resulting in loss of phosphorus. Pressures
greater than atmospheric cause the furnace gases to escape, releasing
fumes and CO in the workroom environment. Furnace gases will escape or
air will be drawn in at openings in the top of the furnace--cracks in
the cement roof, pokeholes, and around the electrodes.

The gas leakage problem can be corrected by making changes which will
reduce pressure fluctuations inside the furnace. It was observed that
removal of minus 10-mesh material from coke was effective in ameliorating
pressure variations. The coke fines were of little benefit in reduction
of the phosphate, but accumulation of this material would create a solid
waste disposal problem. Consequently, fines were fed to the furnace at
a controlled rate.

Data were obtained on the composition of material in the furnaces by
taking samples during shutdowns and analyzing the samples to determine
variations in carbon and $SiO_2$ contents. Figure 8 gives the results
of the analyses of one set of samples. The data show that the carbon
varied from 19 to 304 percent of that needed to reduce the $P_2O_5$. (One
sample had 1377 percent of the carbon needed to reduce the $P_2O_5$, but
this sample probably contained unreacted coke fines which collect on the
surface of the slag.) The $SiO_2$:CaO weight ratio varied from 0.70 to
1.44. The wide variations were attributed to segregation of the materials, and it was observed that segregation lessened when the charge
materials had approximately the same size.

Segregation of materials in solid fertilizer mixtures had caused wide
variations in nutrient contents. Studies showed that segregation can
be corrected by using matched sizes of fertilizer solids. Differences
in particle shape or densities had little influence on the segregation
of components in mixtures.

The phosphate regularly used in the furnace was nodules augmented with beneficiated Florida phosphate as described above. Matched sizes of phosphate, coke, and silica rock were expected to result in a nonsegregating charge mixture inside the furnace. It was postulated that reduction or elimination of segregation would improve phosphorus furnace performance.

The metallurgical coke normally used in the furnace had an average particle size of 0.3- to 0.4-inch, whereas nodulized phosphate had an average size of about 0.8-inch. Silica rock was slightly larger than the nodules. A special purchase of large size metallurgical coke was made, and the material was crushed and screened to prepare material having an average size of 1.0 inch. A special nodulizing run was made whereby Tennessee matrix was briquetted and then calcined in a nodulizing kiln. The phosphate was selected so that it would have a $SiO_2$:CaO ratio of 0.85. No silica rock was needed for fluxing. Average size of the calcined phosphate was about 1.0 inch. The mixture of phosphate and coke was fed to a phosphorus furnace to determine the effect of using these matched sizes of material on the furnace performance.

Figure 9 is the furnace pressure chart for the first day of the test with matched materials. It shows that with the regular charge, furnace pressures were varying ±0.3-inch of water, but with matched sizes of materials, pressure variations decreased to ±0.025-inch of water. Gas analyses indicated that inleakage of air was nearly eliminated.

Figure 10 is sections of the power chart showing the effect of changing to matched charge sizes. With the regular charge the powerload was set on 9.4 MW, but there were deep dips to only 1.0 to 2.0 MW, and numerous smaller dips to 5.0 to 6.0 MW. The average powerload was 13.8 percent less than the preset load of 9.4 MW. No large dips occurred with the matched charge; the average powerload was 1.3 percent less than the set load. These data indicate that the capacity of the furnace can be increased 10-12 percent if matched sizes of the charge materials are used. Electric energy requirement was reduced about 10 percent, but longer tests are needed to make more accurate determination of energy reduction.

The temperature of the gases leaving the furnace held steady at 475°F-- about 200°F less than the average temperature with the segregating charge. Dewpoint of phosphorus in the gas was about 460°F: therefore, little further reduction in gas temperature can be tolerated.

The furnace tests described above were carried out about 18 years ago, but it was impractical to use matched sizes of charge materials on a regular basis. Improved agglomerating methods must be developed to permit the preparation of charge materials with the desired size.

Emission of gases and fumes from molten slag and ferrophosphorus causes a pollution problem at phosphorus furnaces. Both $P_2O_5$ and fluorine compounds volatilize from hot slag. The $P_2O_5$ vapor condenses in air to form foglike particles with an average particle size of about 0.6-micron. This pollution problem was reviewed in a publication[4] which gave the energy required for collecting the particles in a high-energy venturi scrubber. A filtration system augmented with a wet scrubber has been proposed for collection of the emissions at lower energy requirement and smaller capital investment.[5] No specific information is available on the filter-scrubber combination.

## Operation of the Condensing System

A diagram of a phosphorus condensing system is shown in figure 11. Phosphorus is contacted with water at various places--at the precipitator, spray condenser, tubular condenser (the tubes are irrigated with water on the inside), exhauster, telescoping electrode seals, and at phosphorus storage tanks. Soluble phosphorus and fluorine compounds dissolve in the water and the concentrations of these compounds increase as water is recirculated throughout the system. It is necessary to bleed off some of the recirculating water to control the concentration of dissolved salts and suspended solids.

Fluosilicate salts accumulate in the water as sodium and potassium salts. When ammonia is used to neutralize acid in the recirculating water, ammonium fluosilicate will be formed. The fluosilicate salts have relatively low water solubilities, and when saturation is reached they precipitate as tenacious scales on the cooler surfaces in the system. Scales form on cooling coils, piping, pumps, and spray nozzles. It has been found from experience that fluorine concentration of the recirculating water should be held to about 10 grams per liter to prevent fluosilicate scale from forming. A higher fluorine concentration can be tolerated when no neutralizing agent is added to the water.

At saturation, water will contain about 3 ppm of elemental phosphorus. This element readily becomes suspended in the water either as pure elemental phosphorus or as phosphorus sludge. In 1960 the lethal concentration of phosphorus to marine animals was reported by the Tennessee Department of Publication Health[6] to be 0.1 ppm for 48-hour exposure. At TVA, facilities were provided to clarify wastewater containing elemental phosphorus by a combination of a clarifier and a 14-acre settling pond. A publication[7] gives detailed information about the treating system.

Figure 12 gives concentration of phosphorus at the different parts of the treating system. Phosphorus content of water discharged as a waste was in the range of 0.3 to 0.4 ppm. Although the phosphorus concentration at the settling pond overflow exceeded the reported lethal concentration, the water was mixed with plant cooling water and wastewater from other plants before it was discharged into the Tennessee River. The elemental phosphorus concentration was thereby reduced below the toxic limit by a combination of clarification and dilution.

Studies were undertaken by the Fisheries Research Board in Canada to make further determinations of the toxicity of water containing elemental phosphorus. This work was undertaken because of fish kills alleged to have been caused by discharges from a phosphorus plant. The Canadian studies, reported in 1970,[8] disclosed that elemental phosphorus was much more toxic to marine animals than was indicated earlier. The studies indicated that elemental phosphorus content of wastewater must be reduced to 40 ppb, or lower, to make the waste suitable for discharging into a receiving stream. This would require removal of both suspended and dissolved phosphorus--technology for which has not been developed.

### Energy Conservation

## Coke Agglomeration

Several reducing carbons are available for use in phosphorus furnaces.[9] Although metallurgical coke is the preferred carbon source, the quantity

of this material is limited and the quality needs improving.  Coke cost
has increased about five-fold over the past 15 years.  Furthermore,
about 15 percent of the coke is minus 10-mesh fines.

The supply of metallurgical coke may be extended by agglomerating the
fines since the small sized material is ineffective for reduction of
phosphates.  Also, agglomeration of fines will increase the average size
of the reducing carbon because coke is normally only about half as large
as agglomerated phosphate.

Laboratory experiments were conducted in apparatus shown in figure
13  to investigate the agglomeration of coke fines.  The apparatus
consists of a drum 10 inches in diameter by 18 inches long, mounted on
rollers and turned by a 1/4-inch electric drill motor.  The critical
speed of the drum is 84 rpm and in most of the experiments it was
rotated at about half of this speed.

Several binders have been tried but the formulation shown in figure
14  appeared to be the best.  The green agglomerates were hardened
by drying at 248°F (120°C) to give a crushing strength of 5.5 pounds.
In fertilizer practice, granule crushing strengths of 5 pounds are adequate to resist degradation by handling.  Higher strength agglomerates
can be made by using larger proportions of phosphoric acid, and agglomerates were made which had strengths up to 10 pounds.  The agglomerates
are stable at temperatures up to about 1040°F, but may deform under
pressure at higher temperatures.

Ground phosphate should be added in proportions to react with the phosphoric acid and form monocalcium phosphate monohydrate.  The $P_2O_5$ in
both the phosphate and acid is expected to be reduced by carbon in the
furnace to form elemental phosphorus.

Figure 15  illustrates the process of forming coke agglomerates.
Green agglomerates are held together by surface tension forces when the
optimum proportion of liquid phase is present.  After drying, the aqueous
salt solution tends to neck out and form a solid bridge between particles.

Energy savings from agglomeration of coke fines are as follows.

1.  Metallurgical coke requirements can be reduced from 1.5 to 1.3 tons
per ton of elemental phosphorus produced.  Furthermore, other more plentiful carbonaceous materials, such as the smaller sizes of anthracite
coal and low volatile bituminous coal, can be agglomerated and used in
phosphorus furnaces, thereby extending the supply of reducing carbons.

2.  The grade of mixture (percent $P_2O_5$ in phosphate plus silica rock) is
increased by using phosphoric acid as a binder.  Less slag is produced
and loss of energy in the molten slag is reduced.  This energy conservation shows up in lower electric energy requirement to produce elemental
phosphorus.  The relationship between electric energy requirement and
grade has been reported.[1]  For agglomeration of coke fines, the grade
will be increased about 0.5 percent and the saving in electric energy
will be about 154 kWh per ton of phosphorus produced.

3.  Agglomerated fines and screened coke mixture will have a larger
average size than screened coke alone.  It may be possible to match the
sizes of reducing carbon and phosphate, in which case electric energy
requirements would be reduced further.  Savings up to 600 kWh per ton of
phosphorus produced may be possible, but additional data are needed to
make an accurate estimate.

## Phosphate Agglomeration

Exploratory experiments were made to investigate agglomeration of small sized phosphate in the laboratory apparatus. The phosphates used were flotation concentrate and ground Florida pebble--materials containing no clay. Merchant-grade wet-process phosphoric acid was the binder. The agglomerates had a crushing strength of about 9 pounds, which is considered adequate to withstand handling, conveying, and feeding in the phosphorus furnace process. Additional experimentation is needed to determine the optimum agglomerating conditions.

An agglomerated mixture of flotation concentrate and flotation tailing may be a desirable charge material for phosphorus furnaces. The proportion of tailing would be adjusted to make a self-fluxing mixture, and $P_2O_5$ in the waste tailing, which is normally lost, would be recovered.

Heat required to dry green phosphate agglomerates is estimated to be about 800,000 Btu per ton, as compared with a heat requirement for a grate-kiln assembly of about 2.8 million Btu per ton. The energy saving is 2 million Btu per ton of furnaceable phosphate, and total potential energy saving for the industry is about 8.5 trillion Btu per year.

### Pollution Abatement

## Agglomeration of Phosphate by High-

## Temperature Methods

Dust emissions are problems at both nodulizing kilns and grate-kiln calciners, particularly with phosphates containing clay as an impurity. Clay contains submicron size particles which readily become airborne. Particulates are discharged from belt conveyors, elevators, bins, crushers, screens, dryers, and kilns, resulting in workroom concentrations exceeding industrial hygiene standards. Small sized phosphate particles also cause difficult-to-correct environmental problems.

Operators of phosphorus plants must contend with a large number of emission points, and dust loadings of air are high. It is practically impossible to maintain balanced airflow throughout the maze of ductwork conducting the dust-laden air to collection equipment. Ducts become stopped with dust and no dust-laden air will be collected. Energy required for ventilation and recovery of the dust is excessive. Air pollution control equipment must be maintained in good operating condition if environmental standards are to be met--a task essentially impossible to achieve.

Dusts emitted from nodulizing kilns or calcining kilns also cause major environmental problems. Particulates become entrained in the gas stream and fluorine is volatilized from the phosphate because of the high temperatures needed for calcining. The gases must be treated to remove both dust and fluorine. As discussed above, a combination of dry dust collectors followed by a spray tower scrubber was installed at the TVA grate-kiln. The dry collectors were overloaded, resulting in excessive dust loading in the wet scrubber. Increased use of water at the scrubber was inadequate to meet particulate emission standards on a sustained basis. Large volumes of scrubber water must be treated to remove suspended solids and fluorine.

Environmental problems are not likely to be solved by applying improved abatement technology to high-temperature agglomeration. Solutions to these problems may require new phosphate agglomerating methods with lower dust and fluorine emissions.

## Pollution Abatement with Low-Temperature

## Agglomeration

Particulates are emitted as fumes and dusts at fertilizer granulators, but the volume and dust loading of gas is small compared with high-temperature agglomeration. Low pressure drop scrubbers are generally adequate to comply with emission standards, and similar scrubbers should abate particulates at low-temperature phosphate agglomerators. The green agglomerates can be dried with hot gas on a woven wire belt. Particulate emissions from the dryer will be amenable to recovery by a low pressure drop scrubber.

In the manufacture of triple superphosphate about half of the fluorine in the phosphate dust is volatilized in the reaction to form monocalcium phosphate monohydrate. About the same proportion of fluorine volatilization is expected from phosphate dust used in low-temperature agglomeration. Dust scrubbing will remove fluorine from the gas.

Effluent from the scrubber can be put in the phosphorus condensing system as makeup water to recover the fluorine and phosphate values.

## Abatement of Water Pollution at

## Phosphorus Condensers

No practical treatment methods have been devised to reduce elemental phosphorus to concentrations low enough for release into receiving streams. New source performance standards are being withheld until suitable abatement technology is developed.

The wastewater rate is about 1300 gallons per ton of phosphorus produced, and it contains about 1700 ppm of elemental phosphorus. The concentration of elemental phosphorus can be reduced to about 120 ppm by clarification. The bulk of the clarified water can be returned to the phosphorus condensing system, but a stream must be released and replaced with makeup water to control the fluosilicate concentration. It has been found that the elemental phosphorus content of clarified water can be further reduced to about 12 ppm by centrifuging in a stacked-disk centrifuge. Underflow from both the clarifier and centrifuge is treated to recover phosphorus values.

The centrifuged water has elemental phosphorus content low enough to permit the waste to be used in making 11-39-0 suspension fertilizer. In making this fertilizer, merchant-grade wet-process phosphoric acid is neutralized with ammonia in a two-step process; ammonium phosphate precipitates, and a suspending clay is added to keep the solids in suspension. Small amounts (0.1 to 0.5% of the mixture) of fluosilicic acid are added to the phosphoric acid as a crystal modifier to assure that the ammonium phosphate precipitates as crystals small enough for suspension in the fertilizer.

Water is added in both neutralization steps to provide sufficient liquid and to replace water lost by evaporation. It is proposed that the water be replaced by the treated condenser water. When this is done, the waste would supply all the fluosilicic acid and 0.7 percent of the nutrients needed for the fertilizer. First-stage neutralization occurs at the boiling point (220°F) and much of the elemental phosphorus will be oxidized; second-stage neutralizer has an atmospheric cooler to cool the process liquid and additional phosphorus will be oxidized. Furthermore, any elemental phosphorus remaining in the suspension fertilizer will be rapidly oxidized in soil, according to a published article.[10]

An alternative method for the utilization of the liquid waste is to apply it directly to the soil. After centrifuging, the waste can be applied to farmland in a manner similar to that used to dispose of ammonia plant condensate.[11] In this case the phosphorus plant should be located near the disposal site to avoid excessive transportation costs.

The proposed methods for treating and using the wastewater will eliminate the discharge of any elemental phosphorus in aqueous effluent. These methods are offered as solutions to the problem of meeting any new source performance standard calling for limiting phosphorus concentrations to values below the toxic level.

## Research and Development Needs

The present paper offers proposals which will markedly reduce the energy required to produce phosphorus. Some major environmental problems can be corrected without making large expenditures for abatement equipment.

Most ideas for plant improvements must undergo a period of testing or experimentation to assure that the expected benefits will be realized. This is the case here, except that the proposed plant improvements are based on the results of extensive test operation at phosphorus producing units, as well as preliminary laboratory experimentation on a low-temperature agglomeration method.

Following is a broad outline for further developmental work on the various proposals.

1. Make studies of the agglomeration of reducing carbons and small sized phosphate. Laboratory equipment such as that shown in figure 13 would be used in the experimental work. Small sizes of anthracite and low volatile bituminous coal would be included in the studies.

2. Prepare fluid fertilizer mixtures in laboratory apparatus. Clarified wastewater should be used in making clear liquid and suspension fertilizers. Techniques and procedures for investigating processes for making fluid fertilizers have been developed at the TVA National Fertilizer Development Center.

3. Make economic studies. Investment and operating cost estimates are needed for the new agglomerating method and for fluid fertilizer production methods. The estimates would include potential savings from use of matched sizes of carbonaceous materials and phosphates in the furnaces.

4. Make plant tests with agglomerated materials. Supplies of agglomerated carbonaceous materials and small sized phosphate would be prepared in a commercial fertilizer granulator. The agglomerated materials would be fed to a phosphorus furnace to determine whether or not these materials are suitable for use in phosphorus furnaces.

512

# REFERENCES

1.  Barber, J. C. and E. C. Marks. "Phosphorus furnace operations; how are they affected by various types of phosphate charges?" J. Metals 14(12):903-6 (1962).

2.  Stout, E. L. Agglomeration of phosphate fines for furnace use. Chem Eng Report No. 4. Tennessee Valley Authority, Muscle Shoals, Alabama 35660 (1950).

3.  Gartrell, F. E. and J. C. Barber. "Pollution control interrelationships" Chem Eng Prog 62(10):44-47 (1966).

4.  Barber, J. C. "Energy requirements for pollution abatement." Chem Eng Prog 72(12):42-46 (1976).

5.  Chem Proc (news item) p. 150 (May 1979).

6.  Isom, Billy G. "Toxicity of elemental phosphorus." J. Water Pollut Control Fed 32(12):1312-16 (1960).

7.  Barber, J. C. "Waste effluent; treatment and reuse." Chem Eng Prog 65(6):70-73 (1969).

8.  Zitko, V., D. E. Aiken, S. N. Tibbo, K.W.T. Besch and John Murray Anderson. "Toxicity of yellow phosphorus to herring (Clupea harengus), Atlantic salmon (Salmo salar), lobster (Homarus americanus), and beach flea (Gammarus oceanicus)." Can Fis Res Board, J 27(1): 21-29 (1970).

9.  Barber, J. C., E. C. Marks, and G. H. Megar. "Evaluation of reducing carbons for use in electric phosphorus furnaces." Tennessee Valley Authority, Chem Eng Bull No. 4 (July 1960).

10. Bohn, Hinrich L. "Detoxification of white phosphorus in soil." Agri Food Chemis, 18(6):1172-73 (1970).

11. Farris, C. M. "Land application of ammonia plant wastes--a profitable alternative to steam stripping." Paper presented at the Division of Fertilizer and Soil Chemistry, Second Chemical Congress of the North American Continent, American Chemical Society, Las Vegas, Nevada, August 25-28, 1980.

12. Barber, J. C. "Solid wastes from phosphorus production." Chapter VII.2. In Solid Wastes, C. L. Mantell (editor). John Wiley and Sons, Inc., New York, NY (1975).

13. Burt, R. B. and J. C. Barber. Production of elemental phosphorus by the electric-furnace method. Chem Eng Report No. 3. Tennessee Valley Authority, Muscle Shoals, Alabama 35660 (1952).

14. Barber, J. C. and T. D. Farr. "Fluoride recovery from phosphorus production." Chem Eng Prog 66(11):56-62 (1970).

| 1 | $2Ca_3(PO_4)_2 + 10C + 6SiO_2 \longrightarrow P_4 + 10CO + 6(CaO \cdot SiO_2)$ |
|---|---|
| 2 | $Fe_2O_3 + 3C \longrightarrow 2Fe + 3CO$ |
| 3 | $8Fe + P_4 \longrightarrow 4Fe_2P$ |

EQUATIONS FOR THE REDUCTION OF PHOSPHATES

Figure 1

| MATERIAL | TONS PER TON OF PHOSPHORUS PRODUCED | REFERENCE |
|---|---|---|
| PHOSPHATE | 9.5 | BOOK SOLID WASTES, CHAPTER VII.2 "SOLID WASTES FROM PHOSPHORUS PRODUCTION," JOHN WILEY AND SONS, INC., 1975 [12] |
| COKE | 1.5 | DITTO |
| SILICA ROCK | 1.7 | DITTO |
| Total | 12.7 | |

FEED MATERIALS REQUIRED

Figure 2

| BYPRODUCT | TONS PER TON OF PHOSPHORUS PRODUCED | REFERENCE |
|---|---|---|
| SLAG | 8.2[a] | BOOK SOLID WASTES, CHAPTER VII.2, "SOLID WASTES FROM PHOSPHORUS PRODUCTION," JOHN WILEY AND SONS, INC., 1975 [12] |
| FERROPHOSPHORUS | 0.33 | CHEMICAL ENGINEERING REPORT NO. 3, TENNESSEE VALLEY AUTHORITY, 1952 [13] |
| FLUORIDES, AS F | 0.03 | CHEMICAL ENGINEERING PROGRESS 66(11):56-62, NOVEMBER 1970 [14] |
| CARBON MONOXIDE GAS | 2.80[b] | CALCULATED. |
| PRECIPITATOR DUST | 0.05 | CHEMICAL ENGINEERING REPORT NO. 3, TENNESSEE VALLEY AUTHORITY, 1952 [13] |
| TOTAL | 11.41 | |

[a] BYPRODUCT SLAG RATE IS LESS WITH HIGHER GRADE PHOSPHATE.

[b] DRY BASIS

BYPRODUCTS MADE

Figure 3

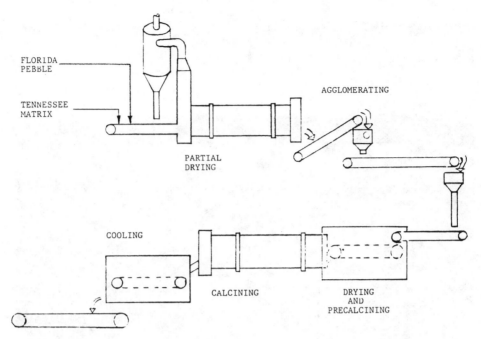

FLORIDA
PEBBLE

TENNESSEE
MATRIX

AGGLOMERATING

PARTIAL
DRYING

COOLING

CALCINING

DRYING
AND
PRECALCINING

GRATE-KILN ASSEMBLY

Figure 4

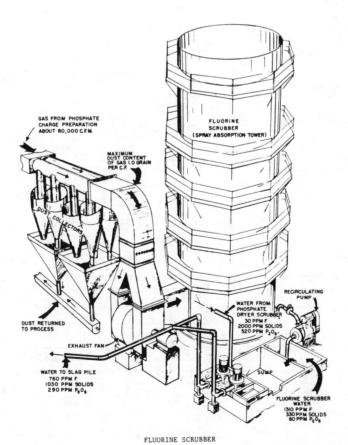

GAS FROM PHOSPHATE
CHARGE PREPARATION
ABOUT 80,000 C.F.M.

MAXIMUM
DUST CONTENT
OF GAS 1.0 GRAIN
PER C.F.

FLUORINE
SCRUBBER
(SPRAY ABSORPTION TOWER)

DUST COLLECTORS

RECIRCULATING
PUMP

WATER FROM
PHOSPHATE
DRYER SCRUBBER
30 PPM F
2000 PPM SOLIDS
520 PPM $P_2O_5$

DUST RETURNED
TO PROCESS

EXHAUST FAN

SUMP

WATER TO SLAG PILE
760 PPM F
1030 PPM SOLIDS
290 PPM $P_2O_5$

FLUORINE SCRUBBER
WATER
1310 PPM F
330 PPM SOLIDS
80 PPM $P_2O_5$

FLUORINE SCRUBBER

Figure 5

POND FOR TREATMENT OF WATER CONTAINING FLUORIDES

Figure 6

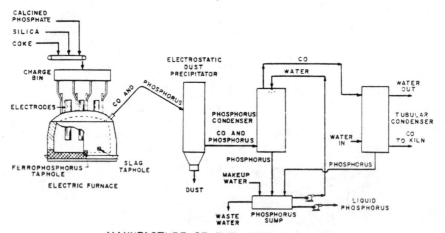

MANUFACTURE OF ELEMENTAL PHOSPHORUS

Figure 7

| Distance from furnace roof | C in sample, % needed to reduce $P_2O_5$ | $SiO_2$:CaO weight ratio |
|---|---|---|
| 4'-0" to 4'-8" | 69 | 0.70 |
| 4'-8" to 5'-3" | 36 | 0.70 |
| 5'-3" to 5'-9" | 19 | 0.77 |
| 5'-9" to 6'-3" | 114 | 0.87 |
| 6'-3" to 6'-9" | 162 | 0.88 |
| 6'-9" to 7'-2" | 139 | 0.80 |
| 7'-2" to 8'-3" | 304 | 0.94 |
| 8'-3" to 9'-8" | 1,377 | 1.44 |

ANALYSES OF SAMPLES REMOVED FROM PHOSPHORUS FURNACE

Figure 8

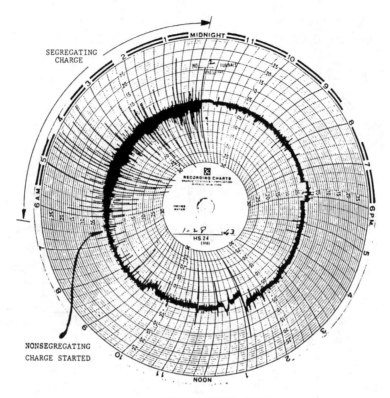

FURNACE PRESSURE CHART

Figure 9

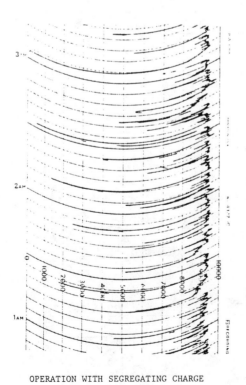

OPERATION WITH SEGREGATING CHARGE          OPERATION WITH NONSEGREGATING CHARGE

FURNACE POWER CHART

Figure 10

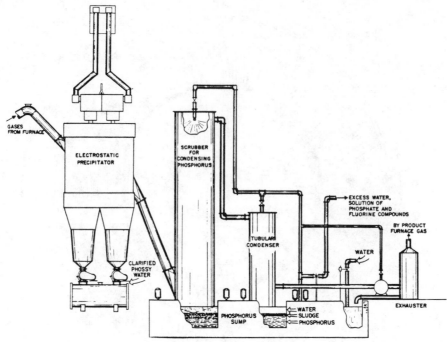

PHOSPHORUS FURNACE CONDENSER

Figure 11

|  | PPM OF ELEMENTAL PHOSPHORUS |
|---|---|
| WATER BLED FROM PHOSPHORUS CONDENSING SYSTEM | 1700 |
| OVERFLOW FROM CLARIFIER | 120 |
| SETTLING POND INLET | 23[a] |
| SETTLING POND DISCHARGE | 0.3-0.4 |

[a] WATER CONTAMINATED WITH ELEMENTAL PHOSPHORUS WAS MIXED WITH PLANT COOLING WATER AND PHOSPHORUS CONCENTRATION WAS DILUTED.

ELEMENTAL PHOSPHORUS CONTENT OF WATER

Figure 12

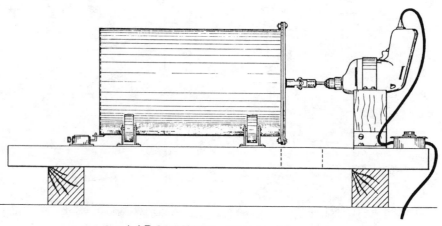

LABORATORY AGGLOMERATOR

Figure 13

| | |
|---|---|
| COKE FINES | 330 GRAMS[a] |
| PHOSPHATE DUST | 42 GRAMS |
| PHOSPHORIC ACID[b] | 63 GRAMS |
| WATER | 85 GRAMS |

[a] WET WEIGHT. COKE FINES CONTAINED 9.1% MOISTURE.

[b] MERCHANT-GRADE WET-PROCESS PHOSPHORIC ACID (53% $P_2O_5$).

FORMULATION FOR AGGLOMERATION OF COKE FINES

Figure 14

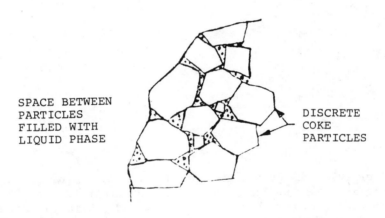

SPACE BETWEEN PARTICLES FILLED WITH LIQUID PHASE

DISCRETE COKE PARTICLES

SECTION OF GREEN AGGLOMERATE

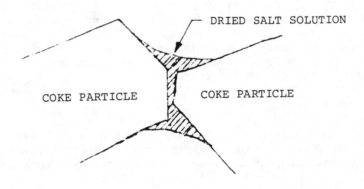

DRIED SALT SOLUTION

COKE PARTICLE      COKE PARTICLE

DRIED AGGLOMERATE

MECHANISM OF COKE AGGLOMERATION PROCESS

Figure 15

519

# HEALTH RISKS OF PHOTOVOLTAIC ENERGY TECHNOLOGIES

**Paul D. Moskowitz**
Brookhaven National Laboratory

## Introduction

The main alternative energy sources for electricity generation in the
United States are coal, nuclear, and solar.  Interest is now focused on
solar alternatives, in part because of widespread public concern about
the health and environmental risks of the competing energy technologies.
The health and environmental risks of solar technologies, including photovoltaic energy systems, are usually considered limited, but it is now
recognized that no energy technology is risk free.

The purpose of this paper is to present an overview of the life cycle
health costs of photovoltaic  energy production.  Health costs are identified by systematically examining all steps in the photovoltaic energy
cycle including material extraction, processing, refining, fabrication,
installation, operation, and disposal to estimate both the type and magnitude of the health risk involved.

## Analysis Methodology

Health effects analysis requires adoption of a comprehensive analytical
framework for preparation of risk estimates.  Overall risks of a technology are estimated through detailed examination of each step in the energy cycle.

Brookhaven's Reference Energy System exemplifies the type of conceptual
framework needed for health impact analysis.[1]  The framework, permitting
comparison of different energy-producing systems, is widely used in analyzing health effects of conventional technologies.  This framework must be
modified to adequately assess technologies such as photovoltaics in which

a substantial fraction of the health costs are not associated with the fuel cycle but rather with the mining of raw materials, and the construction, operation, and disposal of the devices used to convert sunlight into useful energy. Conventional technologies, of course, also have effects associated with such activities, but these have generally been ignored in favor of the large primary fuel cycle impacts.

Health impacts can be traced by using the Reference Material System shown in Figure 1 and Figure 2. Although the framework is simplified, it nevertheless provides a foundation for making impact estimates and identifying potential risk centers.[2] The following information is used to produce estimates of potential health impacts in this framework: (i) end-use material demands; (ii) efficiency coefficients for all processes, e.g., the ratio of material output to material input; (iii) standardized labor productivity estimates for all processes, e.g., the amount of labor required to mine one ton of copper ore; (iv) occupational health and safety coefficients by process, i.e., worker days lost (WDL)/100 man-years (MY); and (v) environmental emission coefficients by process. Specification of a network structure and quantities (i) to (v) suffice to generate some of the occupational health and safety risks and environmental residuals produced by this technology. Other risks, such as occupational and public exposure to toxic chemicals, must be evaluated in light of other variables including dispersion, exposure, and dose-response.

Reference Systems

Four alternative photovoltaic energy systems are examined; (i) silicon n/p single-crystal cells produced by an ingot-growing process; (ii) silicon metal/insulator/semi-conductor cells produced by a ribbon-growing process; (iii) cadmium sulfide back-wall cells produced by a spray deposition process; and (iv) gallium arsenide heterojunction cells produced by a modified ingot-growing process. These alternatives cover a reasonable range of manufacturing options (e.g., ingot vs. spray deposition) and materials (e.g., silicon vs. arsenic) which might be used in future commercialization efforts. The generic design for the reference technologies described in Figure 3 and Figure 4 is based upon a 25 - kW$_{pk}$ decentralized photovoltaic system designed and constructed by the MIT Lincoln Laboratory.[3] This system serves as a preliminary example of a photovoltaic plant, but with three important qualifications regarding its applicability to the design of future installations: (i) structural material requirements greatly overestimate materials likely to be included in future designs; (ii) fabrication technology is rapidly advancing towards thinner and more efficient solar cells, and (iii) the system does not adequately characterize centralized nor is it relevant to important roof-top decentralized applications. Because of the overestimated material requirements, calculated health risks may represent worst case estimates. Liberal estimates used for conversion efficiencies (10 to 14%), operating lifetimes (30 yr), and load factors (100%), however, may balance the overall costs.

Public Health

Public health effects stem principally from pollutants released during various stages of the photovoltaic energy cycle. This initial study explored only effects of arsenic, cadmium, and silicon.

Estimates of the general public's exposure to these pollutants from four major activities (mining and milling, refining, fabrication, and disposal) were calculated by using a Gaussian dispersion model which estimates

annual average pollutant concentrations at ground-level receptor points.[2]
Summaries of the estimated exposures are shown in Figure 5.

Combining these exposure estimates with toxicological information permits
both quantitative and qualitative estimation of some public health risks
of photovoltaic technologies. Although silicon technologies are nearest
to commercialization, prediction of public health risks from exposures to
silicon or silicon monoxide is not possible. The toxicology of silicon
dioxide (silica) is well documented, but it cannot be assumed to represent the toxic effects of elemental silicon and silicon monoxide.[4] In
contrast, information on health effects of arsenic and cadmium compounds
is extensive, and dose-response functions for kidney damage from cadmium[5,6] and lung cancer from arsenic can be constructed. [7,8]

Modeling results (Figure 6) indicate that direct inhalation of cadmium emitted to the atmosphere during the photovoltaic energy cycle will
contribute only a very small increment ( ∼ 2%) of the total kidney burden
from all sources. The kidney is the organ of critical concern. Maximum
kidney cortex concentrations from all sources are well below the concentrations (150 to 300 $\mu g/g$) at which renal damage is expected. Arsenic released during the photovoltaic energy cycle could induce a lung cancer
rate of $10^{-8}$ to $10^{-5}$ deaths/100,000 individuals per year within 50 miles
of photovoltaic-related facilities (Figure 7). If exposures beyond
50 miles are considered because of the non-threshold carcinogenic effect
assumed for arsenic, total U.S. deaths could range from $10^{-3}$ to $10^{-2}$
deaths per year per $10^{12}$ Btu installed capacity produced. The present
lung cancer rate from all causes is ∼43 deaths/100,000 individuals per
year.

Thus, potential public health effects from use of arsenic and cadmium in
the photovoltaic energy cycle appear to be small in comparison with other
known hazards. Public health risks from silicon and silicon monoxide exposure cannot be assessed for lack of toxicologic information at this
time. Exposure measurements and compilations of some basic toxicologic
information are required to improve the accuracy of the arsenic and cadmium damage estimates and to assess risks from silicon exposure.

Occupational Health

The chemical and physical hazards produced in the work-place are the
source of occupational health risks, e.g., effect of trauma to a hand or
carcinogenic hazard of exposure to arsenic. Occupational risks from
chemical hazards were assessed by using approaches similar to those described for public health. As noted, the toxicology of silicon dioxide
(silica) is well documented and clearly shows that hazards exist in dirty
situations like quartz mining. But silicosis, commonly suggested as a
potential hazard in the photovoltaics industry, is entirely preventable;
its prevalence in existing industries has declined in the past decade.

Exposures to silicon monoxide are expected from the different activities
examined, but no toxicologic information exist for this material. Cadmium health effects modeling suggests that chronic exposure in the workplace to levels above 10 $\mu g/m^3$ is likely to produce a cadmium kidney burden ( ∼200 $\mu g/g$) near or exceeding a concentration above which renal
damage can be expected (Figure 6).[2] Results of similar arsenic
modeling efforts are shown in Figure 7. The analysis suggests that
if workers are chronically exposed to inplant arsenic air levels of 10
$\mu g/m^3$, their cancer rates will be 10 to 50% of the background level for
all cancers.[2]

Most occupational mortality and morbidity effects for photovoltaics are probably similar to those encountered in the day-to-day operation of any industrial plant. By using the Reference Material System, risks from material extraction and processing (Figures 12.1.8 and 12.1.9) and from fabrication, installation, operation, and decommissioning (Figures 10 and Figure 12 of the four different photovoltaic devices were examined.[2,9] Among the alternatives, there are only small differences in total health costs. Material supply, installation, and operation appear to contribute substantial damage within the entire photovoltaics energy cycle, while the contribution of the fabrication facilities is small. The relative risk to individual workers at the fabrication facilities ($\sim$50 WDL/100 MY) is smaller than in competing technologies; for example, in coal mining the relative risk to workers is >100 WDL/100 MY. The labor intensiveness of the photovoltaic energy cycle, however, increases the absolute societal cost (risk per worker x number of workers) of producing electricity by photovoltaic energy systems.

Conclusion

In this analysis, we have attempted to develop a consistent framework to crudely estimate the public and occupational health costs of producing electricity via photovoltaic energy technologies. Admittedly, these costs provide only crude estimates of the total risk and ignore other important factors required in the decision making process, e.g., dollar costs. Nevertheless, the analysis places into perspective the potential risks of photovoltaic energy systems.

Acknowledgements

This work was supported by the Division of Photovoltaic Energy Systems, Office of Solar Applications for Buildings, Assistant Secretary for Conservation and Solar Energy, and by the Health and Environmental Risk Analysis Program, Office of Health and Environmental Research, Assistant Secretary for Environment, U.S. Department of Energy, under Contract No. DE-AC02-76CH00016. I thank L.D. Hamilton, H. Fischer, S.C. Morris, K.M. Novak, and M.D. Rowe for their comments and contributions. A. Link and A. Warren assisted in manuscript preparation.

References

1. S.C. Morris, K.N. Novak, and L.D. Hamilton, Databook for the Quantitation of Health Effects from Coal Energy Systems, Brookhaven National Laboratory, Upton, N.Y. 1979.

2. P.D. Moskowitz, L.D. Hamilton, S.C. Morris, K.M. Novak, and M.D. Rowe, Photovoltaic Energy Technologies: Health and Environmental Effects Document, Brookhaven National Laboratory, BNL 51284, Upton, N.Y. 1980. (In press)

3. R.L. Watts, W.E. Gurwell, C.H. Bloomster, S.A. Smith, T.A. Nelson, and W.W. Pawlewicz, Some Potential Material Supply Constraints in the Deployment of Photovoltaic Solar Electric Systems, Pacific Northwest Laboratory, PNL-2971, Richland, WA., 1978.

4. K.M. Novak, Toxicity of Silicon Compounds, Brookhaven National Laboratory, Upton, N.Y., In preparation.

5. K.M. Novak, Toxicity of Cadmium and Its Compounds: A Review of the Literature Aimed at Identifying Dose-Response Relationships, Brook-

haven National Laboratory, Upton, N.Y., In preparation.

6.  C.V. Robinson and K.M. Novak, Programs and Subroutines for Calculating Cadmium Body Burdens Based on a One-Compartment Model, Brookhaven National Laboratory, Upton, N.Y., In preparation.

7.  P.G. Green, Toxicity of Arsenic and Its Compounds, Brookhaven National Laboratory, Upton, N.Y., In preparation.

8.  P. Mushak, W. Galke, V. Hasselblad, and L. Grant, Health Assessment Document for Arsenic, Environmental Criteria and Assessment Office, U.S. Environmental Protection Agency, Research Triangle Park, N.C., 1980.

9.  T. Owens, L. Ungers, and T. Briggs, Estimates of Occupational Safety and Health Impacts Resulting from Large-Scale Production of Major Photovoltaic Technologies, Brookhaven National Laboratory, Upton, N.Y., In preparation.

| Raw Resource | Resource Extraction | Resource Preparation | Processing & Refining | Finishing | Final Product Identification |
|---|---|---|---|---|---|
| PRIMARY MATERIALS | | | | | |
| Copper Ore | Mining 2.70 | Crushing 2.70(0.95)=2.57 | Smelting & Refining 2.57(0.39)=1.00 | | Arsenic 1.00 |
| Zinc Ore | Mining 2.50 | Crushing 2.50(0.95)=2.38 | Smelting & Refining 2.38(0.42)=1.00 | | Cadmium 1.00 |
| Bauxite | Mining 2.50 | Crushing 2.50(0.95)=2.38 | Smelting & Refining 2.38(0.42)=1.00 | | Gallium 1.00 |
| Silica Ore | Mining 1.17 | Crushing 1.17(0.95)=1.11 | Smelting & Refining 1.11(0.90)=1.00 | | Silicon 1.00 |
| SECONDARY MATERIALS | | | | | |
| Iron Ore | Mining 1.67 | Beneficiation 1.67(0.95)=1.59 | Smelting & Steel Making 1.59(0.63)=1.00 | | Iron & Steel 1.00 |
| Bauxite | Mining 1.27 | Beneficiation 1.27(0.95)=1.20 | Smelting 1.20(0.83)=1.00 | Rolling & Finishing 1.0(1.00)=1.0 | Finished Aluminum 1.00 |
| Copper Ore | Mining 2.22 | Refining 2.22(0.95)=2.11 | Smelting 2.11(0.47)=1.00 | Rolling & Finishing 1.0(1.00)=1.0 | Finished Copper 1.00 |
| Clay, limestone Sand, etc. | Mining 1.0 | Crushing 1.0(1.00)=1.0 | Blending, Firing & Grinding 1.0(1.00)=1.00 | | Cement 1.00 / Concrete |
| Sand, limestone, & Soda ash | Mining 1.11 | Crushing 1.11(0.95)=1.06 | Glass Making 1.06(0.95)=1.00 | | Glass 1.00 |
| Natural Gas & Oil | Drilling 1.20 | | Processing 1.20(0.83)=1.00 | | Plastic (Acrylic) 1.00 |

[1]Numbers in (  )    defined as efficiency, i.e., material input x efficiency = material output

SUPPLY REFERENCE MATERIAL SYSTEM – MATERIAL FLOWS (TONS OF REFINED PRODUCT)[1]

FIGURE 1

| Cell Type/ Materials | Component Manuf. & Fabrication | Installation | Operation & Maintenance | Decommissioning and Disposal |
|---|---|---|---|---|
| Silicon-Ingot Silicon 97.4 | Collector & Module 97.14 (0.21) = 20.4 | | | |
| Metals 115 | Energy Conditioning 115 | | | |
| Cement 2666 | Cement Batching 2666 | 2666 | | 2666 |
| Cell Area (Acres) 9.9 | | | 9.9 | |
| Silicon-Ribbon Silicon 13.4 | Collector & Module 13.4 (0.53) = 7.11 | | | |
| Metals 162 | Energy Conditioning | 162 | | |
| Cement 3721 | Cement Batching 3721 | 3721 | | 3721 |
| Cell Area (Acres) 13.8 | | | 13.8 | |
| CdS/CaS Cadmium 0.72 | Collector & Module 0.72 (0.50) = 0.36 | | | |
| Metals 2604 | Energy Conditioning 2604 | | | |
| Cement 3721 | Cement Batching 3721 | 3721 | | 3721 |
| Cell Area (Acres) 13.8 | | | 13.8 | |
| GaAs-Ingot Gallium Arsenide 0.33 | Collector & Module 0.33 (0.33) = 0.11 | | | |
| Metals 2158 | Energy Conditioning 2158 | | | |
| Cement 480 | Cement Batching 480 | 480 | | 480 |
| Cell Area (Acres) 1.9 | | | 1.9 | |

[1]Lines indicate real activities, ...lines shown for continuity only.
[2]Numbers in ( ) defined as efficiency, i.e. material input x efficiency = material output.

END-USE REFERENCE MATERIAL SYSTEM – MATERIAL FLOWS (TONS PER $10^{12}$ BTU)

FIGURE 2

| Technical Characteristics | System | | | |
| --- | --- | --- | --- | --- |
| | Si-N/P | Si-MIS | CdS/CuS | GaAs |
| Nameplate Design (Peak kW) | 5575 | 5575 | 5575 | 5575 |
| Efficiency | 0.14 | 0.10 | 0.10 | 0.14 |
| Packing Factor | 0.80 | 0.80 | 0.80 | 0.80 |
| Cell Area (Acres) | 9.9 | 13.8 | 13.8 | 1.9 |
| Land Area (Acres) | 19.8 | 27.6 | 27.6 | 9.6 |
| Concentration Factor | 1X | 1X | 1X | 500X |
| Annual Insolation ($kWh/m^2$-yr) | 1752 | 1752 | 1752 | 1752 |
| Annual Electric Prod. (kWh)* | $9.76 \times 10^6$ | $9.76 \times 10^6$ | $9.76 \times 10^6$ | $9.76 \times 10^6$ |
| Lifetime Electric Prod. (kWh) | $2.93 \times 10^8$ | $2.93 \times 10^8$ | $2.93 \times 10^8$ | $2.93 \times 10^8$ |
| Lifetime Electric Prod. (Btu) | $1.00 \times 10^{12}$ | $1.00 \times 10^{12}$ | $1.00 \times 10^{12}$ | $1.00 \times 10^{12}$ |

*Assumes 30 year lifetime, and 100% load factor

PHOTOVOLTAIC SYSTEM CHARACTERISTICS

FIGURE 3

| Material | System | | | |
| --- | --- | --- | --- | --- |
| | Si-N/P | Si-MIS | CdS/CuS | GaAs |
| Primary | | | | |
| Arsenic | | | | 0.06 |
| Cadmium | | | 0.36 | |
| Gallium | | | | 0.05 |
| Silicon | 20 | 7 | | |
| Secondary | | | | |
| Acrylic | 3 | 5 | 5 | 168 |
| Aluminum | 1604 | 2316 | 2503 | 907 |
| Cement | 2666 | 3721 | 3721 | 480 |
| Copper | 120 | 160 | 161 | 199 |
| Glass | 0.2 | 0.2 | 599 | 0.2 |
| Iron and Steel | 1260 | 1416 | 1054 | 2155 |

PHOTOVOLTAIC BULK MATERIAL REQUIREMENTS (TONS PER $10^{12}$ BTU OUTPUT)

FIGURE 4

| Technology | Activity | Probable Pollutant | Total Pop. Exposure (person μg/m³)* | Pop. Weighted Ave Conc. (μg/m³) | Histogram of Exposures and Numbers of Persons (μg/m³) | | | | | |
|---|---|---|---|---|---|---|---|---|---|---|
| | | | | | $<10^{-6}$ | $10^{-6}$-$10^{-5}$ | $10^{-5}$-$10^{-4}$ | $10^{-4}$-$10^{-3}$ | $10^{-3}$-$10^{-2}$ | $<10^{-2}$ |
| Silicon N/P | Mine & Mill | $SiO_2$, SiO, $SiO_2$ | $1.4 \times 10^{2}$ | $1.7 \times 10^{-4}$ | – | $7.1 \times 10^{4}$ | $6.1 \times 10^{5}$ | $6.9 \times 10^{4}$ | $3.2 \times 10^{4}$ | – |
| | Refine | Si, SiO, $SiO_2$ | $8.8 \times 10^{0}$ | $1.1 \times 10^{-5}$ | – | $4.2 \times 10^{5}$ | $3.6 \times 10^{5}$ | $2.0 \times 10^{3}$ | – | – |
| | Fabricate | Si, SiO | $2.1 \times 10^{1}$ | $2.7 \times 10^{-5}$ | – | $2.5 \times 10^{5}$ | $5.0 \times 10^{5}$ | $3.6 \times 10^{4}$ | – | – |
| | Dispose | Si, SiO, $SiO_2$ | $7.5 \times 10^{1}$ | $9.5 \times 10^{-5}$ | $5.6 \times 10^{5}$ | $2.0 \times 10^{5}$ | $3.0 \times 10^{4}$ | $2.0 \times 10^{3}$ | – | – |
| Silicon MIS | Mine & Mill | $SiO_2$, SiO, $SiO_2$ | $9.8 \times 10^{2}$ | $1.2 \times 10^{-3}$ | – | – | $8.4 \times 10^{4}$ | $6.2 \times 10^{5}$ | $6.1 \times 10^{4}$ | $2.4 \times 10^{4}$ |
| | Refine | Si, SiO, $SiO_2$ | $6.5 \times 10^{1}$ | $8.2 \times 10^{-5}$ | – | – | $6.0 \times 10^{5}$ | $1.9 \times 10^{5}$ | – | – |
| | Fabricate | Si, SiO | $1.5 \times 10^{2}$ | $1.9 \times 10^{-4}$ | – | – | $4.0 \times 10^{5}$ | $3.6 \times 10^{5}$ | $2.4 \times 10^{4}$ | – |
| | Dispose | Si, SiO, $SiO_2$ | $2.7 \times 10^{0}$ | $3.4 \times 10^{-5}$ | $1.8 \times 10^{5}$ | $5.5 \times 10^{5}$ | $3.2 \times 10^{4}$ | $2.4 \times 10^{5}$ | – | – |
| Cadmium Sulfide | Mine & Mill | CdO | $3.2 \times 10^{1}$ | $4.1 \times 10^{-5}$ | – | $6.1 \times 10^{5}$ | $1.4 \times 10^{5}$ | $3.2 \times 10^{4}$ | – | – |
| | Refine | CdO, $CdCl_2$ | $2.0 \times 10^{1}$ | $2.5 \times 10^{-5}$ | – | – | $7.6 \times 10^{5}$ | $2.6 \times 10^{4}$ | – | – |
| | Fabricate | $CdCl_2$, CdS | $1.1 \times 10^{0}$ | $1.4 \times 10^{-6}$ | $5.2 \times 10^{5}$ | $2.4 \times 10^{5}$ | $2.0 \times 10^{4}$ | – | – | – |
| | Dispose | CdS, CdO | $1.3 \times 10^{-1}$ | $1.7 \times 10^{-7}$ | $7.5 \times 10^{5}$ | $3.2 \times 10^{4}$ | – | – | – | – |
| Gallium Arsenide | Mine & Mill | $As_2O_3$, $As_2O_5$ | $9.9 \times 10^{0}$ | $1.3 \times 10^{-5}$ | $8.4 \times 10^{4}$ | $6.2 \times 10^{5}$ | $5.9 \times 10^{4}$ | $2.6 \times 10^{4}$ | – | – |
| | Refine | $As_2O_3$, $As_2O_5$ | $8.1 \times 10^{0}$ | $1.0 \times 10^{-5}$ | – | $7.6 \times 10^{5}$ | $2.2 \times 10^{4}$ | – | – | – |
| | Fabricate | $As_2O_3$, $As_2O_5$ | $2.8 \times 10^{-1}$ | $3.5 \times 10^{-7}$ | $7.5 \times 10^{5}$ | $3.8 \times 10^{4}$ | – | – | – | – |
| | Dispose | $As_2O_3$, $As_2O_5$ | $6.3 \times 10^{0}$ | $8.0 \times 10^{-6}$ | – | $5.8 \times 10^{5}$ | $1.7 \times 10^{5}$ | $3.0 \times 10^{4}$ | $2.0 \times 10^{3}$ | – |

*Expressed as μg of the elemental substance (i.e., Si, Cd, As)/m³ and not the specific compound.

SUMMARY OF ESTIMATED EXPOSURES ASSOCIATED WITH PHOTOVOLTAIC ENERGY CYCLES

FIGURE 5

| | Public | | | | | | Occupational | | | | | |
|---|---|---|---|---|---|---|---|---|---|---|---|---|
| | Age of Peak Concentration | Source of Cadmium Burden at Peak Size | | | | | Age of Peak Concentration | Source of Cadmium Burden at Peak Size | | | | |
| Activity | | Food | Ambient* Air | Smoking | Occupational Air | Total | | Food | Ambient* Air | Smoking | Occupational Air | Total |
| Mine & Mill | 48 | 43.36 | 0.85 | 10.82 | - | 55 | 48 | 43.36 | 0.85 | 10.82 | 295.53 | 350 |
| Refinery | 47 | 43.36 | 0.04 | 10.82 | - | 54 | 48 | 43.36 | 0.04 | 10.82 | 591.05 | 650 |
| Fabrication | 47 | 43.36 | 0 | 10.82 | - | 54 | 48 | 43.36 | 0 | 10.82 | 147.76 | 200 |
| Disposal | 47 | 43.36 | 0.01 | 10.82 | - | 54 | 48 | 43.36 | -0.01 | 10.82 | 73.88 | 130 |

* Based upon ambient air exposure in the worst sector around each facility.

ESTIMATED CADMIUM CONCENTRATION IN THE RENAL CORTEX ($\mu$G/G) ASSOCIATED WITH THE PHOTOVOLTAIC ENERGY CYCLE

FIGURE 6

| | Public | | | | | | Occupational | | | | | |
|---|---|---|---|---|---|---|---|---|---|---|---|---|
| Activity | Total Pop. | Maximum Conc. ($\mu g/m^3$) | Total Pop. Exp. (Person $\mu g/m^3$) | Pop. Weighted Av. Conc. ($\mu g/m^3$) | Estimated Cancer Deaths per Year | Cancer Death Rate (Deaths/ 100,000 per Year) | Total Pop. | Maximum Conc. ($\mu g/m^3$) | Total Pop. Exp. (Person/ $\mu g/m^3$) | Pop. Weighted Av. Conc. ($\mu g/m^3$) | Estimated Cancer Deaths per Year | Cancer Death Rate (Deaths/ 100,000 per Year) |
| Mine & Mill | $7.8 \times 10^5$ | $4.9 \times 10^{-4}$ | $9.9 \times 10^0$ | $1.3 \times 10^{-5}$ | $1.4\text{-}7.2 \times 10^{-4}$ | $1.8\text{-}9.2 \times 10^{-5}$ | $8.4 \times 10^{-3}$ | $1.0 \times 10^1$ | $4.2 \times 10^0$ | $1.0 \times 10^1$ | $1.2\text{-}6.1 \times 10^{-6}$ | $1.4\text{-}7.3 \times 10^1$ |
| Refinery | $7.8 \times 10^5$ | $3.3 \times 10^{-5}$ | $2.8 \times 10^0$ | $3.6 \times 10^{-6}$ | $0.40\text{-}2.1 \times 10^{-4}$ | $0.51\text{-}2.7 \times 10^{-5}$ | $5.2 \times 10^{-3}$ | $5.0 \times 10^2$ | $2.6 \times 10^0$ | $5 \times 10^2$ | $0.37\text{-}1.9 \times 10^{-4}$ | $.71\text{-}3.7 \times 10^3$ |
| Fabrication | $7.8 \times 10^5$ | $6.0 \times 10^{-6}$ | $2.8 \times 10^{-1}$ | $3.5 \times 10^{-7}$ | $0.39\text{-}2.1 \times 10^{-5}$ | $0.50\text{-}2.6 \times 10^{-6}$ | $5.7 \times 10^1$ | $1.0 \times 10^1$ | $2.8 \times 10^4$ | $1.0 \times 10^1$ | $0.81\text{-}4.2 \times 10^{-2}$ | $1.4\text{-}7.3 \times 10^1$ |
| Disposal | $7.8 \times 10^5$ | $1.2 \times 10^{-7}$ | $6.3 \times 10^{-3}$ | $8.0 \times 10^{-9}$ | $0.90\text{-}4.6 \times 10^{-7}$ | $1.2\text{-}5.9 \times 10^{-8}$ | $5.2 \times 10^{-4}$ | $1.0 \times 10^1$ | $2.6 \times 10^{-1}$ | $1.0 \times 10^1$ | $0.74\text{-}3.8 \times 10^{-7}$ | $1.4\text{-}7/3 \times 10^1$ |

ESTIMATED HEALTH EFFECTS OF ARSENIC RELEASED DURING THE PHOTOVOLTAIC ENERGY CYCLE

(PER $10^{12}$ BTU OUTPUT)

FIGURE 7

| Product/Activity | SIC Code | Material Req. (Tons) | Labor Impacts Labor Prod. (MH/Ton) | Total Labor $(10^2 MY)$ | Health & Safety Incid. Rates Accidents $(WDL/10^2 MY)$ | Fatalities $(Deaths/10^2 MY)$ | Health & Safety Impacts Accidents (WDL/Ton Product) | Fatalities (Deaths/Ton Prod) |
|---|---|---|---|---|---|---|---|---|
| **Arsenic** | | | | | | | | |
| Extraction | 10 | 2.70 | 22.6 | $3.1 \times 10^{-4}$ | 235 | 0.06 | $7.2 \times 10^{-2}$ | $1.8 \times 10^{-5}$ |
| Preparation | 10 | 2.57 | 12.7 | $1.6 \times 10^{-4}$ | 183 | 0.01 | $3.0 \times 10^{-2}$ | $1.6 \times 10^{-6}$ |
| Refining | 3331 | 1.00 | 22.8 | $1.1 \times 10^{-4}$ | 119 | 0.01 | $1.4 \times 10^{-2}$ | $1.1 \times 10^{-6}$ |
| Total | | | | $5.8 \times 10^{-4}$ | | | $1.1 \times 10^{-1}$ | $2.1 \times 10^{-5}$ |
| **Cadmium** | | | | | | | | |
| Extraction | 10 | 2.50 | 1.40 | $1.8 \times 10^{-5}$ | 247 | 0.09 | $4.3 \times 10^{-3}$ | $1.6 \times 10^{-6}$ |
| Preparation | 10 | 2.38 | 12.7 | $1.5 \times 10^{-4}$ | 121 | 0.01 | $1.8 \times 10^{-2}$ | $1.5 \times 10^{-6}$ |
| Refining | 3339 | 1.00 | 27.4 | $1.4 \times 10^{-4}$ | 97 | 0.01 | $1.3 \times 10^{-2}$ | $1.4 \times 10^{-6}$ |
| Total | | | | $3.1 \times 10^{-4}$ | | | $3.5 \times 10^{-2}$ | $4.5 \times 10^{-6}$ |
| **Gallium** | | | | | | | | |
| Extraction | 10 | 2.50 | 0.35 | $4.4 \times 10^{-6}$ | 117 | 0.00 | $5.1 \times 10^{-4}$ | 0.0 |
| Preparation | 10 | 2.38 | 12.7 | $1.5 \times 10^{-4}$ | 1265 | 0.00 | $1.9 \times 10^{-1}$ | 0.0 |
| Refining | 3334 | 1.00 | 37.0 | $1.9 \times 10^{-4}$ | 99 | 0.02 | $1.8 \times 10^{-2}$ | $3.7 \times 10^{-6}$ |
| Total | | | | $3.4 \times 10^{-4}$ | | | $2.1 \times 10^{-1}$ | $3.7 \times 10^{-6}$ |
| **Silicon** | | | | | | | | |
| Extraction | 14 | 1.17 | 0.08 | $4.7 \times 10^{-7}$ | 104 | 0.17 | $4.9 \times 10^{-5}$ | $8 \times 10^{-8}$ |
| Preparation | 14 | 1.11 | 0.47 | $2.6 \times 10^{-6}$ | 94 | 0.06 | $2.5 \times 10^{-4}$ | $1.6 \times 10^{-7}$ |
| Refining | 321 | 1.00 | 13.3 | $6.7 \times 10^{-5}$ | 96 | 0.01 | $6.4 \times 10^{-3}$ | $4.7 \times 10^{-7}$ |
| Total | | | | $7.0 \times 10^{-5}$ | | | $6.7 \times 10^{-3}$ | $7.1 \times 10^{-7}$ |
| **Iron & Steel** | | | | | | | | |
| Extraction | 10 | 1.17 | 0.73 | $4.3 \times 10^{-6}$ | 117 | 0.04 | $5.0 \times 10^{-4}$ | $1.7 \times 10^{-7}$ |
| Preparation | 10 | 1.11 | 12.7 | $7.0 \times 10^{-5}$ | 140 | 0.02 | $9.9 \times 10^{-3}$ | $1.4 \times 10^{-6}$ |
| Refining | 3312 | 1.00 | 17.1 | $8.6 \times 10^{-5}$ | 86 | 0.01 | $7.3 \times 10^{-3}$ | $8.6 \times 10^{-7}$ |
| Total | | | | $1.6 \times 10^{-4}$ | | | $1.8 \times 10^{-2}$ | $2.4 \times 10^{-6}$ |
| **Aluminum** | | | | | | | | |
| Extraction | 10 | 1.27 | 0.35 | $2.2 \times 10^{-6}$ | 117 | 0.00 | $2.6 \times 10^{-4}$ | 0.0 |
| Preparation | 10 | 1.20 | 12.7 | $7.6 \times 10^{-5}$ | 1265 | 0.00 | $9.6 \times 10^{-2}$ | 0.0 |
| Refining | 3334 | 1.00 | 9.17 | $4.6 \times 10^{-5}$ | 99 | 0.02 | $4.6 \times 10^{-3}$ | $9.2 \times 10^{-7}$ |
| Finishing | 3355 | 1.00 | 22.3 | $1.1 \times 10^{-4}$ | 87 | 0.02 | $9.9 \times 10^{-3}$ | $2.2 \times 10^{-6}$ |
| Total | | | | $2.3 \times 10^{-4}$ | | | $1.1 \times 10^{-1}$ | $3.1 \times 10^{-6}$ |
| **Copper** | | | | | | | | |
| Extraction | 10 | 2.22 | 22.6 | $2.5 \times 10^{-4}$ | 235 | 0.06 | $5.9 \times 10^{-2}$ | $1.5 \times 10^{-5}$ |
| Preparation | 10 | 2.11 | 12.7 | $1.3 \times 10^{-4}$ | 183 | 0.01 | $2.5 \times 10^{-2}$ | $1.3 \times 10^{-6}$ |
| Refining | 3331 | 1.00 | 18.7 | $9.4 \times 10^{-5}$ | 119 | 0.01 | $1.1 \times 10^{-2}$ | $9.4 \times 10^{-7}$ |
| Finishing | 3351 | 1.00 | 28.3 | $1.4 \times 10^{-4}$ | 145 | 0.01 | $2.1 \times 10^{-2}$ | $1.4 \times 10^{-6}$ |
| Total | | | | $6.1 \times 10^{-4}$ | | | $1.2 \times 10^{-1}$ | $1.9 \times 10^{-5}$ |
| **Cement** | | | | | | | | |
| Extraction | 14 | 1.00 | 0.55 | $2.8 \times 10^{-6}$ | 145 | 0.07 | $4.1 \times 10^{-4}$ | $1.9 \times 10^{-7}$ |
| Preparation | 14 | 1.00 | – | – | 153  149 | 0.07  0 07 | - | - |
| Refining | 327 | 1.00 | 0.74 | $3.7 \times 10^{-6}$ | 135 | 0.02 | $5.0 \times 10^{-4}$ | $7.4 \times 10^{-8}$ |
| Total | | | | $6.5 \times 10^{-6}$ | | | $9.1 \times 10^{-4}$ | $2.6 \times 10^{-7}$ |
| **Glass** | | | | | | | | |
| Extraction | 14 | 1.11 | 0.08 | $4.4 \times 10^{-7}$ | 104 | 0.17 | $4.6 \times 10^{-5}$ | $7.5 \times 10^{-8}$ |
| Preparation | 14 | 1.06 | 0.47 | $2.5 \times 10^{-6}$ | 94 | 0.00 | $2.3 \times 10^{-4}$ | 0.0 |
| Refining | 321 | 1.00 | 12.7 | $6.4 \times 10^{-5}$ | 96 | 0.007 | $6.1 \times 10^{-3}$ | $4.4 \times 10^{-7}$ |
| Total | | | | $6.7 \times 10^{-5}$ | | | $6.4 \times 10^{-3}$ | $5.2 \times 10^{-7}$ |
| **Acrylic** | | | | | | | | |
| Extraction | 131 | 1.20 | 0.73 | $4.4 \times 10^{-6}$ | 43 | 0.04 | $1.9 \times 10^{-4}$ | $1.8 \times 10^{-7}$ |
| Preparation | | | | | - | - | - | - |
| Refining | 282 | 1.0 | 0.38 | $1.9 \times 10^{-6}$ | 36 | 0.005 | $6.8 \times 10^{-5}$ | $9.5 \times 10^{-9}$ |
| Total | | | | $6.3 \times 10^{-6}$ | | | $2.6 \times 10^{-4}$ | $1.9 \times 10^{-7}$ |

PRIMARY MATERIAL OCCUPATIONAL HEALTH AND SAFETY IMPACTS

FIGURE 8

| Cell Type/ Material | Material Req. (Tons/$10^{12}$ Btu) | Labor Requirements Labor Prod. (MH/Ton) | Labor (MH/$10^{12}$ Btu) | Health & Safety Incid. Rates Accidents (WDL/Ton) | Fatalities (Deaths/Ton) | Health & Safety Impacts Accidents (WDL/$10^{12}$ Btu) | Fatalities (Deaths/$10^{12}$ Btu) |
|---|---|---|---|---|---|---|---|
| **Silicon N/P** | | | | | | | |
| Silicon | 97.4 | $1.4 \times 10^1$ | $1.4 \times 10^3$ | $6.7 \times 10^{-3}$ | $7.1 \times 10^{-7}$ | $6.5 \times 10^{-1}$ | $6.9 \times 10^{-5}$ |
| Aluminum | 1604 | $4.6 \times 10^1$ | $7.4 \times 10^4$ | $1.1 \times 10^{-1}$ | $3.1 \times 10^{-6}$ | $1.8 \times 10^2$ | $5.0 \times 10^{-3}$ |
| Copper | 120 | $1.2 \times 10^2$ | $1.5 \times 10^4$ | $1.2 \times 10^{-1}$ | $1.9 \times 10^{-5}$ | $1.4 \times 10^1$ | $2.3 \times 10^{-3}$ |
| Cement | 2666 | $1.3 \times 10^0$ | $3.5 \times 10^3$ | $9.1 \times 10^{-4}$ | $2.6 \times 10^{-7}$ | $2.4 \times 10^0$ | $6.9 \times 10^{-4}$ |
| Glass | 0.2 | $1.3 \times 10^1$ | $2.7 \times 10^0$ | $6.4 \times 10^{-3}$ | $5.2 \times 10^{-7}$ | $1.3 \times 10^{-3}$ | $1.0 \times 10^{-7}$ |
| Iron & Steel | 1260 | $3.2 \times 10^1$ | $4.0 \times 10^4$ | $1.8 \times 10^{-3}$ | $2.4 \times 10^{-6}$ | $2.3 \times 10^1$ | $3.0 \times 10^{-3}$ |
| Acrylic | 3 | $1.3 \times 10^0$ | $3.9 \times 10^0$ | $2.6 \times 10^{-4}$ | $1.9 \times 10^{-7}$ | $7.8 \times 10^{-4}$ | $5.7 \times 10^{-7}$ |
| Total | | | $1.3 \times 10^5$ | | | $2.2 \times 10^2$ | $1.1 \times 10^{-2}$ |
| **Silicon MIS** | | | | | | | |
| Silicon | 13.4 | $1.4 \times 10^1$ | $1.9 \times 10^2$ | $6.7 \times 10^{-3}$ | $7.1 \times 10^{-7}$ | $9.0 \times 10^{-2}$ | $9.5 \times 10^{-6}$ |
| Aluminum | 2316 | $4.6 \times 10^1$ | $1.1 \times 10^5$ | $1.1 \times 10^{-1}$ | $3.1 \times 10^{-6}$ | $2.5 \times 10^2$ | $7.2 \times 10^{-3}$ |
| Copper | 160 | $1.2 \times 10^2$ | $2.0 \times 10^4$ | $1.2 \times 10^{-1}$ | $1.9 \times 10^{-5}$ | $1.9 \times 10^1$ | $3.0 \times 10^{-3}$ |
| Cement | 3721 | $1.3 \times 10^0$ | $4.8 \times 10^3$ | $9.1 \times 10^{-4}$ | $2.6 \times 10^{-7}$ | $3.4 \times 10^0$ | $9.7 \times 10^{-4}$ |
| Glass | 0.2 | $1.3 \times 10^1$ | $2.7 \times 10^0$ | $6.4 \times 10^{-3}$ | $5.2 \times 10^{-7}$ | $1.3 \times 10^{-3}$ | $1.0 \times 10^{-7}$ |
| Iron & Steel | 1416 | $3.2 \times 10^1$ | $4.5 \times 10^4$ | $1.8 \times 10^{-3}$ | $2.4 \times 10^{-6}$ | $2.6 \times 10^1$ | $3.4 \times 10^{-3}$ |
| Acrylic | 5 | $1.3 \times 10^0$ | $6.3 \times 10^0$ | $2.6 \times 10^{-4}$ | $1.9 \times 10^{-7}$ | $1.3 \times 10^{-3}$ | $9.5 \times 10^{-7}$ |
| Total | | | $1.8 \times 10^5$ | | | $3.0 \times 10^2$ | $1.5 \times 10^{-2}$ |
| **Cadmium Sulfide** | | | | | | | |
| Cadmium | 0.72 | $6.2 \times 10^0$ | $4.5 \times 10^0$ | $3.5 \times 10^{-2}$ | $4.5 \times 10^{-6}$ | $2.5 \times 10^{-2}$ | $3.2 \times 10^{-6}$ |
| Aluminum | 2503 | $4.6 \times 10^1$ | $1.2 \times 10^5$ | $1.1 \times 10^{-1}$ | $3.1 \times 10^{-6}$ | $2.8 \times 10^2$ | $7.8 \times 10^{-3}$ |
| Copper | 161 | $1.2 \times 10^2$ | $2.0 \times 10^4$ | $1.2 \times 10^{-1}$ | $1.9 \times 10^{-5}$ | $1.9 \times 10^1$ | $3.1 \times 10^{-3}$ |
| Cement | 3721 | $1.3 \times 10^0$ | $4.8 \times 10^3$ | $9.1 \times 10^{-4}$ | $2.6 \times 10^{-7}$ | $3.4 \times 10^0$ | $9.7 \times 10^{-4}$ |
| Glass | 599 | $1.3 \times 10^1$ | $8.0 \times 10^3$ | $6.4 \times 10^{-3}$ | $5.2 \times 10^{-7}$ | $3.8 \times 10^0$ | $3.1 \times 10^{-4}$ |
| Iron & Steel | 1054 | $3.2 \times 10^1$ | $3.4 \times 10^4$ | $1.8 \times 10^{-3}$ | $2.4 \times 10^{-6}$ | $1.9 \times 10^1$ | $2.5 \times 10^{-3}$ |
| Acrylic | 5 | $1.3 \times 10^0$ | $6.3 \times 10^0$ | $2.6 \times 10^{-4}$ | $1.9 \times 10^{-7}$ | $1.3 \times 10^{-3}$ | $9.5 \times 10^{-7}$ |
| Total | | | $1.9 \times 10^5$ | | | $3.3 \times 10^2$ | $1.5 \times 10^{-2}$ |
| **Gallium Arsenide** | | | | | | | |
| Gallium | 0.15 | $6.8 \times 10^1$ | $1.0 \times 10^1$ | $2.1 \times 10^{-1}$ | $3.7 \times 10^{-6}$ | $3.2 \times 10^{-2}$ | $5.6 \times 10^{-7}$ |
| Arsenic | 0.18 | $1.2 \times 10^{-2}$ | $2.1 \times 10^1$ | $1.1 \times 10^{-1}$ | $2.1 \times 10^{-5}$ | $2.0 \times 10^{-2}$ | $3.8 \times 10^{-6}$ |
| Aluminum | 907 | $4.6 \times 10^1$ | $4.2 \times 10^4$ | $1.1 \times 10^{-1}$ | $3.1 \times 10^{-6}$ | $1.0 \times 10^2$ | $2.8 \times 10^{-3}$ |
| Copper | 199 | $1.2 \times 10^2$ | $2.4 \times 10^4$ | $1.2 \times 10^{-1}$ | $1.9 \times 10^{-5}$ | $2.4 \times 10^1$ | $3.8 \times 10^{-3}$ |
| Cement | 480 | $1.3 \times 10^0$ | $6.2 \times 10^2$ | $9.1 \times 10^{-4}$ | $2.6 \times 10^{-7}$ | $4.4 \times 10^{-1}$ | $1.2 \times 10^{-4}$ |
| Glass | 0.2 | $1.3 \times 10^1$ | $2.7 \times 10^0$ | $6.4 \times 10^{-3}$ | $5.2 \times 10^{-7}$ | $1.3 \times 10^{-3}$ | $1.0 \times 10^{-7}$ |
| Iron & Steel | 2155 | $3.2 \times 10^1$ | $6.9 \times 10^4$ | $1.8 \times 10^{-3}$ | $2.4 \times 10^{-6}$ | $3.9 \times 10^1$ | $5.2 \times 10^{-3}$ |
| Acrylic | 168 | $1.3 \times 10^0$ | $2.1 \times 10^2$ | $2.6 \times 10^{-4}$ | $1.9 \times 10^{-7}$ | $4.4 \times 10^{-2}$ | $3.2 \times 10^{-5}$ |
| Total | | | $1.4 \times 10^5$ | | | $1.6 \times 10^2$ | $1.2 \times 10^{-2}$ |

MATERIAL SUPPLY OCCUPATIONAL HEALTH AND SAFETY IMPACTS (PER $10^{12}$ BTU)

FIGURE 9

| Cell Type/Activity | SIC Code | Material Req. (Tons) | Labor Requirements | | Health & Safety Incid. Rates | | Health & Safety Impacts | |
|---|---|---|---|---|---|---|---|---|
| | | | Labor Prod. (Mil/Ton) | Total Labor ($10^2$ MY) | Accidents (WDL/$10^2$ MY) | Fatalities (Deaths/$10^2$ MY) | Accidents (WDL/$10^{12}$ Btu) | Fatalities (Deaths/$10^{12}$ Btu) |
| **Silicon-N/P** | | | | | | | | |
| Collector | 3674 | N.A. | ---- | $3.6 \times 10^{-1}$ | 47.5 | 0.002 | $1.8 \times 10^1$ | $7.4 \times 10^{-4}$ |
| Conditioning | 36 | 175 | 280 | $2.45 \times 10^{-1}$ | 50.3 | 0.006 | $1.23 \times 10^1$ | $1.47 \times 10^{-3}$ |
| Cement Batching | 161 | 2666 | 0.09 | $1.20 \times 10^{-3}$ | 105.9 | 0.03 | $1.27 \times 10^{-1}$ | $3.60 \times 10^{-5}$ |
| Installation | 162 | 2666 | 200 | $2.29 \times 10^0$ | 113.2 | 0.03 | $2.60 \times 10^2$ | $6.88 \times 10^{-2}$ |
| Operation & Maintenance[1] | 49 | 40063 | 7.5 | $1.50 \times 10^0$ | 76.5 | 0.02 | $1.15 \times 10^2$ | $3.00 \times 10^{-2}$ |
| Decommissioning | 179 | 2666 | 17 | $2.27 \times 10^{-2}$ | 124.4 | 0.03 | $2.82 \times 10^0$ | $6.80 \times 10^{-4}$ |
| Total | | | | $4.42 \times 10^0$ | | | $3.97 \times 10^2$ | $1.02 \times 10^{-1}$ |
| **Silicon-MIS** | | | | | | | | |
| Collector | 3674 | N.A. | ---- | $2.7 \times 10^{-1}$ | 44.9 | 0.002 | $1.21 \times 10^1$ | $5.7 \times 10^{-4}$ |
| Conditioning | 36 | 175 | 280 | $2.45 \times 10^{-1}$ | 50.3 | 0.006 | $1.23 \times 10^1$ | $1.47 \times 10^{-3}$ |
| Cement Batching | 161 | 3721 | 0.09 | $1.67 \times 10^{-3}$ | 105.9 | 0.03 | $1.77 \times 10^{-1}$ | $5.02 \times 10^{-5}$ |
| Installation | 162 | 3721 | 172 | $3.20 \times 10^0$ | 113.2 | 0.03 | $3.62 \times 10^2$ | $9.60 \times 10^{-2}$ |
| Operation & Maintenance[1] | 49 | 55847 | 7.5 | $2.09 \times 10^0$ | 76.5 | 0.02 | $1.60 \times 10^2$ | $4.19 \times 10^{-2}$ |
| Decommissioning | 179 | 3721 | 1.7 | $3.16 \times 10^{-1}$ | 124.4 | 0.03 | $3.93 \times 10^0$ | $9.49 \times 10^{-4}$ |
| Total | | | | $6.12 \times 10^0$ | | | $5.50 \times 10^2$ | $1.93 \times 10^{-1}$ |
| **Cadmium Sulfide** | | | | | | | | |
| Collector | 3674 | N.A. | ---- | $1.72 \times 10^{-1}$ | 45.8 | 0.002 | $9.10 \times 10^0$ | $5.70 \times 10^{-4}$ |
| Conditioning | 36 | 175 | 280 | $2.45 \times 10^{-1}$ | 50.3 | 0.006 | $1.23 \times 10^1$ | $1.47 \times 10^{-3}$ |
| Cement Batching | 161 | 3721 | 0.09 | $1.67 \times 10^{-3}$ | 105.9 | 0.03 | $1.77 \times 10^{-1}$ | $5.02 \times 10^{-5}$ |
| Installation | 162 | 3721 | 172 | $3.20 \times 10^0$ | 113.2 | 0.03 | $3.62 \times 10^2$ | $9.60 \times 10^{-2}$ |
| Operation & Maintenance[1] | 49 | 55847 | 7.5 | $2.09 \times 10^0$ | 76.5 | 0.02 | $1.60 \times 10^2$ | $4.19 \times 10^{-2}$ |
| Decommissioning | 179 | 3721 | 17 | $3.16 \times 10^{-2}$ | 124.2 | 0.03 | $3.93 \times 10^0$ | $9.49 \times 10^{-4}$ |
| Total | | | | $5.74 \times 10^0$ | | | $5.48 \times 10^2$ | $1.40 \times 10^{-1}$ |
| **Gallium Arsenide** | | | | | | | | |
| Collector | 3674 | N.A. | ---- | $5.7 \times 10^{-1}$ | 49.2 | 0.002 | $2.94 \times 10^1$ | $1.6 \times 10^{-3}$ |
| Conditioning | 36 | 175 | 280 | $2.45 \times 10^{-1}$ | 50.3 | 0.006 | $1.23 \times 10^1$ | $1.47 \times 10^{-3}$ |
| Cement Batching | 161 | 480 | 0.09 | $2.16 \times 10^{-4}$ | 105.9 | 0.03 | $2.29 \times 10^{-2}$ | $6.48 \times 10^{-6}$ |
| Installation | 162 | 480 | 172 | $4.13 \times 10^{-1}$ | 113.2 | 0.03 | $4.67 \times 10^1$ | $1.24 \times 10^{-2}$ |
| Operation & Maintenance[1] | 49 | 7689 | 7.5 | $2.88 \times 10^{-1}$ | 76.5 | 0.02 | $2.20 \times 10^1$ | $5.76 \times 10^{-3}$ |
| Decommissioning | 179 | 480 | 1.7 | $4.08 \times 10^{-2}$ | 124.2 | 0.03 | $5.07 \times 10^{-1}$ | $1.22 \times 10^{-4}$ |
| Total | | | | $1.58 \times 10^0$ | | | $1.10 \times 10^2$ | $2.14 \times 10^{-2}$ |

[1]Operational maintenance measured in $m^2$.

FABRICATION, INSTALLATION, OPERATION, AND DECOMMISSIONING OCCUPATIONAL HEALTH AND SAFETY IMPACTS

FIGURE 10

| Cell Type/ Activity | Total Labor $(10^2$ MY) | Health and Safety Impacts | |
|---|---|---|---|
| | | Accidents (WDL) | Fatalities (Persons) |
| Silicon N/P | | | |
| Material Supply | $6.68 \times 10^{-1}$ | $2.17 \times 10^2$ | $1.1 \times 10^{-2}$ |
| Fabrication | $6.05 \times 10^{-1}$ | $3.03 \times 10^1$ | $2.21 \times 10^{-3}$ |
| Installation | $2.29 \times 10^0$ | $2.60 \times 10^2$ | $6.88 \times 10^{-2}$ |
| O & M | $1.50 \times 10^0$ | $1.15 \times 10^2$ | $3.00 \times 10^{-2}$ |
| Decommission | $2.27 \times 10^{-2}$ | $2.82 \times 10^0$ | $6.80 \times 10^{-4}$ |
| Total | $4.09 \times 10^0$ | $6.25 \times 10^2$ | $1.31 \times 10^{-1}$ |
| Silicon MIS | | | |
| Material Supply | $8.82 \times 10^{-1}$ | $3.03 \times 10^2$ | $1.5 \times 10^{-2}$ |
| Fabrication | $5.15 \times 10^{-1}$ | $2.44 \times 10^1$ | $2.04 \times 10^{-3}$ |
| Installation | $3.20 \times 10^0$ | $3.62 \times 10^2$ | $9.60 \times 10^{-2}$ |
| O & M | $2.09 \times 10^0$ | $1.60 \times 10^2$ | $4.19 \times 10^{-2}$ |
| Decommission | $3.16 \times 10^{-1}$ | $3.93 \times 10^0$ | $9.49 \times 10^{-4}$ |
| Total | $7.01 \times 10^0$ | $8.53 \times 10^2$ | $1.56 \times 10^{-1}$ |
| Cadmium Sulfide | | | |
| Material Supply | $9.07 \times 10^{-1}$ | $3.21 \times 10^1$ | $1.5 \times 10^{-2}$ |
| Fabrication | $4.17 \times 10^{-1}$ | $2.14 \times 10^1$ | $2.04 \times 10^{-3}$ |
| Installation | $3.20 \times 10^0$ | $3.62 \times 10^2$ | $9.60 \times 10^{-2}$ |
| O & M | $2.09 \times 10^0$ | $1.60 \times 10^2$ | $4.19 \times 10^{-2}$ |
| Decommission | $3.16 \times 10^{-2}$ | $3.93 \times 10^0$ | $9.49 \times 10^{-4}$ |
| Total | $6.64 \times 10^0$ | $5.79 \times 10^2$ | $1.55 \times 10^{-1}$ |
| Gallium Arsenide | | | |
| Material Supply | $6.79 \times 10^{-1}$ | $1.63 \times 10^1$ | $1.2 \times 10^{-2}$ |
| Fabrication | $8.35 \times 10^{-1}$ | $4.17 \times 10^1$ | $3.07 \times 10^{-3}$ |
| Installation | $4.13 \times 10^{-1}$ | $4.67 \times 10^1$ | $1.24 \times 10^{-2}$ |
| O & M | $2.88 \times 10^{-1}$ | $2.20 \times 10^1$ | $5.76 \times 10^{-3}$ |
| Decommission | $4.08 \times 10^{-2}$ | $5.07 \times 10^{-1}$ | $1.22 \times 10^{-4}$ |
| Total | $2.26 \times 10^0$ | $1.27 \times 10^2$ | $3.34 \times 10^{-2}$ |

SUMMARY OF THE OCCUPATIONAL HEALTH AND SAFETY IMPACTS
OF PHOTOVOLTAIC ENERGY TECHNOLOGIES
(PER $10^{12}$ BTU OUTPUT)

FIGURE 11

# ENVIRONMENTAL REGULATIONS:
## APPLICABILITY TO ADVANCED PHOTOVOLTAIC CONCEPTS

**D. A. Shaller**
Solar Energy Research Institute

Introduction

Relative to many competing forms of energy conversion technologies, photovoltaic systems are considered among the most environmentally benign. This is especially apparent during the projected 20-30 years of system operation. However, in a life cycle context, which includes manufacture, installation, operation and maintenance, and decommission, photovoltaic system development poses a number of potential environmental, health, and safety issues. Federal environmental regulatory programs will apply to waste streams generated in the development of photovoltaic systems and will ultimately determine the degree of environmental acceptability of the technology.

The Solar Energy Research Institute (SERI) has lead national laboratory responsibility for advanced research and development within the federal photovoltaics program. Advanced materials undergoing research at SERI are being evaluated in light of many criteria, including materials availability and environmental effects. Environmental research efforts in FY80 have focused on two advanced photovoltaic cell materials: copper sulfide/cadmium sulfide and polycrystalline silicon. Information and conclusions presented in this paper are adapted from SERI's 1980 work. A final report is in preparation and will be available in early 1981.

As these materials may be fabricated into photovoltaic cells in numerous ways, four representative processes were selected for analysis: the front wall $Cu_2S/CdS$ approach; a diffused p/n junction in a polycrystalline silicon wafer; spray deposition of a tin oxide junction on a polycrystalline silicon wafer; and the epitaxial deposition of high purity polycrystalline silicon onto a low cost substrate. In very simplified terms, the front wall $Cu_2S/CdS$ approach involves vacuum deposition of CdS onto an electroformed copper foil substrate. A $Cu_2S$ layer is formed and added through a wet chemical dip.

While four representative cell fabrication approaches were considered in the environmental review, only the $Cu_2S/CdS$ approach will be highlighted in this paper. SERI's forthcoming report on the FY80 research will provide a detailed look at all four processes. In this report, the environmental effects of manufacture are presented for each of the cell production options. In addition, the potential effects of installation, operation and maintenance, and decommission are evaluated. The report also discusses the potential for controlling and avoiding any adverse environmental effects identified. Finally, a major section of the analysis is dedicated to an overview of federal environmental, health, and safety programs applicable to each advanced material option. It is this last stage, the regulatory overview as applied to the $Cu_2S/CdS$ option, that is covered in this paper.

The $Cu_2S/CdS$ Process

There are two methods of fabricating copper sulfide/cadmium sulfide thin-film heterojunction cells: the front wall (or front surface) cell and the back wall (or back surface) cell. Development of the front wall $Cu_2S/CdS$ cell is supported by the U.S. Department of Energy (DOE) and the Solar Energy Research Institute through contracts with Westinghouse Corporation and the Institute of Energy Conversion at the University of Delaware. Both have proposed processes for commercial scale production (shown schematically in Figure 1).

Although this cell is constructed of a layer of $Cu_2S$ and CdS, the $Cu_2S$ layer is extremely thin ($0.2\mu m$). The CdS film ($5-30\mu m$) comprises the bulk of the semiconducting region and dominates the process considerations and operations of the cell. The heterojunction is formed from the CdS and $Cu_2S$ layers. The CdS, primarily because of the stoichiometric deficiency of sulfur, is an n-type semiconductor; the $Cu_2S$ is a p-type semiconductor due to a deficiency of copper.

Environmental Regulations and the $Cu_2S/CdS$ Process

Commercial scale production of the $Cu_2S/CdS$ photovoltaic material option will result in various by-product waste streams and associated health and safety risks at one or more stages of the cell production process. These waste streams may affect air, water and land environments depending on the handling, treatment and disposal methods employed. Federal environmental regulatory programs will apply to these waste streams much as they apply to a variety of wastes from other industrial activities.

SERI's FY80 research identified those regulatory programs which will likely apply at each of six discrete stages of the technology's development: 1) materials extraction; 2) materials processing; 3) cell fabrication; 4) installation; 5) decommissioning; and 6) materials, processing inputs and waste transportation (see Figure 2).

The analysis has been limited in part due to the strictly qualitative information available on the materials inputs and waste stream by-products of the $Cu_2S/CdS$ process. However, pollutant emission standards, limitations and exposure thresholds have been identified and can be used to help suggest whether there may be regulatory constraints at the commercialization stage of technology development.

As the cell fabrication stage is unique to the photovoltaic industry, its waste stream characterization and regulatory profile is of great interest to both SERI and the participating industries. Figure 3 presents a summary of the emissions, effluents, and solid wastes expected from front wall $Cu_2S/CdS$ cell fabrication.

## Air Quality

From an air quality standpoint, the fabrication of $Cu_2S/CdS$ cells generates few atmospheric contaminants. Hydrogen sulfide and hydrogen chloride are both emitted during the etching process and in the formation of the $Cu_2S$ barrier. Nitrogen gas is also emitted during the $Cu_2S$ barrier formation process step. Silicon monoxide or tantalum pentoxide particulate matter may be emitted if an antireflective (AR) coating is used. Interconnection of the cells is accomplished by soldering, which emits various acid fluxes, such as zinc chloride, ammonium and stannous chlorides, lead suboxides, and formaldehyde and fluorine fumes. [1] Hydrocarbons are produced during the degreasing done in the substrate preparation stage; however, these may be controlled with ventilation hoods, and activated carbon or other absorbing means. [2]

The particulates which are emitted in AR coating and the hydrocarbons produced by degreasing are both criteria pollutants for which National Ambient Air Quality Standards have been established. $Cu_2S/CdS$ cell manufacturing facilities are not now classified as major stationary sources by the Environmental Protection Agency and therefore do not automatically qualify for Prevention of Significant Deterioration review or Non-attainment rules. It does not appear that such facilities would emit greater than 250 tons per year of a regulated pollutant. Without the creation of a new stationary source category by EPA, cell manufacturing facilities would probably not be subject to either PSD or Non-attainment rules. Additionally, there are no New Source Performance Standards which now apply to such fabrication facilities, nor do any of the National Emission Standards for Hazardous Air Pollutants appear currently applicable to the $Cu_2S/CdS$ process.

## Occupational Health

Worker health and safety standards regulating air quality issued by the Occupational Safety and Health Administration are applicable to the $Cu_2S/CdS$ cell manufacturing industry. Standards for workplace exposure have been developed for many of the atmospheric contaminants generated in production of the cells. Selected time weighted average workplace air contaminant levels are listed in Figure 4.

OSHA also regulates several chemicals as carcinogens. Recently, regulations have been issued to further identify, classify and control potential occupational carcinogens. [3] As part of an initial screening of more than 2,000 substances. OSHA has listed 107 substances as candidates for further review as potential carcinogens. Only a few of these 107 are used as material inputs or created as waste by-products in either of the $Cu_2S/CdS$ or polycrystalline silicon options. However, a number of substances on an earlier OSHA list of potential carcinogens are prominent in the two cell processes. Among these are cadmium, cadmium oxide, cadmium sulfate, cadmium sulfide, sulfuric acid, nickel ammonium , chromium and chromium compounds. [4] Nevertheless, listing

535

does not represent a scientific determination of carcinogenicity.

## Water Quality

Water used during several of the process steps as a rinse solution can become contaminated with metal salts and other wastes. Specific liquid effluents may include: phosphorus detergent, zinc fluoroborate bath solution, which typically consists of zinc fluoroborate, ammonium chloride ammonium fluoroborate, and hydrogen chloride, and methanol from the substrate preparation steps; cadmium sulfide and hydrogen chloride from etching; sodium, copper, and cadmium chlorides generated during barrier formation; and gold, nickel and copper bath solutions used in the metalization step.

These liquid effluents may be discharged directly into receiving waters, indirectly via publicly owned treatment works, or they may be stored and treated as solid wastes. If they are directly discharged, the cell manufacturing facility may be considered a point source subject to the National Pollutant Discharge Elimination System permit program under the Clean Water Act. Under this program, manufacturing operations producing the $Cu_2S/CdS$ cells could be considered a primary industry source under the electronic component manufacturing category. Effluents generated will have to undergo pre-permit application testing for toxic, metals, conventional, non-conventional, and hazardous pollutants as defined by EPA.

No industry specific effluent limitation guidelines have been issued for $Cu_2S/CdS$ cell or module manufacturing operations. However, there are effluent guidelines which apply to electroplating and etching operations performed during the cell manufacturing process.[5] Plating operations may be used for substrate preparation and grid application during metallization. Clean Water Act regulations also control discharges resulting from the electroplating of common metals, such as zinc in the substrate preparation step or copper and nickel in the metallization step. Applicable standards depend upon the size of the regulated facility, the volume of discharge water, and the number of plant employees.

There are no New Source Performance Standards developed to date for the $Cu_2S/CdS$ cell manufacturing industry, nor for any of the specific process steps involved. Additionally, none of the toxic pollutant effluent standards issues to date by EPA appear applicable to the manufacture of these cells.
Some of the liquid effluents generated by the manufacture of the $Cu_2S/CdS$ cells are listed as hazardous substances under the CWA. Ammonium chloride and ammonium fluoroborate, which are generated during substrate preparation, and cadmium chloride resulting from barrier formation, are listed and their discharge is generally limited. Deep well disposal of any liquid effluent produced during the manufacture of $Cu_2S/CdS$ cells must be in compliance with applicable Underground Injection Control (UIC) programs under the Safe Drinking Water Act.

## Solid and Hazardous Wastes

The disposal of nonhazardous waste produced in $Cu_2S/CdS$ manufacture is regulated via state programs established under the Resource Conservation and Recovery Act. Some of the wastes generated in the cell production process qualifies for regulation under the Hazardous Waste Management (HWM) program under RCRA if it is to be disposed of on land or if stored for treatment. Wastes containing any of the toxic constituents listed by EPA are considered hazardous wastes. Wastewater treatment sludges, spent bath solutions, bath sludges, spent stripping and cleaning bath solutions from electroplating operations, and certain spent solvents used in degreasing operations are all listed as hazardous wastes by EPA. Their occurrence in the cell fabrication process would likely trigger application of the stringent handling, reporting and treatment provisions of the HWM program.

## Occupational Safety

In addition to the workplace air quality standards established by OSHA, that agency also issues safety and health regulations. These job safety rules address handling, labeling and warning requirements with respect to materials and prescribe protective gear for workers. Manufacturing operations, such as photovoltaic cell fabrication, would be covered by these additional OSHA stipulations.

## Regulatory Trends

Changes in the environmental regulatory framework may easily modify this analysis. It is not uncommon for variances and other exceptions to regulations to be instituted. Patterns of enforcement are also not well established for many of the programs, as legal challenges and revisions to the regulations occur frequently. These factors will all be important in the determination of specific regulatory applicability at the commercialization stage of each PV material option. Also, alternative manufacturing processes may be adopted at later stages of technology development. This could easily result in the addition or subtraction of regulatory programs from the sets identified.

## Findings and Conclusions

SERI's analysis identified the range of federal environmental, health, and safety regulatory programs which will likely be applicable to one or more stages of $Cu_2S/CdS$ and polycrystalline silicon photovoltaics development. No attempt was made, nor is it indeed possible, to indicate with certainty whether any stage of PV development would experience regulatory compliance difficulties. At best, the analysis indicates the potential applicability of environmental regulatory programs.

Based on the qualitative review of regulations which has been done, and on an understanding of waste stream control options, it is possible to identify the two or three most significant regulatory programs affecting the two advanced PV material options which should receive continuing and more detailed study. The regulatory programs which appear to be of most consequence are the effluent limits and permit procedures of the Clean Water Act, the workplace exposure regulations of the Occupational Safety and Health Act, and the hazardous waste disposal regulations under the Resource Conservation and Recovery Act.

It is important to note that there are no present effluent limitation guidelines which apply specifically to the industrial category of photovoltaics manufacture, whether by cadmium or silicon process. The ultimate size and nature of the emerging PV industry will be the force which triggers industry specific standards. At other stages in the production process for PV cells, the manufacturing, mining, and processing activities are sufficiently common as to already be addressed by regulations on an industry specific basis (i.e., electroplating, glass manufacture, inorganic chemical manufacturing). Figure 5 lists these industries and the regulatory programs which apply.

The processing, fabrication and installation of PV cells under each materials option will be subject to the general workplace exposure and safety standards authority of the Occupational Safety and Health Administration. Standards have been and will continue to be established for specific hazardous chemical substances which must be controlled in the work environment irrespective of the industrial category affected. A stringent cadmium exposure standard for the workplace will have implications for a variety of industries besides $Cu_2S/CdS$ cell production.

OSHA will be considering a variety of factors in individual rulemaking procedures on occupational carcinogen determinations. These include: a) the estimated number of exposed workers; b) the estimated levels of workers' exposure; c) the molecular similarity of the substance to a known carcinogen; and d) the availability of safer substitute substances.

Qualification of a substance under OSHA's rules of evidence for carcinogenicity will trigger regulation at "lowest feasible" levels. [6] This level of control will include the economic costs of compliance. Specific work practices to limit exposure will be determined on an individual chemical basis. If substitutes for the substance exist, the substance could be banned. Suspected carcinogens which do not meet OSHA's rules of evidence may still be regulated in the workplace at less stringent levels.

Under RCRA, there will be a tight regulatory program developed for the "cradle-to-grave" tracking and monitoring of hazardous wastes. To the extent that the PV materials options examined include the use or disposal of hazardous chemicals, there will be regulatory responsibilities to be maintained.

In summary, then, the extent to which $Cu_2S/CdS$ and polycrystalline silicon materials options will be affected by environmental, health, and safety regulations depends on three major determinants:

- the specific quantitative dimensions of waste streams, control options, and discharge sites to be employed;

- the modifications which may occur within waste-generating industrial processes at all stages of PV development and use; and

- the changes, either more or less restrictive, in quantitatively enforced environmental, health, and safety standards.

As progress is achieved in materials R&D and as forecasts may be made of the quantitative dimensions of the PV cell production process, more conclusive findings as to regulatory compliance and associated costs may be reached. The qualitative regulatory review comprising part of SERI's FY 80 research can help guide the direction of subsequent research into these important questions.

References

1. T.M. Briggs and T.W. Owens; Industrial Hygiene Characterization of the Photovoltaic Solar Cell Industry; Report No. N1OSH-80-112; PEDCO Environmental, Inc.; March 1980.

2. K. Lawrence, S. Morgan, D. Schaller and T. Wilczak; Life Cycle Environmental, Health and Safety Effects of Selected Advanced Photovoltaic Material Options; TR-743-799; Solar Energy Research Institute; (publication pending).

3. U.S. Department of Labor, Occupational Safety and Health Administration; "OSHA Names 107 Substances as Candidates for Further Scientific Review;" News; USDL: 80-502; August 12, 1980.

4. U.S. Environmental Protection Agency; A Handbook of Key Federal Regulations and Criteria for Multimedia Environmental Control; EPA-600/7-79-175; August 1979; pp. 128-129.

5. 40 CFR Part 413.

6. "OSHA Develops New Cancer Policy;" Science; Vol. 207; p. 742; (1980).

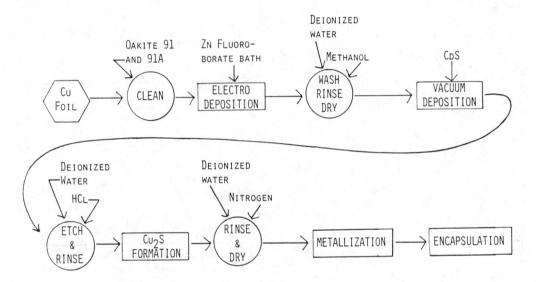

FABRICATION OF FRONT WALL $Cu_2S/CdS$
PHOTOVOLTAIC CELL

FIGURE 1

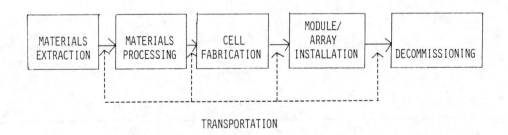

TRANSPORTATION

LIFE CYCLE STAGES OF PHOTOVOLTAICS DEVELOPMENT

FIGURE 2

| | | |
|---|---|---|
| $H_2S$(G) | LIQUID ABSORPTION (AMMONIA) (STERN '77) | S OR $H_2SO_4$ |
| $CdCl_2$ | ADD LIME (SITTIG '76); OR LIME AND $FeSO_4$ (PATTERSON '75) | $Cd(OH)_2$; $Fe(OH)_2$ |
| $HCl$(G) | LIQUID ABSORPTION (+AMMONIA?) (STERN '77) | $NaCl$ |
| $N_2$(G) | VENT | $N_2$ |
| 4A. ALTERNATE $Cu_2S$ FORMATION | | |
| $CdCl$ | ADD LIME (SITTIG '76) OR LIME AND $FeSO_4$ (PATTERSON '75) | $Cd(OH)_2$; $Fe(OH)_2$ |
| $CuCl$ | ADD LIME (SITTIG '76; PATTERSON '75) | $Cu(OH)_2$ |
| 5. GRID | | |
| $Au$ | ION EXCHANGE | |
| $Ni$ | ADD LIME (SITTIG '76; PATTERSON '75) | $Ni(OH)_3$ |
| $Cu$ PLATING BATH | ADD LIME (SITTIG '76; PATTERSON '75) | $Cu(OH)_2$ |

$Cu_2S$/$CdS$ FRONT WALL PROCESS WASTE STREAM AND TREATMENTS

FIGURE 3

| | | | ACCEPTABLE MAXIMUM PEAK ABOVE THE ACCEPTABLE CEILING CONCENTRATION FOR AN 8-HOUR SHIFT | |
|---|---|---|---|---|
| MATERIAL | 8-HOUR TIME WEIGHTED AVERAGE | ACCEPTABLE CEILING CONCENTRATION | CONCENTRATION | MAXIMUM DURATION |
| CADMIUM FUME (Z37,5-1970) | $0.1$ MG./$M^3$ | $0.3$ MG./$M^3$ | | |
| CADMIUM DUST (Z37,5-1970) | $0.2$ MG./$M^3$ | $0.6$ MG./$M^3$ | | |
| CHROMIC ACID AND CHROMATES | | $1$ MG./$10M^3$ | | |
| FORMALDEHYDE (Z37,16-1967) | $3$ P.P.M. | $5$ P.P.M. | $10$ P.P.M. | $30$ MINUTES |
| HYDROGEN FLUORIDE (A37,28-1969) | $3$ P.P.M. | | | |
| HYDROGEN SULFIDE (Z37,2-1966) | | $20$ P.P.M. | $50$ P.P.M. | $10$ MINUTES ONCE ONLY IF NO OTHER MEASURABLE EXPOSURE OCCURS |
| FLUORIDE AS DUST (Z37,28-1969) | $2.5$ MG./$M^3$ | | | |
| MERCURY (Z37,8-1971) | | $1$ MG./$10M^3$ | | |

SOURCE: OSHA

SELECTED TIME WEIGHTED AVERAGE WORKPLACE
AIR CONTAMINANT LEVELS

FIGURE 4

| Industry | Clean Air Act | | Clean Water Act | | | RCRA |
|---|---|---|---|---|---|---|
| | NESHAPS | NSPS | Effluent Limits | Pretreatment Standards | Toxic Standards | Hazardous Waste |
| Cement manufacturing | | X | X | | | |
| Electronic component production | | | | | X | X |
| Electroplating | | | X | X | | X |
| Ferroalloy manufacturing | | X | X | | | X |
| Glass manufacturing | | X | | | | |
| Ink formulating | | X | | | | X |
| Chlorine production | | | | | | X |
| Sulfuric acid plants | | X | | | | |
| Iron and steel manufacturing | | X | X | | | |
| Mineral mining and processing | | | X | | | |
| Nonferrous metals manufacturing | | X | X | X | | |
| Ore mining and dressing | X | | X | | | |
| Paint formulation | | | X | | | X |
| Plastics and synthetics manufacturing | | | X | | | |

Industries Related to PV Cell Production
Specifically Addressed by Federal
Environmental Regulations

Figure 5

| Process Step/Waste | Possible Treatment | Probable Effluents |
|---|---|---|
| 1. Substrate preparation chromate coating | Lower pH with $H_2SO_4$, ad agent ($SO_2$) reduce $Cr^{+3}$, add lime (Patterson, '75; Sittig '76) or ion exchange w/ recycle (Patterson '75) | $Cr(OH)_3$ |
| Oakite 91 and 91A (phosphorus detergent) | Biological, activated sludge, municipal or lime treatment and anaerobic digestion (Callely, '76) | Precipitate (insoluble) |
| 2. Zinc Plating | | |
| $ZnBF_4$ | Add $Ca^{++}$, raise pH (Yehaskel '79) | $CaF_2$, AnOH, $H_3BO_3$ |
| $NH_4Cl$ | Add NaOH (Powers, '76) | NaCl; $NH_3$ |
| $NH_4BF_4$ | Add $Ca^{++}$ (Yehaskel '79) | $CaF_2$, $H_3BO_3$ |
| Licorice root solution | Activated carbon | |
| HCl (to pH 3 to 4) | Raise pH with $Ca(OH)_2$ (Powers '76) | $CaCl_2$ |
| $CH_3OH$ (methanol) | Solvent extraction, fluidized bed, activated carbon (Ross '68, Powers '76) | |
| 3. CdS Deposition and Etch | | |
| HCl(g) | Liquid absorption and neutralization (ammonia) (Stern '77) | NaCl |
| $CdCl_2$ | Add lime (Sittig '76); lime and $FeSO_4$ (Patterson '75) | $Cd(OH)_2$ (insoluble); FE(OH) |
| $H_2S$(g) | Liquid absorption (ammonia) (Stern '77) | S or $H_2SO_4$ |
| 4. $Cu_2S$ Barrier Formation | | |
| NaCl | Ion exchange if concentration excessive | NaCl |
| CuCl | Add lime (Sittig '76; Patterson '75) | $Cu(OH)_2$ |

Figure 6

541

# SOLAR POWER SATELLITE ASSESSMENT

**F. A. Koomanoff**
U.S. Department of Energy

## Background

The SPS energy concept adopted for commonality for all DOE and NASA Concept Development and Evaluation Program (CDEP) evaluations featured a large photovoltaic array and a microwave generation and transmission system.  The entire satellite, Figure 1, would be built in space at its geostationary earth orbit (GEO) altitude with construction supported from a base in low earth orbit (LEO).  Heavy lift launch vehicles (HLLV) would be used to transport material from earth to LEO, and cargo and personnel orbit transfer vehicles (COTV and POTV, respectively) would be used for transportation between LEO and GEO.  A high degree of automated assembly, such as modular construction methods, was presumed during CDEP, but hundreds of space workers would nevertheless be needed to construct and maintain power satellites in space.

The photovoltaic SPS Reference System would use klystron power tubes to convert energy from direct current (dc) electricity to microwaves.  The microwave energy, at a presumed frequency of 2.45 GHz, would be transmitted through space from a slotted waveguide antenna on the satellite to a receiving array of rectifying dipole antennas (rectenna) on earth.  The rectified energy would then be processed conventionally and supplied either as dc or alternating current (ac) electricity to utility grids. It is important to recall that the system does not represent a preferred engineering design; its only function was a common basis for evaluation during CDEP.

## Systems Definition

The SPS-CDEP results show that energy can be collected in space either by photovoltaic processes (as in the Reference System) or thermal processes.

The energy collected by photovoltaic means could be converted to micro-wave energy by either power tubes of solid state electronics, or alter-natively to laser energy by any of several laser methods. Thermal energy also could be converted to either microwave or laser energy.

Microwave conversion was emphasized during the three-year CDEP because of the advanced state of this technology relative to laser technology.

One of the most important objectives of the CDEP process adopted for studying SPS was identifying key technological, environmental and societal issues associated with the concept. This objective was a critical factor in conducting systems definition studies. Unlike conventional approaches, systems definition was oriented toward identifying alternatives which could help resolve key issues rather than to develop a preferred engi-neering concept independent of the likely issues.

There obviously are numerous trade-offs that could be made between en-gineering preferences, system operating characteristics, costs, environ-mental issues and societal concerns to achieve a truly "preferred" sate-llite power system. Although such trade-offs were beyond the scope of CDEP, the process identified the necessary studies and their nature.

Regardless of the configuration which might be selected for a power sate-llite or its precise operating parameters, considerable more work is needed to define SPS transportation requirements. The information from CDEP is not unique to the SPS concept, nor would results of studies be applicable only to SPS. Both would be applicable to any contemplated venture involving large space structures useful for long periods of time.

Environmental Assessment

In recent years, and especially during the last decade, there has been great public sensitivity about environmental quality in the United States. In fact, since 1970 federal agencies have been required by law to consid-er the environmental implications of major actions (including decisions) which may significantly affect man's environment. These considerations must be included as early as possible in the formulation of potential major actions and decision-making.

Environmental sensitivity coupled with significant public concern about technology has made it essential to evaluate the environmental implica-tions of potential actions not only early, but in a straight-forward man-ner. Thus for SPS the CDEP process included a substantial environmental assessment effort with high public visibility.

Many of the potential environmental effects of SPS would be no different than one might reasonably expect from any action which would require sub-stantial use of natural and human resources, a large and diverse indus-trial and manufacturing capability, a significant terrestrial construc-tion effort (rectennas), and a widespread and diverse transportation net-work. That is, many of the environmental effects of SPS would be no dif-ferent from those which have become familiar to contemporary, developed societies. Moreover, most of those effects could be controlled in the same way for SPS as they are for contemporary activities. Those controls include advanced industrial, manufacturing and construction methods, pro-cesses conforming to environmental protection legislation and occupation-al and public health and safety legislation; safe and efficient use of transportation resources; and well-controlled waste disposal and so forth.

There are several potential environmental effects of SPS which are somewhat unique to the concept. With one exception, uniqueness is due in large part to their scale rather than their nature. In other words, these effects can be produced by some of society's current activities but they generally are limited in either geographic or population extent. They are also somewhat unique in terms of energy-producers. That is not to say that they are either less acceptable or more acceptable, however.

The results of CDEP suggest that the following potential environmental effects of SPS may be the most difficult to control, define quantitatively in terms of risk, and/or become especially significant sociopolitical issues:

o Microwave electromagnetic exposure effects on health and ecology

o Electromagnetic radiation effects on astronomy and other deep space research

o Electromagnetic radiation and space transportation exhaust effluent effects on the upper atmosphere

o Health and safety of space workers

Microwave electromagnetic exposure effects on health and ecology have received substantial scientific, political and public interest in many nations in recent years. An SPS equipped for microwave transmission from space to earth would naturally be expected to receive national and international scrutiny.

The CDEP process precluded a quantitative risk assessment of microwave exposure for many reasons. The CDEP did include some qualitative assessment of the potential effect based on the scientific literature and supported by limited research, however.

Power density profiles are used to describe eletromagnetic environment exposure of telecommunications and other electronic equipment. Potential electromagnetic interference problems with both terrestrial and space-based equipment of many types would be possible, but CDEP studies show that conventional mitigating strategies of proven effectiveness (including judicious rectenna siting) could be applied successfully to prevent most cases of direct-coupled interference. Potential problems may be formidable in terms of exposure extent, spectrum management and mitigation cost and management, but no new technological innovations seem apparent.

Interference effects on astronomy and other celestial studies (aeronomy, deep space research, etc.) remain an important concern. Light reflected from power satellites, heat energy and electromagnetic emissions represent undesirable effects on optical and radio astronomers and others. Some candidate ameliorative measures and mitigating strategies have been identified. Definitive work to demonstrate effectiveness and costs requires SPS design information not now available, however.

Transmitting microwave energy from space to earth in amounts suggested by the SPS concept raises important questions about possible electromagnetic interference due to effects on radio propagation. Considerable theoretical studies of radio propagation effects were made during CDEP, and those studies were coordinated with experiments involving operating telecommunications links in the U.S. and the Carribean area. The SPS transmissions were simulated at facilities maintained in the State of Colorado and in Puerto Rico. Sufficient work was completed to indicate that microwave power densities associated with the Reference System probably would not affect radio wave propagation dependent upon ionospheric reflections.

Space vehicle exhaust emissions conceivably could alter the atmosphere in ways which would affect radio propagation. In addition, these emissions might have effects on space travel, weather and climate. There is insufficient operational experience or experimental data for the upper atmosphere to describe these potential effects definitively.

The most unique feature of the SPS concept compared to other energy-producing ventures is the need for protecting the health and ensuring the safety of workers in space. Building and operating power satellites could require hundreds of workers regardless of the degree of automation which might be achieved.

Available information about man's ability to work efficiently and safely in space without suffering debilitating health aftereffects is limited. Few people have experienced life in space, and only relatively simple tasks have been performed. Generally, the current information must be extended to apply to large groups of individuals. A range of biological age and a spectrum of physiological health and psychological make-up must be correlated with current knowledge about space environment effects on health, performance and safety before a venture like SPS can be considered.

Societal Concerns

As noted earlier in this paper, technology and the use of natural and human resources and the possible environmental and societal consequences of doing so receive substantial public scrutiny in the United States today. The same is true in many other nations. The CDEP process was adopted to ensure that the SPS concept would not be developed without a sensitivity for those concerns. Many possible concerns were identified during CDEP, and their likely intensity covers a broad spectrum from purely local to national to international. Two examples serve to illustrate what was learned.

The SPS concept as depicted by the Reference Systems would result in a need for contiguous land areas of a size that may not be acceptable to the public.

Rectenna land requirements for the Reference System are large, and it appears that existing land uses, undesirable terrain features, etc. would preempt about 60% of the land of the contiguous states as potential rectenna sites.

Multiple uses of the land needed for rectennas could be helpful, and the use of offshore areas also might be possible. The latter is especially interesting because the vast majority of the U.S. population is concentrated along the nation's coastlines. The impact of SPS on flyways used by migratory birds must be determined.

The international implications of SPS are best exemplified by the issue of geostationary orbit allocations. No nation could responsibly consider a venture like SPS without obtaining international agreement on this question. Recognizing that GEO is a limited resource, it also is recognized that a power satellite must be designed to occupy minimum space by virtue of occupying minimum spectrum (maximum electromagnetic compatibility). Doing so would enhance the possibility of achieving international agreement on geostationary earth orbit allocations.

Conclusions

The three-year Concept Development and Evaluation Program for SPS conducted jointly by the U.S. Department of Energy and the National Aeronautics and Space Administraion indicates this energy concept, in any of several possible configurations, could provide continuous baseload electricity. It could do so while using an inexhaustible fuel. The concept is international in scope. The basic underlying principles of SPS are different from those of most other new energy technologies, and the environmental effects and the societal concerns associated with this energy concept also are different.

References

1.  -; Satellite Power System (SPS) Concept Development and Evaluation Program Plan, July 1977 - August 1980;U.S. Department of Energy, DOE/ET-0034; February 1978

2.  Glaser, P.E.; Method and Apparatus for Converting Solar Radiation to Electrical Power; U.S. Patent 3781647; December 1973

3.  Koomanoff, F.A. and Sandahl, C.A.; Status of the Satellite Power System Concept Development and Evaluation Program; 30th Congress of the International Astronautical Federation, Munich; September 1979

4.  -; Satellite Power System Concept Development and Evaluation Program Reference System Report; U.S. Department of Energy and National Aeronautics and Space Administration, DOE/ER-0023; October 1978

5.  Brown, W.C.; Power Amplifiers (Tube); The Final Proceedings of the Solar Power Satellite Program Review; U.S. Department of Energy, CONF-800491 pp. 336-339; July 1980

6.  -; National Environmental Policy Act - Implementation of Procedural Provisions; Council on Environmental Quality; 40 CFR 1500-1508 (43 FR 55978); November 1979

7.  Krebs, J.S.; Heynick, L.N.; and Polson, P.; Environmental Assessment for the Satellite Power System Concept Development and Evaluation Program Microwave Health and Ecological Effects; U.S. Department of Energy (In Press)

8.  Davis, K.; Grant, W.B.; Morrison, E.L.; and Juroshek, J.R.; Environmental Assessment for the Satellite Power System Concept Development and Evaluation Program - Electromagnetic Systems Compatibility; U.S. Department of Energy (In Press)

9.  Rush, C.; Environmental Assessment for the Satellite Power System Concept Development and Evaluation Program - Effects of Ionospheric Heating on Telecommunication; U.S. Department of Energy (In Press)

10. Rote, D.; Environmental Assessment for the Satellite Power System Concept Development and Evaluation Program - Atmospheric Effects; U.S. Department of Energy (In Press)

11. White, M.; Environmental Assessment for the Satellite Power System
    Concept Development and Evaluation Program - Nonmicrowave Health and
    Ecological Effects; U.S. Department of Energy (In Press)

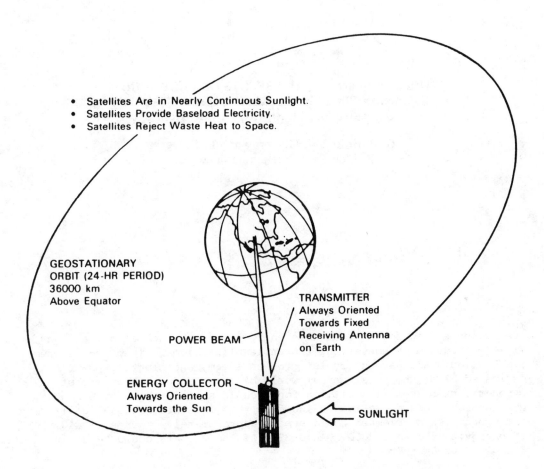

- Satellites Are in Nearly Continuous Sunlight.
- Satellites Provide Baseload Electricity.
- Satellites Reject Waste Heat to Space.

GEOSTATIONARY
ORBIT (24-HR PERIOD)
36000 km
Above Equator

TRANSMITTER
Always Oriented
Towards Fixed
Receiving Antenna
on Earth

POWER BEAM

ENERGY COLLECTOR
Always Oriented
Towards the Sun

SUNLIGHT

FIGURE 1   SATELLITE POWER SYSTEM CONCEPT

# THE ENVIRONMENTAL EFFECTS OF PRODUCING
# BIOMASS IN PINE AND HARDWOOD PLANTATIONS
# FOR RENEWABLE ENERGY RESOURCES

**G. J. Hollod, J. C. Corey** and **E. F. Dyer**
E. I. du Pont de Nemours and Co.

## Introduction

The nation's need for nonrenewable energy sources, such as oil and gas, might be reduced by converting renewable plant biomass into useful chemicals, liquid and gaseous fuels, process steam, and electric power. An estimated 5% of the total annual world biomass production from photosynthetic processes (146 billion tons) could satisfy the entire annual U.S. energy needs.[1] Estimates indicate that feasible national goals from the production of biomass could supply approximately 3 quads (3 $\times 10^{15}$ BTU) of energy by the year 2000. About 85% of the biomass could be derived from wood products.[2]

The annual above-ground growth of biomass in commercial forests in the U.S. in 1978 was calculated at 580 million dry tons.[3] The useful sources of wood biomass include wood and bark residues, logging residues, noncommercial or surplus stands of timber, and thinning from commercial timber stands. If all these sources of wood biomass were available, they could provide 3 to 6 quads (3-6 x $10^{15}$ BTU) per year of renewable energy resources. The ultimate value of wood materials as energy resources depends on the content of the wood product, the use of the wood with a minimum of objectionable emissions, and the net energy expended in preparing it for use. Some of the various programs for utilizing the renewable energy resources of wood products have previously been reviewed.[4,6]

If the U.S. is going to exploit the forests for energy and organic resources, then the subsequent effects on the forest ecosystem must be considered. The long-term objective of the biomass fuels research at the

548

Savannah River Plant (SRP) is to evaluate the environmental effects of enhancing tree production with sewage sludge.  These effects include the impact on the nutrient and mineral cycle, the hydrologic cycle, soil quality, and the disease rate in forest ecosystems.  Therefore, the Savannah River Laboratory is studying the land application of sewage sludge for its nutrient and organic matter content to increase soil fertility in pine and hardwood plantations.

Biomass production in pine and hardwood plantations could be improved by land application of sewage sludge because of its high nutrient content. The fertilizer capacity of sewage sludge could increase forest productivity and save approximately 660 million  dollars per year in fertilizer cost.  The energy-related expenses of fertilizer production account for 60-70% of this savings.[7]  The addition of nutrients and organic matters from the sludge to the soil will decrease the impact of continual tree removal on the nutrient and carbon budget in soils.[8]

Experimental Procedures

The experimental plots are on the Department of Energy's 300-square-mile Savannah River Plant (SRP) site near Aiken, South Carolina (Figure 1). Because of forestry management activities at SRP, 100,000 acres of pine plantations of different ages are available for evaluating biomass production.  The wide diversity of soil types (20) on the site is useful for studying the retention and movement of sludge-derived nutrients, metals, and organics through soil profiles representative of the southeastern U.S.

The experimental plots are located in loblolly pine stands and in a hardwood plantation. The sludge will be applied to provide nitrogen at an equivalent rate of 0 lb/acre, 300 lb/acre, and 600 lb/acre.

The loblolly pine plantations have planting dates of 1953, 1972, 1978, and 1981 and are located on light and medium textured soils.  Each pine study area has 9 plots on a randomized block design with three replications of three treatments.  The plot dimensions  are 150 square feet (1/2 acre) with an interior measurement plot for intensive environmental and biological sampling.

Prior to sludge application, the stand conditions were characterized by measurements of tree height, diameter at breast height, basal area, stem volume, and incidence of disease.  Composite vegetation, litter layers, and subsoil samples were collected to determine initial conditions.  The soil samples were taken at depths of 0-3 in., 3-6 in., 12-15 in., and 24-30 in. from 10 random points in each plot and combined to form a composite average plot sample from each depth.  The soil samples were air dried and crushed to pass a 10 mesh screen.  Texture was determined by the hydrometer method, organic matter content by wet oxidation, and exchangeable bases by exchange with normal ammonium acetate at pH 7 followed by analyses using atomic absorption.  The vegetation, litter, and composite soil samples will be analyzed for cadmium, chromium, zinc, lead, copper, mercury, manganese,  and nickel.

Fritted glass tubular connectors are being used to collect soil solution samples.[9]  Two pairs of these tubes have been placed on each 1/2 acre plot at depths of 20 in. and 40 in.  Solution samples are collected seasonally and after some rain events.  Three deep-water wells (60-120 ft) have been installed around the perimeter of each study area for monitoring any groundwater response to sludge amendments.

The hardwood plantation is located on 16 acres, which was planted with five species of hardwood trees on 4 ft x 8 ft centers in February 1980. The species were red maple (Acer rubrum), sweet gum (Liquid-amber styraci flua L.), American sycamore (Plantanus occidentalis L.), black locust (Robinia pseudoacacia), and black alder (Alnus glutisosa). Seedlings were planted in a random block design to evaluate biomass production by coppice growth of trees treated with and without sewage sludge. Heavy metals, nutrients, and toxic organics in vegetation, the soil profile, and water samples will be monitored following application of sewage sludge. Treatment effects on nutrient uptake by coppice growth will be evaluated by sampling various plant parts.

The sewage sludge is supplied by two area waste water treatment plants. The Horse Creek Pollution Control Facility (HCPCF), located in North Augusta, SC, has a capacity of 20 million gallons per day and a sludge production of 40 wet tons/week. The sludge is aerobically digested, heat-treated, and dewatered (30% solids). The second source of sewage sludge is the Augusta Municipal Waste Water Treatment Plant (AMWWTP) located in Augusta, GA. This facility produces 50,000 gallons per day of anaerobically digested sewage sludge (3.5% solids).

The HCPCF sludge is applied with conventional farm equipment including an all-purpose farm tractor, a front-end loader, a flail spreader with hydraulic lid opener for sludge delivery from either side, and a manure spreader. The flail will be used in the thinned pine plantations, and the rear delivery spreader in the younger pine stands and in the hardwood plantations. The AMWWTP sludge is subsurface injected using a commercially available Terra-gator (Registered Trademark of Ag-Chem).

Discussion

The principal objective of this program is to evaluate the environmental effects of using sewage sludge as a fertilizer and soil conditioner to increase biomass production in pine and hardwood plantations. A cost-benefit analysis will be made by comparing the benefits from increases in wood fiber production with the costs associated with sludge handling.

Environmental Effects

The impacts to the system may be categorically identified by the specific silvicultural manipulations that are implemented in the production of the wood products. In this study, the land application of sewage sludge to increase tree productivity will primarily affect the hydrologic and nutrient cycle, distribution of heavy metals, and transport of organics in the tree plantations. The emphasis of the program is to predict the environmental effects of using sewage sludge by developing an ecological model which can integrate submodels developed for each of the above mentioned parameters.

One of the primary impacts of harvesting wood in managed tree plantations is the removal of nutrients which are essential to the productivity of the site. If some type of fertilization program is not used to balance the nutrient budget, then continued harvesting will eventually result in a significant loss in the productivity of the site. Utilization of the nutrients and organic matter in sewage sludge will increase soil fertility and aid in maintaining site productivity.

A preliminary nutrient budget was calculated for harvest of tree stems and application of sewage sludge on two 27-year-old loblolly pine experimental plots on the SRP site. Removal of all loblolly pine stems per acre (180 trees) would result in a depletion of 122 lb N, 18 lb P, and 91 lb K per acre (Tables 1 and 2). Application of AMWWTP sewage sludge at a rate of 1 dry ton per acre adds 110 lb N, 11 lb P, and 4 lb K per acre (Table 2). The nutrient balance on the experimental plots for complete stem removal and application of sewage sludge (5.45 dry tons/acre) would result in a net gain of 438 lb N and 42 lb P and a net loss of 69 lb K per acre.

TABLE 1

Average Nutrient Compostion of 27-Year-Old Loblolly Pine Stems and Sewage Sludge*

|  | Nutrient, dry wt % | | |
| --- | --- | --- | --- |
|  | N | P | K |
| Loblolly Pine ** | 0.18 | 0.03 | 0.13 |
| Augusta, GA, Sewage Sludge *** | 5.50 | 0.55 | 0.22 |

* Stems include bolewood and bolebark.

** Reference 18.

*** Reference 19.

Nutrient cycling data collected from the experimental plots will be used to determine the availability and uptake of N/P/K for the trees at different application rates in order to maximize the fertilization capacity of the sewage sludge. Micronutrient and heavy metal mass balances will also be constructed for tree stem removal and sewage sludge application. These parameters are dependent on the soil characteristics and will vary from site to site.

TABLE 2

Nutrient Budget for Removal of Loblolly Pine Stems and Application of Sewage Sludge*

|  | Nutrients, lb/acre | | |
| --- | --- | --- | --- |
|  | N | P | K |
| Nutrients Removed by Tree Harvest ** | 122 | 18 | 91 |
| Nutrients Added from Application of Sewage Sludge (dry wt) | 110 | 11 | 4 |

* Calculations are based on removal of 180 27-year-old loblolly pine stems (bark + bole) per acre and application of 1 ton/acre of Augusta, GA, sewage sludge.

** Nutrient removal rates are based on values in Table 1 and the following calculations for SRP experimental site loblolly pine trees.

a. Ovendry density of loblolly pine:

$$\rho = (1 + m/100)(Sp.Gr.)(\rho~H_2O) \div (1 - S/100)$$

where m is the moisture content of the wood, $\rho$ is density, and S is the percent shrinkage by volume of wood due to removal of water (shrinkage values taken from Reference 20).

$$\rho = (1+0.10)(0.47)(62.43~lb/ft^3) \div (1 - 0.07)$$
$$\rho = 34.7~lb/ft^3$$

b. Volume and weight of loblolly pine stems (60 ft height and 9 in. diameter at base height) allowing 7% reduction in volume after drying to 10% moisture content (tree volumes taken from Reference 21).

$$\rho = (11.6~ft^3/tree) \times (0.93) \times (34.7~lb/ft^3) \times (180~trees/acre) \times (1~ton/2000~lb)$$

$$\rho = 34~tons/acre$$

The presence of heavy metals in the sewage sludge may be the ultimate constraint for use of sludge in biomass production. Heavy metals may have a beneficial effect on trees as micronutrients,[10] but when certain concentrations in the environment are exceeded, or when very large amounts are taken up, toxic effects have been observed in trees.[11]

The three physical variables that control the movement and availability of heavy metals and nutrients to trees are soil type, stand age, and secondary successional cover crops. The analysis of the soil solution samples collected from lysimeters located in different soil types and age stands will provide information about metal and nutrient availability to trees. Analysis of plant parts will determine the accumulation rate of heavy metals and importance of levels of sludge treatments.

There are two general groups of organic chemicals present in sewage sludge that will be studied. The first group of organics, which includes humic and fulvic acids, proteins, carbohydrates, and lipid material, constitutes a major fraction of the organics in sewage sludge. This group is important because it influences, via complexation and adsorption, chemical speciation and thus the transport of heavy metals and nutrients in soil.[12,13] In addition, these organics will change the cation exchange capacity.[14] The fulvic acid fraction[15] is the most important water-soluble organic fraction of sewage sludge, with respect to interactions with clay minerals and metal cations in soil solution. The levels and movement of the second group of organics, the refractory organics, may limit the environmental acceptance of sludge treatments. These organics are a result of man's use of plasticizers, pesticides, and fossil fuels. Polychlorinated biphenyls, phthalates, and polycyclic aromatic hydrocarbon are found in ppm concentrations in wastewater treatment plant effluents.[16] The importance of these compounds for limiting land application of sludge is not well documented.

Benefit-Cost Analyses

The costs of sludge delivery and handling must be compared to the benefits of increased wood production to economically evaluate the usefulness of land application of sewage sludge in biomass production. All of the energy inputs involved in planting, harvesting, transporting, fuel produc-

552

tion, and recycling of byproduct streams will be considered. The primary objective is to produce more renewable fuel substitutes than fossil fuels consumed in the system.

The benefit-cost analysis entails more than cost accounting of tree plantation operation and the resulting useable wood products. It includes ecological research directed at understanding and quantifying relationships between treatments, climate, nutrient cycling, hydrology, heavy metal cycling, and biomass productivity in context with the broader economic issues. Economic comparisons between environmental impacts and fuel-related expenses will be made using calories as a common unit.[17]

Equations will be developed which describe the costs of sludge transportation to sites, application, site preparation, labor, capital, and tree harvest. The sensitivity of the benefit-cost analysis to each of these constraints can be assessed in relation to the productivity of wood. The constraints will each generally yield a single value (calories will be the common unit), but the values can fluctuate with changes in costs of fossil fuels, wages, etc. These sliding scales will be incorporated into the benefit-cost analysis.

Given the state-of-the-art in breeding productive strains of trees, climate will influence the rate of production regardless of the supply of available nutrients. Results of the plot studies should identify an optimal rate of sludge application; however, metabolic constraints for the conversion of nutrients in sludge to tree biomass will ultimately place a limit on productivity. Depending upon concentration of heavy metals in the sludge, a feedback between optimal rates of nutrient supply vs. toxic accumulation of heavy metals could develop. This plus potential groundwater contamination with metals, nitrates, and organic compounds might impose legal, as well as economic constraints on optimal application rates. Potentially hazardous levels of heavy metal concentrations in consumer organisms, especially deer which are hunted by the public on the SRP site, may impose additional constraints.

To perform the benefit-cost analyses, benefits accrued through increased fiber production under varying amendment regimes must be quantified. These benefits (converted to units of energy) include the energetic equivalent of the fiber used for lumber, pulp, or biomass, and positive effects in the local economy. Where necessary, economic consultants will be asked to work with the basic research team on the plant site to generate appropriate methods for the benefit-cost analyses.

Conclusion

This study will evaluate the environmental effects that growing trees for energy have on the nutrient and mineral cycle, hydrologic cycle, soil quality, and disease rate in forest ecosystems. Ecological models will be developed to predict the effects tree harvest and application of sewage sludge have on the fertility and productivity of pine and hardwood plantations.

References

1. Riley, G.A. "The Carbon Metabolism and Photo-Synthetic Efficiency of the Earth as a Whole." Am. Sci. 32, 129-134 (1964).

2. U.S. Department of Energy. "Prediction of Energy Derived from Biomass." Report to the Senate Committee on Commerce, Science, and

Transportation (1980).

3.  Burwell, C.C. "Solar Biomass Energy: an Overview of U.S. Potential." _Science_ 199 (10), 1041-1048 (1978).

4.  Bach, W., W. Manshard, W.H. Matthews, and H. Brow. _Renewable Energy Resources_. Pergamon Press (1980).

5.  Hiser, M.L. _Wood Energy_. Ann Arbor Science, Ann Arbor, MI (1978).

6.  Vigerstad, T.J., and J.M. Sharp. "Assessment of the Availability of Wood as a Fuel for Industry in SouthCarolina." S.C. Energy Research Institute, Columbia, SC. (1979).

7.  Pritchett, W.L. _Properties and Management of Forest Soils._ John Wiley and Sons, New York, NY (1979).

8.  Patric, J.H. "Effects of Wood Products Harvest on Forest Soil and Water Relations." _J. Environ. Qual._ 9(1), 78-80 (1980).

9.  Long, F.L. "A Glass Filter Soil Solution Sampler." _Soil Sci. Soc. Am. J._ 42, 834-835 (1978).

10. Mortvedt, J.J., P.M. Giordano, and W.K. Lindsay. "Micronutrients in Agriculture." _Soil Sci. Soc. of Am.,_ Madison,WI. 666 pp (1972).

11. Davis, J.B., and R.L. Barnes. "Effects of Soil Applied Fluoride and Lead on Growth of Loblolly Pine and Red Maple." _Environ. Poll_ 35-44 (1973).

12. Holtzclaw, K.M., and G. Sposito. "Analytical Properties of the Soluble, Metal-Complexing Fractions in Sludge-Soil Mixtures: III. Unaltered Anionic Surfactants in Fulvic Acid." _Soil Sci. Soc. Am.J._ 42, 607-611 (1978).

13. Sposito, G., K.M. Holtzclaw, and C.S. Le Vesque-Madore. "Calcium Ion Complexation by Fulvic Acid Extracted from Sewage Sludge-Soil Mixtures." _Soil. Sci. Soc. Am. J._ 42, 600-606 (1977).

14. Knox, K., and P.H. Jones. "Complexation Characteristics of Sanitary Landfill Leachates." _Water Res._ 13, 839-846 (1979).

15. Holtzclaw, K.M., G. Sposito, and G.R. Bradford. "Analytical Properties of the Soluble, Metal-Complexing Fractions in Sludge-Soil Mixtures: I. Extraction and Purification of Fulvic Acid." _Soil Sci.Soc. Am.J._ 40, 254-258 (1976).

16. Pahren, H.R., J.B. Lucas, J.A. Ryan, and G.K. Dotson. "Health Risks Associated with Land Application of Municipal Sludge." _JWPCF_ 51(11), 1588-2601 (1979).

17. Odum, H.T. _Environment, Power and Society._ John Wiley and Sons, New York, NY (1971).

18. Metz, L.J., and C.G. Wells. "Weight and Nutrient Content of the Aboveground Parts of Some Loblolly Pines." USDA Forest Serv. Res. Pap. SE-17, 20 pp., Southeast. Forest Exp. Sta., Asheville, NC (1965).

19.  "Report on Composition of Augusta Municipal Waste Water Treatment
     Plant Sewage Sludge."  Carr and Associates Laboratory (Columbia,
     SC) (1980).

20.  Peck, E.C. "Shrinkage of Boards of Douglas Fir, Western Yellow Pine
     and Southern Pines."  Amer. Lumberman 2774, 52-54 (1928).

21.  Romancier, R.M.  "Weight and Volume of Plantation Grown Loblolly
     Pine."  USDA Forest Serv. Res. Note 161, 2 pp., Southeast.  Forest
     Exp. Sta., Asheville, NC (1961).

This paper was prepared in connection with work under Contract No.
DE-AC09-76SR00001 with the U.S. Department of Energy.  By acceptance of this paper, the publisher and/or recipient acknowledges
the U.S. Government's right to retain a nonexclusive, royalty-free
license in and to any copyright covering this paper, along with
the right to reproduce and to authorize others to reproduce all or
part of the copyrighted paper.

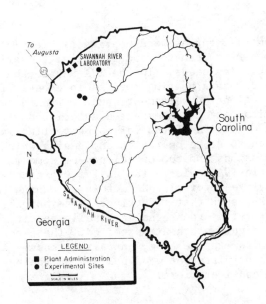

FIGURE 1.  Savannah River Plant Site

# FACTORS AFFECTING WOOD HEATER
# EMISSIONS AND THERMAL PERFORMANCE

**J. P. Harper** and **C. V. Knight**
Tennessee Valley Authority

Introduction

Emissions from residential wood heating are receiving greater attention
as more homeowners switch to wood as a fuel to satisfy all or part of
their winter heating needs.  In the TVA Region the percentage of residences using wood as a heating fuel is in excess of 20-percent in some
areas.  As a regional development agency, TVA is concerned about the
potential environmental consequences associated with residential wood
heating.  The objective of the following testing was to identify those
factors related to stove operation and design which may significantly
affect the emissions and the efficiency of some residential wood heaters
commonly used in TVA Region.  The analysis presented shows that emissions
and efficiency are interrelated factors governing the overall performance
of wood heaters; such that reductions in emissions, for example carbon
monoxide and total hydrocarbons, typically result in an increase in the
efficiency of eight wood heaters tested.  The paper not only presents the
data obtained from the testing, but also proposes a performance index for
the comparison of different wood heaters on the basis of both efficiency
and emissions considerations.

Methods of Analysis

Figure 1 depicts the experimental layout of the stove and the emissions
monitoring equipment utilized during the testing.  The stove and chimney
were supported on a platform which is placed on a digital scale.  The
scale provides a continuous measurement of the mass rate of change of the
wood being burned.  Emission measurements of the flue gas composition,
were made on a continuous basis.  Carbon monoxide, carbon dioxide, total
hydrocarbons (as methane $CH_4$), oxygen, sulfur dioxide, and nitrogen oxide
were measured near the exit of the stove.  Values for each of these

parameters, as well as stove temperature and flue gas temperature, were recorded from the instrumentation for every minute of the stove's operation. The use of continuous monitoring equipment in evaluating the emissions and thermal performance of wood heaters is very important because of the transient burn conditions always existing in the wood heater. Also, this approach differs from previous work (DeAngelis et. al., 1980) (Dyer et. al., 1978) (Shelton et. al., 1978) (Butcher and Sorenson, 1979) which used wet-Orsat and grab-type sampling in the measurement of flue gas composition. The continuous method of measurement offers greater consistency and timeliness in data collection which is needed to identify interrelationships between the rate of combustion of the wood, the emissions, and the instantaneous efficiency.

The operating procedure during the test was to first predetermine the appropriate draft settings to obtain conditions representative of low, medium, and high burn rate conditions. Manufacturers recommendations or pretests were utilized in obtaining the best draft setting for establishing these three burn rates.

The wood charge for the tests was specified at one pound per cubic foot of combustion chamber volume for low burn, two pounds per cubic foot of combustion chamber volume for medium burn, and three pounds of cubic foot of combustion chamber volume for high burns. The wood used was either Douglas Fir brands (Fireplace Institute, 1979) or 4" triangular oak sections. The Douglas Fir brands are commonly utilized for safety testing of wood heaters and have been proposed by the Fireplace Institute and the ASHRAE 106P Committee for efficiency testing. The 4" triangular oak sections were taken as a fuel more similar to that commonly burned in a wood heater, therefore giving a closer approximation of reat world conditions. Both woods used were oven dried to approximately six-percent moisture content.

First a precharge similar in weight and type to the test fuel was placed in the stove to condition it to the proper temperature. Next, a second charge was placed into the heater to commence the test measurements. Testing proceeded from low, to medium, to high burn rates with precharges being burned between each test of record.

The instantaneous data was then input into a computer program which calculated both the instantaneous and overall efficiency of the wood heater and also the instantaneous and average emission factors for the wood heater. The efficiency program utilized an indirect stack loss approach for calculating efficiency (Shelton et. al., 1978). The flue gas composition data is used to balance a theoretical combustion equation in the computer program. The combustion equation in turn gives the air/fuel ratio which is then utilized to calculate the sensible and latent heat losses and chemical heat losses realized during the test.

Efficiency and Emissions Data

Table 1 summarizes the efficiency and emissions test results obtained for the group of wood heaters tested in this project. Literature values are also presented for comparison purposes. Figure 2 presents a plot of the efficiency versus burn rate for all eight wood heaters tested. Figures 3, 4, and 5 are plots of carbon monoxide, total hydrocarbons, and nitrogen oxide versus burn rate for wood heaters 4 through 8. These Figures, when compared with efficiency data in Figure 2 illustrates a major point of this paper, namely, that efficiency and emissions are interrelated. The general trend exhibited for the wood heaters tested is that as the efficiency increases the emissions of carbon monoxide, total hydrocarbons, and nitrogen oxide generally go down.

The significance of this observation is that a criterion can be developed which could provide a comparative basis for ranking wood heaters not only the basis of thermal efficiency but also on the basis of lower emissions. Later in the paper a performance index is proposed utilizing this concept which could provide a basis for comparing wood heaters both on their thermal and environmental performance.

Effective Fuel Size

One of the more significant operational variables which we found affected both the efficiency and the emissions of the wood heaters was the fuel size. The effect on efficiency is illustrated in Figure 2. The straight lines on the graph (representing wood heater 1, 2, and 3) depict the efficiency relationship obtained when fir brands are utilized as the test fuel. A straight linear relationship decreasing with increasing burn rate is observed. Contrary to this observation, when 4" triangular oak sections are utilized as fuel, the optimal efficiency indicated on the graph lies somewhere between low and the medium burn rates. The surface to volume ratio is thought to account for this difference, since testing with oak kindling also showed a similar alteration in the efficiency when used as a fuel.

A series of efficiency tests alternatively burning the 4" triangular oak sections and the fir brands done at high burn rate on wood heaters 4 through 8 showed that the observed difference was in fact due to a decrease in the combustion efficiency between the two fuels for a particular stove (see Table 2). The major determinant of this reduction in combustion efficiency was excess carbon monoxide production which on the average is significantly higher when fir brands are burned (see Table 3). This observation on fuel surface/volume effects also illustrates that efficiency and emissions interrelationships can provide useful insights on how best to operate wood heating appliances.

The reason for surface area being a primary determinant may be that the excessive burning surface area in the stove requires larger quantities of oxygen than the wood heater's design is capable of supplying directly into the combustion zone to prevent pyrolysis from ocurring. Pyrolysis would result with incomplete combustion products being generated. Thus, when using a fuel with a high surface area, a larger proportion of the wood is heated above $400^{\circ}F$, and pyrolysis or thermal chemical degradation of the wood tends to become self-sustaining. The system then operates more as a gasifier rather than a combustion chamber resulting in significantly higher emissions of carbon monoxide.

Effective Design

The impact of design on either the emissions or the thermal performance of wood heaters is very difficult to ascertain. The large number of design variables and the limited capability to do sufficient testing to provide statistical basis for comparing current designs makes it difficult to identify the characteristic effect of each design parameter. However, we have identified variables we feel important to the thermal performance of wood design and have made an attempt to interpret the results of our test data regarding these variables. The variables include the presence of fire brick, whether the air is preheated, whether the device has a grate, and whether the device contains baffles in the combustion chamber. Each of these factors was thought to be beneficial with respect to improving the thermal efficiency of the wood heating device. In general, the more of these design components each stove had, the more efficient the stove was. No single wood heater, however, had all the major design components we considered to be significant. The better performing heaters typically had either baffles or grates, with the existence of a preheated air supply improving either one of these parameters. Unfortunately, without a greater statistical base further interpretation of our test results would be highly speculative at this time.

Wood Heater Performance Index

The need for reducing the emissions in wood heaters and simultaneously improving the efficiency is self-evident. A similar type of effort was undertaken in the 1920's and 30's with regard to industrial solid-fuel boiler systems which resulted in the mid-30's with highly efficient solid-fuel combustion systems. It is thought that a similar sort of program could occur now in the 1980's with regard to improvement in the design of residential solid-fuel heating devices.

In this paper, we propose the following index to be utilized for a comparative evaluation of wood heaters:

$$\text{Wood Heater Performance Index} = \frac{\text{Thermal Efficiency}}{\text{Source Severity}}$$

This index provides a function which increases with increasing thermal efficiency and decreases with increasing emissions. The source severity (Chalekode and Blackwood, 1978) (DeAngelis, et. al., 1979) was chosen as the environmental factor for evaluating wood heaters because it provides a standard basis of comparison for different emission factors. The source severity is a measure of the maximum ground level concentration of a particular pollutant obtained for a given emission factor and standard set of operating conditions, to the maximum allowable primary ambient air quality standard of that pollutant or its threshold limit value. By using the source severity we also have a normalized basis for analyzing the cumulative effect of different pollutant sources. Figures 6, 7, and 8 illustrate how the wood heater performance index ranks the different stoves tested on the basis of carbon monoxide emissions, total hydrocarbons emissions, and the sum of carbon monoxide, total hydrocarbon and nitrogen oxide emissions. Thus, the proposed index not only allows one to come up with a relative comparison among the performance of wood heaters based on a singular emission factors, but also a comparative ranking of different stoves based on the sum of all pollutant sources.

The use of this index in itself is not so important but rather the importance of the proposed index is that consideration needs to be given in the development of wood heaters not only to improving their thermal efficiencies, but also in terms of reducing their pollution output. This index could provide a standard basis of comparison against which alternative designs can be compared on the overall basis of both efficiency and emissions.

Summary

The primary conclusion of this report from the environmental viewpoint, is by reducing emissions from wood heaters it is also possible to improve the efficiency of these devices. Therefore, the need exists to do both simultaneous emissions and efficiency testing. Secondly, fuel wood size may be a major operating variable governing the efficiency and emission of residential wood heaters. Thirdly, design considerations, primarily those affecting air flow into the combustion zone and flue gas resonance time in the wood heater, tend to be present in the more efficient stoves. Fourth, a comparative basis for evaluating different wood heaters, a performance index, was proposed. This performance index equalled the ratio of the efficiency to the source severity. This performance index could provide a basis for rank ordering those wood heaters which are the most highly efficient and least polluting devices. Many of the environmental problems associated with residential wood combustion may be lessened with the development of stoves which are significantly more efficient.

References

1. DeAngelis, D. G. et. al., "Preliminary Characteristics of Emissions from Wood-Fired Residential Combustion Equipment," EPA-600/7-80-040, March, 1980.

2. Dyer, D. F. et. al., "Improving the Efficiency, Safety, and Utility of Wood Burning Unite--Volume I" Quarterly Report No. WB-4, Contract No. ERDA ECT180552, Auburn University, 1978.

3. Shelton, J. W. et. al., "Wood Stove Testing Methods and Some Preliminary Experimental Results," ASHRAE Transactions Vol. 48, Part 1 (1978).

4. Butcher, S. S. and E. M. Sorenson; "A Study of Wood Stove Particulate Emissions," Journal of Air Pollution Control Association, V. 29(7): 724-728, July (1979).

5. Fireplace Institute, "Test Standard for Rating Wood-Fired, Closed Combustion-Chamber, Heating Appliances," Chicago, Illinois (1979).

6. DeAngelis, D. G. et. al., "Source Assessment: Residential Combustion of Coal," EPA-600/2-79-019a, January, 1979, pp. 111-122.

7. Chalekode, P. K. and T. R. Blackwood, "Source Assessment: Coal Refuse Piles, Abandoned Homes and Outcrops, State of the Air," EPA-600/2-78-004v, July, 1978, pp.35.

TABLE 1   Preliminary Wood Heater Emissions Data From Literature and TVA Testing Project.

| | TVA WOOD HEATER TESTS | | | | | | LITERATURE SOURCES[1] | |
| | OAK SECTION 4"x4" TRIANGULAR | | | FIR BRANDS | | | | |
| EMISSIONS FACTORS | AVERAGE g/kg (Lbs/10^6 BTU OUT) | RANGE g/kg (Lbs/10^6 BTU OUT) | STD. DEV. g/kg (Lbs/10^6 BTU OUT) | AVERAGE g/kg (Lbs.10^6 BTU OUT) | RANGE g/kg (Lbs/10^6 BTU OUT) | STD. DEV. g/kg (Lbs/10^6 BTU OUT) | AVERAGE g/kg | RANGE g/kg |
|---|---|---|---|---|---|---|---|---|
| CARBON MONOXIDE (CO) | 92.31 (22.13) | 26.06-185.5 (5.66-37.26) | 43.13 (10.46) | 196.27 (43.74) | 67.10-310.26 (15.7-73.8) | 68.45 (16.65) | 181.3 | 91-370 |
| TOTAL HYDROCARBONS (THC) | 28.36 (6.36) | 8.55-78.26 (1.96-19.95) | 17.87 (4.58) | 29.10 (6.39) | 11.30-56.15 (2.63-11.72) | 10.80 (2.3) | 12.6 | 0.3-44.4 |
| SULFUR DIOXIDE (SO$_2$) | 1.06 (0.24) | 0.49-2.82 (.069-.423) | 0.64 (0.17) | 0.78 (0.16) | 0.36-1.78 (.075-0.37) | 0.33 (0.07) | 0.2 | 0.16-0.24 |
| NITROUS OXIDE (NO) | 2.51 (0.63) | 1.27-4.22 (0.24-1.86) | 0.78 (0.37) | 2.83 (0.62) | 1.03-4.13 (0.20-0.91) | 0.87 (0.18) | 0.49 | 0.2-0.8 |
| FILTERABLE [2] PARTICULATE | 1.481 (0.34) | 0.23-3.50 (.044-.89) | 1.01 (0.26) | 4.37 (.963) | 0.33-8.36) (.063-1.62) | 2.39 (.488) | 3.6 | 1.8-7.0 |
| TOTAL POLYCYLIC ORGANIC MATERIAL (POM) [3] | .095 (0.020) | 0.0003-.314 (0.0006-.065) | .114 (.023) | .296 (.065) | 0.021-.865 (0.004-.189) | 0.27 (.059) | 0.27 | 0.19-0.37 |
| GASEOUS POM [3] (XAD-2 TRAP) | .026 (.0056) | .00003-.1793 (.000006-.385) | .054 (.012) | .091 (0.021) | 0.002-0.267 (0.0004-.064) | .098 (.022) | | |
| FILTERABLE POM [3] (METHOD 5) | .069 (0.015) | .0024-.311 (.00048-.064) | .108 (.022) | .206 (.044) | .019-.786 (0.004-.171) | .266 (.057) | | |
| THERMAL EFFICIENCY (%) | 57.8 | 47.2-65.7 | 5.14 | 56.03 | 47.7-64.6 | 4.94 | 55.0 | 39.4-76.9 |

[1] De Angelis et al., 1979
[2] Particulate Mass on Filter
[3] Lower Range Estimate Established by Fluorescence Screening for POM.

TABLE 2:   OVERALL THERMAL EFFICIENCY,
COMBUSTION EFFICIENCY AND HEAT TRANSFER EFFICIENCY FOR
FIR BRANDS AND TRIANGULAR OAK 4" SECTIONS

| WOOD HEATER | FUEL | BURN RATE | OVERALL THERMAL EFFICIENCY | COMBUSTION EFFICIENCY | HEAT TRANSFER EFFICIENCY |
|---|---|---|---|---|---|
| # 4 | Fir Brands | High | 59% | 77% | 76% |
| # 4 | 4" Oak Sections | High | 66% | 94% | 70% |
| # 5 | Fir Brands | High | 51% | 87% | 58% |
| # 5 | 4" Oak Sections | High | 58% | 95% | 61% |
| # 6 | Fir Brands | High | 52% | 77% | 68% |
| # 6 | 4" Oak Sections | High | 58% | 93% | 62% |
| # 7 | Fir Brands | High | 54% | 82% | 66% |
| # 7 | 4" Oak Sections | High | 59% | 89% | 66% |
| # 8 | Fir Brands | High | 48% | 80% | 60% |
| # 8 | 4" Oak Sections | High | 54% | 87% | 62% |

TABLE 3:   CO EMISSION FACTORS FOR
OAK AND FIR BRAND TESTS

| BURN RATE | CO EMISSION FACTORS (LB/TON WOOD) | |
| | 4" TRIANGULAR OAK SECTIONS | FIR BRANDS |
|---|---|---|
| Low | 217 | 275 |
| Medium | 166 | 344 |
| High | 191 | 366 |

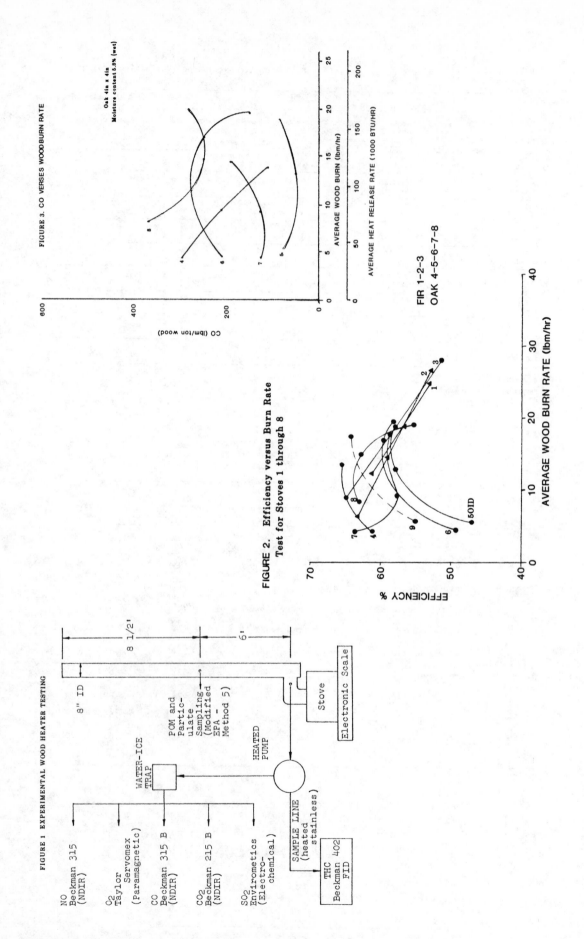

FIGURE 3. CO VERSES WOODBURN RATE

Oak 4in x 4in
Moisture content 6.8% (wet)

CO (lbm/ton wood)

600

400

200

0

AVERAGE WOOD BURN (lbm/hr)

AVERAGE HEAT RELEASE RATE (1000 BTU/HR)

FIR 1-2-3
OAK 4-5-6-7-8

FIGURE 2. Efficiency versus Burn Rate
Test for Stoves 1 through 8

EFFICIENCY %

70

60

50

40

AVERAGE WOOD BURN RATE (lbm/hr)

5OID

FIGURE 1 EXPERIMENTAL WOOD HEATER TESTING

8 1/2'

8" ID

6'

POM and Partic-
ulate
Sampling
(Modified
EPA -
Method 5)

WATER-ICE
TRAP

HEATED
PUMP

Stove

Electronic Scale

NO
Beckman 315
(NDIR)

O2
Taylor
Servomex
(Paramagnetic)

CO
Beckman 315 B
(NDIR)

CO2
Beckman 215 B
(NDIR)

SO2
Envirometics
(Electro-
chemical)

SAMPLE LINE
(heated
stainless)

THC
Beckman 402
FID

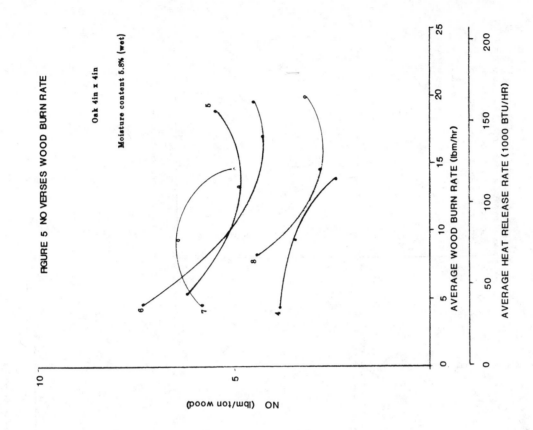

FIGURE 5 NO VERSES WOOD BURN RATE

Oak 4in x 4in

Moisture content 5.8% (wet)

NO (lbm/ton wood)

AVERAGE WOOD BURN RATE (lbm/hr)

AVERAGE HEAT RELEASE RATE (1000 BTU/HR)

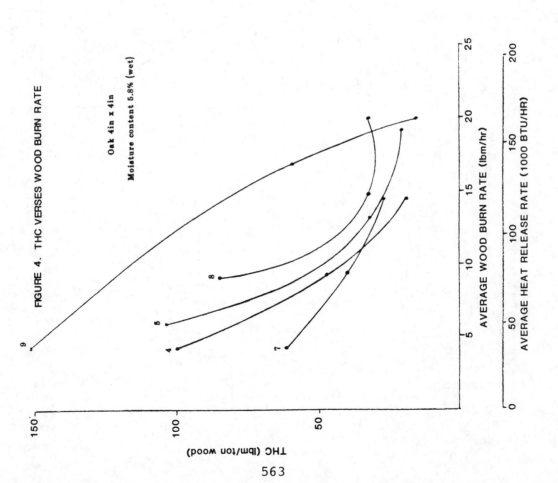

FIGURE 4. THC VERSES WOOD BURN RATE

Oak 4in x 4in

Moisture content 5.8% (wet)

THC (lbm/ton wood)

AVERAGE WOOD BURN RATE (lbm/hr)

AVERAGE HEAT RELEASE RATE (1000 BTU/HR)

563

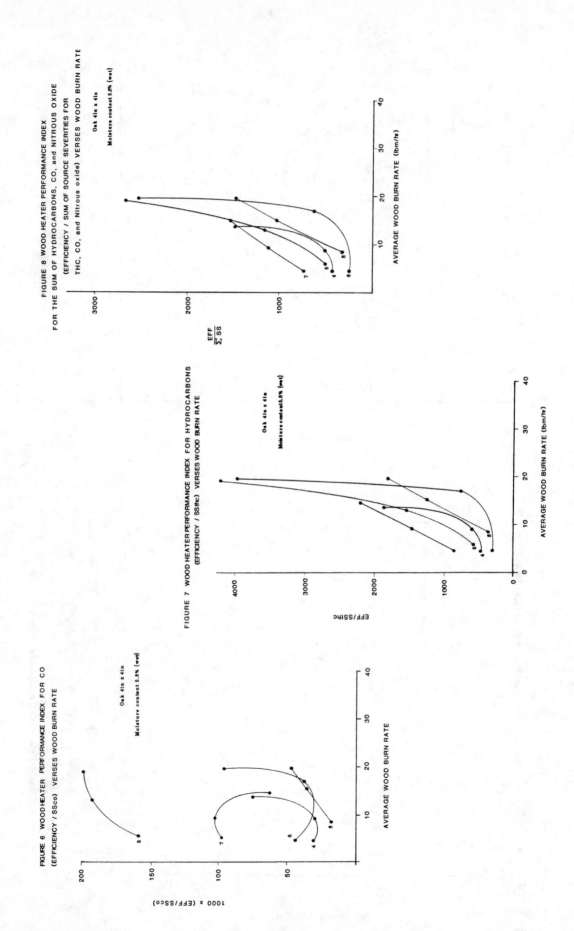

FIGURE 8 WOOD HEATER PERFORMANCE INDEX
FOR THE SUM OF HYDROCARBONS, CO, and NITROUS OXIDE
(EFFICIENCY / SUM OF SOURCE SEVERITIES FOR
THC, CO, and Nitrous oxide) VERSES WOOD BURN RATE

Oak 4in x 4in

Moisture content 5.8% (wet)

FIGURE 7 WOOD HEATER PERFORMANCE INDEX FOR HYDROCARBONS
(EFFICIENCY / SSthc) VERSES WOOD BURN RATE

Oak 4in x 4in

Moisture content 5.8% (wet)

FIGURE 6 WOOD HEATER PERFORMANCE INDEX FOR CO
(EFFICIENCY / SSco) VERSES WOOD BURN RATE

Oak 4in x 4in

Moisture content 5.8% (wet)

564

# SOLAR ENERGY AND THE SEARCH FOR EMISSION OFFSETS

**Barbara Euser**
Solar Energy Research Institute

## I.  Introduction

Before General Motors could build a long-planned Oklahoma City assembly plant, a search was undertaken to identify industrial polluters in the area.  This search revealed a serious air pollution problem:  the failure of local oil companies to cover their petroleum storage tanks was resulting in the discharge of thousands of tons of hydrocarbons per year.  The Oklahoma City Chamber of Commerce persuaded the oil companies to install floating roofs on the storage tanks[1,2], thereby reducing the pollution they caused, and construction of the assembly plant was allowed to proceed.

The inducement for this activity was a legal one:  the requirements of the Clean Air Act Amendments.  Section 129 of the 1977 Amendments[3] provides that, before a company may construct a major polluting plant in an area that has not met any of five national ambient air quality standards (NAAQS)—a nonattainment area—it must reduce air pollution in the area by more than the amount the new plant will add.  This procedure is called "offsetting."[4]  By reducing hydrocarbon emissions by 5,316 tons per year, GM was able to offset the air pollution of its new plant, projected to be 3,340 tons per year.[2]  As a result, the automaker obtained the necessary preconstruction permit.

The offset policy, a creation of the Environmental Protection Agency (EPA), gives industry the opportunity to develop its own plans for meeting clean air requirements.  The policy has dispelled early fears that the Clean Air Act would ban industrial growth in nonattainment areas.  As the states take over the responsibility for meeting the air quality standards from the EPA,[5,7] indications are that most will continue the offset approach in some form.

Our focus is on solar energy as an alternative means of offsetting. Because solar energy is nonpolluting, any reduction in emissions that results from the installation of solar devices can, in theory at least, be offset against the pollution from conventional energy sources. This alternative will be particularly useful where the pool of available offset candidates may have been all but exhausted, or where easily—and cheaply—obtained emissions offsets may be in short supply. We will examine the legal and economic issues raised by such a proposal.

II. Overview of the Offset Policy
        A.  Framework of the Clean Air Act

A detailed explanation of the Clean Air Act is beyond the scope of this article.[5,7] Briefly, national ambient air quality standards (NAAQS) are set by the EPA; and the states develop plans, approved by the EPA, to implement the standards.[5] The country is divided into 247 air quality regions.[7] A region may be in compliance with the NAAQS for some pollutants and not for others. Construction or modification of major sources of air pollution depends on permits granted after review.[5,7,8] In regions which meet the standard for a particular pollutant, review is for "prevention of significant deterioration" (PSD).[5,7,9,10] The purpose of PSD review is to allow industrial expansion in those areas without degrading the quality of their air. In nonattainment areas, the purpose of the review is to allow growth while assuring compliance with plans to achieve the standard.

Primary responsibility for implementing the Clean Air Act rests with the states. The Act directs state agencies to develop and adopt state implementation plans (SIPs) setting forth control measures sufficient to achieve compliance with the NAAQS standards.[11]

        B.  Background of the Offset Policy

The offset policy arose in connection with the EPA's implementation of rules governing preconstruction review of new or modified sources in nonattainment areas.[12] In applying these regulations, the EPA found itself in a political dilemma. On the one hand, the Clean Air Act, as interpreted by the courts,[13] "subordinate[d] demands for industrial expansion to the need for attaining and maintaining clean air."[14] On the other, economic stagnation seemed to many an intolerably high price to pay for clean air.[5,7,14]

The offset policy emerged as an attempt to reconcile these conflicting demands.[14] As one commentator put it: "The essence of the Offset Policy is that major new growth should be allowed in nonattainment areas only if air quality is improved as a result of that growth."[7]

Originally set forth in an EPA interpretative ruling,[15] the offset policy allowed construction of major new sources of pollution and modification of major existing sources in nonattainment areas where greater than one-for-one emission offsets were achieved.[14]

The premise underlying the policy, one commentator writes, "is that even where severe abatement measures have already been required, untapped opportunities exist for further emission reductions."[5] Offsets, then, "must represent emission reductions that otherwise would not be required."[5]

566

The offset concept was incorporated, with changes, into Section 129 of the 1977 Clean Air Act Amendments.[16] On January 16, 1979, the EPA revised the original interpretative ruling;[17] and, in September 1979, the agency proposed additional changes.[18]

Response to the offset policy has been generally positive. The program encourages innovation and economic efficiency on the part of industry, adding flexibility to the seemingly rigid commands of the Clean Air Act and allowing industrial growth.

### C. Scope of the Offset Policy

Sources subject to nonattainment review are "major" stationary sources of air pollution—those "which directly emit, or have the potential to emit, one hundred tons per year or more of any air pollutant,"[19] after accounting for any air pollution control equipment that may be used.[20] If a modification of a source of this size increases emissions at all, the modification is subject to review.[21] The EPA may, however, exempt increases which are too small for the agency to be able to regulate[21] or which are de minimis,[21] that is, of inconsequential size.

Sources covered by the offset policy are those emitting particulates, sulfur dioxide ($SO_2$), nitrogen oxide ($NO_x$), volatile organic compounds (hydrocarbons which produce ozone), or carbon monoxide.[5,7,22]

The EPA regulations limit emission offsets to those of the same type, that is, "intrapollutant" emission offsets.[23] If a new source will emit, for example, hydrocarbons, it will have to find offsets in existing sources that also emit hydrocarbons.

This limitation may be a serious barrier to solar offsetting. If, for example, solar power is offset against emissions from coal-fired industrial boilers, offsets would be limited to sulfur dioxide and particulates, the major byproducts of coal burning. Accordingly, only proposed new or modified plants that would emit these pollutants could take advantage of solar offsets.

The EPA has been criticized for its refusal to allow interpollutant offsets,[6] a policy apparently based on the difficulty of comparing pollutants.[6] The limitation may be moderated as the states assume responsibility for administering the program under their revised implementation plans. This will not, of course, be an issue in those regions that are nonattainmment areas for only one NAAQS.

Sufficient offsets must be obtained from existing sources in the area of the proposed source to provide a "positive net air quality benefit" in the affected area.[5,23] This test is intended to ensure that, on balance, an air quality benefit is achieved in the general area affected by the new source.[24] Whether this presents an additional requirement above the net reduction in emissions is not clear,[24] but an offset for a new (or modified) source may be required to be fairly close to the new source—the EPA is using a rule of thumb of 5 miles—or the required ratio of offsets may increase as the offsetting is done further from the new source.[25]

### D. Offsetting Under State Implementation Plans

The 1977 Amendments require the states to revise their implementation plans for nonattainment areas. While these state implementation plans

were to have been approved by the EPA by July 1, 1979,[7,26,27] few plans
have yet been approved.[28] Where there is no revised plan conforming to
the 1977 Amendments, the Act imposes a ban on construction or modification
of major new sources within the nonattainment area.[27,29]

Whenever it is filed and approved, a SIP becomes the governing law for
all existing sources within the nonattainment area and all new or modified sources affecting the area.[17,30] That is, the state takes over the
problem of growth in nonattainment areas. States have a choice of retaining the offset approach or of reducing emissions far enough below the
standard to allow for growth without case-by-case offsets.[5,27,31] If the
growth allowance is used up or if none is created, the state may issue
permits only by using case-by-case offsets.

Will the states continue using the offset approach? It will be difficult
for states to reduce emissions enough to make room for growth, so it
seems likely that the offset approach will remain in frequent use. Solar
energy can play a role both in the expansion of state offset programs
and in the creation of growth allowance margins.

### E. The "Bubble"

Related to offsetting is the "bubble" concept, used in the PSD procedure.[21,32,33] The bubble functions like an offset: reduced emissions in
one source at a plant or even at other plants are traded for emissions at
the new or modified source.[32,34,35,36,37] The original idea behind the
bubble was to treat the various buildings which make up an industrial
plant as a single source, as if all their smokestacks were under a giant
bubble with one outlet.[1,36,37] Offsetting increases and decreases in
emissions from various plant components entail no regulatory consequences.[36,38] The bubble concept allows a manufacturer to select the
most cost-effective mix of control technology.

Although the EPA has never included the bubble in its offset policy,[37]
it has advised the states that they may use the concept in their revised
SIPs for areas which are expected to attain the air quality standards
by 1982.[37,39,40,41]

### III. A Solar Offset Program

Solar energy has potential applications in any pollution control program,
but its most promising use would seem to be in nonattainment areas. It
is in those areas that new and modified sources are likely to be most
stringently controlled; Hence very expensive pollution control may be
needed, boosting the value of solar energy as a substitute.

### A. Offset Possibilities

The first offset possibility, when the new source is an expansion of an
existing plant, is, as Quarles points out, an internal offset: the manufacturer installs tighter controls on existing operations. This might
include use of innovative technology or controls "so costly that they
would not normally be required."[5]

A new or modified source might alternatively "strike a deal with another
large industrial plant in the area to apply extra pollution control
measures, presumably with a sum of money changing hands."[5] "[A]n applicant may simply purchase an existing facility in order to clean it up
or close it down, particularly an old, obsolete plant with large pollut-

ant emissions."[5]   A market for pollution rights may develop.[1,5]   In fact, however, almost all offsets have been internal, that is, intraplant.[42,43]

In any of these instances, solar energy could be substituted for pollution controls.  In the last instance, its use could avoid the possibility that construction of a new plant would cause a new decrease in jobs by avoiding the closing of an existing plant.  Solar installations could be used to reduce emissions from new or modified sources, thus reducing or avoiding the need for offsetting altogether, particularly if the bubble concept were adopted in a SIP.

Substituting nonpolluting solar energy for polluting fuel-burning will result in a net decrease in emissions represented by this formula:

$$\frac{\text{Emissions (tons)}}{\text{Energy Produced}} \times \text{Net Decrease in Energy Production} = \text{Emission Offset Credit}$$

## B.   Solar Energy Applications

Solar collectors could be placed on individual residences to offset oil-burning or powerplant emissions, although administrative difficulties are likely to limit this application.[44]   Providing solar heating or cooling for office complexes or entire industrial parks is likely to be more practical and could produce more easily measurable offsets.  In fact, however, the use of solar energy for industrial process heat (IPH) is likely to be the most frequent application for offsetting.

## C.   The Economics of Offsetting

Two sets of cost information must be considered in determining the economic feasibility of a solar offset program:  the cost of pollution control plus conventional energy production on one side and the cost of converting the sun's energy to use on the other.

A comparison of estimates of the cost of controlling the final 10% of emissions, plus conventional IPH, with the cost of solar IPH, is set forth in Table I.  This is a hypothetical case rather than an example of the costs involved.[45]   Figure I illustrates some of the assumptions in Table I.  Figure 2 shows that the marginal cost of cleaning up the last 10% or so of emissions increases sharply.

As noted, Table I assumes a single, daylight shift;  conventional backup equipment is needed, although seldom used, producing negligible fuel costs but no saving on capital.  That is, the conventional energy production equipment is installed with both systems.  The only difference in capital costs between the two systems is for pollution control equipment in one system and solar equipment in the other.  Much of major industry works two or three shifts or uses processes that run around the clock, reducing the likely role of solar energy for IPH to 30-40% at best in these industries.  But there are many plants, mostly small and medium size, that run only one shift of 8 or 10 hours, five or six days a week.[46] In such plants, with the proper amount of solar radiation, use of conventional backup power would be expected to be very infrequent.  A further assumption is made:  that pollution controls are not required for very intermittent use of conventional backup for the solar energy system.[47]

Assuming a certain level of control has been established by virtually all large plants in a given nonattainment area (90% in this example), the cost of installing additional control devices for the purpose of obtaining offsets would be disproportionately high.  We hypothesize that, at some

point in that final increment (Point A in Figure 2), investment in solar
energy as an alternative offsetting technique will become cost-competitive
in the long term.

### D. The Economics of Solar Industrial Applications

Industry is the largest energy user in the United States, consuming about
37% of the energy used.[48]  It is estimated that about 50% of industrial
demand is for industrial process heat (IPH), thermal energy used in
manufacturing processes.[49]

Although solar thermal energy can be supplied at almost any temperature
required by industry, it is now produced most cost-effectively at temperatures of less than 550°F.  Considering the terminal temperature requirements of process heating in industry and the theoretical demand for energy
for preheating to 550°F for higher temperature requirements, about half
of IPH demand is at or below 550°F.[50]  Commercially available solar
equipment is now capable of operating at temperatures up to 550°F.  Like
conventional process heat, solar IPH can be supplied directly or through
a heat transfer fluid such as hot water;  hot, dry air;  heat transfer
oil;  or low-pressure steam.

There are many potential applications for solar energy in industry.
Solar energy may be used to preheat water for oil-fired boilers producing
steam.    It may be used to heat water used directly in industrial processes, as in food processing or textile dyeing,[51] possibly offsetting
emissions from oil or coal.  Solar energy may be used to dry foods such
as onions, garlic, raisins,  and prunes,[52] replacing natural gas or oil.

Table 2 lists several projects underway to test the feasibility of solar
IPH.  These projects have demonstrated benefits which will be of increasing importance to industrial management.  In addition to pollution
control, they include: (1) alleviation of costly plant shutdowns resulting from fuel supply interruptions, (2) avoidance of rapidly rising construction costs resulting partly from pollution control requirements,
(3) avoidance of rapidly escalating fuel costs, and (4) public relations
benefits resulting from reduction of air pollution and conservation of
nonrenewable energy resources.[50]

Solar IPH is not now economically feasible when compared with less expensive alternatives such as coal, oil, or natural gas.  When, however,
the cost of pollution control equipment for the last increments of control is added to the cost of conventional process heat production, solar
energy will likely be competitive in the long term for some applications.
Given the massive size of the IPH energy demand estimated as 10.60 quads
in 1978,[49] even a small portion of this demand is substantial.

As suggested by Figure 2, the cost of cleaning up the final 10% increment
is likely to be double or even triple previous unit costs.  This is
precisely the situation that will confront industrial managers in many
nonattainment areas:  adding to pollution control at a prohibitive price.

As shown by Table I,  it is estimated that additional research and
development and the introduction of mass production techniques will reduce the cost of solar industrial systems markedly, perhaps by as much
as 50%.  Similarly, changes in tax policy may create significant incentives.  At present, fuel costs can be deducted from taxable income, while
solar energy, because it is not explicitly "purchased," does not qualify
as a tax deductible expense.[53] Changing the tax laws to provide a deduc-

570

tion for fuel saved or to permit accelerated depreciation could bring solar energy costs closer to the combined cost of pollution control and energy.

The upshot is that solar energy, while not presently an economic alternative to pollution control equipment, may become so in the future. Although strict economic analysis may not dictate adoption for some time, solar energy may nevertheless be attractive in certain areas of the country and for certain industries.

### E. Procedures for Offsetting

Should a company decide to go ahead with a solar offset program, what is the proper course of action? The answer depends to a large extent on the nature of the company's proposed operations. If relatively minor design or operational modifications are proposed, internal offsets will be most practical, particularly if the bubble concept has been adopted in the SIP. If major expansion or new construction is desired, a larger, external offset may be necessary. Growth allowance offsets, particularly if created through a government-sponsored cost-sharing arrangement, may be feasible.

Offsets may be initiated either by the owner of the proposed source or by the appropriate state or local government.[54] In a state of community-initiated offset, the state or local government commits itself to reduce emissions from existing sources enough to outweigh the impact of the new or modified source. In either case, the offset must be submitted as a formal SIP revision.[54]

Ultimately, government-initiated offsets may be the vehicle through which solar energy is blended with the offset program. Because of the high initial investment required for solar energy, a cost-sharing arrangement involving both industry and government would be desirable. Both entities would benefit from such an arrangement—industry by being allowed to locate where it chooses, government by the additional revenues and jobs accruing from industrial growth. Unlike dollars invested in pollution control equipment, solar-invested dollars would produce a tangible economic benefit—energy.

## IV. Conclusion

The concept of cleaning up the air through solar technology has intrinsic appeal. Without betraying the cause of clean air, demands for industrial growth and energy self-sufficiency can be answered. Without undermining the operation of the marketplace, incentives for solar energy investment are created. The concept represents, in a very real sense, the best of both worlds.

But intrinsic appeal does not always mean economic good sense. Until additional financial incentives for industrial investment in solar energy are created, solar offsetting will remain a conceptually sound, yet economically unrealistic, idea. Once these incentives are created and it is demonstrated that solar energy can work in an industrial setting, solar offsets can and should be encouraged. Until then, decisionmakers in both industry and government should study the potential of the proposal with an eye to the future.

# REFERENCES

1. Timothy B. Clark, "New Approaches to Regulatory Reform--Letting the Market Do The Job," *National Journal*, vol. 11 no. 32, August 11, 1979; "Dirty Deals," *Forbes*, vol. 121 no. 7, April 3, 1978.

2. Kang Kun Wu, Benny D. Cranor, "The Air Quality Impact of the Proposed General Motors Assembly Plant," papaer presented at 71st Air Pollution Control Association Annual Meeting and Exhibition, Houston, TX, June 25-30, 1978 (unpublished).

3. Pub. L No. 95-95, § 129 (a), 91 Stat. 745, as amended by Pub. L No. 95-190, §§ 14 (b)(2), 14 (b)(3) (1977), 91 Stat. 1404, 42 U.S.C.A. § 7502 note (non-attainment Areas), incorporating by reference EPA Interpretative Ruling, 41 Fed. Reg. 55,524 (December 21, 1976)

4. Since July 1, 1979, the offset policy has been in effect in only very limited circumstances, having been replaced, for the most part, by a growth ban in non-attainment areas, 42 U.S.C.A. § 7410 (a)(2)(I) (1978 Pam.), or revised state implementation plans. 44 Fed. Reg. 3,275 (January 16, 1979).

5. Quarles, *Federal Regulation of New Industrial Plants*, 10 ENVIR. REP. Monograph No. 28 (May 4, 1979).

6. "EPA's Emission Offset Policy," *Environmental Science & Technology*, vol. 12 no. 9, September 1978.

7. Bradley I. Raffle, *Prevention of Significant Deterioration and Nonattainment under the Clean Air Act -- A Comprehensive Review*, 10 ENVIR. REP. Monograph No. 27 (May 4, 1979).

8. 42 U.S.C.A. §§ 7502(b)(6), 7503, 7502 note (Nonattainment Areas, State Implementation Plan Revision), 7475 (1978) Pam.).

9. 42 U.S.C.A. §§ 7470-7471 (1978 Pam.).

10. 40 C.F.R. § 52.21 (1978).

11. 42 U.S.C.A. §§ 7410, 7502 (1978 Pam.).

12. 40 C.F.R. § 51.18 (1978 Pam.).

13. Union Electric Co. v. EPA, 427 U.S. 246 (1976); Train v. NRDC, 421 U.S. 60 (1975); Sierra Club v. EPA, 540 F.2d 1114 (D.C. Cir. 1976); Sierra Club v. Ruckelshaus, 344 F. Supp. 253 (D.D.C. 1972), aff'd per curiam, 2 ENVT'L L. REP. 20656 (D.C. Cir. 1972), aff'd by an evenly divided court sub nom. Fri v. Sierra Club, 412 U.S. 541 (1973).

14. 7 ENVT'L L. REP. 10029 (1977).

15. 41 Fed. Reg. 55,524 (December 21, 1976).

16. 42 U.S.C.A. §§ 7502 note, 7503 (1978 Pam.).

17. 44 Fed. Reg. 3,274 (January 16, 1979).

18. 44 Fed. Reg. 51,924 (September 5, 1979).

19. 42 U.S.C.A. § 7602(j) (1978 Pam.).

20. Alabama Power Co. v. Costle, ____F.2d____ 13 ENVIR. REP. CAS.
    1993, 2003-05 (D.C. Cir. 1979); cf. 44 Fed. Reg. 51,956 (September 5, 1979) (proposed). These rules were proposed after a preliminary decision in the case in June, 606 F.2d 1068 (D.C. Cir. 1979). Although the decision directly affected only the PSD rules, some of the same definitions also govern the nonattainment rules; so the EPA proposed new rules for both PSD and offsetting to comply with the court's ruling. See 44 Fed. Reg. 51,925 (September 5, 1979).

21. Alabama Power Co. v. Costle, ____F.2d____ 13 ENVIR. REP. CAS.
    1993.

22. 44 Fed. Reg. 3,282 (January 16, 1979), 40 C.F.R. Part 51, App. S;

23. 44 Fed. Reg. 3,284 (January 16, 1979).

24. 44 Fed. Reg. 3,279 (January 16, 1979).

25. Telephone conversation, Joel Schwartz, policy analyst, Policy Planning Division, EPA, Washington, D.C., February 5, 1980.

26. 42 U.S.C.A. § 7502(a)(1) (1978 Pam.).

27. See 44 Fed. Reg. 3,275 (January 16, 1979);

28. As of mid-January, the EPA had fully approved one SIP, disapproved one, and conditionally approved eight. Telephone conversation with Steve Kuhrtz, Special Assistant to the Assistant Administrator for Air, Noise and Radiation, EPA, Washington, D.C., January 17, 1980.

29. 42 U.S.C.A. § 7410(a)(2)(I) (1978 Pam.).

30. U.S.C.A. § 7502(a)(1).

31. U.S.C.A. § 7503(1).

32. 44 Fed. Reg. 3,741 (January 18, 1979).

33. 44 Fed. Reg. 51,927, 51,948 (September 5, 1979) (proposed).

34. See 44 Fed. Reg. 51,926 (September 5, 1979) (proposed).

35. Fed. Reg. 71,780, 71,783 (December 11, 1979).

36. 9 ENVT'L L. REP. 10027 (1979).

37. 10 ENVIR. REP. 1591 (1979).

38. 8 ENVT'L L. REP. 10052 (1978).

39. 44 Fed. Reg. 71,781, 71,782 (December 11, 1979).

40. 44 Fed. Reg 51,926, 51,957-58 (September 5, 1979).

41. 44 Fed. Reg. 3,742 (January 18, 1979).

42. Wes Vivian, William Hall, An Empirical Examination of U.S. Market Trading in Air Pollution Offsets, Ann Arbor: University of Michigan, December 1979 (Draft), 8.

43. William H. Foskett, Emission Offset Policy at Work: A Summary Analysis of Eight Cases, Washington, D.C.: Performance Development Institute (for National Bureau of Standards), 1979.

44. Aside from the difficulty for industry of arranging a large number of such small, decentralized solar installations, the difficulty of verification is likely to make the EPA cautious in authroizing SIPs to allow offsets from uncontrolled sources. Telephone conversation, Joel Schwartz, note 25 above.

45. For cost comparisons in a specific case, see Peter de Leon, Kenneth C. Brown, et al., Solar Technology Application to Enhanced Oil Recovery, Golden, CO: SERI, SERI/TR-352-392 (UC-59B), December 1979.

46. Telephone conversations, Ken Brown, senior engineer, Solar Energy Research Institute, Golden, Colorado, February 4, 7, 1980. Brown notes that a recent, small survey turned up three such single-shift plants, a poultry processor in Dallas, a clothing manufacturer in Amarillo, Texas, and a metal fabricator in Tennessee.

47. While this assumption is questionable now, changes in the regulatory scheme in the next few years may make it less so. The most likely forthcoming change is elimination of the 24-hour short-term emission standard in favor of a 30-day standard. Telephone conversation, Joel Schwartz, note 25 above.

48. In 1978, industry used an estimated 29.228 quads, of an estimated U.S. energy use of 78.393 quads. Monthly Energy Review (Energy Information Administration), November 1979, DOE/EIA-0035/9(79), NTISUB/E/127-009, at 8. A quad is a quadrillion ($10^{15}$) Btus.

49. The IPH portion of industrial demand has been variously estimated, but DOE data suggest that 47% of demand in 1974 was for IPH. Energy Information Administration, End Use Energy Consumption Data Base - Series 1 Tables, Washington, D.C.: EIA, December 1977 (Draft), vol. 1.

50. Solar Energy Research Institute, Putting the Sun to Work in Industry, Golden, CO: SERI, September 1979 (2d ed.).

51. See James B. Trice, "Textile Dyeing," in Proceedings: Solar Industrial Process Heat Conference, October 18-20, 1978, vol. 1, Golden, CO: SERI, SERI/TP-49-065.

52. See E.J. Carnegie, "Prune and Raisin Drying," in *id.*; Gerald R. Grinn, "Soybean Drying," in *id.*; P. D. Seirer, Jr., "Onion Drying," in *id.*

53. See Herbert F. Schuler, Duane F. Rost, Gene Ameduri, "Federal Government Taxes Sun's Energy at 48% Rate," Solar Engineering, vol. 4 no. 3, March 1979.

54. 44 Fed. Reg. 3,285 (January 16, 1979).

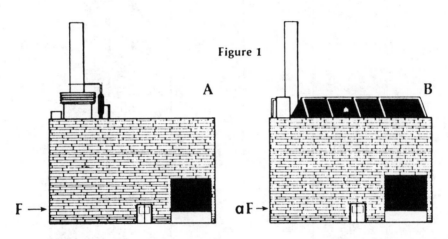

**Figure 1**

A

B

Factory A adds pollution control equipment which reduces polution

F = Fuel
Total Operating Cost = Cost of P + Cost of F

Factory B adds solar energy system which reduces polution and provides energy

Total Operating Cost =
Cost of S + (Cost of $\alpha$F) ~ O

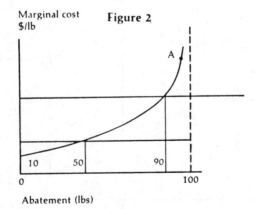

Marginal cost
$/lb     **Figure 2**

Abatement (lbs)

Graph—Adapted from D. N. Dewees, C. K. Everson, W. A. Sims, *Economic Analysis of Environmental Polices* (Univ. of Toronto Press, 1975)

Table 1

| Thermal Power Plant 200 MW | Capital Cost ($/kW$_t$) | Operating and Maintenance Cost (O + M) (¢/kWh$_t$) | Levelized Cost $\left[\dfrac{\text{Capital Cost} \times 0.10}{\text{Annual Output}} + O + M\right]$ c) (¢/kWh$_t$) | Energy Cost (¢/kWh$_t$) | Total Plant Operating Costs (¢/kWh$_t$) f) |
|---|---|---|---|---|---|
| Pollution Controls (SO$_2$) a) | 90 | 0.55 | 0.86 | 3.0 d) | 3.86 g) |
| Solar Energy | Near Term - 3,000 b)<br>Long Term - 1,000 b) | 0.10<br>0.10 | 10.27<br>3.52 | 0 e)<br>0 e) | 10.37<br>3.52 |

a) Limestone scrubbing;
b) Assume no saving on cost of capital of conventional energy production system, used as occasional backup for solar system;
c) Assume 10% fixed charge rate (FCR); annual output = 584,000,000 kWh$_t$, based on a single, daytime shift;
d) 3¢/kWh$_t$ = $8.78/MBtu, equivalent to purchase price of $44/bbl of oil where oil is used at 80% conversion efficiency;
e) Assume output of solar system is 584,000,000 kWh$_t$; additional fuel costs to match conventional output then near zero;
f) Calculations done on before-tax basis; i.e., no allowance made for deductibility of fuel cost as an operating expense nor for 20% investment tax credit on industrial installation of solar equipment.
g) Total plant operating costs = levelized cost of pollution control + energy cost.

**Table 2**

**Solar IPH Field Tests**

| Location | Process | Collectors | Owner | Status |
|---|---|---|---|---|
| **Hot Water (140°-212°F)** | | | | |
| Sacramento, Calif. | can washing | flat-plate & parabolic trough | Campbell Soup Co. | operational (April 1978) |
| Harrisburg, Pa. | concrete block curing | multiple reflector | York Building Products | operational (Sept. 1978) |
| LaFrance, S.C. | textile dyeing | evacuated tube | Riegel Textile Corp. | operational (June 1978) |
| **Hot Air (140°-212°F)** | | | | |
| Fresno, Calif. | fruit drying | flat-plate | Lamanuzzi & Pantaleo Foods | operational (May 1978) |
| Canton, Miss. | kiln drying of lumber | flat-plate | LaCour Kiln Services, Inc. | operational (Nov. 1977) |
| Decatur, Ala. | soybean drying | flat-plate | Gold Kist, Inc. | operational (May 1978) |
| Gilroy, Calif. | onion drying | evacuated tube | Gilroy Foods, Inc. | operational (Sept. 1978) |
| **Low Temperature Steam (212°-350°F)** | | | | |
| Fairfax, Ala. | fabric drying | parabolic trough | WestPoint Pepperell | operational (Sept. 1978) |
| Sherman, Tex. | gauze bleaching | parabolic trough | Johnson & Johnson | operational (Nov. 1979) |
| Pasadena, Calif. | laundry | parabolic trough | Home Cleaning & Laundry | construction |
| Bradenton, Fla. | orange juice pasteurization | evacuated tube | Tropicana Products, Inc. | construction |
| **Intermediate Temperature Steam (350°-550°F)** | | | | |
| Mobile, Ala. | oil heating | parabolic trough | Ergon, Inc. | design only |
| Dalton, Ga. | latex production | multiple reflector | Dow Chemical | construction |
| Newberry Springs, Calif. | hectorite processing | parabolic trough | Nat'l Lead industries | design only |
| Hobbs, N. Mex. | oil refinery | parabolic trough | Southern Union Co. | construction |
| San Antonio, Tex. | brewery | parabolic trough | Lone Star Brewing Co. | design only |
| Henderson, Nev. | chlorine manufacturing | parabolic trough | Stauffer Chemical Co. | design only |
| Ontario, Ore. | potato processing | parabolic trough | Ore-Ida Co. | construction |
| **Privately Funded** | | | | |
| Youngstown, Ohio | aluminum anodizing | fixed half parabolic | General Extrusions, Inc. | operational (Sept. 1977) |
| Jacksonville, Fla. | beer pasteurization | evacuated tube | Anheuser-Busch, Inc. | operational (Feb. 1978) |

PROGRAM FOR THE
SEVENTH NATIONAL CONFERENCE
ON ENERGY AND THE ENVIRONMENT


WORKSHOPS:  Sunday, November 30, P.M.

W.1  Rulemaking and its Effect on Energy Development, *Ray Kary*

W.2  Resource Conservation and Recovery -- An Industrial Perspective, *Lori Spencer*

W.3  Fine Particulate Control for Industrial and Utility Boilers, *Louis Theodore, Anthony Buonicore*

W.4  Environmental Control Implications of Coal Use, *Paul Farber, David Livengood*


PLENARY SESSION:  Monday, December 1, A.M.

    Chairman:  *Joseph Reynolds*
               Chairman, Chemical Engineering
               Manhattan College
               Bronx, N.Y.

1.1  Keynote Address, *Ted Williams*, Acting Director, Arizona Health Dept.

1.2  Mount St. Helens - 1980, *Donal R. Mullineaux*

1.3  Cultural Impacts of Energy Development on the Northern Cheyenne, *Ted Rising Sun*

SESSION 2:  CONSEQUENCES OF ENERGY DEVELOPMENT ON AMERICAN INDIAN
            RESERVATIONS - Monday, December 1, P.M.

    Chairmen:  *Alfred Q. Colton*          *Harold W. Tso*
               Manager, Environmental      Director, Environmental Pro-
                 Services Dept.              tection Commission,
               Salt River Project          Navajo Nation

2.1  Environmental Issues Regarding the Baca Geothermal Project, *Harold Sando*

2.2  Environmental Analysis Activities for Energy Resource Development on Indian Lands, *George Thomas*

2.3  Impact of Uranium Mining on Local Residences, *Harold W. Tso*

2.4  Socioeconomic Impact Assessment of Energy Development on Indian Reservations, *A.T. Anderson*

2.5  Panel Discussion:  Solar-Renewable Energy Development on Indian Reservations, *Dean B. Suagee, Richard Curley* and *Alan Parker*

SESSION 3:  ENERGY UTILIZATION AND IMPACT ON VISIBILITY - Monday, December 1, P.M.

Chairmen:  *Raymond Kary*                          *Donald Moon*
           Manager, Environmental          Supervisory Environmental
                 Dept.                              Analyst,
           Arizona Public Service          Salt River Project

3.1  How Visibility Regulations Will Affect Energy Development in the West, *Terry L. Thoem* and *David B. Joseph*

3.2  Visibility of a $NO_2$ Plume in Northern Arizona, *Neil S. Berman*

3.3  A Sulfate Visibility Relationship Derived from Field Data in New York State, *John M. Stansfield* and *James B. Homolya*

3.4  Statutory and Regulatory Aspects of the Visibility Issue, *Michael L. Teague*

3.5  Panel Discussion:  Can Visibility Be Regulated Successfully? *Edwin Roberts, Prem S. Bhardwaja, Nils Larson, Michael Teague* and *Terry L. Thoem*

SESSION 4:  CONSIDERATIONS IN THE DEVELOPMENT OF OIL SHALE - Monday, December 1, P.M.

Chairmen:  *Gary Baughman*                   *Marla M. Moody*
           Senior Project Engineer,     Hydrologist, Rio Blanco
           Colorado School of Mines         Oil Shale Company
              Research Institute

4.1  Overview of Oil Shale Development, *Thomas A. Sladek*

4.2  Water Availability and Requirements for Western Oil Shale Development, *G.A. Miller*

4.3  Air Quality Impact of Oil Shale Development, *Terry L. Thoem*

4.4  Ecological Effects of Oil Shale Development: Problems, Perspectives, and Approaches, *Thomas E. Hakonson*

4.5  Estimation and Mitigation of Socioeconomic Impacts of Western Oil Shale Development, *Dee R. Wernette*

4.6  Oil Recovery and Recycle Water Treatment for In Situ Oil Sand Production, *J. Nagendran* and *S.E. Hrudey*

SESSION 5:  MATERIALS: EXCESSES, SHORTAGES OR DISRUPTIONS - Tuesday, December 2, A.M.

Chairmen:  *Michael J. Deutch*              *Edmund R. Rolinski*
           Consulting Engineer and     Professor of Chemical
                Economist                      Engineering,
                                        University of Dayton

5.1  Current Problems in Nonfuel Materials Policy: Priorities, Trade-offs, Institutional Constraints, and Countermeasures to Cope with Shortages in Supply and Disruption in Essential Production Services, *Michael J. Deutch*

5.2 Environmental Consequences of a Coal Conversion Industry in Western Kentucky, *Harry G. Enoch*

5.3 Resource Conservation and Air Pollution Control: A Case Study of Compatibility, *Anthony J. Buonicore, Joseph Billotti* and *Panaj R. Desai*

5.4 The Air Force Prepares for Synthetic Jet Fuel, *Charles L. Delaney* and *Herbert R. Lander*

SESSION 6: EMERGING INNOVATIVE PROCESS TECHNOLOGY - Tuesday, December 2, A.M. and P.M.

Chairmen: *Anthony J. Buonicore*  *Imre Zwiebel*
Vice-President and  Chairman, Dept. of Chemical
General Manager,  Engineering
York Research Corp.  Arizona State University

6.1 New Process Technology: Energy Conservation Potential and Environmental Impacts, *Charles L. Kusik* and *James I. Stevens, Harry E. Bostian, Herbert S. Skovronek*

6.2 Environmental, Operational, and Economic Aspects of Thirteen Selected Energy Technologies, *Larry Hoffman*

6.3 Environmental Considerations and Comparisons of Alternate Generation Systems, *Kennard F. Kosky* and *Jackson B. Sosebee*

6.4 Environmental Assessment and Site-Selection Methodology for Selected Energy Storage Systems, *Louis J. DiMento*

6.5 An Environmental Assessment of Coal-Using Prime Movers for Residential/Commercial Total Energy Systems, *M.L. Jain, J.C. Bratis* and *T.J. Marciniak*

6.6 Environmental Effects of a Successful Underground Coal Gasification Test in a Deep, Thin-Seam Bituminous Coal, *J.W. Martin* and *A.J. Liberatore*

6.7 Impact of Different Environmental Constraint Levels on an Indirect Coal Liquefaction Process, *P.J. Johnson* and *Suman P.N. Singh*

6.8 Water Quality and Solid Waste Management for a Proposed TVA Atmospheric Fluidized-Bed Combustion Demonstration Plant, *T.Y. Julian Chu, D.B. Cox, John W. Shipp, Jr.* and *L.H. Woosley, Jr.*

6.9 Environmental Considerations of Cogeneration-Based District Heating, *Elliot P. Levine* and *Danilo J. Santini*

6.10 A Standardized Costing Method for Emerging Technologies, *G.R. Miller* and *M.S. Massoudi*

SESSION 7:   REGULATORY IMPACT - Tuesday, December 2, P.M.

Chairmen:   *Myron Gottlieb*          *Roger Raufer*
            Chief, Fossil and In-      Manager, Environmental
               exhaustible Technology,    Studies,
            U.S. Department of Energy  ETA Engineering

7.1   Environmental Overview for Emerging Coal Technologies,
      *James E. Mann* and *Thomas C. Ruppel*

7.2   The Impact of the Resource Conservation and Recovery Act (RCRA) on
      Coal-Fired Facilities, *Dave Burstein*

7.3   PSD and Energy Development: The First Five Years, *D.B. Garvey*
      and *S.B. Moser*

7.4   Impact of Recent Energy and Environmental Legislation on the
      Petroleum Refining Industry, *Doan L. Phung*

7.5   Regulatory Responses to the Problem of Acidic Precipitation,
      *Ralph Luken*

7.6   The Uranium Mystique: A Search for Identity, *Stephen Schermerhorn*


SESSION 8: EMERGING INNOVATIVE CONTROL TECHNOLOGY - Tuesday,
           December 2, P.M.

Chairmen:   *Cecil Lue-Hing*          *Craig Toussaint*
            Director for Research      Program Manager, Environmental
               & Development              Protection Programs,
            Metropolitan Sanitary      U.S. Department of Energy
            District of Greater
            Chicago

8.1   Design of Environmental Control Systems for Coal Synthetic Fuels
      Plants, *Edward C. Mangold*

8.2   Methodology for Comparison of Alternative Processes Using Life-
      Cycle Consumption, *J.C. Uhrmacher*

8.3   Simultaneous Heat Recovery and Pollution Control with Fluidized-
      Bed Heat Exchangers,   *John Vogel* and *Paul Grogan*

8.4   Use of Nahcolite for Coal-Fired Power Plants, *Dennis E. Lapp* and
      *Eric Samuel, Theodore G. Brna, Navin D. Shah* and *Ben Weichman,
      Ronald L. Ostop*

8.5   High Temperature Particulate Removal by Granular Bed Filtration,
      *Larry L. Moresco, Jerry Cooper* and *John Guillory*

8.6   Hydrocarbon Emission Control with Energy Savings and a Positive
      Payback, *James H. Mueller*

8.7   Flue Gas Desulfurization in Molten Salt Electrochemical Cell:
      Preliminary Experiments, *Omar E. Abdel-Salam*

SESSION 9: FEASIBILITY OF SYNTHETIC FUELS DEVELOPMENT - Wednesday,
          December 3, A.M. and P.M.

   Chairman:   *Stephen L. Brown*
               Director, Center for
                  Resources and En-
                  vironmental Systems
                  Studies,
               S.R.I. International

9.1  DOE Alternative Fuels Solicitations, *Michael E. Card*

9.2  A Perspective on the Economic Readiness of Methods for Producing
     Coal Liquids, *Ronald L. Dickenson, Dale R. Simbeck* and *A. James
     Moll*

9.3  Synthetic Fuels and the Environment, *Minh-Triet Lethi*

9.4  Potential Environmental Impacts of Synthetic Fuels Production,
     *Peter W. House*

9.5  Synfuel Development in the West: National Versus Local Perspective,
     *Douglas C. Larson*

9.6  Selecting Sites for Synfuels Plants, *Charles A. D'Ambra* and
     *Gaylord M. Northup*

9.7 Environmental Development Capacity for Siting Synfuels Industries,
     *Robert V. Steele* and *F. Jerome Hinkle*

9.8 Potential Health Problems from Coal Conversion Technologies,
     *R.A. Wadden* and *A.L. Trabert*

9.9 A Human Health Assessment Model for Pollutants Discharged from
     Emerging Energy Technologies, *Amiram Roffman* and *Joseph A. Maser*


SESSION 10: INDUSTRIAL ENERGY USAGE AND CONSERVATION - Wednesday,
            December 3, A.M.

   Chairmen:   *David B. Nelson*        *James C. Wade*
               Monsanto Research Corp.   Dept. of Agricultural
                                            Economics,
                                         University of Arizona

10.1 Industrial Energy Conservation-What Measure for Efficiency?
     *D.B. Wilson*

10.2 The Environmental Desirability of Nuclear Process Heat,
     *Leon Green, Jr.*

10.3 Industrial Cogeneration: Economic Feasibility of Selected Examples,
     *C.L. Kusik* and *L.K. Fox, J.R. Hamm* and *K.D. Weeks, I. King* and
     *V.P. Buscemi*

10.4 Productivity and Thermal Energy Storage Technology, *Ben-Chieh Liu,
     Joe Asbury, Bob Giese, Jarilaos Stavrou* and *Chuck Maslowski*

10.5 Process Models: Analytical Tools for Managing Industrial Energy Systems, *Stephen O. Howe, David A. Pilati* and *Chip Balzer, F.T. Sparrow*

10.6 An Approach to Measuring the Potential for Energy Conservation in the Industrial Sector, *Samir Salama* and *Harold Kalkstein*


SESSION 11:  MINING AND SMELTING OPERATIONS - Wednesday, December 3, P.M.

   Chairmen:  *Nils Larson*                    *Earle B. Amey*
              Chief, Bureau of Air            Staff Chemical Engineer,
                 Quality Control,            U.S. Bureau of Mines
              Arizona Dept. of Health

11.1 Effects of Uranium Mining and Milling on Surface Water in New Mexico, *Lynn L. Brandvold, Donald K. Brandvold* and *Carl J. Popp*

11.2 Evaluation and Control of Permeability Damage During Carbonate Solution Mining of Uranium, *Terry R. Guilinger, I.H. Silberberg* and *Robert S. Schecter*

11.3 Environmental and Energy Considerations of Foreign Nonferrous Smelter Technology, *A. Christian Worrell III, Thomas K. Corwin* and *Mary A. Taft, John O. Burckle*

11.4 Energy Conservation and Pollution Abatement at Phosphorus Furnaces, *James C. Barber*


SESSION 12:  CONSIDERATIONS IN THE DEVELOPMENT OF SOLAR ENERGY - Wednesday, December 3, P.M.

   Chairmen:  *Lewis Goidell*                 *Carl L. Strojan*
              Environmental Scientist,        Senior Environmental Scientist,
              Savannah River Operations       Solar Energy Research Institute
                 Office,
              U.S. Department of Energy

12.1 Health Risks of Photovoltaic Energy Technologies, *Paul D. Moskowitz*

12.2 Environmental Regulations: Applicability to Advanced Photovoltaic Concepts, *D.A. Shaller*

12.3 Solar Power Satellite Assessment, *F.A. Koomanoff*

12.4 The Environmental Effects of Producing Biomass in Pine and Hardwood Plantations for Renewable Energy Resources, *G.J. Hollod, J.C. Corey* and *E.F. Dyer*

12.5 Factors Affecting Wood Heater Emissions and Thermal Performance, *J.P. Harper* and *C.V. Knight*

12.6 Solar Energy and the Search for Emission Offsets, *Barbara Euser*

# AWARDS

All sessions and papers were critically evaluated by a panel of five judges for the purpose of determining the session and papers that deserved special recognition. The following persons were nominated by the panel for their outstanding contributions to the Conference:

Best Session:

Awarded to:

**Stephen L. Brown**
Chairman of Session 9: Feasibility of Synthetic Fuels Development

Best Paper and Presentation[1]:

Awarded to:

**Charles L. Kusik**
New Process Technology: Energy Conservation Potential and Environmental Impacts (Co-authors: James I. Stevens, Harry E. Bostian and Herbert S. Skrovronek)

Honorable Mention:

**Stephen L. Brown**[2]
Environmental Development Capacity for Siting Synfuels Industries (Co-authors: Robert V. Steele and F. Jerome Hinkle)

Honorable Mention:

**Kennard F. Kosky**
Environmental Considerations and Comparisons of Alternate Generation Systems (Co-author: Jackson B. Sosebee)

Honorable Mention:

**Larry L. Moresco**
High-Temperature Particulate Removal by Granular Bed Filtration (Co-authors: Jerry Cooper and John Guillory)


The Awards Committee consisted of:

Thayer Masoner, Jacobs Engineering Group, Inc. (Chairman)
Hank Coffer, C. K. GeoEnergy
M. Rita Howe, Pathfinder Publishing Services
Torsten Rothman, Enviro Control, Inc.
Eugene M. Wewerka, Los Alamos Scientific Laboratory

---

[1] Papers were judged on the basis of both content and presentation.

[2] This paper was presented by Stephen L. Brown, who, although a contributor to the paper, was not listed in the program as an author.

THE SEVENTH NATIONAL CONFERENCE ON ENERGY AND THE ENVIRONMENT

List of Attendees

Peter F. Allard, P.E.
Chief Chemical Engineer
Engineers Testing Laboratories
P.O. Box 21387
Phoenix, Arizona 85036

James A. Amend, P.E.
Consulting Engineer
3037 Del Mar Avenue
Yuma, Arizona 85364

Earle B. Amery
Staff General Engineer
Bureau of Mines
1508 Crofton Parkway
Crofton, Md. 21114

A.T. Anderson, President
AISES
35 Porter Avenue
Middlebury, Ct. 06770

George Aulenbacher
Multimineral Corp.
330 North Belt East
Houston, Texas 77060

James C. Barber
James C. Barber and Associates
Suite 118, Courtview Towers
Florence, Al. 35630

Gary L. Baughman
Sr. Project Engineer
Colorado School of Mines
Research Institute
P.O. Box 112
Golden, Colorado

Dale L. Beck, Vice President
Engineering Design Const. Ltd.
3030 S. Rural Road
Tempe, AZ. 85282

Mike A. Beckwith, Scientist
Battelle-Northwest Lab.
P.O. Box 999
Battelle Blvd.
Richland, Washington 99352

Tracey Bell, Envir. Engineer
A 251
401 Chestnut Street
Chattanooga, Tenn. 37401

Neil S. Berman, Professor
Arizona State University
Dept. of Chemical Engineering
Tempe, Arizona 85281

Prem S. Bhardwaja,
Sr. Environmental Analyst
Salt River Project
P.O. Box 1980
Phoenix, Arizona 85001

Iona Black, Chemistry Instructor
Ballou S/M H.S.
2420 16 Street
N.W. #404
Washington, D.C. 20009

James O. Blankenship
Forest Service, USDA
1221 Buttonwood Drive
Ft. Collins, Colorado 80525

John F. Boland, Jr., Attorney
Evans, Kitchell & Jenckes
363 North First Avenue
Phoenix, Arizona 85003

Larry Bowles, Marketing Manager
Ecological Services
P.O. Box 225621 M/S 349
Dallas, Texas 75265

Lynn A. Brandvold, Chemist
New Mexico Bureau of Mines and
Mineral Resources
Workman Center, NMIMT
Socorro, New Mexico 87801

Prof. Donald K. Brandvold
New Mexico Institute of Mining
Technology
Dept. of Chemical, NMIMT
Socorro, New Mexico 87801

Carolyn Bread, Community Planner
San Carlos Apache Tribe
P.O. Box O
San Carlos, Arizona 85550

Lynne A. Bridson, Air Quality
Analyst
Pima County Air Quality Cont.
151 W. Congress Street
Tucson, Arizona 85701

Udell T. Brown, Planning Director
San Carlos Apache Tribe
P.O. Box O
San Carlos, Arizona 85550

Keith J. Brown, President
North American Weather Consultant
1141 East 3900 So. #A130
Salt Lake City, Utah 84117

Stephen L. Brown, Director Cress
SRI International
333 Ravenswood Ave.
Menlo Park, Calif. 94025

Dr. J. Clement Burdick, III
Mgr. of Permitting & Siting
TRC Environmental Consultants
8775 East Orchard Rd. Suite 816
Englewood, Colorado 80111

Anthony J. Buonicore, Vice Pres.
York Service Company
One Research Drive
Stamford, Conn. 06906

David Burstein, Design Mgr.
Engineering Science
57 Executive Park South, #590
Atlanta, Georgia 30329

Janet M. Cain, Acting Chief,
Environmental Planning
Minnesota Pollution Control
Agency
1935 West Co Road B2
Roseville, Mn. 55113

Robert B. Candeloria
Environmental Engineer
Salt River Project
P.O. Box W
Page, Arizona

J.A. Catalano, President
AEROCOMP, Inc.
3303 Harbor Blvd.
Suite F-4
Costa Mesa, Calif. 92626

Michael A. Chartock
Research Fellow, Sci. &
Public Policy
University of Oklahoma
601 Elm Str. #431
Norman, Oklahoma 73019

T.Y. Julian Chu, Envir.Specialist
Tennessee Valley Authority
NRB-N
Norris, Tenn. 37828

Mr. Hank Coffer, Judge
C.K. GeoEnergy
3376 So. Eastern Ave.
Suite 145
Las Vegas, Nevada 89109

Alfred Q. Colton, Mgr. Envir. Serv.
Salt River Project
P.O. Box 1900
Phoenix, Arizona 85001

Vincent D. Coppolecchia
Gibbs and Hill Inc.
393 Seventh Avenue
New York, New York 10001

Leo Craton, Dir. of Marketing
HDR Sciences
824 Anacapa
Santa Barbara, Calif. 93101

Lawrence Crisafulli, P.H. Engineer
Maricopa Company
Health Dept.
1845 East Roosevelt
Phoenix, Arizona

Walter H. Croft, P.E.
Consulting Engineer
3021 North 15 Drive
Phoenix, Arizona 85015

Charles Crowley, Chief Env.Serv. Br.
Rural Electric Administration
USDA-REA, Agri-South
Room 2860
Washington, D.C. 20250

Charles D'Ambra, Assoc. Scientist
Center for Environment & Man Inc.
275 Windsor Street
Hartford, Conn. 06120

Dr. Gregory A. Daneke, Dir. of Env.
Energy Planning Program
University of Arizona (BPA Bldg.)
Tucson, Arizona 85721

James E. Dean, Envir. Coordinator
Bureau of Land Management Rm.700
1600 Broadway
Denver, Colo. 80202

Ann De Bano, Energy Chairperson
League of Women Voters of East
Maricopa
507 E. Fremont Drive
Tempe, Arizona 85282

Dr. Michael J. Deutch
Consulting Engineer
2820 32 Street N.W.
Washington, D.C. 20008

Ronald L. Dickenson, Principal
Synthetic Fuels Associates
2 Palo Alto Square
Suite 528
Palo Alto, Calif. 94304

Louis J. DiMento, Sr. Envir.
Planner
NUS Corporation
4 Research Place
Rockville, Md. 20850

Stephen Dixon, Proj.Engineer
Hittman Associates Inc.
9190 Red Branch Road
Columbia, Maryland 21045

Thomas Douville, Director
Mesa County Health Dept.
515 Patterson Road
Grand Junction, Colo. 81501

Wayne R. Duchemin
Transportation/Air Quality
Planner
Cape Cod Regional Transit
Authority
275 Mill Way - P.O.Box 318
Barnstable, Ma. 02630

Deborah Duffy, Policy Editor
McGraw Hill's Coal Week
621 National Press Bldg.
Washington, D.C. 20045

Michael Dyer, Engineer
Salt River Project
P.O. Box 1980
Phoenix, Arizona 85001

Gerald K. Eddlemon, Res.Assoc.
Union Carbide (ORNL)
Oak Ridge National Lab.
Bldg. 1505, P.O. Box X
Oak Ridge, Tenn. 37830

Steve Ehrman
Salt River Project
P.O. Box 1980
Phoenix, Arizona 85001

Al Elliott, Chemistry Instructor
USAF Academy
4303 G
Colorado Springs, Colo. 80840

Harry Enoch, Asst.Dir. of Tech.
Assessment
Kentucky Dept. of Energy
Box 11888
Lexington, Ky. 40578

Bert Enserink, Director
Dynamic Science Inc.
1850 W. Pinnacle Peak Road
Phoenix, Arizona 85027

Barbara J. Euser, Managing Editor
Solar Law Reporter
Solar Energy Research Institute
1617 Cole Blvd.
Golden, Colorado 80401

Paul Farber, Program Manager
Argonne National Lab
Bldg. 362
9700 S. Cass Avenue
Argonne, Illinois 60439

Dennis Farrell
Science Reporter
Phoenix Gazette
Phoenix, Arizona

Roger Ferland, Asst.Attorney Gen'l
State of Arizona
1740 West Adams St. Rm. 211
Phoenix, Arizona 85007

Richard C. Flory, Marketing Mgr.
TRC Environmental Consultants
8775 E. Orchard Rd. Suite 816
Englewood, Colo. 80111

Lee Fox, Chemical Engineer
Arthur D. Little Inc.
Acorn Park
Cambridge, Mass. 02140

Carl A. Fox
Research Scientist
Southern California Edison
P.O. Box 800
Rosemead, Calif. 91770

Harold E. Francis, P.E.
P.O. Box 1649
Lake Havasu City, Arizona 86403

Peter Frascino, Supervising
Environmental Engineer
United Eng. & Constructors Inc.
100 Summer St.
Boston, Mass. 02110

Arnold Frautnick, Engineer
Salt River Project
P.O. Box 1980
Phoenix, Arizona 85001

Bayne Freeland
The Phoenix Gazette
Phoenix, Arizona

William F. Fuller, Mgr. CE/SM
Beckman Instruments
2500 Harbor Blvd.
Fullerton, Calif. 92634

Jim Furlow
Dames & Moore
234 N. Central - Suite 111A
Phoenix, Arizona 85004

Richard Ganzel, Associate Prof.
University of Nevada, Reno
Political Science Department
Reno, Nevada 89557

Doris Garvey, Scientific Assoc.
Argonne National Laboratory
9700 South Cass Ave. Bldg. 12
Argonne, Illinois 60439

Stephen L. Gash, Mgr. Gov't
and Envir. Affairs
Marathon Resources
1515 Arapahoe Street - Suite 1300
Denver, Colorado 80202

Robert W. George II
Foster-Miller Association
350 Second Avenue
Waltham, Ma. 02154

Kurt Gernerd
Utah State University
1058 Crescant Drive
Logan, Utah 84321

Lewis C. Goidel, Envir. Scientist
U.S. Department of Energy
Savannah River Operations Office
P.O. Box A
Aiken, South Carolina 29801

Mel Gordon, Chairman
PAC N.W. River Basin
1 Columbia River
Vancouver, Washington 98669

Allen Gordon, Envir.Engineer
Stone & Webster Engineering Corp.
3 Executive Campus - Box 5200
Cherry Hill, N.J. 08034

Del Green, President
Del Green Associates Inc.
1155C Chess Drive - Suite A
Foster City, Calif. 94404

Leon Green, Jr. Consultant
2101 Connecticut Avenue N.W.
Washington, D.C. 20008

Kevin Greene
Citizens for a Better Environment
59 East Van Buren, Suite 1600
Chicago, Illinois 60605

Terry Guilinger, Graduate Student
University of Texas at Austin
Petroleum Engineering Bldg. 211
Austin, Texas 78712

G.T. Gutierrez, Principal
Gutierrez-Palmerberg Inc.
2159 W. Sharon Avenue
Phoenix, Arizona

James L. Guyton, Mgr.
Arizona Dept. of Health Service
1740 W. Adams Street
Phoenix, Arizona 85023

Dennis Haase, Proj. Meteorologist
Dames and Moore
234 N. Central - Suite 111-A
Phoenix, Arizona 85004

Thomas S. Hakonson, Staff Member
Los Alamos Scientific Lab.
P.O. Box 1663
Los Alamos, N.M.

Frederic C. Hamburg, Mgr.
Atmosphere Science Div.
Radiation Mgmt. Corp.
6917 Arlington Rd., Suite 206
Bethesda, Md. 20014

Rex Hamilton, Dir. of Marketing
Law Engineering Testing Co.
2749 Delk Road
Marietta, Georgia 30067

Bill Heckman, Coordinator
Arapahoe Community College
5900 S. Santa Fe
Littleton, Colo. 80120

Michael Heerschap, Envir. Spec.
Hills County Environmental
Protection Commission
1900 Ninth Avenue
Tampa, Florida 33605

Dr. E.R. Hendrickson
Chairman of the Booard
Environmental Science & Eng. Inc.
P.O. Box ESE
Gainesville, Florida 32602

Lawrence Hoffman, President
Hoffman-Muntner Corp.
8750 Georgia Avenue #E134
Silver Spring, Md. 20910

Gregory J. Hollod, Research Engr.
E.I. DuPont Company
P.O. Box 1100 Rt. 5
Aiken, South Carolina 29801

Dr. Steven O. Howe, Asst. Scientist
Brookhaven National Laboratory
Bldg. 475
Upton, New York 11973

M. Rita Howe
Pathfinder Publishing Services
3890 Mowry Avenue
Suite G
Fremont, Calif. 94538

Arthur E. Hudson, Mgr.
TRC Environmental Consultants
8775 East Orchard Road - Suite 816
Englewood, Colorado 80111

Peter Hyde, Chemist
Pima County Air Quality
151 W. Congress Street
Tucson, Arizona 85707

R. Fred Iacobelli, Chief
Bureau of Vehicular Emissions Insp.
Arizona Dept. of Health Services
600 N. 40 Street
Phoenix, Arizona 85008

Arun G. Jhaveri, Energy Mgmt. Spec.
John Graham and Company
1110 Third Avenue
Seattle, Washington 98101

Nancy L. Johnson, Program Analyst-
Policy and Evaluation
Department of Energy
PE-84, Mail Stop 7E-088 FORSTL
Washington, D.C. 20585

Archie Johnston, Mgr. Qual./Envir.
Affairs
Douglas Oil Company
14700 Downey Avenue
Paramount, Calif. 90723

Larry R. Johnson, Asst. Economist
Argonne National Laboratory
9700 South Cass Ave. EES-12
Argonne, Illinois 60439

Richard L. Johnson, Engr. Mgr.
Arizona Public Services Co.
Four Corners Power Plant
Fruitland, N.M. 87416

Ross M. Johnston, Mgr. Envir. Sales
Marblehead Lime Company
300 W. Washington St. #1010
Chicago, Illinois 60606

Thomas W. Kalinowski, Asst.Lab. Dir.
Pacific Environmental Laboratory
657 Howard Street
San Francisco, Calif. 94105

Harold Kalkstein, Proj. Mgr.
Energy & Environmental Analysis Inc.
1111 North 19 Street
Arlington, Va. 22201

Raymond E. Kary, Mgr.Envir.Dept.
Arizona Public Service Co.
P.O. Box 21666
Phoenix, Arizona 85036

Akira Kato, Student
University of Minnesota
1100 Como Ave. S.E. #2
Minneapolis, Minn. 55414

E.K. Kauper, Dir.
Metro Monitoring
436 N. Barranca Avenue
Covina, Calif. 91723

Vic Kebely, Proj. Mgr.
The Aerospace Corp.
2260 E. El Segundo Blvd.
D5/1107
El Segundo, Calif. 90245

Karen Kelley, Sociologist
National Park Service
18th & C St. N.W. Room 3021
Washington, D.C. 20040

Gene Keluche
7700 East McCormick Pkwy.
Scottsdale, Arizona 85258

Jocelyn Kempe, Gov't Affairs Rep.
Chevron, USA
595 Market Street
San Francisco, Calif.

Hari Khanna, Sr. Planner
Arizona Dept. of Transportation
206 S. 17th Avenue, Room 310B
Phoenix, Arizona 85007

Paul Klores, Vice Pres.
Public Affairs
Valley National Bank of Arizona
P.O. Box 71
Phoenix, Arizona 85001

Dr. Charles V. Knight
Consultant to TVA
Tennesseee Valley Authority
6921 Starlite Rd.
Hixson, Tenn. 37343

Harold J. Kholmann, Sr.Vice Pres.
Hydrotechnic Corporation
1250 Broadway
New York, New York 10001

Frederick A. Koomanoff, Dir.SPS Proj.
U.S. Department of Energy
ER-14   MS G-256
Washington, D.C. 20545

Kennard Kosky, Assoc. Vice Pres.
Environmental Science & Engineering
P.O. Box ESG
Gainesville, Florida 32602

Vijendra P. Kothari
Chemical Engieer
Department of Energy DOE/METC
P.O. Box 880
Morgantown, West Virginia 26505

Shyang-Lai Kung, System Engineer
MITRE Corp./Metrek Division
1820 Dolley Madison Blvd.
Mail Stop W249
McLean, Virginia 22101

C.L. Kusik, Sr. Professional Staff
Arthur D. Little Inc.
20 Acorn Park
Cambridge, Mass. 02140

Dennis Lapp
Buell Division Envirotech Corp.
200 North 7th Street
Lebanon, Penn. 17042

Dr. Kerby E. LaPrade
East Texas State University
Commerce, Texas 75428

Doug Larson, Exec. Dir.
Western Interstate Energy Board
3333 Quebeck - Suite 2500
Denver, Colorado 80207

Nils I. Larson, Chief
Bureau Air Quality Control
Arizona Dept. of Health Services
1740 W. Adams
Phoenix, Arizona 85007

John Lerohl, Sr. Envir. Scientist
H.D.R. Sciences
1020 North Fairfax St.
Suite 201
Alexandria, Va. 22314

Minh-Triet Lethi, Energy Policy
EPA
401 M Street S.W.
Washington, D.C. 20460

Elliott P. Levine,Envir.Scientist
Argonne National Lab.
Energy & Environmental Systems Div.
Argonne, Illinois 60439

Arthur J. Liberatore, Proj. Mgr.
Department of Energy
P.O. Box 880
Morgantown, West Va. 26505

Malcolm Lindsay, Economist
Wisc. Division of State Energy
101 S. Webster Street
Madison, Wisc. 53703

Ben-Chieh Liu, Program Mgr.
Argonne National Lab.
9700 South Cass Ave.
Bldg. 12
Argonne, Illinois 60439

C.D. Livengood, Envir. Sys.Engr.
Argonne National Lab.
9700 South Cass Ave.
Bldg. 362
Argonne, Illinois 60439

Keith Long, Ass't Proj. Mgr.
Stearns-Roger
P.O. Box 7165
University Street
Grand Forks, N.D. 58202

Ralph A. Luken, Economist
U.S. EPA
401 M Street S.W.
Washington, D.C. 20460

Edward P. Lynch, Chem. Engr.
Argonne National Lab.
9700 South Cass Ave.
Argonne, Illinois 60439

Stephen V. McBrien, Tech. Staff
MITRE Corp.
1820 Dolley Madison Blvd.
McLean, Va. 22102

H. Roy McBroom, Chief
U.S. Bureau of Land Management
6088 S. Clayton
Littleton, Colo. 80121

William N. McCarthy, Jr., Chem.Engr.
U.S. EPA
10813 Vista Road
Columbia, Md. 21044

Edward C. Mangold, Env.Sys.Scientist
MITRE Corp.
1820 Dolly Madison Blvd.
McLean, Va. 22102

Thomas J. Marciniak, Mgr
Argonne National Lab.
Energy & Env. Systems Div.
9700 South Cass Avenue
Argonne, Illinois 60439

Dan Marlowe, Sr. Chem. Engineer
Lurgi Corp.
1 Davis Drive
Belmont, Calif. 94002

G. Marmer, Envir. Scientist
Argonne National Laboratory
9700 South Cass Ave.
Argonne, Illinois 60439

Kenneth M. Martin, P.E.,Mech. Engr.
John Carollo Engineers
3308 North 3rd Street
Phoenix, Arizona 85012

S. Arthur Martinez
Associate Dean of Instruction
Coastline Community College
13521 Edwards Street
Westminster, Calif. 92683

Shannan Marty, Legislative Intern
Salt River Project
P.O. Box 1980
Phoenix, Arizona 85001

Joseph A. Maser, Sr. Scientist
Energy Impact Associates
P.O. Box 1899
Pittsburg, Penn. 15230

Thayer Masoner Judge
Jacobs Engineering Group Inc.
251 S. Lake
Pasadena, Calif. 91101

Dr. M.S. Massoudi, Sr. Engineer
Teknekron
2118 Milvia St.
Berkely, Calif. 94704

David C. Maurer,
Urban Affairs Mgr.
Phoenix Chamber of Commerce
34 W. Monroe Street
Phoenix, Arizona 85003

Carl J. Mercer, Supervisor
Source Testing
Arizona Department of Health
1740 W. Adams
Phoenix, Arizona 85007

Jay Messer, Research Asst. Prof.
Utah State University
UMC 82
Logan, Utah 84322

Glen A. Miller
USGS - AOSO
131 North 6th
Grand Junction, Colo. 81501

Peter J. Miller, Tech.
Support Mgr.
Texas Instruments Inc.
P.O. Box 225621, MS 349
Dallas, Texas 75265

Gerald R. Miller, Engineer
Teknekron
2118 Milvia Street
Berkeley, Calif. 94704

Robert F. Minnitti
Director-School Facilities
Superintendent of Public Instr.
Old Capitol Bldg.
Olympia, Wa. 98504

Andrew C. Montz, Prog. Mgr.
Radian Corp.
8501 Mo-Pac Blvd.
P.O. Box 9948
Austin, Texas 78766

Donald W. Moon, Supervisor
Environmental Analyst
Salt River Project
P.O. Box 1980
Phoenix, Arizona 85001

Larry L. Moresco, Ph.D.
Project Scientist
Combustion Power Co. Inc.
1346 Willow Road
Menlo Park, Ca. 94025

Paul V. Morgan, Vice Pres. &
General Manager
NUS Corporation
#2 Palo Alto Square - Suite 624
Palo Alto, Calif. 94304

Paul D. Moskowitz, Assoc.Scientist
Brookhaven National Lab.
Bldg. 475
Upton, New York 11973

James H. Mueller, President
REECO Inc.
P.O. Box 600
520 Speedwell Avenue
Morris Plains, N.J. 07950

Donal R. Mullineaux, Geologist
U.S Geological Survey
MS 903, Box 25046, Fed. Center
Denver, Colo. 80225

Keshava S. Murthy, Staff Scientist
Battelle Northwest Lab.
Battelle Blvd.
Richland, Washington 99352

Jay Nagendran, Engineer
Alberta Environment
9820 106th Street
Edmonton, Alberta, Canada

Charles C. Nathan, Dir.
New Mexico Institute of Mining &
Technology
Socorro, New Mexico 87301

Michael A. Neher, Health Engineer
Arizona Dept. of Health Service
1213 E. El Camino
Phoenix, Arizona 85020

Eric T. Nelson, Eng.
Gilbert Associates
Box 1498
Reading, Pa. 19603

Julian M. Nielsen, Dept. Mgr.
Battelle-Northwest Lab.
1611 Sunset St.
Richland, Wa. 99352

Ramona Noline, Secretary
San Carlos Apache Tribe
Box O
San Carlos, Arizona 85550

Eric A. Nordhausen, Sr. Envir.
Mountain States Engineers
P.O. Box 17960
Tucson, Arizona 85731

Frank O'Donnell - Press
Associate Editor
Coal Outlook/Synfuels Week
1730 K Street N.W.
Suite 713
Washington, D.C. 20006

Veronica O'Donnell
Salt River Project
Box 1980
Phoenix, Arizona 85001

Nancy Osborne
Executive Asst. to Sec.
Department for Natural
Resources & Environmental
Protection
Frankfort, Kentucky

Ragner Overby, Envir. Affairs
Specialist
World Bank
1818 H Street N.W.
Washington, D.C. 20433

Christine Papageorgis, Proj. Mgr.
Princeton Aqua Science
789 Jersey Ave.
P.O. Box 151
New Brunswick, N.J. 08902

Marc Papai, Staff Engr.
Radian Corp.
7927 Jones Branch Drive
Suite 600
McLean, Va. 22102

E.P. Parry, Dir.
Rockwell EMSC
2421 W. Hillcrest
Newbury Park, Calif. 91320

Susan G. Peterson, Env.Planning
Environmental Affairs
Standard Oil, California
44 Montgomery St.
San Francisco, Calif. 94119

L.J. Pearsall, Equip. Analyst
City of Phoenix
2441 S. 22nd Avenue
Phoenix, Arizona 85009

Dennis Perrone, Sales Engineer
Dresser Industries
1630 Newell Avenue
Walnut Creek, Calif. 94596

James W. Peterson, Supervisor
Texas Eastern Transmission Corp.
P.O. Box 2521
Houston, Texas 77001

Richard Petrenka, Planning/Spec.Proj.
Marcopa County Bureau of A.P.C.
1845 E. Roosevelt
Phoenix, Arizona 85202

Doan L. Phung, Sr. Scientist
Institute for Energy Analysis/ORAU
P.O. Box 117
Oak Ridge, Tenn. 37830

A. "Tony" Pietsch, Sr. Proj.Engr.
Air Research Mfg. Co. of Arizona
P.O. BOX 5217
Phoenix, Arizona 85010

Richard A. Porter, Ecological Serv.
Texas Instrument Inc.
P.O. Box 22564
MS 349
Dallas, Texas 75265

Donald D. Potter, Maj. USAF
Group Leader, Fuels Analysis R & D
U.S. Air Force AFWAL/POSF
Wright Patterson AFB
Ohio 45385

Roger Raufer, Mgr. Envir. Studies
ETA Engineering Inc.
415 E. Plaza Drive
Westmont, Illinois 60559

Glenda Rauscher
Salt River Project
P.O. Box 1980
Phoenix, Arizona 85001

Dr. John A. Reagan
University of Arizona
Engr. Bldg. #20
Tucson, Arizona 85721

Tom Renckly, Student
Arizona State University
1217 W. Pebble Beach Drive
Tempe, Arizona 85282

Dr. Joseph Reynolds, Chairman
Chemical Engineering Dept.
Manhattan College
Bronx, New York 10471

Thomas H. Rhodes, Envir.Sr. Advisor
Exxon Chemical Americas
P.O. Box 3272
13501 Katy Freeway-77079)
Houston, Texas 77001

L.W. Rickert, Inf. Ctr. Analyst
Oak Ridge National Lab.
Box X
Oak Ridge, Tenn. 37830

Ed M. Roberts, Supervisor
APS
P.O. Box 21666
Phoenix, Arizona

Dr. Amiram Roffman, Sr. Consultant
Energy Impact Associates Inc.
P.O. Box 1899
Pittsburgh, Penn. 15230

Sharron E. Rogers, Principal Sci.
Battelle Northwest Lab.
200 Park Drive
Box 12056
Research Triangle Park,N.C. 27709

Dr. Edmund Rolinski
Department of Chemical Engr.
University of Dayton
Dayton, Ohio 45469

Hope S. Roman, Vice Pres.
M.R. West Marketing Research
221 East Indianola Ave.
Phoenix, Arizona 85012

Henry Rosenfield
Environmentalist
Texas Eastern Transmission Corp.
P.O. Box 2521
Houston, Texas 77001

Tor Rothman, P.E.
Enviro Control Inc.
11300 Rockville Pike
Rockville, Md. 20852

G. Starr Rounds, Attorney
Evans, Kitchel & Jenckes
363 North First Avenue
Phoenix, Arizona 85003

Dennis Ruddy
U.S. EPA
401 M Street S.W.
Washington, D.C. 20460

Thomas C. Ruppel, Chemical Engr.
U.S. Department of Energy
P.O. Box 10940
Pittsburgh, Pa. 15236

S. James Ryckman Jr., Env. Engr.
USAF/AFLC Wright Patterson AFB
1225 Lytle Lane #2
Dayton, Ohio 45409

Samir Salama, Sr. Professional
Energy & Environmental Analysis Inc.
1111 North 19th Street
Arlington, Va. 22209

Omar E. Abdel-Salam
Cairo University
Chemical Engineering Dept.
Giza, Cairo, Egypt

Dr. Norman Sather
Argonne National Laboratory
9700 South Cass Avenue
Argonne, Illinois 60439

Herbert J. Sawyer, Dir.Tech. Serv.
Del Green Associates Inc.
1155C Chess Drive - Suite A
Foster City, Calif. 94404

David A. Schaller, Sr. Envir.Sci.
S E R I
1617 Cole Blvd.
Golden, Colorado 80401

Steve Schermerhorn, President
Impact Ltd.
1409 Larimer Square
Denver, Colorado

Yale M. Schiffman, Group Leader
MITRE Corp.
1820 Dolley Madison Blvd.
McLean, Va. 22102

Donna Schomburg, Planning Analyst
Exxon Coal, USA
P.O. Box 2180
Houston, Texas 77001

Harold S. Schneider, Mgr.Envir.Proj.
Envirosphere Company
130 Newport Center Drive
Newport Beach, California 92660

R. Bruce Scott, Consultant
Chemical Material Mgr.
Honeywell
P.O. Box 6000
Phoenix, Arizona

Dr. N.D. Shah, Dir.
Multi Mineral Corp.
715 Horizon Drive #380
Grand Junction, Colo. 81501

Tom Shillington, Res. Assoc.
Montana State University
Dept. of Agric. Economics
Boseman, Mt. 59717

Jack A. Siegfried
Civil Engr. Asst. ADOT
Environmental Planning Service
205 S. 17th Avenue Room 240
Phoenix, Arizona 85007

Joseph F. Silvery, Mgr.
Envirosphere Company
130 Newport Center Drive
Newport Beach, Calif. 92660

Suman P.N. Singh,
Staff Member
Oak Ridge National Lab.
Room 224, Bldg. 4500N
P.O. Box X
Oak Ridge, Tenn. 37830

J.L. Shapiro
Chief Nuclear & Envi.Engr.
Bechtel Power Corp.
12400 East Imperial Hwy
Norwalk, Calif. 90650

Tom Sladek, Dir. Energy Div.
CSMRI
P.O. Box 112
Golden, Colo. 80401

Susan E. Small, Grad. Student
University of New Mexico
1406½ Carlisle S.E.
Albuquerque, N.M. 87106

J.M. Sorge, Sr. Consultant
ERTEC Inc.
23 Belmont Drive
Somerset, N.J. 08873

Lori Spencer, Cosultant
Spencer Environmental Consultants
2313 Old Columbiana Road
Birmingham, Alabama 35216

Donald L. Stahlfeld, Staff Engr.
Gulf Research & Development
P.O. Drawer 2038
Pittsburgh, Pa. 15230

John M. Stansfield
Brown & Root
11490 Harwin #1115
Houston, Texas 77072

L.E. Steele, Research Mgr.
Naval Research Laboratory
7624 Highland St.
Springfield, Va. 22150

C.F. Steiner, Vice Pres.
Engineering Science
600 Bancroft Way
Berkeley, Calif. 94710

Joseph G. Stites, Ph.D.
Asst. to General Mgr.
Air Correction Div. UOP Inc.
101 Merritt-7  P.O.Box 5440
Norwalk, Conn. 06856

Dean B. Suagee, Envir.Prot.Spec.
U.S. Bureau of Indian Affairs
BIA Envir. Services (204)
1951 Constitution Ave. N.W.
Washington, D.C. 20245

Fred N. Tabak, County Supervisor
Milwauke? Company
Board of Supervisors
Room 201 Courthouse
Milwaukee, Wi. 53233

Michael L. Teague
Hunton & Williams
Box 1535
Richmond, Va. 23212

Dr. L. Theodore
Manhattan College
Manhattan College Pkwy
Chemical Engineering Dept.
Bronx, N.Y. 10471

Terry Thoem, Dir., Energy Office
EPA
1860 Lincoln Street
Denver, Colo. 80203

George Thomas
Council on Energy Resource Tribes
5660 South Syracuse Circle
Plaza North - Suite 206
Englewood, Colo. 80111

Dr. Ellen Thomas, Res. Assoc.
Arizona State University
Dept. of Geology
Tempe, Arizona 85281

Robert J. Thompson
Energy Consultant
1514 E. Driftwood Drive
Tempe, Arizona 85263

Millicent E. Tissair
Science Teacher
Fairfield Public School
96 Inwood Rd.
Trumbell, Conn. 06611

Craig R. Toussaint, Mgr.
U.S. DOE
Rt. 2, P.O. Box 99
Frederick, Md. 21701

Angela L. Trabert
Environmental Scientist
816 16th Street
Wilmette, Illinois 60091

P.P. Turner, Chief-APB
EPA-RTP-IERL
405 Clayton Road
Chapel Hill, N.C. 27514

Stephen Utter
U.S. Bureau of Mines
Denver Research Ctr.
Bldg. 20
Denver Federal Ctr.
Colorado 80225

Dr. Grey Verner
Occidental Chemical
P.O. Box 1185
Houston, Texas 77001

Pete Vesecky, Proj. Engr.
Evans, Kuhn & Assoc.
4350 E. Camelback Rd.
Suite 200-A
Phoenix, Arizona 85016

G. John Vogel, President
GJV Corp.
168 Chandler Ave.
Elmhurst, Illinois 60126

Dan Vossler
Environmental Coordinator
Room 205A Ct. House Annex
San Luis Obispo, Calif. 93401

Prof. Richard A. Wadden
University of Illinois
School of Public Health
P.O. Box 6998
Chicago, Illinois 60680

Prof. James C. Wade
University of Arizona
Dept. of Ag. Economics
Tucson, Arizona 85715

Dee R. Wernette, Asst.Env.Scientist
Argonne National Lab.
Bldg. 12
9700 South Cass Ave.
Argonne, Illinois 60439

Mr. Eugene M. Wewerka    Judge
U.S. DOE
Los Alamos Science Lab.
Envir. Science Group-MS-734
Los Alamos, N.M. 87545

John White
Salt River Project
P.O. Box 1980
Phoenix, Arizona 85001

Sandi Wiedenbaum, Envir.Scientist
Argonne National Laboratory
9700 South Cass Ave.
Argonne, Illinois 60439

Howard P. Willett, Group  Vice Pres.
Peabody Process Systems
835 Hope Street
Stamford, Conn. 06907

Prof. James D. Williams
Citrus College
P.O. Box 632
Glendora, Calif. 91740

Pamela Williamson, Public Health
Engr. Assistant
Arizona Dept. of Health Services
1740 W. Adams
Phoenix, Arizona 85007

Steven Wolf, Sr. Proj. Mgr.
WAPORA Inc.
211 E. 43 Street
New York, N.Y. 10017

Jane C. Woods
3242 S. Magda
Tucson, Arizona 85730

Paul H. Woods, Sr. Engineer
Fred C. Hart Associates
530 Fifth Ave.
New York, N.Y. 10036

A. Christian Worrell, III
Environmental Scientist
PEDCO Environmental Inc.
11499 Chester Road
Cincinnati, Ohio 45246

Wen C. Yu
Principal Engineer
UOP/SDC
7929 West Park Drive
McLean, Va. 22102

Julia Zilkanich, Pur. Agent
Proc. & Purchasing Section
U.S. Department of Energy
Morgantown Energy Technology Ctr.
Morgantown, W.V. 26505

Jon Zuck
Battelle Northwest
P.O. Box 999
Richland, Washington 99352

Imre Zwiebel, Chairman,
Chemical Engr. Dept.
Arizona State University
Tempe, Arizona 85281

591

# INDEX

electrolytic flotation  68-71
electron-beam irradiation  275
electrostatic precipitator  275; 288; 322-323; 433
emission coefficient  191
emission factor  148-149
Energy Conversion Alternatives Study (ECAS)  122
Energy Policy Statement  51
Energy Policy and Conservation Act  241
Energy Security Act  320-331; 351; 357; 364; 372
Energy Supply and Environmental Co-ordination Act  241
England  120
Entitlement Program  24
Environmental Impact Statements (EIS)  91; 136-138; 211; 364; 368
environmental assessment  136-137; 211
Environmental Protection Agency  13-17; 52-55; 107-108; 118; 120; 147; 149; 165; 209-210; 213-216; 220; 230; 243; 287-289; 340; 345-348; 351; 358; 360; 469; 535-537; 565-568
ethanol  54; 95; 347
Export Control Act  75

F

Federal Coal Leasing Program  361-362
Federal Power Commission  201
Finland  107
Flat Tops Wilderness Area  52; 55
Florida  231; 504
flue gas desulfurization (FGD)  119; 122; 194; 287-293; 321-325; 433
fluidized-bed combustion  119; 151-152; 176-181; 254; 272-277; 497
France  107; 420; 427
fuel cell  122; 200; 202

G

gasket curing ovens  99-100
gasohol  355
Georgia  231,233
Geothermal Demonstration Program  1
geothermal power  200; 202; 273
Germany  107; 120; 336; 420; 427
Grand Canyon  235
grandfathering  52
Grants Mineral Belt  467-468; 471
granular filtration  304-307
Great Plains Gasification Project  329; 357
Green River  44; 46; 93

H

H-Coal Process  91
Hidden Mountain Dam  471
high-sulfur coal  119; 161
Human Exposure and Health Effect Model (HEHEM)  403-409
hydrogasification  120-121
hydrologic disturbance  163-166

I

Illinois  233; 305-306; 341; 375; 382-387
impact avoidance  58-63
impact elements  368
impact mitigation  58-63
in-situ recovery technologies  48; 66-71; 121-122; 161-166; 351-352; 394; 427
in-situ solution mining  477-485
incineration  100; 267-268
INCO oxygen flash smelting  497
increments (consumption)  233-235; 358; 377; 570
incremental production  108-109
Indiana  231; 234; 385
Indians, tribal society  5-8
-, underdevelopment  5
-, water rights  378
indirect coal liquefaction  See: coal liquefaction
industrial cogeneration  432-436
industrial preparedness base  79; 86
industrial process heat (IPH)  569-571
industry models  457-462
integral vistas  15; 38; 40-41
Intermountain Power Project  233; 360-361
isomerization  243

J

Japan  79; 107; 336; 427; 494-496
Jemez Region  1-4
Jones Act  241

K

Kansas  230; 234
Kentucky  75; 90-95; 176-181; 231; 234; 368; 379; 385
Kivcet Process  498
Klystron power tubes  542

L

"laissez faire philosophy"  80
landfill  108; 194; 212; 225; 268; 288

Laramie River  362
laser technology  543
leaching  212; 250
life cycle energy analysis  263-269
Louisiana  139; 234; 342
low-sulfur coal  232
lowest achievable emission rate (LAER)
    243

## M

magnetohydrodynamics (MHD)  122-
    123
Mandatory Oil Import Program (MOIP)
    241
Maryland  379
Materials Policy Bill  79
methanol  120; 171; 336; 338; 340-341;
    368; 427
methyl mercury  7
Mexico  47
mill tailings  250-253
Mississippi  139
Missouri  230
Mitsubishi Process  495-496
Montana  13; 232; 234; 358-359
Morrison Formation  468

## N

nahcolite  287-293
National Ambient Air Quality Standards
    (NAAQS)  147; 212-213; 230; 235;
    312; 535; 565
national emission standards (for hazard-
    ous air pollutants)  110; 535
National Energy Conservation Act  241
National Energy Plan  220; 420
National Environmental Policy Act  147;
    211
National Primary Interim Drinking
    Water Standard  178
national visibility goal  41
NATO  79; 87
Navajo Indians  378
New Mexico  1-4; 359; 378-384; 421;
    467
New Source Performance Standards
    (NSPS)  53-54; 147-149; 231; 243;
    366; 535-536
New York  25-31
nitrogen dioxide  14-15; 19-22; 376
nitrogen oxides  93; 109-111; 119; 123;
    125-126; 128-129; 147; 149-152;
    191-192; 243; 272-278; 345; 428;
    567
    See also: nitrogen dioxide
"No Growth Syndrome"  80
noise impacts  193; 210; 240
North Carolina  354

North Dakota  13; 331; 357-360
Northern Great Plains  90

## O

Oak Ridge National Laboratory  170-
    172; 177; 200
offsetting  565-571
    See also: "bubble" concept,
    increments
Ohio River Basin  93; 176; 178-181; 234
oil shale  13; 44-48; 50-55; 121-122;
    242; 287; 345-346; 348; 352; 354;
    357-359; 365; 375
Oklahoma  234; 565
OPEC  80; 87
Overthrust Belt  13; 360; 362

## P

Parachute Creek  46
particulates  14-17; 19; 21; 51; 109;
    123; 125; 127; 129; 165; 231; 243;
    275-276; 304-307; 322; 346; 358;
    433; 505; 535
Pennsylvania  211
phenol  164-166; 255; 395-396
photochemical oxidants  93; 147
photovoltaic energy  200; 202; 520-523;
    533-538; 542
Piceance Basin  44; 46; 48; 51; 59; 287;
    354
plumes  4; 15-17; 20-22; 37; 40; 165-
    166; 253
Pollution Control Guidance Documents
    (PCGD)  54
Powder River  341; 358-359
pressurized fluidized bed (PFB)  119
Prevention of Significant Deterioration
    (PSD)  14; 16; 52-53; 147; 212-
    213; 366; 374; 376; 535; 566; 568
process models  See: industry models
Prototype Oil Shale Leasing Program  50
pyrolysis  120-121; 254-255

## R

radioactivity  210; 250-253
Reasonably Available Control Tech-
    nology (RACT)  243
Redondo Peak  1-2
reduction
    nonselective -  274-275
    selective -  274-275
reduction in intensity  19-22
reforming  240; 243
regenerative incineration  312-317
Resource Conservation and Recovery
    Act  110; 147; 209-212; 214; 220-
    226; 243; 536-538